Insects as food and feed:

from production to consumption

INSECTS AS FOOD AND FEED:
from production to consumption

edited by:

Arnold van Huis

Jeffery K. Tomberlin

Wageningen Academic Publishers

Open Access e-learning

This book is accompanied by Open Access e-learning that has been made possible by the **Uyttenboogaart-Eliasen Foundation** (UES).

The e-learning materials are located at https://e-insects.wageningenacademic.com. These materials include high resolution macro photographs and 360 degrees view photographs of several edible insect species, relevant information on the species that are being shown, questions to test your knowledge on edible insects, and information explaining the technique used to make the photographs.

EAN: 9789086862962
e-EAN: 9789086868490
ISBN: 978-90-8686-296-2
e-ISBN: 978-90-8686-849-0
DOI: 10.3920/978-90-8686-849-0

Cover photo of *Locusta migratoria* made by Hans Smid bugsinthepicture.com

First published, 2017

Table of

contents

Regulation, ethics and promotion

Future prospects

Preface

In tropical countries, insects have been a food source for a very long time, and the local population has an intricate knowledge about which species can be eaten, when and how they can be harvested, and which methods to use for conserving, preparing and consuming the insects. In the western world, this knowledge is absent as insects were never considered as food. However, insects are farmed by private companies as feed for pets, birds, fish, reptiles, etc. This utilisation is only a small niche market and the volumes produced are rather limited.

There is an increasing awareness that with the growing world population and increasing welfare that business as usual for the production of animal proteins in not an option. The land area available on our planet to satisfy the increasing demand for animal proteins is not enough. Besides, there are concerns about the environmental footprint of producing livestock (greenhouse gas and ammonia emissions), about infectious diseases of vertebrate animals that can naturally be transmitted to humans (zoonosis). For those reasons, society is receptive to the idea of searching alternative protein sources. Although it was known in the western world that insects are eaten in the tropics, only now they are considered as an option to be used both as human food and animal feed. Only since the last ten years there is an increasing interest to farm insects as a new source of protein for humans and for animals. This in particular when considering the lower environmental impact when compared to the conventional protein products.

However, to go from a niche market to producing edible insects at an industrial level is new. The only examples of mass rearing insects are those used in pest control. One example is the sterile insect technique in which large numbers of sterile insects are released into the wild. The sterile males compete with wild males to mate with the females. Females that mate with a sterile male produce no offspring, thus reducing the next generation's population. Insects that have been controlled (eradicated) this way are screw-worm flies, fruit flies and tsetse flies. Biological control companies have experience in rearing insects to control agricultural

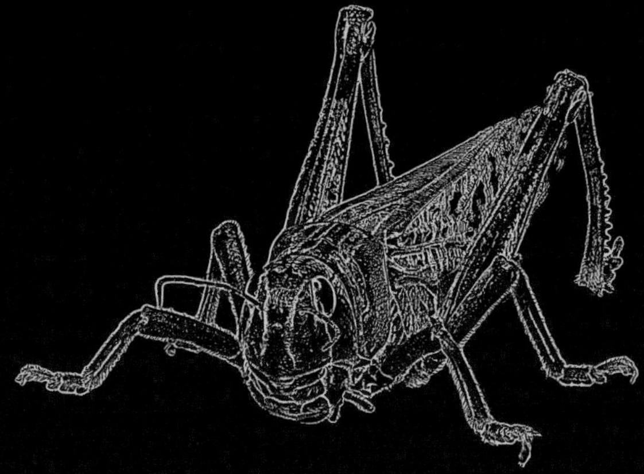

pests (i.e. natural enemies, such as predators and parasitoids). However, the companies that produce insects as feed for animals are rather small with a lot of labour involved making it too expensive to be competitive to conventional feed products.

So, the field of rearing insects for either human food or animal feed is new. Companies are engaged in automating the system in order to bring the price down. However, there are many challenges. To make it interesting to rear insects, such approaches have to be done sustainably and environmental impact should compare with the farming of common production animals. What kind of production design and what kind of facilities are needed to farm the insects in an optimal way? Can we fine tune the rearing technique to respond to the biological requirements of the target species? What about using different strains of insects and how can we maintain a high quality of the insects (genetics). What kind of diets are we going to use taking into account the price and the large influence substrates have on the development of the insect and on the nutritional value of the end product. Of particular interest is the use of organic side-streams as this would largely contribute to a circular economy. Other factors of importance to the nutritional quality are for example the abiotic conditions and the stage of the insects to be harvested. Also, we have a lot of knowledge about entomopathogens to kill pest insects, but very limited knowledge about diseases and pests that occur in insect rearing. Wat about insect welfare? Does the killing of thousands of insects for food or feed morally equal the slaughtering of one cow? Then after harvesting the larvae, what is done with the substrate and how to process the insects? How to deal with chitin, protein and oil? Finally, the product needs to marketed. When insects are used as food then consumer attitudes play an important role and what kind of strategies to use to convince the consumer? Another very important issue in such an innovative sector is the legislative framework which is often not conducive. What kind of enabling conditions are required for the edible insect industry to thrive?

Finding answers to all these questions served as the impetus for bringing together a diverse group of researchers and practitioners from around the world to develop this book dealing with all aspects of farming insects and developing methods for efficiently bringing insect-related products to the consumer.

In order to improve the quality of the book most chapters have been subjected to a peer review process. We tried to find the best independent subject experts around the globe to do so. Most chapters have been checked for accuracy and readability by at least one reviewer, most often two and sometimes even three reviewers. The detailed feedback, from a range of expert reviewers, ensures that that we deliver relevant, cutting edge, and high quality information useful for students, teachers, practitioners, and researchers. We wish to thank all reviewers, listed on the next page for the willingness to check the chapters and provide their valuable comments.

The editors and Wageningen Academic Publishers want to thank the Uyttenboogaart-Eliasen Foundation (UES; http://tinyurl.com/y77pkj6b) in the Netherlands for their generous grant to develop the e-learning module attached to this book (https://e-insects.wageningenacademic.com). The grant also allowed the incorporation of specially made high quality photographs both for the book and the e-learning module. The module enables readers to test whether they have learned from the book and to get feedback.

Arnold van Huis and Jeffery K. Tomberlin

Reviewers

Based on the insightful comments and suggestions provided by reviewers, the authors addressed the suggested revisions, which greatly improved the quality of the book. We wish to express our sincere gratitude to the following reviewers:

Paul Aman
University of Liège – AgricultureIsLife, Gembloux Agro-Bio Tech, Passage des Déportés 2, 5030 Gembloux, Belgium

Simone Belluco
Istituto Zooprofilattico Sperimentale delle Venezie, Food Safety Department, Risk Analysis Division, Viale dell'Università 10, 35020 Legnaro (PD), Italy

M. Eric Benbow
Michigan State University, Department of Entomology, Department of Osteopathic Medical Specialties, 288 Farm Lane, East Lansing, MI 48824, USA

Jonathan A. Cammack
Texas A&M University, Department of Entomology, TAMU 2475, College Station, TX 77843-2475, USA

Marcel Dicke
Wageningen University & Research, Laboratory of Entomology, P.O. Box 16, 6700 AA Wageningen, the Netherlands

Stefan Diener
Eawag: Swiss Federal Institute of Aquatic Science and Technology, P.O. Box 611, 8600 Dübendorf, Switzerland

Darja Dobermann
Rothamsted Research, BCP, West Common, Harpenden AL5 2JQ, United Kingdom; University of Nottingham, Division of Nutritional Sciences, North Laboratory Building, Sutton Bonington LE12 5RD, United Kingdom

James C. Dunford
McGuire Center for Lepidoptera and Biodiversity, Florida Museum of Natural History, University of Florida, Gainesville, FL 32611-2710, USA

Cordelia I. Ebenebe
Nnamdi Azikiwe University, Department of Animal Science and Technology, P.M.B. 5025 Awka 307720010, Anambra State, Nigeria

René Cerritos Flores
Universidad Nacional Autónoma de México, Department of Experimental Medicine, Experimental Immunology Laboratory, Dr. Balmis 148 Col. Doctores, C.P. 06720 Mexico City, Mexico

Asnath M. Fuah
Bogor Agricultural University, Faculty of Animal Science, Department of Animal Production and Technology, Agatis Street, Bogor 16680, Indonesia.

Laura Gasco
University of Turin, Department of Agricultural, Forest and Food Sciences, Largo P. Braccini 2, 10095 Grugliasco, Torino, Italy

Peter W.G. Groot Koerkamp
Wageningen University & Research, Farm Technology Group and Wageningen Livestock Research, P.O. Box 16, 6700 AA Wageningen, the Netherlands

Jonas House
University of Sheffield, Department of Geography, Winter Street, Sheffield, S10 2TN, United Kingdom

Anna Jansson
Swedish University of Agricultural Sciences, Department of Anatomy, Physiology and Biochemistry, P.O. Box 7011, 75007 Uppsala, Sweden

Simon Knutsson
Foundational Research Institute, Hardenbergstraße 9, 10623 Berlin, Germany

Jeffrey A. Lockwood
University of Wyoming, Department of Philosophy, College of Arts & Sciences, Dept. 3392, 1000 E. University Ave, Laramie, WY 82071, USA

Andrew Müller
Humboldt-Universität zu Berlin, Department of Social Sciences, Universitätsstraβe 3B, 10117 Berlin, Germany

Christina Nielsen-LeRoux
INRA, UMR-1319 Micalis – AgroParisTech, University Paris-Saclay, Génétique Microbienne & Environnement, 4 Avenue Jean Jaures, 78350 Jouy en Josas, France

Dennis Oonincx
Wageningen University & Research, Laboratory of Entomology, P.O. Box 16, 6700 AA Wageningen, the Netherlands

Charlotte Payne
University of Cambridge, Department of Zoology, David Attenborough Building, New Museums site, Cambridge, CB2 3QY, United Kingdom

Christine J. Picard
Indiana University Purdue University Indianapolis (IUPUI), Department of Biology, 723 W. Michigan Street, Indianapolis, IN 46202, USA

Giulia Poma
University of Antwerp, Department of Pharmaceutical Sciences, Toxicological Center, Universiteitsplein 1, 2610 Wilrijk, Belgium

Roberta Salomone
University of Messina, Department of Economics, Piazza Pugliatti 1, 98122 Messina, Italy

Matan Shelomi
Max Planck Institute of Chemical Ecology, Department of Entomology, Hans-Knöll-Str. 8, 07743 Jena, Germany

Baneshwar Singh
Virginia Commonwealth University, Department of Forensic Science, 1015 Floyd Avenue, Richmond, VA 23284, USA

Renate C. Smallegange
http://www.renatesmallegange.nl

Sergiy Smetana
German Institute of Food Technologies (DIL-e.V.), Prof. von Klitzing Str. 7, 49610 Quakenbrück, Germany

Arnoud Togtema
Wageningen Food & Biobased Research, Fresh Food & Chains, P.O. Box 17, 6700 AA Wageningen, the Netherlands

Martin Tschirner
Leibniz Institute of Freshwater Ecology and Inland Fisheries, Department Ecophysiology and Aquaculture, Müggelseedamm 310, 12587 Berlin, Germany

Gordon Tween
United States Department of Agriculture, Animal and Plant Health Inspection Service, International Programs, 4700 River Road, Riverdale, MD 20737, USA

Just Vlak
Wageningen University & Research, Laboratory of Virology, P.O. Box 16, 6700 AA Wageningen, the Netherlands

Photo

credits

The photographs at the beginning of each section and of each chapter have been made and are copyrighted by Hans Smid (bugsinthepicture.com), except for the photograph in Chapter 2, which was made by Nina Fatouros (bugsinthepicture.com). The photographs of Hans Smid were optimized for increased depth of field by making multiple images taken at various distances with a camera (Canon M5, Tokyo, Japan) mounted on a Stackshot automated macro rail (Cognisys, Traverse city, MI, USA). Resulting stacks of images were processed in Zerene stacker software vs 1.04 (Zerene systems, Richland, WA, USA). The following insect species are shown:

Sections

- Introduction — *Lucilia sericata*
- Tropical production systems — Freeze-dried locusts
- Production technology and management — Freeze-dried *Alphitobus diaperinus*
- Industrial production systems — *Hermetia illucens*
- Nutritional quality and processing — Freeze-dried crickets
- Regulation, ethics and promotion — *Lucilia sericata*
- Future prospects — *Calliphora vomitoria*

Chapters

- 1 *Lucilia sericata*
- 2 *Trichogramma brassicae* parasitizing an egg of *Pieris brassicae*
- 3 *Pachnoda marginata*
- 4 *Imbrasia belina* (sun-dried)
- 5 *Gryllus bimaculatus*
- 6 *Rhynchophorus ferrugineus*
- 7 *Locusta migratoria*
- 8 *Alphitobus diaperinus*
- 9 *Gryllodes sigillatus*
- 10 *Zophobas morio*
- 11 *Calliphora vomitoria*
- 12 *Hermetia illucens*
- 13 *Musca domestica*
- 14 *Tenebrio molitor*
- 15 *Acheta domesticus*
- 16 *Blaptica dubia*
- 17 *Galleria mellonella*
- 18 *Omphisa fuscidentalis* (dried larvae)
- 19 *Gryllus assimilis*
- 20 *Apis mellifera*
- 21 Freeze-dried insects for human consumption
- 22 *Bombyx mori* (canned pupa)

Introduction

1.

Lucilia sericata

The potential of insects as food and feed

A. van Huis[1*] and J.K. Tomberlin[2]
[1]Laboratory of Entomology, Wageningen University, P.O. Box 16, 6700 AA Wageningen, the Netherlands; [2]Department of Entomology, Texas A&M University, TAMU 2475, College Station, TX 77843-2475, USA; arnold.vanhuis@wur.nl

Abstract

Insect production has gained recognition globally as a viable economy in terms of protein production for food and feed. While many tropical countries have practiced entomophagy for millennia, western nations are just now tapping into this tremendous resource. The purpose of this text is to review the many facets related to insect production ranging from examining the diversity of arthropods currently being mass-produced, ethics of such systems, to industrialized processes currently employed to meet the growing demand. While the architecture from production of the insect to packaging is still in its infancy, this growing sector of agriculture is presented with a unique opportunity to address potential hurdles prior to their encounter through communication, education, and practice.

Keywords: edible insects, insects as food and feed, animal protein, alternative protein sources, environment, organic side streams

1.1 Introduction

This book provides you with information for those who rear, or intend to rear, insects either for human consumption or for animal feed. Rearing is equivalent to producing or farming insects. The word 'breeding' is also often used. However, we follow the definition of 'animal breeding' used by Encyclopaedia Britannica, viz: 'controlled propagation of domestic animals in order to improve desirable qualities', and the following was added: 'humanity has been modifying domesticated animals to better suit human needs for centuries' (https://www.britannica.com/science/animal-breeding). In a textbook the following definition was used: 'animal breeding involves the selective breeding of domestic animals with the intention to improve desirable (and heritable) qualities in the next generation' (Oldenbroek and Waaij, 2014). This book deals with all aspects related to rearing and marketing edible insects and only one chapter (Chapter 9: Jensen *et al.*, 2017) is reserved for breeding.

1.2 Future of animal protein

According to the '2012 revision of the official United Nations population estimates and projections', the world population of 7.2 billion in mid-2013 is projected to increase by almost one billion people within the next twelve years, reaching 8.1 billion in 2025, and to further increase to 9.6 billion in 2050 and 10.9 billion by 2100 (United Nations, 2013). Aggregate meat consumption increased almost 60% between 1990 and 2009, from 175,665 thousand tonnes to 278,863 thousand tonnes, driven in part by the growing world population (Henchion *et al.*, 2014). Meat consumption will increase because of the growing world population in the period 2005/2007 to 2050. However, the increase will be larger because per capita meat consumption is predicted to increase in this period by 26%. This projection is 50% in developing countries. When China and Brazil are excluded, who both account for 56% of meat consumption in developing countries, the increase is 76% (Table 1.1). This calculation indicates that other factors, in addition to population growth, are influencing demand. This outcome is very likely due to increased welfare. For example, South Asia's middle class will go from under one million in 2012 (less than 0.1% of the population) to some 660 million (29% of the population) in 2050 (Table 6.3 in Van der Mensbrugghe *et al.*, 2011). Global population growth, rising incomes and increased urbanization in developing countries stimulate the growth for animal food source also because populations adopt more Westernized diets (Turk, 2016).

Table 1.1. Changes in meat[1] (carcass weight) consumption (kg/person/year) in the world and by country groups (Alexandratos and Bruinsma, 2012).

	2005/2007	2030	2050
World	39	45	49
Developing countries	28	36	42
(minus China and Brazil)	17	23	30
Developed countries	80	87	91

[1] Meat includes bovine, ovine, poultry and pig-meat.

Of total available agricultural land, about 68% (3.3 billion ha) is under permanent meadows and pastures (http://www.fao.org/faostat). Another 10% (500,000 ha) is used as cropland to produce animal feed (Steinfeld *et al.*, 2006); so in total, 78% of all agricultural land is used to produce livestock. Total meat production from 2005/2007 to 2050 is expected to increase from 258 to 455 million tonnes (57%) (Alexandratos and Bruinsma, 2012). How can this be accommodated when so much land is already used for livestock? When no additional land is being added for pasture, marginal lands are now the frontiers and will become quickly exhausted (Turk, 2016). Growth of crop yields has slowed considerably. Studies have shown global crop production needs to double by 2050 to meet the projected demands from rising population, diet shifts, and increasing biofuels consumption. Production of four key global crops, maize, rice, wheat, and soybean, that currently produce nearly two-thirds of global agricultural calories, will increase between 38 and 67% by 2050, which is far below what is needed (Ray *et al.*, 2013). Is the lower growth potential, together with modest increases in cultivated land, sufficient to meet the increased requirements? Climate change is another risk, which may affect the production potentials of agricultural resources in many areas of the world. These factors together indicate we need to find solutions to satisfy the demand for animal protein.

1.3 Livestock, environment and health

1.3.1 Greenhouse gases

The report 'Livestock's long shadow: environmental issues and options' was released by the Food and Agriculture Organization of the United Nations (FAO) on 29 November 2006 (Steinfeld *et al.*, 2006). The report assessed the full impact of the livestock sector on environmental problems, along with potential technical and policy approaches to mitigation. The livestock sector emerged as one of the top two or three most significant contributors to serious environmental problems. For example, livestock globally emits more greenhouse gases (GHGs) than all transport systems worldwide. Other problems caused by the livestock sector were: a major contributor to ammonia emissions, deforestation, soil erosion, desertification, loss of plant biodiversity, negative impact on public health, and water pollution.

Gerber *et al.* (2013) consider livestock a significant contributor to climate change, because of the emission of GHGs, which absorb and emit radiation within the thermal infrared range. The relevant GHGs in the earth's atmosphere are carbon dioxide (CO_2), methane and nitrous oxide, denoted by carbon dioxide equivalents (CO_2-eq). They estimated emissions at 7.1 gigatonnes CO_2-eq per annum, representing 14.5% of human-induced GHG emissions. Cattle are the main contributor to the sector's emissions with about 4.6 gigatonnes CO_2-eq, representing 65% of sector emissions. Herrero *et al.* (2015) estimated the contribution of cattle even larger, between 65 and 78% of the emissions. If we look at livestock products, beef is responsible for 41% of the sector's emissions, while swine meat and poultry meat and eggs contribute 9 and 8%, respectively (Table 1.2). The largest process contributors are feed production and processing (45%) and enteric fermentation by methane (39%) (Table 1.2). Methane and nitrous oxide emissions from manure storage and processing represent about 10% of the sector's emissions.

Table 1.2. Percentage of greenhouse gas emissions from livestock sector per product and per process (Gerber *et al.*, 2013).

Products	%	Processes	%
Beef	41	feed production and processing[1]	45
Cattle milk	20	enteric fermentation	39
Swine meat	9	manure storage and processing	10
Poultry meat and eggs	8	processing and transportation of animal products	6

[1] Included in feed production, the expansion of pasture and feed crops into forests accounts for about 9% of the sector's emissions.

Methane remains in the atmosphere for 9-15 years and is 23 times more effective in trapping heat than carbon dioxide. Concentration of methane have increased by about 150% since pre-industrial times (Steinfeld *et al.*, 2006). Nitrous oxide is present in the atmosphere with extremely small amounts but is 289 times more effective than carbon dioxide in trapping heat and has a very long atmospheric lifetime (Table 1.3). Methane is the largest contributor (44%), followed by carbon dioxide and nitrous oxide, each about 27%.

1.3.2 Ammonia

Beusen *et al.* (2008) estimated that the contribution from livestock production to global agricultural ammonia emission is 59-71% (16-27 Tg/y with a mean of 21 Tg/y, i.e. 21×10^{12} g); 31-55% of this emission comes from animal houses and storage systems, 17-37% from grazing animals and 23-38% from the spreading of animal manure. Atmospheric ammonia causes acidification and eutrophication of the ecosystems. Global emissions of ammonia have increased by a factor of two to five since preindustrial times, and they are projected to continue to rise over the next 100 years (Zhu *et al.*, 2015).

1.3.3 Other problems associated with the livestock industry

A number of books highlight the concerns associated with livestock production. First of all the book by FAO entitled 'Livestock long shadow' (Steinfeld *et al.*, 2006) mentioning a number of environmental and public health problems. After that a number of other books have appeared (Bramble and Fischer, 2016; D'Silva and Webster, 2010; Foer, 2009; Lymbery, 2014),

Table 1.3. Anthropogenic CO_2 emissions: global warming potential (GWP), anthropogenic emissions in gigatons CO_2-eq per annum (Gt-CO_2-eq/y), percentage emission per gas; and percentage of the agricultural sector's emissions of a specific gas (Gerber *et al.*, 2013; IPCC, 2007).

Greenhouse gas	GWP	Gt CO_2-eq/y	% of emissions of a specific greenhouse gas	% of sector's emissions
Carbon dioxide	1×	2	5	27
Methane	23×	3.1	44	44
Nitrous oxide	289×	2	53	27

which deal with ethical and policy implications of livestock production and consumption. Among the environmental impacts of livestock production are: huge water and land use; cereals being used as animal fodder, which could feed billions of people; cattle ranching and soy expansion as major drivers of deforestation in Brazil; global warming by greenhouse gas emissions; and loss of biodiversity. The overall message is often that we should eat less meat or become vegetarian. Also, in the books the ethics of eating meat is discussed: are animals harmed or benefited by our practice of raising and killing them for food? Health issues are mentioned as the widespread use of antimicrobials in food animals contributes to bacterial drug resistance and human infection.

Pathogens, responsible for a number of diseases are transmitted from vertebrate animals to humans (zoonotic infectious diseases), account for an estimated 60% of all human infectious diseases (Klous *et al.*, 2016). Livestock associated infectious diseases are a major threat to human health and they mainly occur in densely human populated areas, and therefore, having a considerable public health impact.

1.3.4 Need to replace fishmeal

In 2014, the aquaculture sector's contribution to the supply of fish for human consumption overtook that of wild-caught fish for the first time, and now provides half of all fish for human consumption (FAO, 2016). Aquaculture provided only 7% fish for human consumption in 1974; this share had increased to 26% in 1994 and 39% in 2004. In 2014, world aquaculture production of fish accounted for 44.1% of total production from capture fisheries and aquaculture (including for non-food uses) (Figure 1.1).

In 2014, 21 million tonnes was destined for non-food products, of which 76% (15.8 million tonnes) was reduced to fishmeal and fish oil. Fishmeal is a commercial product mostly made from fish containing a high percentage of bones and oil, and is usually deemed not suitable for direct human consumption. A small portion is made from the bones and offal left over from processing fish used for human consumption. The larger portion is manufactured from

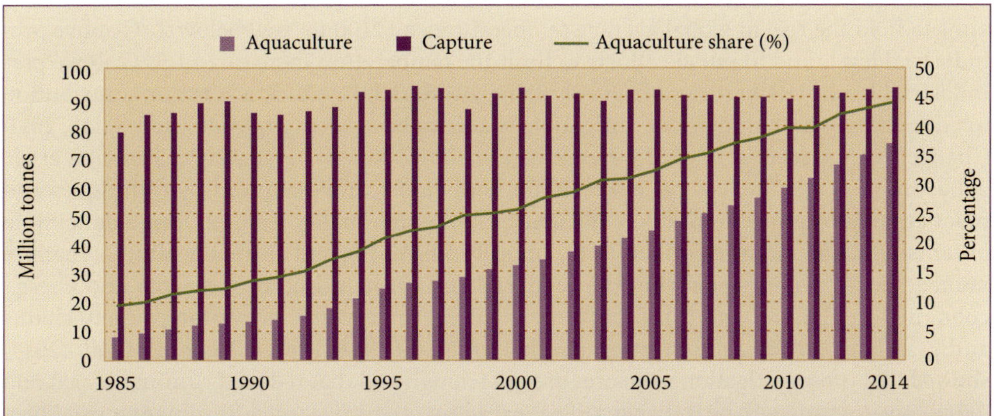

Figure 1.1. Share of aquaculture in total production of aquatic animal (including non-food uses) (FAO, 2016; p. 20).

fish stocks of wild-caught, small marine fish. Many different species are used for fishmeal and fish-oil production, with oily fish, especially anchoveta, the main group of species utilized. Fishmeal is the crude flour obtained after milling and drying fish or fish parts, while fish oil is usually a clear brown/yellow liquid obtained through the pressing of the cooked fatty fish.

Fishmeal and fishoil are used to feed poultry, swine, and farmed fish. Fishmeal production peaked in 1994 and has followed an overall declining trend since then (FAO, 2016). Prices of fishmeal have remained below 1000 US$/metric tonnes from 1987 to 2006, and after 2012 have peaked above 2,000 US$/metric tonnes (http://tinyurl.com/nxlajav). This partly due to overexploitation of fish stocks. Today, 31% of commercially important assessed marine fish stocks worldwide are overfished (FAO, 2016). Fishmeal used to be a cheap commodity included in animal feed. It is regarded as an excellent source of proteins, it is easily digestible, has an excellent composition of essential amino acids, and long chain omega-3 fatty acids, such as eicosapentaenoic acid and docosahexaenoic acid; it is also a good source of vitamins (riboflavin, niacin, vitamins A and D) and minerals (calcium, phosphorus, iron, zinc, selenium and iodine) product (Olsen and Hasan, 2012). The rapidly growing aquaculture industry using fishmeal as a major protein source in compounded feed has strongly contributed to increased demand and prices for this product, and for that reason massive research is conducted to identify alternative more sustainable plant and animal protein and lipid sources (Leschin-Hoar, 2016; Olsen and Hasan, 2012; Rana *et al.*, 2009; Tacon *et al.*, 2009).

1.4 How to tackle the meat crises?

There are a number of strategies possible to tackle the meat crisis (Théwis and Galiù, 2012): (1) the livestock sector itself can devise strategies to curb GHG and ammonia emissions, the so-called mitigation strategies; (2) cutting on meat consumption; (3) developing alternate methods for producing meat (e.g. cultured meat); and (4) develop alternate protein sources.

1.4.1 Mitigation strategies

To curb emissions, it is particular important to adhere to the 2015 Paris Agreement of the United Nations Framework Convention on Climate Change (UNFCCC). The agreement aims to hold the rise in global average temperatures by 2100 to 'well below 2 °C above pre-industrial levels and to pursue efforts to limit the temperature increase to 1.5 °C above pre-industrial levels' (Richards *et al.*, 2015). Being aware that livestock is a major contribution to GHG emissions, a number of mitigation strategies have been proposed (Boer *et al.*, 2011; Change, 2015; Gerber *et al.*, 2013; Herrero *et al.*, 2016; Ndegwa *et al.*, 2008; Paustian *et al.*, 2006; Smith *et al.*, 2008; Vergé *et al.*, 2007). Emissions can be reduced by techniques and practices that improve production efficiencies: (1) better quality feed; (2) feed balancing to lower enteric and manure emissions; (3) improved breeding and animal health; (4) manure management practices ensuring the recovery and (5) recycling of nutrients and energy contained in manure; and (6) improvements in energy use efficiency along supply chains and grassland carbon sequestration (Gerber *et al.*, 2013). However, Hedenus *et al.* (2014) showed that those mitigation measures are not enough and that reduced ruminant meat and dairy consumption will be indispensable for reaching the 2 °C target. Ruminant meats (beef and lamb) have emissions per gram of protein are about 250 times those of legumes, and also much higher than those of poultry and swine (Tilman and Clark, 2014). To shift away from ruminants (e.g. cattle and sheep) to monogastrics (swine and poultry) has been suggested as solutions (Gerber *et al.*, 2013; McAlpine *et al.*, 2009). Also putting a tax on in particular ruminant meat has been proposed (Revell, 2015).

1.4.2 Eating less meat

In order to reduce global agricultural GHG emissions, reduce land clearing and resultant species extinctions, it seems obvious that reducing meat consumption would be the most sensible solution (Davis *et al.*, 2016; Hedenus *et al.*, 2014; Schösler *et al.*, 2012; Tilman and Clark, 2014; Wollenberg *et al.*, 2016).

1.4.3 Cultured meat

In vitro culturing of meat has been proposed as one of the alternatives for livestock meat production (Post, 2012). On August 5, 2013, a hamburger prototype made from cultured, or *in vitro*, meat was tasted at a well-publicized event in London (Post, 2014). The meat used for preparation of this hamburger was not grown in an animal, but rather from bovine skeletal muscle stem cells in Dr Mark Post's laboratory at Maastricht University in the Netherlands. The process entails the generation of bio-artificial muscles from satellite cells. Such cells generated should mimic meat in visual appearance, smell, texture, and of course, taste. Loose myosatellite cells can be cultured on a substrate, and mature muscle cells can be harvested after differentiation and processing them into various meat products. The effective culture of skeletal muscle seems to be possible with current technology generating an acceptable mimic of meat tissue. Main issues to consider with this process are: scalability of the production process, quality control of mammalian cell/tissue cultures, maintaining sterility in the culture, prevention of contamination or disease and the controlled breeding of stem cell donor animals. A great deal of research is still needed to establish a sustainable *in vitro* meat culturing system on an industrial scale (Fayaz Bhat and Fayaz, 2011). Using a life cycle assessment, the environmental impacts of cultured meat production is substantially lower than those of conventionally produced meat (Tuomisto and Teixeira de Mattos, 2011): compared to conventionally produced European meat, cultured meat involves approximately 7-45% lower energy, 78-96% lower GHG emissions, 99% lower land use, and 82-96% lower water use depending on the product compared (Tuomisto and Teixeira de Mattos, 2011). This evaluation was confirmed by Mattick *et al.* (2015), who also found that *in vitro* biomass cultivation could require smaller quantities of agricultural inputs and land than livestock, although uncertainty ranges are large.

1.4.4 Alternative protein sources

Dietary change, in areas with affluent diet, is one of the options to reach environmental goals. One group of researchers reviewed 14 peer-reviewed journal articles assessing the GHG emissions and land use demand of in total 49 dietary scenarios (Hallström *et al.*, 2015). This strategy is able to reduce GHG emissions and land use demand up to 50% compared to the current diet. Reason to study more in detail which dietary changes are possible.

The search for alternative and more sustainable protein sources and recently a whole book was published on sustainable protein sources, among those in particular many plant protein sources (Nadathur *et al.*, 2016). We will only discuss upcoming sources of protein which are canola/rapeseed, mycoprotein, heterotropic algae and of course insects (Table 1.4).

Table 1.4. Overview of common food proteins from various sources (Haque *et al.*, 2016).

Source	Protein
Animal	muscle proteins (myofibrillar, sarcoplasmic, stroma proteins)
	blood proteins (hemoglobin, plasma protein)
	connective tissue proteins (collagen, elastin)
	milk proteins (whey proteins, caseins, lactoferrin)
	egg proteins (egg albumin, yolk)
Botanical	cereals proteins (gluten, zein, barley, oats, rice proteins)
	proteins from legumes and pulses (protein isolates from peas, soybeans, lupines, lentils)
	proteins from oilseeds (protein isolates from rapeseed, canola, cottonseed, peanut, chia seed, flaxseed)
	proteins from tubers (potato protein)
	proteins from leaves (alfalfa leaf protein concentrate)
Algal	proteins from green and blue-green seaweed (spirulina, anabaena,nostoc, ulva, enteromorpha)
Microbial	fungal proteins (mycoproteins)

Canola/rapeseed

Rapeseed, or the canola plant, belongs to the Brassicaceae family. This crop is a food grade oil-producing plant that has become a major oilseed in the last 40 years (Wanasundara *et al.*, 2016). Canola meal at present is primarily used as animal feed. The product is a competitive animal feed source owing to its high protein value and has an amino acid profile very competitive with other plant proteins currently on the market. The food protein source has a high bioavailability and digestibility (Campbell *et al.*, 2016). Suitable production techniques are needed that can isolate proteins appropriate for human food application, and the product needs to fit within the regulatory frameworks required for novel protein sources (Wanasundara *et al.*, 2016). One of the biggest challenges determining the success of canola protein as a food ingredient is being able to bring a product to the market at a competitive price.

Mycoprotein

To find an optimum strain for microbial proteins, approximately 3,000 fungi were screened, and *Fusarium venenatum* was selected as the best (Ritchie *et al.*, 2017). The cell walls of the hyphae (cells) are the source of dietary fibre (chitin and glucan), and the cell membranes are the source of polyunsaturated fat and the cytoplasm is the source of high-quality protein (Malav *et al.*, 2015). The fungus has a favourable fatty acid profile and is cholesterol free with a fibre content that is comparable to other vegetarian protein sources. By continuous fermentation of this fungal species, a product is obtained that has a fibrous texture analogous to that of meat, which can be flavoured and further textured to resemble meat. Mycoprotein is produced and commercialized by Marlow Foods in the UK, and sold under the brand name Quorn (Ritchie *et al.*, 2017). These products serve a narrow, premium, vegetarian market, and only significant price reductions will expand the market. Biotechnological innovations may be able to half the price.

Algae

Algae include microalgae and seaweeds. They have increasingly important applications as both food ingredients and animal feed (Packer *et al.*, 2016). There are some 350,000 species of algae but only a few species are currently domesticated. The unexploited potential for use as food or in feeds is therefore large.

Microalgae are unicellular. They can achieve high areal yields and can be grown on non-arable land. Because of the unique lipid composition and the high protein content, they are considered a next generation bio-based/food feedstock (Eppink *et al.*, 2017). The closed-system growth process allows for a consistent and clean production. Microalgae grown on sugarcane feedstock can produce a highly sustainable source of protein (Klamcynska and Mooney, 2016). To develop a sustainable and economically feasible process all (fragile) biomass components should be used. Therefore, an integrated continuous biorefinery technology concept for microalgae processing is needed.

Macroalgae (seaweed and kelp in English) are multicellular. Since macroalgae can be grown in water (oceans and lakes), they will not compete with land-based crops, and thus will not be in competition with human foods. Their composition is highly variable, depending on the species, time of collection and habitat, and on external conditions such as water temperature, light intensity and nutrient concentration in water. Seaweeds tend to accumulate heavy metals (arsenic), iodine, and other minerals, and feeding such seaweeds could deteriorate animal and human health (Packer *et al.*, 2016). Regular monitoring of minerals in seaweeds would prevent toxic and other undesirable situations. There are about 10,000 species of seaweeds, but only about ten are of interest for use as animal feed. Seaweeds have seen a renewed interest as feed ingredients since the 1960s, when Norway started producing seaweed meal from kelp (Makkar *et al.*, 2016). Seaweeds include brown algae (Phaeophyceae), red algae (Rhodophyceae) and green algae (Chlorophyceae).

Insects and other arthropods will be covered in the next pages.

1.5 Edible insects

1.5.1 Species and occurrence

Jongema (2017) produced a list of edible insect species around the world; 2,111 different species belonging to the following groups: beetle larvae (31%), caterpillars (17%), wasps, bees and ants (15%), crickets, grasshoppers and locusts (13%), true bugs (11%), dragonflies (3%), termites (3%), flies (2%), cockroaches (2%), spiders (1%) and others (2%) (Figure 1.2). A world map has also been produced on the occurrence of edible insects based on refereed literature data. However, caution has to be taken as the number of species often reflects research intensity rather than actual figures (e.g. Mexico figures with 395 species, very likely to the activity of Julieta Ramos Elorduy, who wrote more than 150 publications on edible insects) (Pino Moreno, 2016).

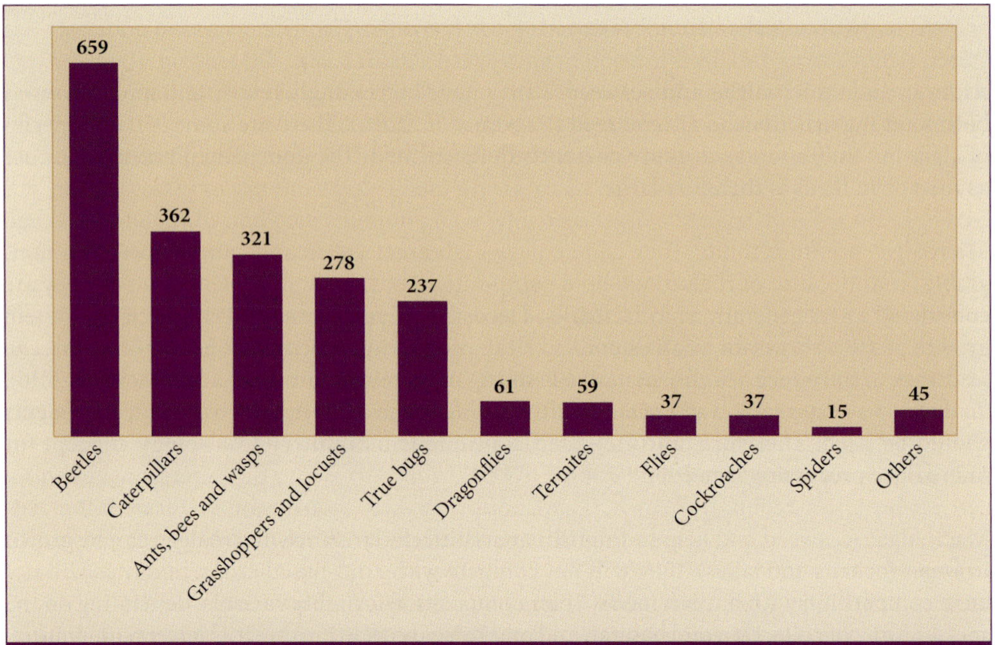

Figure 1.2. Number of recorded edible insect species per group in the world (Jongema, 2017).

1.5.2 History

An historical account of eating insects in western civilisation is given by Evans *et al.* (2015). They did so because the authors were puzzled by significance of the word 'entomophagy'. The term was used by researchers without a background of eating insects in their own culture. The best example is the book 'Peuples entomophages et insects comestibles' (Entomophagous people and edible insects) by Emile Bergier (Bergier, 1941), which clearly shows that some ethnic groups mainly in tropical countries had the 'strange' habit of eating insects. Such an overview was given also by Bodenheimer (1951) in his book 'Insects as human food; a chapter of the ecology of man.' Afterwards, DeFoliart (2012) gave such an overview in his on line bibliography 'The human use of insects as a food resource: a bibliographic account in progress.' The question is why insects are eaten in the tropics but not in temperate zones? Van Huis (2017a) gives several reasons: insects are available in the tropics throughout the year, they are larger and often aggregated facilitating harvesting, people are closer into contact with nature, and westerners do not value insects very much. The main reason is: insects can be harvested much easier. However, there are examples in the western world of insects being eaten, when insects are abundant. One is the cockchafer *Melolontha* spp. (Coleoptera: Scarabaeidae). This species was a major problem in agriculture and forestry and in the pre-industrialized era. The main mechanism to control their numbers was to collect and kill the adult beetles, thereby interrupting the cycle, which is why they were considered as food in France and Germany (Van Huis *et al.*, 2013; p. 55). Another example is that of `the Rocky Mountain locust (*Melanoplus spretus*), an extinct species of locust that ranged through the western half of the United States and some western portions of Canada until the end of the 19th century. Swarms have been sighted covering 510,000 km^2, greater than the area of California. Charles Valentine Riley was appointed as the first State Entomologist for Missouri in 1868. He dealt with the locust plagues from 1873 to 1877 and was considered one of the leading entomologists in the United States. He recommended to eat the insect, and five pages of the commission's first report were devoted to consuming locusts (Lockwood, 2004; p. 116).

1.5.3 From harvesting from nature to rearing

Insects were/are harvested from nature, and there are only few examples where the availability of edible insect populations were encouraged, also called semi-domestication. Several examples are given by Van Itterbeeck and Van Huis (2012): providing egg laying sites of reed and grasses for aquatic Hemiptera in lakes of Mexico; manipulating the habitat to increase edible caterpillars in Africa; and cutting deliberately palm trees to encourage palm weevils to lay their eggs in Africa and Latin America.

However, is harvesting from nature an option if insects are becoming more popular as either food or feed? There are already examples where the food resource is threatened due to overexploitation, habitat changes and environmental contamination (Ramos-Elorduy, 2006). There are several ways to farm the insects, depending on the quantities needed. In Thailand insects are reared in small farms; for example, some 20,000 farms produce 7,500 tonnes of crickets a year for home consumption and sale at the local market(Hanboonsong *et al.*, 2013). However, it can also be done large scale in which large automated facilities produce tonnes of insects a week. Large scale insect rearing is known in pest management. One is the insect-sterile technique (SIT) in which millions of insects are produced and released to disrupt population dynamics of the pest (i.e. sterile males mate with females resulting in sterile eggs deposited).

Experience in rearing insects in large numbers are in biological control programmes, either by introduction, e.g. the control of the cassava mealybug by parasitoids in Africa (Neuenschwander, 2001), by augmentation (Van Lenteren, 2000) or by sterile insect technique (SIT) (Klassen, 2005). In particular, SIT requires very large numbers of insects, often flies such as the Mediterranean fruit fly, screw flies, and tsetse flies. However, insects are also reared as food for reptiles, birds and zoos. These are niche markets and do not concern huge numbers. To feed crew on a long duration flight on a spacecraft, insects as human food have a number of advantages. They can be reared in automated systems. The first attempts to rear the insects in reactors were with storage insects as a protein source (Box 1.1). However, now there are a number of companies worldwide engaging in large scale rearing of black soldier flies, house flies, mealworms and crickets.

Box 1.1. Automated insect systems derived from food models for long duration flights on a spacecraft.

Aboard a spacecraft on long-duration travels, stocking food is not really an option. How to grow and harvest food on the craft considering the limited space and required low energy supply. The insects selected need to be those that can convert low quality feed to high value proteins. Robert Kok (see Chapter 8: Kok, 2017) was one of the first who explored this idea and he used as model organism the drugstore beetle, *Stegobium paniceum* (Kok, 1983). He wrote in this article: '...it will be considerably simpler to program cybernauts and androids to handle insect colonies than to herd cattle or sheep'. With several co-authors, he further developed this idea with *Tribolium confusum* and he showed that it was feasible to mass produce insects for human consumption by industrial methods (Kok *et al.*, 1988). He prepared several foods from the pupae such as bread, spaghetti sauce and dog wieners. Volunteers found them palatable.

Katayama *et al.* (2008) for the same purpose explored several other species, which do not compete with human in terms of food resources, but convert inedible biomass or waste into edible food for humans. The species they choose were the silkworm *Bombyx mori*, the hawk-moth *Agrius convolvuli*, the drugstore beetle *Stegobium paniceum*, and the termite *Macrotermes subhyalinus*. Their idea was that in a space-based agro-ecosystem, insects can recycle materials, process waste, serve as food and feed, and pollinate plants. The silkworm was further investigated providing them mulberry and stem lettuce leaves (Tong *et al.*, 2011) and it was even possible to process the silk into an edible protein for humans (Yang *et al.*, 2009). For missions to Mars, a round trip taking 2.5 years, the above-mentioned insects were part (Katayama *et al.*, 2005, 2008; Yamashita *et al.*, 2009). The yellow mealworm (*Tenebrio molitor*) also proved to be an interesting candidate. The insect was tested in a closed integrative experimental facility, the Lunar Palace 1, using a 105-day multi-crew closed integrative bioregenerative life support systems (Li *et al.*, 2015). Jones (2015) calculated that the daily mealworm harvest that satisfies the protein requirement of a stable crew population of 160 on a craft is approximated to be ~162,000.

1.6 Why edible insects?

Reasons will be given why insects as a food and feed source is a serious alternative to conventional meat products and feedstocks, such as fishmeal. First the nutritional value of insects will be discussed.

1.6.1 Nutrition

Due to the large variety of insect species that can be eaten, more than 2,000, drawing generalisations can be quite difficult. Besides, there are many factors that have an influence on nutritional values: such as life stage harvested, diet, and environmental factors such as temperature, humidity, and photoperiod (Finke and Oonincx, 2014). Rumpold and Schlüter (2013) reviewed the nutritional composition of a large number of edible insect species and concluded that the data were subject to a large variation that many insect species meet amino acid requirements for humans, are high in monounsaturated fatty acids and polyunsaturated fatty acids, and rich in several micronutrients such as copper, iron, magnesium, manganese, phosphorous, selenium, and zinc as well as riboflavin, pantothenic acid, biotin, and in some cases folic acid. Payne *et al.* (2016) determined the nutrient value score of crickets and mealworms had a significant higher nutritional value than the commonly consumed meats beef and chicken. In their analyses, not a single comparison showed the insect species (crickets, honeybee, silkworm, mopane caterpillar, mealworms, palm weevil larvae) to be nutritionally inferior to beef, chicken or pork. The high iron and zinc content of many insect species (Latunde-Dada *et al.*, 2016) is interesting considering the high incidence of anaemia and zinc deficiencies in the world. For example, in the Democratic Republic of Congo infants who consumed a caterpillar cereal had higher haemoglobin concentration and fewer were anaemic (Bauserman *et al.*, 2015).

Chitin, the exoskeleton of insects, has shown to have an interesting role, in particular with regard to immune responses (Lee *et al.*, 2008; Muzzarelli, 2010; Reese *et al.*, 2007), among other in relation to asthma and allergy (Brinchmann *et al.*, 2011). Using chitin from shrimps, Khempaka *et al.* (2011) fed broilers with up to 15% shrimp meal and the result was increased populations of intestinal *Lactobacillus* and decreased intestinal *Escherichia coli* and cecal *Salmonella*. Humans likely have chitinase in their gastric fluids (Paoletti *et al.*, 2007).

1.6.2 Less greenhouse gases and ammonia emissions

Insects produce less ammonia and GHG than conventional livestock (Oonincx *et al.*, 2010). Using a life cycle analyses, production of one kg of edible protein from milk, chicken, swine, or beef resulted in higher GHG emissions than mealworms and it required much more land, while similar amounts of energy were required (Oonincx and De Boer, 2012). An extensive overview of environmental impact of producing edible insects in comparison to common production animals is given in Chapter 3 of this book (Oonincx, 2017).

1.6.3 High feed conversion efficiency

Insects are poikilotherm (cold blooded), so they do not need to maintain a body temperature, contrary to homeothermic animals, which is probably why insects have higher feed conversion efficiencies than common production animals. Using literature references, Van Huis (2013) mentioned the following feed conversion ratio's (kg feed/kg liveweight) comparing crickets with conventional livestock: crickets 1.7; poultry 2.5; swine 5; and beef 10. Collavo *et al.* (2005) estimated feed conversion ratio's (feed consumed/biomass gained) from literature values to be for: crickets 1.1-1.7 (last figure was their own experiment); poultry 2.1-2.9; pigs 3.2-3.6; cattle 6.3-6.7. If edible weight would be calculated than the crickets are even doing better because at least 80% of the body can be consumed as to only 40-55% of the livestock species. Oonincx *et al.* (2010) indicated that CO_2 production expressed as g/kg mass gain indicated higher feed conversion efficiencies for insects. Van Broekhoven *et al.* (2015) calculated feed conversion efficiencies for three mealworm species (weight ingested food/weight gained) using different quality diets and found: *Tenebrio molitor* 3.4-6.1, *Alphitobius diaperinus* 3.0-24.6; and *Zophobas atratus* 2.7-5.6. Oonincx *et al.* (2015a) also used high and low quality diets and found the following values (weight ingested food/weight gained); crickets 2.3-10; black soldier fly 1.8-2.6; and the yellow mealworm 3.9-19.1. The last two studies indicated that the quality of the diet has a large influence. That is why Lundy and Parrella (2015) for the house cricket, indicated that only when organic side-streams can be found of relatively high-quality, it may compete with livestock products.

1.6.4 Ability to convert organic side streams

FAO (2011) estimated that 'roughly one-third of food produced for human consumption is lost or wasted globally, which amounts to about 1.3 billion tonnes per year'. The value of the lost and wasted food has been estimated to be 750 billion US$ (The Economist, 2014). Most of it is vegetables (23%), followed by meat (21%), fruits (19%) and cereals (18%) (Figure 1.3). Besides, this loss and waste is associated with approximately 173 billion cubic meters of water consumption per year and 198 million hectares of cropland per year (Lipinski *et al.*, 2013).

The use of side streams to grow insects is of course a very interesting option. Most considered as insects as feed are the black soldier fly and the house fly. The capacity to live on organic side streams seems to be somewhat larger for the black soldier fly. Both have quite different life cycles in which the advantage of the house fly is their tremendous reproductive capacity (Figure 1.4). However also other fly species such as the blow fly, *Lucilia sericata*, and face fly, *Musca autumnalis*, can be considered (Čičková *et al.*, 2015). For insects for human consumption the yellow mealworm and crickets can also be grown on organic side-streams, in particular fruit and vegetable remains as well as dried distillers grains with solubles can be considered. An overview of these species grown on different types of organic side streams is given in Table 1.5. Also, a list is given of references indicating which insect species can be fed to pigs, poultry, to animals in general and to a number of fish species (Table 1.6). As alternatives for fish meal plant proteins (soya protein, corn gluten, pea meal and wheat gluten) have been and will probably continue to be the main choice when replacing fishmeal in aquaculture. However, these products have limitations due several nutritional drawbacks compared to fishmeal (Lock *et al.*, 2015; Olsen and Hasan, 2012) particularly in diets for carnivorous species, which are not adapted to plant feed. Besides, there is a relative low content of proteins, the amino acid profile is unbalanced, and the fibre content is high. Anti-nutritional components are present, reducing digestion or absorption of nutrients, counteract the function of vitamins and may even induce toxicity. Plant proteins also compete with use for human consumption.

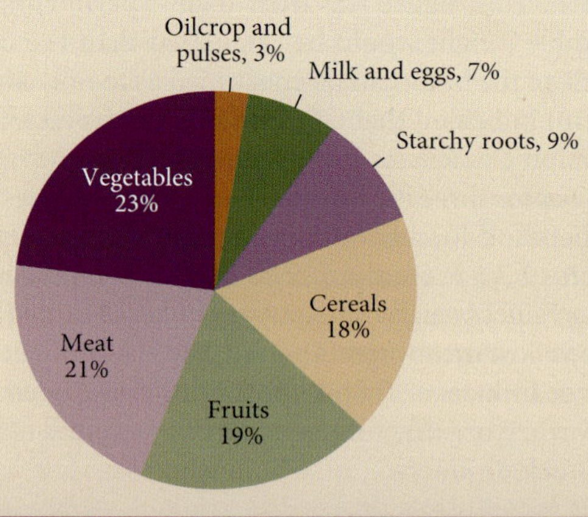

Figure 1.3. Global economic costs of food wastage (750 billion US$), by commodity, 2007 (The Economist, 2014).

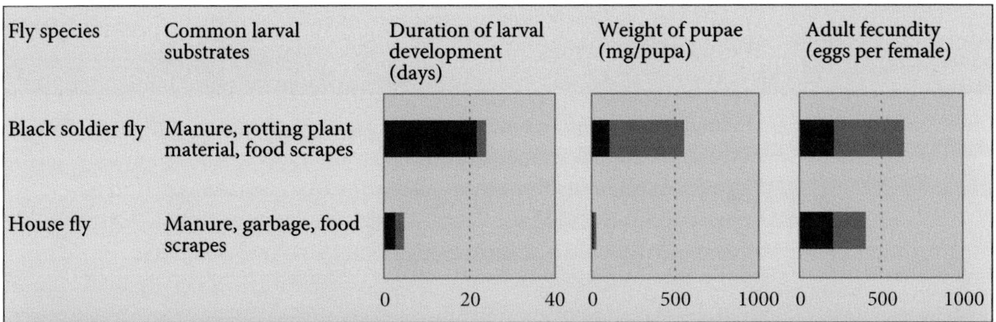

Figure 1.4. Comparison of life histories (under optimum conditions) of some fly species used for biodegradation (Ĉiĉková *et al.*, 2015).

Table 1.5. Insects species able to convert certain types of organic waste streams and references.

Insect species	Organic side streams and references
Black soldier fly	• organic waste (Čičková *et al.*, 2015; Diener *et al.*, 2011b; Leong *et al.*, 2015; Oonincx *et al.*, 2015a; Pastor *et al.*, 2015; Surendra *et al.*, 2016)
	• manure (Lalander *et al.*, 2015; Li *et al.*, 2011; Newton *et al.*, 2005a; Oonincx *et al.*, 2015b; Zhou *et al.*, 2013); pig manure (Newton *et al.*, 2005b; Wang *et al.*, 2013); cattle manure (Liu *et al.*, 2008; Myers *et al.*, 2008; Rehman *et al.*, 2017); chicken manure (Erickson *et al.*, 2004; Sheppard, 1983; Sheppard *et al.*, 1994); kitchen waste (Driemeyer, 2016)
	• coffee pulp (Lardé, 1989, 1990)
	• vegetables (Pineda Mejia, 2015; Rehman *et al.*, 2017; Spranghers *et al.*, 2016; Supriyatna *et al.*, 2016)
	• catering waste (Jeon *et al.*, 2011; Spranghers *et al.*, 2016; Zheng *et al.*, 2012a,b)
	• municipal organic waste (Diener *et al.*, 2009, 2011a; Gabler, 2014)
	• straw (Manurung *et al.*, 2016; Nicks *et al.*, 2003; Zheng *et al.*, 2012a)
	• dried distillers grains with solubles (Spranghers *et al.*, 2016; Webster *et al.*, 2015)
	• sorghum and cowpea (Tinder *et al.*, 2017)
House fly	• organic waste (Čičková *et al.*, 2015; Pastor *et al.*, 2015; Ramos-Elorduy and Morales, 1989)
	• manure (Shah *et al.*, 2016); pig manure (Čičková *et al.*, 2012a,b; Roffeis *et al.*, 2015; Wang *et al.*, 2013; Zhang *et al.*, 2012); poultry (El Boushy, 1991; Teotia and Miller, 1974); cattle (Hussein *et al.*, 2017)
	• municipal organic waste (Ocio *et al.*, 1979)
Yellow mealworm	• organic waste (Oonincx *et al.*, 2015a)
	• vegetables (Ramos-Elorduy *et al.*, 2002; Van Broekhoven *et al.*, 2015)
	• dried distillers grains with solubles (Van Broekhoven *et al.*, 2015)
Crickets	• organic waste (Oonincx *et al.*, 2015a)
	• vegetables/cereals (Caparros Megido *et al.*, 2015; Miech *et al.*, 2016)
	• weeds (Miech *et al.*, 2016)
Cydia pomonella	• wastewater sludges (Brar *et al.*, 2008)

Table 1.6. References of insect species fed to animals, and more specifically fed to poultry and pigs as well as a number of fish, molluscs and crustacean species.

Insect species	Animal species[1] and references
General	general (Kenis *et al.*, 2014; Moon and Lee, 2015; Sánchez-Muros *et al.*, 2016); poultry and pigs (Makkar *et al.*, 2014; Veldkamp and Bosch, 2015; Veldkamp *et al.*, 2012); fish (Barroso *et al.*, 2014; Riddick, 2013, 2014; Tran *et al.*, 2015)
Coleoptera	
• Yellow mealworm (*Tenebrio molitor*)	poultry (Alhamidi, 2017; Ballitoc and Sun, 2013; Biasato *et al.*, 2016; Bovera *et al.*, 2015; De Marco *et al.*, 2015; Islam and Yang, 2016; Loponte *et al.*, 2017; Ramos-Elorduy *et al.*, 2002); pigs (Jin *et al.*, 2016); fish and shrimp: *Ameiurus melas* (Roncarati *et al.*, 2015); *Clarias gariepinus* (Ng *et al.*, 2001); *Litopenaeus vannamei* (Panini *et al.*, 2017); *Oreochromis niloticus* (Sánchez-Muros *et al.*, 2015); *Pagellus bogaraveo* (Laconisi *et al.*, 2017)
• Super mealworm (*Zophobas morio*)	poultry (Islam and Yang, 2016)
• Lesser mealworm (*Alphitobius diaperinus*)	poultry (Despins and Axtell, 1995)
Diptera	
• Black soldier fly (*Hermetia illucens*)	poultry (Cullere *et al.*, 2016; De Marco *et al.*, 2015; Elwert *et al.*, 2010; Gaffigan, 2017; Leiber *et al.*, 2015; Loponte *et al.*, 2017; Oluokun, 2000); pigs (Baiocco *et al.*, 2016); fish and shrimp: *Cyprinus carpio* var. Jian (Li *et al.*, 2017); *Dicentrarchus labrax* (López, 2017; Magalhães *et al.*, 2017); *Ictalurus punctatus* (Bondari and Sheppard, 1987); *L. vannamei* (Cummins Jr *et al.*, 2017); *Oncorhynchus mykiss* (Sealey *et al.*, 2011; St-Hilaire *et al.*, 2007); *Oreochromis aureus* (Bondari and Sheppard, 1987); *Salmo salar* (Lock *et al.*, 2015)
• House fly (*Musca domestica*)	poultry (Adeniji, 2007; Hwangbo *et al.*, 2009; Pieterse *et al.*, 2013); fish and mollusc: *Achatina* spp. (Mbunwen *et al.*, 2011); *C. gariepinus* (Aniebo *et al.*, 2009; Arong and Eyo, 2017; Fasakin *et al.*, 2003; Idowu *et al.*, 2003; Kareem and Ogunremi, 2012); *Heteroclarias* (Ekelemu, 2015); *O. mykiss* (St-Hilaire *et al.*, 2007); *O. niloticus* (Chenyambuga *et al.*, 2017; Ogunjil *et al.*, 2006; Wang *et al.*, in press)
• Maggots (not specified; likely house fly)	general (Bouafou, 2011); poultry (Adesina *et al.*, 2011; Akpodiete and Inoni, 2000; Awoniyi *et al.*, 2003; Ncobela and Chimonyo, 2015; Téguia *et al.*, 2002); fish: *Clarias anguillaris* (Madu and Ufodike, 2003)
Other flies	
• Asian latrine fly (*Cryosoma megacephala*)	fish: *Oreochromis* sp. (Sing *et al.*, 2014); *O. niloticus* (Akhter, 2015)
• Green bottle fly (*Lucilia sericata*)	fish: *Sparus aurata* (De Haro *et al.*, 2015)

>>>

Table 1.6. Continued.	
Insect species	**Animal species[1] and references**
Orthoptera	
• Grasshopper, locust	poultry (Adeyemo *et al.*, 2008; Daudu *et al.*, 2015; Longe *et al.*, 2007; Ncobela and Chimonyo, 2015; Ojewola *et al.*, 2005; Wang, 2007); pigs (Hemsted, 1947); fish: *C. gariepinus* (Alegbeleye *et al.*, 2011)
• House cricket (*Acheta domesticus*)	poultry (Nakagaki *et al.*, 1987); fish: *C. gariepinus* (Norhidayah, 2016)
• *Sitophilus zeamais*	poultry (López Vergé, 2013)
Lepidoptera	
• Silkworm	poultry (Ijaiya and Eko, 2009; Ncobela and Chimonyo, 2015); pigs (Medhi *et al.*, 2016); fish: *C. gariepinus* (Kurbanov *et al.*, 2015)
• *Cirina butyrospermi*	fish: *C. gariepinus* (Anvo *et al.*, 2016)
Isoptera (termites)	poultry (Munyuli Bin Mushambanyi and Balezi, 2002); fish: *C. gariepinus* (Arong and Eyo, 2017)
Dyctioptera (cockroaches)	poultry (Munyuli Bin Mushambanyi and Balezi, 2002)

[1] Common names of fish, mollusc and crustaceaen species: *Achatina* spp. (African giant snail); *A. melas* (common catfish); *C. anguillaris* (catfish); *C. gariepinus* (African catfish); *C. carpio* var. Jian (Jian carp); *D. labrax* (European seabass); *Heteroclarias* (hybrid of *C. gariepinus* × *Heterobranchus bidorsalis*); *I. punctatus* (channel catfish); *L. vannamei* (Pacific white shrimp); *O. mykiss* (rainbow trout); *Oreochromis* sp. (red tilapia); *O. aureus* (blue tilapia); *O. niloticus* (Nile tilapia); *P. bogaraveo* (blackspot sea bream); *S. salar* (Atlantic salmon); *S. aurata* (gilthead sea bream).

1.7 From producing to consuming

We will below give a brief overview of the different chapters. An introduction (this chapter) is given in Chapter 1 'The potential of insects as food and feed'.

A negative attitude towards insects is deeply embedded in Western cultures (Looy and Wood, 2015). They are often considered as alien, threatening, dangerous, and disgusting. This cultural preconception will have implications when promoting insect as food. Therefore, we start the book by explaining the significance of insects in our ecosystems. We all know that insect produce honey and silk, but insects also provide important ecosystem services such as pollination, dung and carrion removal and they control pests as biological control agents. Besides, insects are important food for birds and fish. Considering the increasing importance of aquaculture and the search for alternative protein feeds for poultry and pig industry, insects as feed can play an important role. In the western world during the last ten years they are increasingly recognized as a potential food source. This all will be explained in Chapter 2 (Dicke, 2017).

The question is when insects are being farmed as food and feed how this compares to conventional livestock species. In Chapter 3, the direct emissions of greenhouse gasses and ammonia from insect production are discussed and compared to conventional production animals (Oonincx, 2017). To assess environmental impacts a live cycle assessment (LCA) is often used and is the technique associates all the stages of a product's life. In insect farming, there are very few of such assessments. It has been done for the yellow mealworm and the results are presented for energy use, global warming potential, land use and water use. Food waste can be used as insect feed but also for biogas production and when this alternative use is taken into account in the assessment, we call this a consequential LCA. The environment impact of feed utilization is discussed for conventional production animals and compared with insects. The environmental impact of honey and silk production is also discussed.

In the tropics, it is common to use insect as food and feed. Insects are harvested from nature and are used either for domestic consumption or put on the market. However, most edible insect species depend on host plants, which entails that they are only seasonally available. But, in tropical zones edible insects are available throughout the year, although the species composition varies and the quantities harvested differ. The only way to make some insect species available throughout the year both in tropical and temperate zones is to farm them. In Chapter 4-7, we have chosen a few species for which simple production systems have been developed. After an introduction (Chapter 4: Van Huis, 2017b), examples of crickets, palm weevils and grasshoppers are given. We have chosen crickets for which in southeast Asia and in East Africa simple and sustainable rearing systems are being developed (Chapter 5: Caparros Megido *et al.*, 2017). We have chosen palm weevils (*Rynchophorous* spp.) of which the larvae are very popular as food throughout the tropics. In Africa, Latin America and Asia different species are eaten and we give an example from Africa (Chapter 6: Muafor *et al.*, 2017). The last example is using grasshoppers, mainly as feed in India (Chapter 7: Haldar and Malakar, 2017).

Mass-producing insects has been done for three different reasons: (1) biological control companies that produce predators, parasitoids and predatory mites to control agricultural pests; (2) sterile insect techniques in which sterile males, when released into the wild, mate with females, which then produce no offspring; (3) as pet feed. Now it is proposed to rear insects as food and feed. Rearing at an industrial scale means producing tonnes of insects a day. In Chapter 8 (Kok, 2017) a number of issues that have to be dealt with are discussed: organisms and feeds, process types, reactor types, environmental requirements, heat and mass transfer, materials handling, process control, air conditioning, waste management, cleaning, strain breeding, culture maintenance, population synchronicity, and safety.

Mass- and laboratory-rearing of insects is a key component of several integrated pest management strategies. However, it can have dramatic effects on insect performance via laboratory adaptation, inbreeding depression, inadvertent selection or through direct rearing effects (e.g. crowding and artificial diets) (Sørensen *et al.*, 2012). Chapter 9 focusses on how to maintain a genetically sound breeding stock while exploiting the genetic potential of a given species, and to utilize both genetic and environmental conditions for optimizing the production (Jensen *et al.*, 2017). A major challenge in production systems is to produce robust animals producing homogenous products. The authors use knowledge from research in population genetics and animal breeding.

Many insect rearing companies fear pests and diseases in their production system. This also because not much is known about the causing organisms which can be predators, parasitoids, and microorganisms (viruses, fungi, bacteria, protozoa, etc.). There is not much expertise, except for those entomopathogens that are used in biological control of agricultural pests. Also how to avoid and prevent that diseases enter the production facility and if it happens what should be done to combat it. Chapter 10 gives a number of recommendations involving: the physical characteristics of the production facilities, the rearing conditions, the monitoring of possible malfunctions, hygiene, genetic variation of the insect stock, the quality of the substrates used, and how to apply for assistance (Eilenberg *et al.*, 2017).

Conditions to produce insects in large numbers differ among species. In Chapter 11-15 four insect species are discussed: two fly species used as feed, the black soldier fly (Tomberlin and Cammack, 2017) and the house fly (Tomberlin *et al.*, 2017); and two species often used as human food, mealworms (Berezina, 2017) and crickets (Mott, 2017).

How to achieve an insect that is nutritionally adequate for the target species, such as poultry, swine, fish and humans? In Chapter 16 (Finke and Oonincx, 2017) this is discussed using a variety of insect species: the protein content and the amino acid profile, the protein digestibility, fat content and the fatty acid profile, carbohydrates, fibre and chitin, essential minerals and vitamin content. It will also be discussed what the effect of diet is on the nutrient composition of the target animal. Other factors that have an effect on the nutritional composition of the insects are: environmental conditions (temperature, light, and humidity) and processing such as protein extraction and cooking.

Food and feed processing of insects is to obtain products with a desired chemical and nutrient composition, digestibility and bioavailability of components, texture, size, colour and taste. At the same time it is necessary to achieve a long shelf-life, ensure product safety, and achieve optimum product quality. The production process need to be sustainable and cost-effective. In Chapter 17 examples are given of decontamination, drying and comminution methods as well as subsequent fractionation methods for protein, fat or chitin recovery (Rumpold *et al.*, 2017). The impact of the different processing steps on product properties is discussed and examples for whole insect processing pathways are presented.

Because insects have in the past hardly been considered to be used both as food and feed, it is only now with the increased attention, that it is realized that legislation is lagging behind developments and that a number of 'legal gaps' need to be addressed. In Chapter 18 these issues are discussed by using the European Union as an example (Meijer and Van der Fels-Klerx, 2017). A number of food safety issues are addressed such as biological hazards (microbes), hazardous chemical, residues of veterinary drugs and persistent organic pollutants and allergens. This can be influenced by type of substrate, insect species, life stage of the insects upon harvest, and the nature of subsequent processing. What are the legal requirements and will they be loosened in future?

The question about insect welfare was hardly asked when dealing with pest management but now that insects are considered as food and feed the topic seems to become relevant. In particular whether the insects can be considered as conscious organisms. Do welfare consideration in livestock welfare apply to insect production system? In welfare considerations how does the killing of one cow compare to the killing of thousands of insects? Do insects possess integrity and can it be violated through large scale production systems? Can insects experience emotions like pain and fear? These issues will be discussed in Chapter 19 (Röcklinsberg *et al.*, 2017).

Enabling conditions deals with a set of institutional (formal or informal) arrangements among stakeholders, often within the national context of a given country or region. Enabling conditions at the country level include: (1) a tight collaboration between government, private sector and academic institutions; (2) a strong message that convinces private and public sector of the potential of the endeavour; and (3) an awareness of urgency to find alternative protein sources. Actors at the national and international level can create enabling conditions, also for consumers. Scientific articles, cookbooks, the press, social media, workshops and conferences all play a role in creating a conducive environment for the sector of insects as food and feed to develop. What were the enabling conditions for a wide range of different stakeholders is dealt with in Chapter 20 (Vantomme, 2017).

How to promote insect products and convince consumers about the benefits, how are insects perceived by the general public. An overview is given per region and country in Chapter 21 (Shockley *et al.*, 2017).

With the farming of edible insects, there are a number of issues that need to be considered. Which insect species and/or insect strain to be used and for which purpose, feed or food? How to avoid genetic problems in the insect rearing? How to obtain insect with the appropriate nutritional composition considering the target species? Which kind or rearing conditions to get optimal development and a low mortality? How to avoid pest and disease problems? When farming the insect is it necessary to take insect welfare into account? How to process the insects? How to ensure food safety and which are the legal requirements? How to convince consumers to eat insect products? What are enabling condition and how to market the insects. This will be discussed in Chapter 22 (Van Huis and Tomberlin, 2017).

References

Adeniji, A.A., 2007. Effect of replacing groundnut cake with maggot meal in the diet of broilers. International Journal of Poultry Science 6: 822-825.

Adesina, M., Adejinmi, O., Omole, A., Fayenuwo, J. and Osunkeye, O., 2011. Performance of broilers' finishers fed graded levels of Cassava peel-maggot meal-based diet mixtures. Journal of Agriculture, Forestry and the Social Sciences 9: 226-231.

Adeyemo, G.O., Longe, O.G. and Lawal, H.A., 2008. Effects of feeding desert locust meal (*Schistocerca gregaria*) on performance and haematology of broilers. In: Book of Abstracts, Tropentag 2008, Hohenheim, Germany. Available at: http://www.tropentag.de/2008/proceedings/proceedings.pdf.

Akhter, Z., 2015. Utilization and growth perfomance ofg Nile tilapia fed Asian latrine fly maggots. MSc thesis, Norwegian University of Life Sciences, Faculty of Veterinary Medicine and BioSciences, Department of Animal and Aquacultural Sciences, Ås, Norway.

Akpodiete, O.J. and Inoni, O.E., 2000. Economics of production of broiler chickens fed maggot meal as replacement for fish meal. Nigerian Journal of Animal Production 27: 59-63.

Alegbeleye, W.O., Obasa, S.O., Olude, O.O., Otubu, K. and Jimoh, W., 2011. Preliminary evaluation of the nutritive value of the variegated grasshopper (*Zonocerus variegatus* L.) for African catfish *Clarias gariepinus* (Burchell. 1822) fingerlings. Aquaculture Research 43: 412-420.

Alexandratos, N. and Bruinsma, J., 2012. World agriculture towards 2030/2050: the 2012 revision. Global Perspective Studies Team. ESA Working Paper No. 12-03. Agricultural Development Economics Division, Food and Agriculture Organization of the United Nations, Rome, Italy.

Alhamidi, M.B., 2017. Energi metabolis tepung ulat Hongkong (*Tenebrio molitor* l.), tepung ikan dan meat bone meal pada broiler [Metabolizable energy of mealworm meal (*Tenebrio molitor* l.), fish meal and meat bone meal in broilers]. Departemen Ilmu Nutrisi dan Teknologi Pakan, Fakultas Peternakan, Institut Pertanian, Bogor, Indonesia.

Aniebo, A.O., Erondu, E.S. and Owen, O.J., 2009. Replacement of fish meal with maggot meal in African catfish (*Clarias gariepinus*) diets [Sustitución de harina de pescado con harina de larvas en dietas para el bagre Africano (*Clarias gariepinus*)]. Revista Científica UDO Agrícola 9: 653-656.

Anvo, M.P.M., Toguyéni, A., Otchoumou, A.K., Zoungrana-Kaboré, C.Y. and Kouamelan, E.P., 2016. Evaluation of Cirina butyrospermi caterpillar's meal as an alternative protein source in *Clarias gariepinus* (Burchell, 1822) larvae feeding. International Journal of Fisheries and Aquatic Studies 4: 88-94.

Arong, G.A. and Eyo, V.O., 2017. Evaluation of house fly (*Musca domestica*) maggot meal and termite (*Macrotermes subhyalinus*) meal as supplementary feed for African catfish *Clarias gariepinus* (Burchell, 1822). International Journal of Entomology and Nematology 3: 42-50.

Awoniyi, T.A.M., Aletor, V.A. and Aina, J.M., 2003. Performance of broiler-chickens fed on maggot meal in place of fishmeal. International Journal of Poultry Science 2: 271-274.

Baiocco, L., Bortoluzzi, C., Kruft, H., Schlingmann, B., Seesink, I. and Zweekhorst, I., 2016. Black soldier flies as an alternative protein source in pig feed in Kenya. Agriprotenza, MSc project course Academic Consultancy Training (ACT). Wageningen University, Wageningen, the Netherlands.

Ballitoc, D.A. and Sun, S., 2013. Ground yellow mealworms (*Tenebrio molitor* L.) feed supplementation improves growth performance and carcass yield characteristics in broilers. Open Sience Repository Agriculture, 14 pp.

Barroso, F.G., de Haro, C., Sánchez-Muros, M.-J., Venegas, E., Martínez-Sánchez, A. and Pérez-Bañón, C., 2014. The potential of various insect species for use as food for fish. Aquaculture 422-423: 193-201.

Bauserman, M., Lokangaka, A., Gado, J., Close, K., Wallace, D., Kodondi, K.-K., Tshefu, A. and Bose, C., 2015. A cluster-randomized trial determining the efficacy of caterpillar cereal as a locally available and sustainable complementary food to prevent stunting and anaemia. Public Health Nutrition 18: 1785-1792.

Berezina, N., 2017. Mealworms, promising beetles for the insect industry. In: Van Huis, A. and Tomberlin, J.K. (eds.) Insects as food and feed: from production to consumption. Wageningen Academic Publishers, Wageningen, the Netherlands, pp. 258-269.

Bergier, E., 1941. Peuples entomophages et insectes comestibles: etude sur les moeurs de l'homme et de l'insecte. Imprimerie Rulliere Freres, Avignon, France.

Beusen, A.H.W., Bouwman, A.F., Heuberger, P.S.C., Van Drecht, G. and Van Der Hoek, K.W., 2008. Bottom-up uncertainty estimates of global ammonia emissions from global agricultural production systems. Atmospheric Environment 42: 6067-6077.

Biasato, I., De Marco, M., Rotolo, L., Renna, M., Lussiana, C., Dabbou, S., Capucchio, M.T., Biasibetti, E., Costa, P., Gai, F., Pozzo, L., Dezzutto, D., Bergagna, S., Martínez, S., Tarantola, M., Gasco, L. and Schiavone, A., 2016. Effects of dietary *Tenebrio molitor* meal inclusion in free-range chickens. Journal of Animal Physiology and Animal Nutrition 100: 1104-1112.

Bodenheimer, F.S., 1951. Insects as human food; a chapter of the ecology of man. Dr. W. Junk, Publishers, The Hague, the Netherlands, 352 pp.

Boer, I.D., Cederberg, C., Eady, S., Gollnow, S., Kristensen, T., Macleod, M., Meul, M., Nemecek, T., Phong, L., Thoma, G., Van de Werf, H., Williams, A. and Zonderland-Thomassen, M., 2011. Greenhouse gas mitigation in animal production: towards an integrated life cycle sustainability assessment. Current Opinion in Environmental Sustainability 3: 423-431.

Bondari, K. and Sheppard, D.C., 1987. Soldier fly, *Hermetia illucens* L., larvae as feed for channel catfish, *Ictalurus punctatus* (Rafinesque), and blue tilapia, *Oreochromis aureus* (Steindachner). Aquaculture and Fisheries Management 18: 209-220.

Bouafou, K.G.M., 2011. Revue bibliographique sur les asticots et leur emploi dans l'alimentation animale [Literature review on maggots and their use in animal feed]. Journal of Animal and Plant Sciences 12: 1543-1551.

Bovera, F., Piccolo, G., Gasco, L., Marono, S., Loponte, R., Vassalotti, G., Mastellone, V., Lombardi, P., Attia, Y.A. and Nizza, A., 2015. Yellow mealworm larvae (*Tenebrio molitor* L.) as possible alternative to soybean meal in broiler diets. British Poultry Science 56: 569-575.

Bramble, B. and Fischer, B., 2016. The moral complexities of eating meat. Oxford University Press, New York, NY, USA.

Brar, S.K., Verma, M., Tyagi, R.D., Valéroand, J.R. and Surampalli, R.Y., 2008. Wastewater sludges as novel growth substrates for rearing codling moth larvae. World Journal of Microbiology and Biotechnology 24: 2849-2857.

Brinchmann, B.C., Bayat, M., Brøgger, T., Muttuvelu, D.V., Tjønneland, A. and Sigsgaard, T., 2011. A possible role of chitin in the pathogenesis of asthma and allergy. Annals of Agricultural and Environmental Medicine 18: 7-12.

Campbell, L., Rempel, C.B. and Wanasundara, J.P.D., 2016. Canola/rapeseed protein: future opportunities and directions. Workshop Proceedings of IRC 2015. Plants 5(2): 17.

Caparros Megido, R., Alabi, T., Nieus, C., Blecker, C., Danthine, S., Bogaert, J., Haubruge, É. and Francis, F., 2015. Optimisation of a cheap and residential small-scale production of edible crickets with local by-products as an alternative protein-rich human food source in Ratanakiri Province, Cambodia. Journal of the Science of Food and Agriculture 96: 627-632.

Caparros Megido, R., Haubruge, E. and Francis, F., 2017. Small-scale production of crickets and impact on rural livelihoods. In: Van Huis, A. and Tomberlin, J.K. (eds.) Insects as food and feed: from production to consumption. Wageningen Academic Publishers, Wageningen, the Netherlands, pp. 100-111.

Change, I.P.o.C., 2015. Climate change 2014: mitigation of climate change, 3. Cambridge University Press, Cambridge, UK.

Chenyambuga, S.W., Madalla, N. and Ally, T., 2017. Evaluation of invertebrates as protein sources in Nile tilapia (*Oreochromis niloticus*) diets. Sustainable seed technology and nutrient input systems/experiment/13SFT01PU. Aquaculture Research Facility, Department of Animal Science and Production at Sokoine University of Agriculture (SUA), Morogoro, Tanzania.

Čičková, H., Pastor, B., Kozánek, M., Martínez-Sánchez, A., Rojo, S. and Takáč, P., 2012a. Biodegradation of pig manure by the housefly, *Musca domestica*: a viable ecological strategy for pig manure management. Plos One 7: e32798.

Čičková, H., Kozánek, M., Morávek, I. and Takáč, P., 2012b. A behavioral method for separation of house fly (Diptera: Muscidae) larvae from processed pig manure. Journal of Economic Entomology 105: 62-66.

Čičková, H., Newton, G.L., Lacy, R.C. and Kozánek, M., 2015. The use of fly larvae for organic waste treatment. Waste Management 35: 68-80.

Collavo, A., Glew, R.H., Huang, Y.-S., Chuang, L.-T., Bosse, R. and Paoletti, M.G., 2005. House cricket small-scale farming. In: Paoletti, M.G. (ed.) Ecological implications of minilivestock: potential of insects, rodents, frogs and snails. CRC press, Boca Raton, FL, USA, pp. 519-544.

Cullere, M., Tasoniero, G., Giaccone, V., Miotti-Scapin, R., Claeys, E., Smet, S.D. and Zotte, A.D., 2016. Black soldier fly as dietary protein source for broiler quails: apparent digestibility, excreta microbial load, feed choice, performance, carcass and meat traits. Animal 10: 1923-1930.

Cummins Jr, V.C., Rawles, S.D., Thompson, K.R., Velasquez, A., Kobayashi, Y., Hager, J. and Webster, C.D., 2017. Evaluation of black soldier fly (*Hermetia illucens*) larvae meal as partial or total replacement of marine fish meal in practical diets for Pacific white shrimp (*Litopenaeus vannamei*). Aquaculture 473: 337-344.

D'Silva, J. and Webster, J., 2010. The meat crisis: developing more sustainable production and consumption. Earthscan, London, UK, 328 pp.

Daudu, O., Yusuf, T., Olugbemi, T. and Odegbile, O., 2015. Effect of grasshopper meal diets on performance of broiler chickens. In: Athanasiadou, S., Richardson, R.I., Chaudhry, A.S., Rooke, J.A., Chikunya, S., Rose, M.T., Denwood, M., Steele, M.A., Flockhart, J., Rymer, C., Flynn, R., Wall, E., Gibbons, J., Wolf, B.T., Houdijk, J.G.M., Wylie, A.R.G., Kenny, D.A., Mather, A., Mayes, R.W., McNeilly, T. and Nash, D.M. (eds.) Proceedings of the British Society of Animal Sciencein association with AVTRW, CFER and EBLEX. Advances in Animal Biosciences 6(2): 104.

Davis, K.F., Gephart, J.A., Emery, K.A., Leach, A.M., Galloway, J.N. and D'Odorico, P., 2016. Meeting future food demand with current agricultural resources. Global Environmental Change 39: 125-132.

De Haro, C., Ramos Bueno, R.P., Barroso, F.G., Sánchez Muros, M.J., Rincón Cervera, M.Á. and Guil-Guerrero, J.L., 2015. Insect larvae as feed ingredient selectively increase arachidonic acid content in farmed gilthead sea bream (*Sparus aurata* L.). Aquaculture Research 47: 2881-2887.

De Marco, M., Martínez, S., Hernandez, F., Madrid, J., Gai, F., Rotolo, L., Belforti, M., Bergero, D., Katz, H., Dabbou, S., Kovitvadhi, A., Zoccarato, I., Gasco, L. and Schiavone, A., 2015. Nutritional value of two insect larval meals (*Tenebrio molitor* and *Hermetia illucens*) for broiler chickens: apparent nutrient digestibility, apparent ileal amino acid digestibility and apparent metabolisable energy. Animal Feed Science and Technology 209: 211-218.

DeFoliart, G., 2012. The human use of insects as a food resource: a bibliographic account in progress. University of Wisconsin, Madison, WI, USA.

Despins, J.L. and Axtell, R.C., 1995. Feeding behavior and growth of broiler chicks fed larvae of the darkling beetle, *Alphitobius diaperinus*. Poultry Science 74: 331-336.

Dicke, M., 2017. Ecosystem services of insects. In: Van Huis, A. and Tomberlin, J.K. (eds.) Insects as food and feed: from production to consumption. Wageningen Academic Publishers, Wageningen, the Netherlands, pp. 60-76.

Diener, S., Gutiérrez, F.R., Zurbrügg, C. and Tockner, K., 2009. Are larvae of the black soldier fly – *Hermetia illucens* – a financially viable option for organic waste management in Costa Rica? Proceedings Sardinia 2009, 12th International Waste Management and Landfill Symposium. October 5-9, 2009. S. Margherita di Pula, Cagliari, Italy.

Diener, S., Studt Solano, N., Roa Gutiérrez, F., Zurbrügg, C. and Tockner, K., 2011a. Biological treatment of municipal organic waste using black soldier fly larvae. Waste and Biomass Valorization 2: 357-363.

Diener, S., Zurbrügg, C., Gutiérrez, F.R., Nguyen, D.H., Morel, A., Koottatep, T. and Tockner, K., 2011b. Black soldier fly larvae for organic waste treatment – prospects and constraints. In: Alamgir, M., Bari, Q.H., Rafizul, I.M., Islam, S.M.T., Sarkar, G. and Howlader, M.K. (eds.) Proceedings of the WasteSafe 2011 – 2nd International Conference on Solid Waste Management in the Developing Countries, February 13-15, 2011, Khulna, Bangladesh, pp 1-8.

Driemeyer, H., 2016. Evaluation of black soldier fly (*Hermetia illucens*) larvae as an alternative protein source in pig creep diets in relation to production, blood and manure microbiology parameters. MSc thesis, Faculty of AgriScience at Stellenbosch University, Stellenbosch, South Africa.

Eilenberg, J., Gasque, S.N. and Ros, V.I.D., 2017. Natural enemies in insect production systems. In: Van Huis, A. and Tomberlin, J.K. (eds.) Insects as food and feed: from production to consumption. Wageningen Academic Publishers, Wageningen, the Netherlands, pp. 200-223.

Ekelemu, J.K., 2015. Cost-benefit analysis and growth response of heteroclarias fingerlings fed diets containing graded levels of maggot meal as replacement for fish meal. Journal of Agriculture and Veterinary Sciences 2: 144-148.

El Boushy, A.R., 1991. House-fly pupae as poultry manure converters for animal feed: a review. Bioresource Technology 38: 45-49.

Elwert, C., Knips, I. and Katz, P., 2010. A novel protein source: maggot meal of the black soldier fly (*Hermetia illucens*) in broiler feed. In: Gierus, M., Kluth, H., Bulang, M. and Kluge, H. (eds.) Tagung Schweine- und Geflügelernährung. Institut für Agrar- und Ernährungswissenschaften, Universität Halle-Wittenberg, Lutherstadt Wittenberg, pp. 140-142.

Eppink, M.H.M., Olivieri, G., Reith, H., Van den Berg, C., Barbosa, M.J. and Wijffels, R.H., 2017. From current algae products to future biorefinery practices: a review. Springer Berlin Heidelberg, Berlin/Heidelberg, Germany, pp. 1-25.

Erickson, M.C., Islam, M., Sheppard, C., Liao, J. and Doyle, M.P., 2004. Reduction of *Escherichia coli* O157:H7 and *Salmonella enterica* serovar Enteritidis in chicken manure by larvae of the black soldier fly. Journal of Food Protection 67: 685-690.

Evans, J., Alemu, M.H., Flore, R., Frøst, M.B., Halloran, A., Jensen, A.B., Maciel-Vergara, G., Meyer-Rochow, V.B., Münke-Svendsen, C., Olsen, S.B., Payne, C., Roos, N., Rozin, P., Tan, H.S.G., Van Huis, A., Vantomme, P. and Eilenberg, J., 2015. 'Entomophagy': an evolving terminology in need of review. Journal of Insects as Food and Feed 1: 293-305.

Finke, M.D. and Oonincx, D.G.A.B., 2017. Nutrient content of insects. In: Van Huis, A. and Tomberlin, J.K. (eds.) Insects as food and feed: from production to consumption. Wageningen Academic Publishers, Wageningen, the Netherlands, pp. 290-316.

Fasakin, E.A., Balogun, A.M. and Ajayi, O.O., 2003. Evaluation of full-fat and defatted maggot meals in the feeding of clariid catfish *Clarias gariepinus* fingerlings. Aquaculture Research 34: 733-738.

Fayaz Bhat, Z. and Fayaz, H., 2011. Prospectus of cultured meat – advancing meat alternatives. Journal of Food Science and Technology 48: 125-140.

Finke, M.D. and Oonincx, D., 2014. Insects as food for insectivores. In: Juan Morales-Ramos, M.G.R. and Shapiro-Ilan, D. (ed.) Mass production of beneficial organisms. Academic Press, San Diego, CA, USA, pp. 583-616.

Foer, J.S., 2009. Eating animals. Little, Brown and Comapny, New York, NY, USA, 341 pp.

Food and Agriculture Organisation (FAO), 2011. Global food losses and food waste – extent, causes and prevention. FAO, Rome, Italy. Available at: http://www.fao.org/docrep/014/mb060e/mb060e00.pdf.

Food and Agriculture Organisation (FAO), 2016. The state of world fisheries and aquaculture: contributing to food security and nutrition for all. FAO, Rome, Italy. Available at: http://www.fao.org/3/a-i5555e.pdf.

Gabler, F., 2014. Using black soldier fly for waste recycling and effective *Salmonella* spp. reduction. Project thesis Sveriges lantbruksuniversitet Swedish University of Agricultural Sciences, Department of Energy and Technology. Uppsala, Sweden.

Gaffigan, M., 2017. Is insect protein a sustainable alternative to soy and fishmeal in poultry feed? Undergraduate Honors theses, 1344. University of Colorado, Boulder, CO, USA.

Gerber, P.J., Steinfeld, H., Henderson, B., Mottet, A., Opio, C., Dijkman, J., Falcucci, A. and Tempio, G., 2013. Tackling climate change through livestock – a global assessment of emissions and mitigation opportunities. FAO, Rome, Italy.

Haldar, P. and Malakar, C., 2017. Farming of grasshoppers for feed and food. In: Van Huis, A. and Tomberlin, J.K. (eds.) Insects as food and feed: from production to consumption. Wageningen Academic Publishers, Wageningen, the Netherlands, pp. 126-139.

Hallström, E., Carlsson-Kanyama, A. and Börjesson, P., 2015. Environmental impact of dietary change: a systematic review. Journal of Cleaner Production 91: 1-11.

Hanboonsong, Y., Jamjanya, T. and Durst, P.B., 2013. Six-legged livestock: edible insect farming, collection and marketing in Thailand. Food and Agriculture Organization of the United Nations, Regional Office for Asia and the Pacific, Bangkok, Thailand.

Haque, Md.A., Timilsena, Y.P. and Adhikari, B., 2016. Food proteins, structure, and function. In: Reference module in food science. Elsevier, Amsterdam, the Netherlands.

Hedenus, F., Wirsenius, S. and Johansson, D.A., 2014. The importance of reduced meat and dairy consumption for meeting stringent climate change targets. Climatic Change 124: 79-91.

Hemsted, W.R.T., 1947. Locusts as a protein supplement for pigs. East African Agricultural Journal 12: 225-226.

Henchion, M., McCarthy, M., Resconi, V.C. and Troy, D., 2014. Meat consumption: trends and quality matters. Meat Science 98: 561-568.

Herrero, M., Henderson, B., Havlik, P., Thornton, P.K., Conant, R.T., Smith, P., Wirsenius, S., Hristov, A.N., Gerber, P., Gill, M., Butterbach-Bahl, K., Valin, H., Garnett, T. and Stehfest, E., 2016. Greenhouse gas mitigation potentials in the livestock sector. Nature Climate Change 6: 452-461.

Herrero, M., Wirsenius, S., Henderson, B., Rigolot, C., Thornton, P., Havlík, P., De Boer, I. and Gerber, P.J., 2015. Livestock and the environment: what have we learned in the past decade? Annual Review of Environment and Resources 40: 177-202.

Hussein, M., Pillai, V.V., Goddard, J.M., Park, H.G., Kothapalli, K.S., Ross, D.A., Ketterings, Q.M., Brenna, J.T., Milstein, M.B., Marquis, H., Johnson, P.A., Nyrop, J.P. and Selvaraj, V., 2017. Sustainable production of housefly (*Musca domestica*) larvae as a protein-rich feed ingredient by utilizing cattle manure. Plos ONE 12: e0171708.

Hwangbo, J., Hong, E.C., Jang, A., Kang, H.K., Oh, J.S., Kim, B.W. and Park, B.S., 2009. Utilization of house fly-maggots, a feed supplement in the production of broiler chickens. Journal of Environmental Biology 30: 609-614.

Iaconisi, V., Marono, S., Parisi, G., Gasco, L., Genovese, L., Maricchiolo, G., Bovera, F. and Piccolo, G., 2017. Dietary inclusion of *Tenebrio molitor* larvae meal: effects on growth performance and final quality treats of blackspot sea bream (*Pagellus bogaraveo*). Aquaculture 476: 49-58.

Idowu, A.B., Amusan, A.A.S. and Oyediran, A.G., 2003. The response of *Clarias gariepinus* fingerlings (Burchell 1822) to the diet containing housefly maggot (*Musca domestica*) (L). Nigerian Journal of Animal Production 30: 139-144.

Ijaiya, A.T. and Eko, E.O., 2009. Effect of replacing dietary fish meal with silkworm (*Anaphe infracta*) caterpillar meal on performance, carcass characteristics and haematological parameters of finishing broiler chicken. Pakistan Journal of Nutrition 8: 850-855.

Intergovernmental Panel on Climate Change (IPCC), 2007. Climate change 2007: mitigation. contribution of Working Group III to the fourth assessment report of the Intergovernmental Panel on Climate Change, Cambridge University Press, Cambridge, UK.

Islam, M.M. and Yang, C.-J., 2016. Efficacy of mealworm and super mealworm larvae probiotics as an alternative to antibiotics challenged orally with *Salmonella* and *E. coli* infection in broiler chicks. Poultry Science 96: 27-34.

Jensen, K., Kristensen, T.N., Heckmann, L.-H.L. and Sørensen, J.G., 2017. Breeding and maintaining high-quality insects. In: Van Huis, A. and Tomberlin, J.K. (eds.) Insects as food and feed: from production to consumption. Wageningen Academic Publishers, Wageningen, the Netherlands, pp. 174-198.

Jeon, H., Park, S., Choi, J., Jeong, G., Lee, S.B., Choi, Y. and Lee, S.J., 2011. The intestinal bacterial community in the food waste-reducing larvae of *Hermetia illucens*. Current Microbiology 62: 1390-1399.

Jin, X.H., Heo, P.S., Hong, J.S., Kim, N.J. and Kim, Y.Y., 2016. Supplementation of dried mealworm (*Tenebrio molitor* larva) on growth performance, nutrient digestibility and blood profiles in weaning pigs. Asian Australas. Journal of Animal Science 29: 979-986.

Jones, R.S., 2015. Space diet: daily mealworm (*Tenebrio molitor*) harvest on a multigenerational spaceship. Journal of Interdisciplinary Science Topics 4.

Jongema, Y., 2017. List of edible insect species of the world. Wageningen University, Wageningen, the Netherlands. Available at: http://tinyurl.com/zb9u3b9.

Kareem, A.O. and Ogunremi, J.B., 2012. Growth performance of *Clarias gariepinus* fed compounded rations and maggots. Journal of Environmental Issues and Agriculture 4: 1-5.

Katayama, N., Ishikawa, Y., Takaoki, M., Yamashita, M., Nakayama, S., Kiguchi, K., Kok, R., Wada, H. and Mitsuhashi, J., 2008. Entomophagy: a key to space agriculture. Advances in Space Research 41: 701-705.

Katayama, N., Yamashita, M., Wada, H. and Mitsuhashi, J., 2005. Entomophagy as part of a space diet for habitation on Mars. Journal of Space Technology and Science 21: 27-38.

Kenis, M., Koné, N., Chrysostome, C.A.A.M., Devic, E., Koko, G.K.D., Clottey, V.A., Nacambo, S. and Mensah, G.A., 2014. Insects used for animal feed in West Africa. Entomologia 2: 107-114.

Khempaka, S., Chitsatchapong, C. and Molee, W., 2011. Effect of chitin and protein constituents in shrimp head meal on growth performance, nutrient digestibility, intestinal microbial populations, volatile fatty acids, and ammonia production in broilers. Journal of Applied Poultry Research 20: 1-11.

Klamcynska, B. and Mooney, W.D., 2016. Heterotrophic algae: a scalable and sustainable protein source. In: Nadathur, S.R., Givaudan, F., Wanasundara, J.P.D. and Scanlin, L. (eds.) Sustainable protein sources. Elsevier, Amsterdam, the Netherlands, pp. 327-339.

Klassen, W., 2005. Area-wide integrated pest management and the sterile insect technique. In: Dyck, V.A., Hendrichs, J. and Robinson, A.S. (eds.) Sterile insect technique: principles and practice in area-wide integrated pest management. Springer Netherlands, Dordrecht, the Netherlands, pp. 39-68.

Klous, G., Huss, A., Heederik, D.J.J. and Coutinho, R.A., 2016. Human-livestock contacts and their relationship to transmission of zoonotic pathogens, a systematic review of literature. One Health 2: 65-76.

Kok, R., 1983. The production of insects for human food. Canadian Institute of Food Science and Technology Journal 16: 5-18.

Kok, R., 2017. Insect production and facility design. In: Van Huis, A. and Tomberlin, J.K. (eds.) Insects as food and feed: from production to consumption. Wageningen Academic Publishers, Wageningen, the Netherlands, pp. 142-172.

Kok, R., Lomaliza, K. and Shivhare, U.S., 1988. The design and performance of an insect farm/chemical reactor for human food production. Canadian Agricultural Engineering 30: 307-317.

Kurbanov, A.R., Milusheva, R.Y., Rashidova, S.S. and Kamilov, B., 2015. Effect of replacement of fish meal with silkworm (*Bombyx mori*) pupa protein on the growth of *Clarias gariepinus* fingerling. International Journal of Fisheries and Aquatic Studies 2: 25-27.

Lalander, C.H., Fidjeland, J., Diener, S., Eriksson, S. and Vinnerås, B., 2015. High waste-to-biomass conversion and efficient *Salmonella* spp. reduction using black soldier fly for waste recycling. Agronomy for Sustainable Development 35: 261-271.

Lardé, G., 1990. Recycling of coffee pulp by *Hermetia illucens* (Diptera, Stratiomyidae) larvae. Biological Wastes 33: 307-310.

Lardé, G., 1989. Investigation on some factors affecting larval growth in a coffee-pulp bed. Biological Wastes 30: 11-19.

Latunde-Dada, G.O., Yang, W. and Vera Aviles, M., 2016. *In vitro* iron availability from insects and Sirloin beef. Journal of Agricultural and Food Chemistry 64: 8420-8424.

Lee, C.G., Silva, C.A.D., Lee, J.-Y., Hartl, D. and Elias, J.A., 2008. Chitin regulation of immune responses: an old molecule with new roles. Current Opinion in Immunology 20: 684-689.

Leiber, F., Gelencsér, T., Stamer, A., Amsler, Z., Wohlfahrt, J., Früh, B. and Maurer, V., 2015. Insect and legume-based protein sources to replace soybean cake in an organic broiler diet: effects on growth performance and physical meat quality. Renewable Agriculture and Food Systems 32: 21-27.

Leong, S.Y., Kutty, S.R.M., Malakahmad, A. and Tan, C.K., 2015. Feasibility study of biodiesel production using lipids of *Hermetia illucens* larva fed with organic waste. Waste Management 47(A): 84-90.

Leschin-Hoar, C., 2016. Buggin' out: tapping the potential of insect meal in aquaculture. Global Aquaculture Advocate, Portsmouth, NH, USA. Available at: http://tinyurl.com/y7w5tdqf.

Li, L., Xie, B., Dong, C., Hu, D., Wang, M., Liu, G. and Liu, H., 2015. Rearing *Tenebrio molitor* L. (Coleptera: Tenebrionidae) in the 'Lunar Palace 1' during a 105-day multi-crew closed integrative BLSS experiment. Life Sciences in Space Research 7: 9-14.

Li, Q., Zheng, L., Cai, H., Garza, E., Yu, Z. and Zhou, S., 2011. From organic waste to biodiesel: black soldier fly, *Hermetia illucens*, makes it feasible. Fuel 90: 1545-1548.

Li, S., Ji, H., Zhang, B., Zhou, J. and Yu, H., 2017. Defatted black soldier fly (*Hermetia illucens*) larvae meal in diets for juvenile Jian carp (*Cyprinus carpio* var. Jian): growth performance, antioxidant enzyme activities, digestive enzyme activities, intestine and hepatopancreas histological structure. Aquaculture 477: 62-70.

Lipinski, B., Hanson, C., Waite, R., Searchinger, T., Lomax, J. and Kitinoja, L., 2013. Reducing food loss and waste: creating a sustainable food future instalment 2. Working Paper I, World Resources Institute, Washington, DC, USA.

Liu, Q., Tomberlin, J.K., Brady, J.A., Sanford, M.R. and Yu, Z., 2008. Black soldier fly (Diptera: Stratiomyidae) larvae reduce *Escherichia coli* in dairy manure. Environmental Entomology 37: 1525-1530.

Lock, E.R., Arsiwalla, T. and Waagbø, R., 2015. Insect larvae meal as an alternative source of nutrients in the diet of Atlantic salmon (*Salmo salar*) postsmolt. Aquaculture Nutrition 22: 1202-1213.

Lockwood, J.A., 2004. Locust: the devastating rise and disappearance of the insect that shaped the American frontier. Basic Books, New York, NY, USA.

Longe, O., Adeyemo, G. and Akintayo, L., 2007. Effects of feeding desert locust meal (*Schistocerca gregaria*) on performance and haematology of broilers. Tropentag 2007, October 9-11, Witzenhausen, Germany.

Looy, H. and Wood, J.R., 2015. Imagination, hospitality, and affection: the unique legacy of food insects? Animal Frontiers 5: 8-13.

López, A.S., 2017. Potential of pre-pupae meal of the Black Soldier Fly (*Hermetia illucens*) as fish meal substitute: effect on growth performance and digestibility in European sea bass (*Dicentrarchus labrax*). MSc thesis, Faculdade de Siências, Universidade do Porto, Porto, Portugal.

López Vergé, S., 2013. Utilization of *Sitophilus zeamais* (Motschulsky) larvae as a dietary supplement for the production of broiler chickens. MSc thesis, Universitat Autònoma de Barcelona, Barcelona, Spain.

Loponte, R., Nizza, S., Bovera, F., De Riu, N., Fliegerova, K., Lombardi, P., Vassalotti, G., Mastellone, V., Nizza, A. and Moniello, G., 2017. Growth performance, blood profiles and carcass traits of Barbary partridge (*Alectoris barbara*) fed two different insect larvae meals (*Tenebrio molitor* and *Hermetia illucens*). Research in Veterinary Science 115: 183-188.

Lundy, M.E. and Parrella, M.P., 2015. Crickets are not a free lunch: protein capture from scalable organic side-streams via high-density populations of *Acheta domesticus*. Plos ONE 10: e0118785.

Lymbery, P., 2014. Farmageddon: the true cost of cheap meat. Bloomsbury Publishing, London, USA.

Madu, C.T. and Ufodike, E.B.C., 2003. Growth and survival of catfish (*Clarias anguillaris*) juveniles fed live tilapia and maggot as unconventional diets. Journal of Aquatic Sciences 18: 47-52.

Magalhães, R., Sánchez-López, A., Leal, R.S., Martínez-Llorens, S., Oliva-Teles, A. and Peres, H., 2017. Black soldier fly (*Hermetia illucens*) pre-pupae meal as a fish meal replacement in diets for European seabass (*Dicentrarchus labrax*). Aquaculture 476: 79-85.

Makkar, H.P.S., Tran, G., Heuzé, V. and Ankers, P., 2014. State-of-the-art on use of insects as animal feed. Animal Feed Science and Technology 197: 1-33.

Makkar, H.P.S., Tran, G., Heuzé, V., Giger-Reverdin, S., Lessire, M., Lebas, F. and Ankers, P., 2016. Seaweeds for livestock diets: a review. Animal Feed Science and Technology 212: 1-17.

Malav, O.P., Talukder, S., Gokulakrishnan, P. and Chand, S., 2015. Meat analog: a review. Critical Reviews in Food Science and Nutrition 55: 1241-1245.

Manurung, R., Supriatna, A., Esyanthi, R.R. and Putra, R.E., 2016. Bioconversion of rice straw waste by black soldier fly larvae (*Hermetia illucens* L.): optimal feed rate for biomass production. Journal of Entomology and Zoology Studies 4: 1036-1041.

Mattick, C.S., Landis, A.E., Allenby, B.R. and Genovese, N.J., 2015. Anticipatory life cycle analysis of *in vitro* biomass cultivation for cultured meat production in the United States. Environmental Science and Technology 49: 11941-11949.

Mbunwen, F.N.H., Onyimonyi, A.E., Nwoga, C.C. and Musongong, G.A., 2011. Biological value of maggot meal as a replacement for fishmeal in the diets of african giant snail (*Achatina* spp.). Hatchings Journal of Life Sciences 5: 821-825.

McAlpine, C.A., Etter, A., Fearnside, P.M., Seabrook, L. and Laurance, W.F., 2009. Increasing world consumption of beef as a driver of regional and global change: a call for policy action based on evidence from Queensland (Australia), Colombia and Brazil. Global Environmental Change 19: 21-33.

Medhi, D., Bhuyan, R., Ahmed, H.F. and Saikia, B.N., 2016. Effect of inclusion of silk worm pupae meal in the diet with or without enzyme supplementation on nutrient utilization and growth performance in crossbred pigs. Indian Journal of Animal Nutrition 33: 336-339.

Meijer, N. and Van der Fels-Klerx, H.J., 2017. Health risks and EU regulatory framework. In: Van Huis, A. and Tomberlin, J.K. (eds.) Insects as food and feed: from production to consumption. Wageningen Academic Publishers, Wageningen, the Netherlands, pp. 346-362.

Miech, P., Berggren, Å., Lindberg, J.E., Chhay, T., Khieu, B. and Jansson, A., 2016. Growth and survival of reared Cambodian field crickets (*Teleogryllus testaceus*) fed weeds, agricultural and food industry by-products. Journal of Insects as Food and Feed 2: 285-292.

Moon, S.J. and Lee, J.W., 2015. Current views on insect feed and its future. Entomological Research 45: 283-285.

Mott, G., 2017. Cricket rearing. In: Van Huis, A. and Tomberlin, J.K. (eds.) Insects as food and feed: from production to consumption. Wageningen Academic Publishers, Wageningen, the Netherlands, pp. 270-287.

Muafor, F.J., Gnetegha, A.A., Dounias, E., Le Gall, P. and Levang, P., 2017. Palm weevil farming contributing to food security in sub-Saharan Africa. In: Van Huis, A. and Tomberlin, J.K. (eds.) Insects as food and feed: from production to consumption. Wageningen Academic Publishers, Wageningen, the Netherlands, pp. 112-125.

Munyuli Bin Mushambanyi, T. and Balezi, N., 2002. Utilisation des blattes et des termites comme substituts potentiels de la farine de viande dans l'alimentation des poulets de chair au Sud-Kivu, République Démocratique du Congo. Tropicultura 20: 10-16.

Muzzarelli, R.A.A., 2010. Chitins and chitosans as immunoadjuvants and non-allergenic drug carriers. Marine Drugs 8: 292-312.

Myers, H.M., Tomberlin, J.K., Lambert, B.D. and Kattes, D., 2008. Development of black soldier fly (Diptera: Stratiomyidae) larvae fed dairy manure. Environmental Entomology 37: 11-15.

Nadathur, S.R., Flavors, G., Wanasundara, J.P.D. and Scanlin, L., 2016. Sustainable protein sources. Elsevier, Amsterdam, the Netherlands.

Nakagaki, B.J., Sunde, M.L. and Defoliart, G.R., 1987. Protein quality of the house cricket, *Acheta domesticus*, when fed to broiler chicks. Poultry Science 66: 1367-1371.

Ncobela, C.N. and Chimonyo, M., 2015. Potential of using non-conventional animal protein sources for sustainable intensification of scavenging village chickens: a review. Animal Feed Science and Technology 208: 1-11.

Ndegwa, P.M., Hristov, A.N., Arogo, J. and Sheffield, R.E., 2008. A review of ammonia emission mitigation techniques for concentrated animal operations. Biosystems Engineering 100: 453-469.

Neuenschwander, P., 2001. Biological control of the cassava mealybug in Africa: a review. Biological Control 21: 214-229.

Newton, G.L., Sheppard, D.C., Watson, D.W., Burtle, G.J., Dove, C.R., Tomberlin, J.K. and Thelen, E.E., 2005a. The black soldier fly, *Hermetia illucens*, as a manure management/resource recovery tool. In: State of the Science, Animal Manure and Waste Management. January 5-7, San Antonio, TX, USA.

Newton, L., Sheppard, C., Watson, D.W. and Burtle, G., 2005b. Using the black soldier fly, *Hermetia illucens*, as a value-added tool for the management of swine manure. Report for Mike Williams, director of the Animal and Poultry Waste Management Center, North Carolina State University, Raleigh, NC, USA. Agreements between the Nc Attorney General, Smithfield Foods, and Premium Standard Farms, and Frontline Farmers.

Ng, W.K., Liew, F.L., Ang, L.P. and Won, K.W., 2001. Potential of mealworm (*Tenebrio molitor*) as an alternative protein source in practical diets for African catfish, *Clarias gariepinus*. Aquaculture Research 32: 273-280.

Nicks, B., Laitat, M., Vandenheede, M., Desiron, A., Verhaeghe, C. and Canart, B., 2003. Emissions of ammonia, nitrous oxide, methane, carbon dioxide and water vapor in the raising of weaned pigs on straw-based and sawdust-based deep litters. Animal Research 52: 299-308.

Norhidayah, M.T., 2016. Cricket meal as an alternative to fishmeal in diets for African catfish (*Clarias gariepinus*). PhD thesis, Faculty of Science, University of Malaya, Kuala Lumpur, Malaysia.

Ocio, E., Viñaras, R. and Rey, J.M., 1979. Housefly larvae meal grown on municipal organic waste as a source of protein in poultry diets. Animal Feed Science and Technology 4: 227-231.

Ogunjil, J., Kloas, W., Wirth, M., Schulz, C. and Rennert, B., 2006. Housefly maggot meal (Magmeal): an emerging substitute of fishmeal in tilapia diets. Conference on International Agricultural Research for Development. Deutscher Tropentag. October 11-13, 2006, Stuttgart-Hohenheim, Germany.

Ojewola, G.S., Okoye, F.C. and Ukoha, O.A., 2005. Comparative utilization of three animal protein sources by broiler chickens. International Journal of Poultry Science 4: 462-467.

Oldenbroek, K. and Van der Waaij, L., 2014. Textbook animal breeding: animal breeding and genetics for BSc students. Centre for Genetic Resources and Animal Breeding and Genomics Group, Wageningen University and Research Centre, Wageningen, the Netherlands.

Olsen, R.L. and Hasan, M.R., 2012. A limited supply of fishmeal: impact on future increases in global aquaculture production. Trends in Food Science and Technology 27: 120-128.

Oluokun, J., 2000. Upgrading the nutritive value of full-fat soyabeans meal for broiler production with either fishmeal or black soldier fly larvae meal (*Hermetia illucens*). Nigerian Journal of Animal Science 3(2).

Oonincx, D.G.A.B., 2017. Environmental impact of insect production. In: Van Huis, A. and Tomberlin, J.K. (eds.) Insects as food and feed: from production to consumption. Wageningen Academic Publishers, Wageningen, the Netherlands, pp. 78-93.

Oonincx, D.G.A.B. and De Boer, I.J.M., 2012. Environmental impact of the production of mealworms as a protein source for humans – a life cycle assessment. Plos ONE 7: e51145.

Oonincx, D.G.A.B., Van Broekhoven, S., Van Huis, A. and Van Loon, J.J.A., 2015a. Feed conversion, survival and development, and composition of four insect species on diets composed of food by-products. PloS ONE 10: e0144601.

Oonincx, D.G.A.B., Van Huis, A. and Van Loon, J.J.A., 2015b. Nutrient utilisation by black soldier flies fed with chicken, pig, or cow manure. Journal of Insects as Food and Feed 1: 131-139.

Oonincx, D.G.A.B., Van Itterbeeck, J., Heetkamp, M.J.W., Van den Brand, H., Van Loon, J.J.A. and Van Huis, A., 2010. An exploration on greenhouse gas and ammonia production by insect species suitable for animal or human consumption. PLoS ONE 5: e14445.

Packer, M.A., Harris, G.C. and Adams, S.L., 2016. Food and feed applications of algae. In: Bux, F. and Chisti, Y. (eds.) Algae biotechnology: products and processes. Springer International Publishing, Cham, Switzerland, pp. 217-247.

Panini, R.L., Freitas, L.E.L., Guimarães, A.M., Rios, C., Da Silva, M.F.O., Vieira, F.N., Fracalossi, D.M., Samuels, R.I., Prudêncio, E.S., Silva, C.P. and Amboni, R.D.M.C., 2017. Potential use of mealworms as an alternative protein source for Pacific white shrimp: digestibility and performance. Aquaculture 473: 115-120.

Paoletti, M.G., Norberto, L., Damini, R. and Musumeci, S., 2007. Human gastric juice contains chitinase that can degrade chitin. Annals of Nutrition and Metabolism 51: 244-251.

Pastor, B., Velasquez, Y., Gobbi, P. and Rojo, S., 2015. Conversion of organic wastes into fly larval biomass: bottlenecks and challenges. Journal of Insects as Food and Feed 1: 179-193.

Paustian, K., Antle, J., Sheehand, J. and Paul, E., 2006. Agriculture's role in greenhouse gas mitigation. PEW center on global climate change. Pew Center on Global Climate Change, Washington, DC, USA.

Payne, C.L.R., Scarborough, P., Rayner, M. and Nonaka, K., 2016. Are edible insects more or less 'healthy' than commonly consumed meats? A comparison using two nutrient profiling models developed to combat over- and undernutrition. European Journal of Clinical Nutrition 70: 285-291.

Pieterse, E., Pretorius, Q., Hoffman, L.C. and Drew, D.W., 2013. The carcass quality, meat quality and sensory characteristics of broilers raised on diets containing either *Musca domestica* larvae meal, fish meal or soya bean meal as the main protein source. Animal Production Science 54: 622-628.

Pineda Mejia, J.R., 2015. Performance of black soldier fly, *Hermetia illucens* (Diptera: Stratiomyidae) larvae fed organic waste diets based on a specific protein: digestible carbohydrate ratio. MSc thesis, Wageningen University, Wageningen, the Nehterlands.

Pino Moreno, J.M., 2016. The career of Dr J. Ramos-Elorduy Blásquez. Journal of Insects as Food and Feed 2: 3-14.

Post, M.J., 2012. Cultured meat from stem cells: challenges and prospects. Meat Science 92: 297-301.

Post, M.J., 2014. Cultured beef: medical technology to produce food. Journal of the Science of Food and Agriculture 94: 1039-1041.

Ramos-Elorduy, J., 2006. Threatened edible insects in Hidalgo, Mexico and some measures to preserve them. Journal of Ethnobiology and Ethnomedicine 2: 51.

Ramos-Elorduy, J., Gonzalez, E.A., Hernandez, A.R. and Pino, J.M., 2002. Use of *Tenebrio molitor* (Coleoptera: Tenebrionidae) to recycle organic wastes and as feed for broiler chickens. Journal of Economic Entomology 95: 214-220.

Ramos-Elorduy, J. and Morales, J.M.P., 1989. The efficiency of the insect (*Musca domestica* L.) in recycling organic wastes as a source of protein. In: Houghton, D.R. and Smith, R.N. (eds.) Biodeterioration 7: 805-810.

Rana, K.J., Siriwardena, S. and Hasan, M.R., 2009. Impact of rising feed ingredient prices on aquafeeds and aquaculture production. Food and Agriculture Organisation, Rome, Italy.

Ray, D.K., Mueller, N.D., West, P.C. and Foley, J.A., 2013. Yield trends are insufficient to double global crop production by 2050. PloS ONE 8: e66428.

Reese, T.A., Liang, H.-E., Tager, A.M., Luster, A.D., Rooijen, N.V., Voehringer, D. and Locksley, R.M., 2007. Chitin induces accumulation in tissue of innate immune cells associated with allergy. Nature 447: 92-96.

Rehman, K.U., Rehman, A., Cai, M., Zheng, L., Xiao, X., Somroo, A.A., Wang, H., Li, W., Yu, Z. and Zhang, J., 2017. Conversion of mixtures of dairy manure and soybean curd residue by black soldier fly larvae (*Hermetia illucens* L.). Journal of Cleaner Production 154: 366-373.

Revell, B.J., 2015. One man's meat…2050? Ruminations on future meat demand in the context of global warming. Journal of Agricultural Economics 66: 573-614.

Richards, M., Bruun, T., Campbell, B., Huyer, S., Kuntze, V., Gregersen, L., Madsen, S., Oldvig, M. and Vasileiou, I., 2015. How countries plan to address agricultural adaptation and mitigation: an analysis of intended nationally determined contributions. CGIAR Research Program on Climate Change, Agriculture and Food Security (CCAFS), Copenhagen, Denmark.

Riddick, E., 2013. Insect protein as a partial replacement of fishmeal in the diets of juvenile fish and crustaceans. In: Morales-Ramos, J.A., Guadalupe Rojas, M. and Shapiro-Ilan, D.I. (eds.) Mass production of beneficial organisms: invertebrates and entomopathogens. Elsevier, Amsterdam, the Netherlands, pp. 565-582.

Riddick, E.W., 2014. Insect protein as a partial replacement for fishmeal in the diets of juvenile fish and crustaceans. In: Morales-Ramos, J.A., Guadalupe Rojas, M. and Shapiro-Ilan, D.I. (eds.) Mass production of beneficial organisms. Academic Press, San Diego, CA, USA, pp. 565-582.

Ritchie, H., Laird, J. and Ritchie, D., 2017. 3f bio: halving the cost of mycoprotein through integrated fermentation processes. Industrial Biotechnology 13: 29-31.

Röcklinsberg, H., Gamborg, C. and Gjerris, M., 2017. Ethical issues in insect production. In: Van Huis, A. and Tomberlin, J.K. (eds.) Insects as food and feed: from production to consumption. Wageningen Academic Publishers, Wageningen, the Netherlands, pp. 364-379.

Roffeis, M., Muys, B., Almeida, J., Mathijs, E., Achten, W.M.J., Pastor, B., Velásquez, Y., Martinez-Sanchez, A.I. and Rojo, S., 2015. Pig manure treatment with housefly (*Musca domestica*) rearing – an environmental life cycle assessment. Journal of Insects as Food and Feed 1: 195-214.

Roncarati, A., Gasco, L., Parisi, G. and Terova, G., 2015. Growth performance of common catfish (*Ameiurus melas* Raf.) fingerlings fed mealworm (*Tenebrio molitor*) diet. Journal of Insects as Food and Feed 1: 233-240.

Rumpold, B.A., Bußler, S., Jäger, H. and Schlüter, O., 2017. Insect processing. In: Van Huis, A. and Tomberlin, J.K. (eds.) Insects as food and feed: from production to consumption. Wageningen Academic Publishers, Wageningen, the Netherlands, pp. 318-342.

Rumpold, B.A. and Schlüter, O.K., 2013. Nutritional composition and safety aspects of edible insects. Molecular Nutrition and Food Research 57: 802-823.

Sánchez-Muros, M.J., Barroso, F.G. and De Haro, C., 2016. Brief summary of insect usage as an industrial animal feed/feed ingredient. In: Dossey, A.T., Morales-Ramos, J.A. and Rojas, M.G. (eds.) Insects as sustainable food ingredients. Academic Press, San Diego, CA, USA, pp. 273-309.

Sánchez-Muros, M.J., De Haro, C., Sanz, A., Trenzado, C.E., Villareces, S. and Barroso, F.G., 2015. Nutritional evaluation of *Tenebrio molitor* meal as fishmeal substitute for tilapia (*Oreochromis niloticus*) diet. Aquaculture Nutrition 22: 943-955.

Schösler, H., De Boer, J. and Boersema, J.J., 2012. Can we cut out the meat of the dish? Constructing consumer-oriented pathways towards meat substitution. Appetite 58: 39-47.

Sealey, W.M., Gaylord, T.G., Barrows, F.T., Tomberlin, J.K., McGuire, M.A., Ross, C. and St-Hilaire, S., 2011. Sensory analysis of rainbow trout, *Oncorhynchus mykiss*, fed enriched black soldier fly prepupae, *Hermetia illucens*. Journal of the World Aquaculture Society 42: 34-45.

Shah, R.M., Azhar, F., Shad, S.A., Walker, W.B., Azeem, M. and Binyameen, M., 2016. Effects of different animal manures on attraction and reproductive behaviors of common house fly, *Musca domestica* L. Parasitology Research 115: 3585-3598.

Sheppard, C.D., Larry Newton, G., Thompson, S.A. and Savage, S., 1994. A value added manure management system using the black soldier fly. Bioresource Technology 50: 275-279.

Sheppard, D.C., 1983. House fly and lesser fly control utilizing the black soldier fly in manure management systems for caged laying hens. Environmental Entomology 12: 1439-1442.

Shockley, M., Allen, R.N. and Gracer, D., 2017. Product development and promotion. In: Van Huis, A. and Tomberlin, J.K. (eds.) Insects as food and feed: from production to consumption. Wageningen Academic Publishers, Wageningen, the Netherlands, pp. 398-427.

Sing, K., Kamarudin, M., Wilson, J. and Azirun, M.S., 2014. Evaluation of blowfly (*Chrysomya megacephala*) maggot meal as an effective, sustainable replacement for fishmeal in the diet of farmed juvenile red tilapia (*Oreochromis* sp.). Pakistan Veterinary Journal 34: 288-292.

Smith, P., Martino, D., Cai, Z., Gwary, D., Janzen, H., Kumar, P., McCarl, B., Ogle, S., O'Mara, F., Rice, C., Scholes, B., Sirotenko, O., Howden, M., McAllister, T., Pan, G., Romanenkov, V., Schneider, U., Towprayoon, S., Wattenbach, M. and Smith, J., 2008. Greenhouse gas mitigation in agriculture. Philosophical Transactions of the Royal Society B-Biological Sciences 363: 789-813.

Sørensen, J.G., Addison, M.F. and Terblanche, J.S., 2012. Mass-rearing of insects for pest management: challenges, synergies and advances from evolutionary physiology. Crop Protection 38: 87-94.

Spranghers, T., Ottoboni, M., Klootwijk, C., Ovyn, A., Deboosere, S., De Meulenaer, B., Michiels, J., Eeckhout, M., De Clercq, P. and De Smet, S., 2016. Nutritional composition of black soldier fly (*Hermetia illucens*) prepupae reared on different organic waste substrates. Journal of the Science of Food and Agriculture 97: 2594-2600.

St-Hilaire, S., Sheppard, C., Tomberlin, J.K., Irving, S., Newton, L., McGuire, M.A., Mosley, E.E., Hardy, R.W. and Sealey, W., 2007. Fly prepupae as a feedstuff for rainbow trout, *Oncorhynchus mykiss*. Journal of the World Aquaculture Society 38: 59-67.

Steinfeld, H., Gerber, P., Wassenaar, T., Castel, V. and Rosales, M., 2006. Livestock's long shadow. Environmental issues and options. Food and Agriculture Organisation, Rome, Italy, 319 pp.

Supriyatna, A., Manurung, R., Esyanti, R.R. and Putra, R.E., 2016. Growth of black soldier larvae fed on cassava peel wastes. An agriculture waste. Journal of Entomology and Zoology Studies 4: 161-165.

Surendra, K.C., Olivier, R., Tomberlin, J.K., Jha, R. and Khanal, S.K., 2016. Bioconversion of organic wastes into biodiesel and animal feed via insect farming. Renewable Energy 98: 197-202.

Tacon, A.G.J., Metian, M., Turchini, G.M. and De Silva, S.S., 2009. Responsible aquaculture and trophic level implications to global fish supply. Reviews in Fisheries Science 18: 94-105.

Téguia, A., Mpoame, M. and Mba, J.A.O., 2002. The production performance of broiler birds as affected by the replacement of fish meal by maggot meal in the starter and finisher diets. Tropicultura 20: 187-192.

Teotia, J.S. and Miller, B.F., 1974. Nutritive content of house fly pupae and manure residue 1. British Poultry Science 15: 177-182.

The Economist, 2014. Food loss and its intersection with food security. Global food security index 2014. Special report of the Intelligence Unit of The Economist, London, UK.

Théwis, A. and Galiù, A.-M., 2012. Livestock production: prospects over the next decades and alternative protein bio-resources. AgroLife Scientific Journal 1: 29-38.

Tilman, D. and Clark, M., 2014. Global diets link environmental sustainability and human health. Nature 515: 518-522.

Tinder, A.C., Puckett, R.T., Turner, N.D., Cammack, J.A. and Tomberlin, J.K., 2017. Bioconversion of sorghum and cowpea by black soldier fly (*Hermetia illucens* L.) larvae for alternative protein production. Journal of Insects as Food and Feed 3: 121-130.

Tomberlin, J.K., 2017. Industrialization of insect farming. In: Van Huis, A. and Tomberlin, J.K. (eds.) Insects as food and feed: from production to consumption. Wageningen Academic Publishers, Wageningen, the Netherlands, pp. 226-228.

Tomberlin, J.K. and Cammack, J.A., 2017. Black soldier fly: biology and mass production. In: Van Huis, A. and Tomberlin, J.K. (eds.) Insects as food and feed: from production to consumption. Wageningen Academic Publishers, Wageningen, the Netherlands, pp. 230-246.

Tomberlin, J.K., Holcomb, C. and Van Leeuwen, S.K., 2017. Industrialization of house fly production for livestock feed. In: Van Huis, A. and Tomberlin, J.K. (eds.) Insects as food and feed: from production to consumption. Wageningen Academic Publishers, Wageningen, the Netherlands, pp. 248-257.

Tong, L., Yu, X. and Liu, H., 2011. Insect food for astronauts: gas exchange in silkworms fed on mulberry and lettuce and the nutritional value of these insects for human consumption during deep space flights. Bulletin of Entomological Research 101: 613-622.

Tran, G., Heuzé, V. and Makkar, H.P.S., 2015. Insects in fish diets. Animal Frontiers 5: 37-44.

Tuomisto, H.L. and Teixeira de Mattos, M.J., 2011. Environmental impacts of cultured meat production. Environmental Science and Technology 45: 6117-6123.

Turk, J., 2016. Meeting projected food demands by 2050: understanding and enhancing the role of grazing ruminants. Journal of Animal Science 94: 53-62.

United Nations, 2013. World population prospects: the 2012 revision. Key findings and advance tables. Working Paper No. ESA/P/WP.227. United Nations, Department of Economic and Social Affairs, Population Division, New York, NY, USA.

Van Broekhoven, S., Oonincx, D.G.A.B., Van Huis, A. and Van Loon, J.J.A., 2015. Growth performance and feed conversion efficiency of three edible mealworm species (Coleoptera: Tenebrionidae) on diets composed of organic by-products. Journal of Insect Physiology 73: 1-10.

Van der Mensbrugghe, D., Osorio-Rodarte, I., Burns, A. and Baffes, J., 2011. Macroeconomic environment, commodity markets: a longer term outlook. Paper prepared for the Expert Meeting on how to feed the world in 2050. June 24-26, 2009. FAO, Rome, Italy.

Van Huis, A., 2013. Potential of insects as food and feed in assuring food security. Annual Review of Entomology 58: 563-583.

Van Huis, A., 2017a. Insects as human food. In: Rômulo, A. and De Albuquerque, U.P. (eds.) Ethnozoology: animals in our lives. Elsevier, Amsterdam, the Netherlands.

Van Huis, A., 2017b. Introducing small production systems for edible insects. In: Van Huis, A. and Tomberlin, J.K. (eds.) Insects as food and feed: from production to consumption. Wageningen Academic Publishers, Wageningen, the Netherlands, pp. 96-99.

Van Huis, A., Itterbeeck, J.V., Klunder, H., Mertens, E., Halloran, A., Muir, G. and Vantomme, P., 2013. Edible insects: future prospects for food and feed security. FAO Forestry Paper 171. FAO, Rome, Italy and Wageningen University and Research Centre, Wageningen, the Netherlands, 187 pp. Available at: http://www.fao.org/docrep/018/i3253e/i3253e.pdf.

Van Huis, A. and Tomberlin, J.K., 2017. Future prospects of insects as food and feed. In: Van Huis, A. and Tomberlin, J.K. (eds.) Insects as food and feed: from production to consumption. Wageningen Academic Publishers, Wageningen, the Netherlands, pp. 430-445.

Van Itterbeeck, J. and Van Huis, A., 2012. Environmental manipulation for edible insect procurement: a historical perspective. Journal of Ethnobiology and Ethnomedicine 8: 1-19.

Van Lenteren, J.C., 2000. Success in biological control of arthropods by augmentation of natural enemies. In: Gurr, G. and Wratten, S. (eds.) Biological control: measures of success. Springer Netherlands, Dordrecht, the Netherlands, pp. 77-103.

Vantomme, P., 2017. Creating an enabling environment. In: Van Huis, A. and Tomberlin, J.K. (eds.) Insects as food and feed: from production to consumption. Wageningen Academic Publishers, Wageningen, the Netherlands, pp. 380-397.

Veldkamp, T. and Bosch, G., 2015. Insects: a protein-rich feed ingredient in pig and poultry diets. Animal Frontiers 5: 45-50.

Veldkamp, T., Van Duinkerken, G., Van Huis, A., Lakemond, C.M.M., Ottevanger, E. and Boekel, M.A.J.S., 2012. Insects as a sustainable feed ingredient in pig and poultry diets – a feasibility study. Wageningen UR Livestock Research, Report 638, Wageningen, the Netherlands.

Vergé, X.P.C., De Krimpe, C. and Desjardins, R.L., 2007. Agricultural production, greenhouse gas emissions and mitigation potential. Agricultural and Forest Meteorology 142: 255-269.

Wanasundara, J.P.D., Tan, S., Alashi, A.M., Pudel, F. and Blanchard, C., 2016. Proteins from canola/rapeseed: current status. In: Nadathur, S.R., Flavors, G., Wanasundara, J.P.D. and Scanlin, L. (eds.) Sustainable protein sources. Elsevier, Amsterdam, the Netherlands, pp. 285-304.

Wang, D., Zhai, S.W., Zhang, C.X., Zhang, Q. and Chena, H., 2007. Nutrition value of the Chinese grasshopper *Acrida cinerea* (Thunberg) for broilers. Animal Feed Science and Technology 135: 66-74.

Wang, H., Zhang, Z., Czapar, G., Winkler, M.K. and Zheng, J., 2013. A full-scale house fly (Diptera: Muscidae) larvae bioconversion system for value-added swine manure reduction. Waste Management and Research 31: 223-231.

Wang, L., Li, J., Jin, J.N., Zhu, F., Roffeis, M. and Zhang, X.Z., in press. A comprehensive evaluation of replacing fishmeal with housefly (*Musca domestica*) maggot meal in the diet of Nile tilapia (*Oreochromis niloticus*): growth performance, flesh quality, innate immunity and water environment. Aquaculture Nutrition, DOI: https://doi.org/10.1111/anu.12466.

Webster, C.D., Rawles, S.D., Koch, J.F., Thompson, K.R., Kobayashi, Y., Gannam, A.L., Twibell, R.G. and Hyde, N.M., 2015. Bio-Ag reutilization of distiller's dried grains with solubles (DDGS) as a substrate for black soldier fly larvae, *Hermetia illucens*, along with poultry by-product meal and soybean meal, as total replacement of fish meal in diets for Nile tilapia, *Oreochromis niloticus*. Aquaculture Nutrition 22: 976-988.

Wollenberg, E., Richards, M., Smith, P., Havlík, P., Obersteiner, M., Tubiello, F.N., Herold, M., Gerber, P., Carter, S., Reisinger, A., Van Vuuren, D., Dickie, A., Neufeldt, H., Sander, B.O., Wassmann, R., Sommer, R., Amonette, J.E., Falcucci, A., Herrero, M., Opio, C., Roman-Cuesta, R., Stehfest, E., Westhoek, H., Ortiz-Monasterio, I., Sapkota, T., Rufino, M.C., Thornton, P.K., Verchot, L., West, P.C., Soussana, J.F., Baedeker, T., Sadler, M., Vermeulen, S. and Campbell, B.M., 2016. Reducing emissions from agriculture to meet the 2 °C target. Global Change Biology 22: 3859-3864.

Yamashita, M., Hashimoto, H. and Wada, H., 2009. On-site resources availability for space agriculture on Mars. In: Badescu, V. (ed.) Mars: prospective energy and material resources. Springer-Verlag, Berlin, Germany, pp. 517-542.

Yang, Y., Tang, L., Tong, L. and Liu, H., 2009. Silkworms culture as a source of protein for humans in space. Advances in Space Research 43: 1236-1242.

Zhang, Z., Wang, H., Zhu, J., Suneethi, S. and Zheng, J., 2012. Swine manure vermicomposting via housefly larvae (*Musca domestica*): the dynamics of biochemical and microbial features. Bioresource Technology 118: 563-571.

Zheng, L., Hou, Y., Li, W., Yang, S., Li, Q. and Yu, Z., 2012a. Biodiesel production from rice straw and restaurant waste employing black soldier fly assisted by microbes. Energy 47: 225-229.

Zheng, L., Qing, L., Jibin, Z. and Ziniu, Y., 2012b. Double the biodiesel yield: rearing black soldier fly larvae, *Hermetia illucens*, on solid residual fraction of restaurant waste after grease extraction for biodiesel production. Renewable Energy 41: 75-79.

Zhou, F., Tomberlin, J.K., Zheng, L., Yu, Z. and Zhang, J., 2013. Developmental and waste reduction plasticity of three black soldier fly strains (Diptera: Stratiomyidae) raised on different livestock manures. Journal of Medical Entomology 50: 1224-1230.

Zhu, L., Henze, D., Bash, J., Jeong, G.R., Cady-Pereira, K., Shephard, M., Luo, M., Paulot, F. and Capps, S., 2015. Global evaluation of ammonia bidirectional exchange and livestock diurnal variation schemes. Atmospheric Chemistry and Physics 15: 12823-12843.

2.

Trichogramma brassicae parasitizing an egg of *Pieris brassicae*

Ecosystem services of insects

M. Dicke
Laboratory of Entomology, Wageningen University, P.O. Box 16, 6700 AA Wageningen, the Netherlands; marcel.dicke@wur.nl

Abstract

Being the most diverse form of eukaryotic life on earth, insects represent a wide biological range found on all continents and filling a rich palette of niches. Thus, insects provide important ecosystem services and are vital components of most ecosystems on our planet. Many of these ecosystem services are obscure to urbanized humanity, yet any human being should be aware of the important contribution to life on earth that is provided by the extremely speciose class of insects. Apart from ecosystem services from naturally occurring insects, some insect species are explicitly exploited for their biological properties and products that are harvested. This chapter highlights some of the insect products that are harvested by humankind as well as the most important ecosystem services provided by insects.

Keywords: insect diversity, food additives, medicine, pollination, detritivores

2.1 Introduction

Life on earth displays a dazzling diversity, ranging from unicellular species such as bacteria, typically a few micrometres in size, to multicellular species such as blue whales measuring up to 30 metres and 170 metric tonnes and giant sequoia trees measuring up to 95 metres and 2,000 metric tonnes. The total number of species on earth is estimated to be ca 5-15 million species (Dirzo and Raven, 2003). Insects account for approximately 66% of all known animal species, with about 1 million described species (Zhang, 2011). New insect species are named on a daily basis and the total number is estimated to be several fold higher than the currently documented insect diversity. Estimates have been made of the total number of insect individuals, which amount to 10^{18}-10^{19} individuals. If every insect were to weigh one milligramme, which is the weight of a small ant, then for each human being, there are 150 to 1,500 kilos of insects on this planet, indicating that insects are more abundant than humans in terms of biomass. Insects are represented by 39 orders such as the Coleoptera (beetles), Lepidoptera (moths and butterflies), Odonata (dragonflies and damselflies), Orthoptera (crickets and locusts), and Diptera (flies and mosquitoes). Among insects, the most speciose group is represented by the beetles with ca 387,000 described species (Zhang, 2011), ranging from tiny species such as a ptiliid beetle of only 0.325 mm (Polilov, 2015) to the cerambycid *Titanus giganteus* that is 500 times larger, measuring up to almost 17 cm (William, 2001). Yet, the longest insects include walking sticks measuring up to 55 cm (Bresseel and Constant, 2014).

Although they all have three pairs of legs and most of them have wings in the adult stage, insects display wide variation among species in almost any aspect of their biology (Gullan and Cranston, 2010). Insects are abundantly present in terrestrial ecosystems where they may occur in the soil, aboveground as well as flying in the air. They are to a lesser extent also present in aquatic environments as well, albeit with fewer species than in terrestrial ecosystems: only 9% of all described insect species are aquatic (having at least one aquatic life stage) (Mitterboeck *et al.*, 2016).

Many species have a complete metamorphosis, which provides juveniles and adults with very different phenotypes. This includes e.g. the beetles, butterflies and moths, and flies and mosquitoes. These are the most speciose insect orders and, thus, the vast majority of insects has this developmental mode. Complete metamorphosis, including a pupation in which the juvenile undergoes a full reorganisation, allows species to exploit very different niches. For instance, butterflies feed on plant nectar while their juveniles, caterpillars, are usually feeding on plant leaves and flowers (Schoonhoven *et al.*, 2005), with caterpillars of some species even being carnivorous and feeding on other insects (Sheppard *et al.*, 2004). However, other groups such as locusts and crickets, and true bugs display incomplete metamorphosis in which the juveniles resemble the adults quite well, even displaying developing wings already.

Insect reproduction also exhibits a diverse range of strategies. In almost any insect order, species can be found that reproduce asexually with females producing unfertilized, yet viable, eggs (Gullan and Cranston, 2010). The best-known example of asexually reproducing species are the aphids that use asexual reproduction during a large part of the season during which the females parthenogenetically produce larvae instead of eggs, alternated with a sexual phase where the females produce overwintering eggs. In the Hymenoptera sexual reproduction is mostly characterized by females being diploid, resulting from fertilized eggs and males being haploid from unfertilized eggs. In many insect species reproduction may be manipulated by endosymbiotic bacteria, *Wolbachi pipientis*, that can induce their host to reproduce asexually, or in some species can even turn a genetic male into a functional female (Stouthamer *et al.*, 1999). The number of eggs laid in a life-time ranges from less than 10 for tsetse flies, that retain eggs individually in the uterus to develop there during the first 3 larval instars (Takken, 1980), to several millions for some eusocial ant species (Brueland, 1995).

Approximately 50% of all known insect species are herbivorous (Schoonhoven *et al.*, 2005), whereas the rest comprise carnivores and detritivores. Thus, this speciose group of animals represents very different trophic niches and, consequently, a wide range of ecological functions in their natural ecosystems. Herbivorous species are especially found among the Lepidoptera (moths and butterflies), Orthoptera (locusts and grasshoppers), Phasmida (walking sticks), Hemiptera (aphids and true bugs), and Coleoptera. Most herbivorous insects are specialists that feed on plants in one or a few related plant families; less than 10% of herbivorous species feed on plants from more than three families (Schoonhoven *et al.*, 2005). Insect herbivores feed on plants that are usually well defended through an array of secondary plant metabolites of various toxicities, such as alkaloids, glucosinolates, and terpenoids. Insect herbivores have evolved intricate mechanisms of detoxifying or sequestering plant secondary metabolites through a co-evolutionary arms race (Schoonhoven *et al.*, 2005). As a consequence, even plant species that defend themselves with highly toxic metabolites are not fully free of insect herbivores. Carnivorous species are especially found among the Odonata (dragonflies and damselflies), Hymenoptera (wasps and ants), Neuroptera (lacewings), Dictyoptera (praying mantids) and Coleoptera. Detritivores are common among the Diptera (flies), Coleoptera. Detritivores live from unpredictable, short-lived resources, usually in the presence of a diversity of competing microbial decomposers. Thus, these species usually feed in microbially rich habitats such as decaying plant or animal tissues. A limited number of arthropods act as vectors of pathogens of plants and animals. Aphids and thrips are well-known vectors of a large variety of viruses to plants (Schoonhoven *et al.*, 2005), and mosquitoes and ticks can transmit various diseases including *Plasmodium* protozoa, West Nile virus, Zika virus, and *Borrelia* bacteria (Takken and Knols, 2007).

Due to their omnipresence on the planet, humans have lived together with insects throughout their evolutionary history. Despite an apparent lack of charisma, insects are perceived in various ways, ranging from irritation to inspiration (Huigens *et al.*, 2006; Lemelin, 2013). In many cases, negative connotations are associated with a lack of knowledge. Altogether, our exposure to insects has resulted in the exploitation of insects for various purposes in human societies. Moreover, as an abundant life form, insects as a whole provide essential ecosystem services that benefit humankind.

2.2 Products of insects

Humankind has explored Nature ever since hominoids evolved. This was true for hunting, for selecting food, as well as to protect their bodies, homes, crops, etc. Insects are known to be used as food by great apes (Deblauwe *et al.*, 2003) and early hominoids (McGrew, 2014). This is especially true for insects from the orders Coleoptera, Hymenoptera, Isoptera, Lepidoptera, and Orthoptera, which have also been used as food. In addition, there is ample archaeological evidence of early humans collecting honey from wild honeybees (Gould and Gould, 1988). Moreover, archaeological evidence shows that insects have been used as medicine, for rituals, in art, in oral tradition, and in technology throughout human evolution (Sutton, 1995) and this continues until the present day (Berenbaum, 1995; Dicke, 2000; Huigens *et al.*, 2006; Van Huis *et al.*, 2013).

2.2.1 Honey

Honey has been recognized as a high-value product for centuries. Archaeological records show that honey has been used to produce fermented beverages as early as the 7th millennium BC in China (McGovern *et al.*, 2004) while archaeological records of cave paintings in Spain, India, Australia, and Southern Africa indicate a similar history of honey collection from natural sources (Boukraâ, 2013; Crittenden, 2011). Honey was the most important source of sugars until sugarcane was used on a large scale and it is still a highly-valued product that indigenous people risk their lives for (Venkataraman *et al.*, 2013). Honey is collected from *Apis* species as well as stingless bees (Meliponini) (Rao *et al.*, 2016). Yet, honey was likely more than only a sugar source as natural honey is often consumed in combination with honeybee larvae that provide proteins, fats and minerals (Crittenden, 2011). However, apart from being a source of sugars, honey also has a long history in medical applications in e.g. China, the Middle East, South America and Egypt (Boukraâ 2013; Lev, 2003; Rao *et al.*, 2016). For instance, it is used to combat inflammation and to support tissue regeneration (Boukraâ, 2013) and flavonoids in honey as well as propolis, another honeybee product, appear to mediate this beneficial effect (Havsteen, 2002). Honey has even made it into language in a very special way, i.e. to address one's loved one.

2.2.2 Silk

Silkworm culture for the production of silk has a long history of circa 5,000 years when the Chinese started to rear *Bombyx mori* moths. Silk has been a highly-valued insect product and its origin remained a well-kept secret for a long time. For centuries, silk clothing was worn exclusively at the Emperor's court, and the Chinese were able to keep the exclusive secret of silk until about the year 600 CE. In China, revelation of the secret of silk was punishable by death. Silk was extensively traded over the Silk Road that was opened in 139 BC (Zhang *et al.*, 2008). It was a highly-valued product whose use was strictly regulated initially, being reserved to royalty and bishops. First production of silk in Europe started during the Renaissance in the Italian city-states and afterwards sericulture spread northwards to France and other countries. Sericulture thus originated from Chinese emperor exclusiveness and had a major impact on European markets and culture (Lemire and Riello, 2008).

Silk moth cocoons are made from a single thread that can reach up to 900 m and protect the silkworm pupa during the transition from caterpillar to moth. The silkworm *B. mori* has been truly domesticated to the extent that the moths are no longer able to fly. Sericulture is an important industry in especially China, the major producer, India and several other countries. Apart from its use in textiles, silk's properties also incited other uses, such as various medical applications (Lawrence, 2014).

2.2.3 Shellac

Shellac is an excretion product from the scale insect *Kerria lacca* (Figure 2.1) that has been used extensively as a varnish, textile dye, insulating material in electrical applications and for making gramophone discs. The insect has been reared since the 16th century in East Asia and its excretion product used as wood finish, until the 1930s (Ghosh *et al.*, 2015; Van Huis *et al.*, 2014). At present shellac is still used as a glazing agent on pills and on sweets and fruits or for pharmaceutical application, such as extended drug release (Pearnchob *et al.*, 2003). When used in food and sweets it is usually indicated on the label under the name E904 (http://www.food-info.net/uk/e/e904.htm).

2.2.4 Cochineal

The scale insect *Dactylopius coccus* is the source of a red dye, cochineal or carminic acid (Figure 2.2). The insect produces carminic acid, which consists of an anthraquinone linked to a glucose unit, to deter predators. The *Dactylopius* genus originates from the America's and they are herbivores of *Opuntia* cacti (Campana *et al.*, 2015). The Maya and Aztecs used this dye and annual tribute payments were sometimes done in the form of cochineal. In the antiquity, a related compound, 1-methyl-2-carboxy-3,5,6,8-tetrahydroxyanthraquinone or kermes dye,

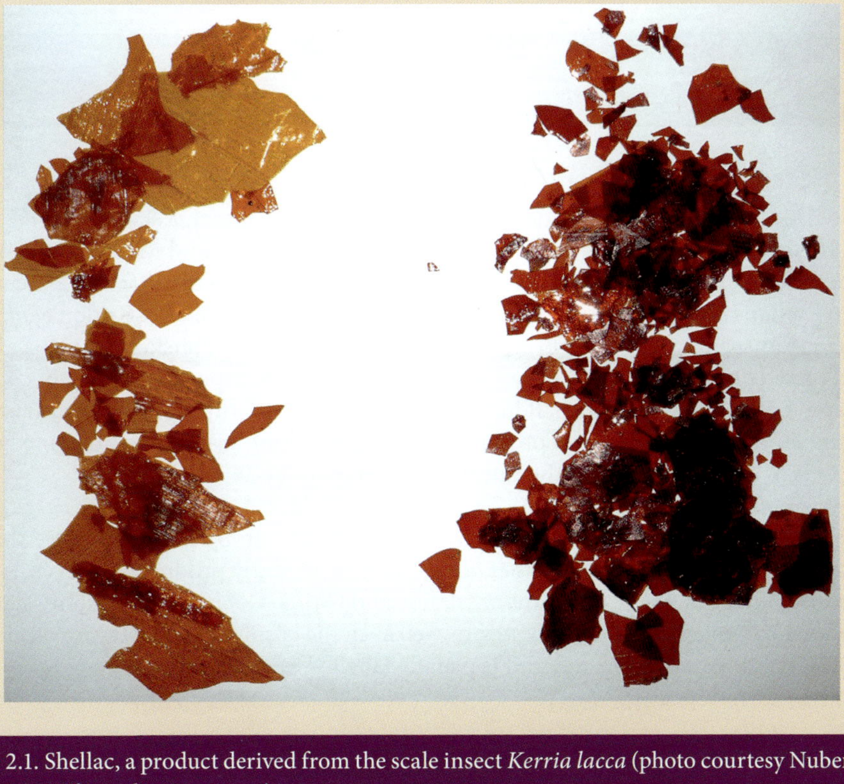

Figure 2.1. Shellac, a product derived from the scale insect *Kerria lacca* (photo courtesy Nuberger13, Wikimedia Commons).

Figure 2.2. Dried female cochineal insects (*Dactylopius coccus*) (photo courtesy Sprinky, Wikimedia Commons).

was also used in Europe. This was derived from another scale insect, *Kermes vermilio*, which lives on *Quercus coccifera* oaks that originate from the Near East and Mediterranean basin. It has been found in, for instance, funeral figurines, found in Hellenistic (3rd to 2nd centuries BC) and Roman (1st to 2nd centuries) tombs in Thessaloniki (Greece) (Fostiridou *et al.*, 2016). The European scale insect, *Porphyrophora polonica*, known as Polish carmine scale, also produces a red dye that is related to carminic acid and was an important source of cochineal in Europe (Lech and Jarosz, 2016). It has been identified in 16th through 18th century textiles (Lech and Jarosz, 2016). European production of red dyes from scale insects declined rapidly after Mexican cochineal entered the European market in the 16th century.

Cochineal is currently used as a dye of food as well and can be found in more than 100 products in a supermarket (Van Huis *et al.*, 2014). It is currently produced in several regions of the world, including Peru, the Canary Islands and China. Cochineal is a highly stable dye and is therefore popular in food industry, as well as in the cosmetics industry for e.g. lipstick, face powders, rouges, and blushes and is usually labelled as E120.

2.3 Ecosystem services

Ecosystem services represent the benefits that humans derive from ecosystem functions. Insects are abundant members of ecosystems that represent a wide biological diversity and, consequently, a wide diversity of ecosystem functions. For instance, insects have important effects on carbon and nutrient cycling by modifying detritus dynamics with consequences for ecosystem functioning (Yang and Gratton, 2014). Their migration, and thus biomass displacement, is not only taking place at short distances (Yang and Gratton, 2014), but mass seasonal migration also takes place over longer distances (Hu *et al.*, 2016), with consequences for the flow of biomass, carbon, nitrogen, phosphorous, etc. Many of the ecosystem functions provided by insects directly or indirectly benefit humans. These ecosystem services make significant contributions to human economy. For instance, ecosystem services provided by naturally occurring insects in the USA have been conservatively estimated to amount to approximately 57 billion US dollars annually (Losey and Vaughan, 2006). This estimate, derived in 2006, was close to the costs incurred to the US economy by the Iraq war in that same year. Insects generally provide such ecosystem services and Losey and Vaughan (2006) analysed four of these for which sufficient data were available: (1) dung burial; (2) pest control; (3) pollination; and (4) recreation and commercial fisheries. The benefits derived by the US economy from these ecosystem services were estimated to be 380 million, 4.49 billion, 3.07 billion and 49.9 billion US dollars, respectively. These four ecosystem services are also among the global list of ecosystem services developed by Costanza *et al.* (1997).

2.3.1 Pollination

About two thirds of all plant species rely on insects for pollination. This relates especially to the activities of bees, beetles, butterflies and flies (Schoonhoven *et al.*, 2005). Thus, plant diversity on earth depends to a large extent on pollination services by insects. And, consequently, animal biodiversity indirectly relies on insect pollinators as well, especially because of the dominance of specialist insect herbivore species (Schoonhoven *et al.*, 2005). This is also reflected in our crops. Approximately 72% of the world's crops are dependent on insects for pollination. For instance, insect pollinators are essential for reproduction of watermelon, melons, gourds, pumpkins, kiwifruit, passionfruit, cashew, macadamia, cocoa, and vanilla. Furthermore, insect pollinators are important for various other crops, such as starfruit, apple, almond, strawberry, avocado, plum, peach, apricot, cherry, cucumber, coconut, mango, soybean, cotton, sunflower, broad beans, and coffee (Gallai *et al.*, 2009; Klein *et al.*, 2007; Losey and Vaughan, 2006). There is ample attention in the media for enhanced mortality in the honeybee *Apis mellifera*, but it should be realized that the number of insect pollinators is much larger, including hundreds of species of solitary bees, bumblebees, flies, etc. (Garibaldi *et al.*, 2013). Moreover, in several crops, wild bee species are more important for pollination than the honeybee *A. mellifera* (Garibaldi *et al.*, 2013). Investing in ecological improvement of agriculture is rewarding as it can create synchronous biodiversity and yield outcomes (Garibaldi *et al.*, 2016). In the study by Losey and Vaughan (2006), pollination accounted for about 3 billion US dollars in the USA. On a world scale, pollination services are estimated to yield approximately 9.5% of all crop production (Gallai *et al.*, 2009). Benefits from pollination are not only represented in terms of yield, but also to an important extent in terms of fruit quality (Garratt *et al.*, 2014).

2.3.2 Dung burial

Cattle produce large amounts of dung, e.g. 9,000 kilos per head per year (Losey and Vaughan, 2006), which may be used as food by various insects, including pest flies and cattle parasites. Dung beetles are important insects that remove dung by burying it to provide food for their offspring, thus improving soil fertility. Dung beetles are specialized to some degree. This was learned from the introduction of cattle into Australia in the 18th century. Australian dung beetles were adjusted to the Australian fauna and were not capable of handling cattle dung. The introduction of cattle, therefore, resulted in serious fouling of rangeland, nitrogen volatilization and the propagation of pest flies such as *Musca vetustissima* (Tyndalebiscoe and Vogt, 1991). In a large-scale programme, dung beetles from Africa (Figure 2.3) and Europe were screened and introduced into Australia to successfully control *M. vetustissima* problems (Davis, 1996; Tyndalebiscoe and Vogt, 1991).

Also in the United Kingdom, dung beetles are documented to provide important benefits to cattle industry, amounting to £ 367 million each year, or ca £ 40 per cow per year (Beynon *et al.*, 2015). This study further estimates that protecting dung beetles under agri-environment schemes, such as by stopping largely unnecessary anti-helminthic treatments of cattle, could save another £ 40 million per year (Beynon *et al.*, 2015). In addition to the interference with rangeland fouling and suppression of fly pests, dung beetles can also play an important role in nutrient cycling and plant growth. Given their sensitivity to environmental disturbance

Figure 2.3. Dung beetle ecosystem services are appreciated in South Africa (photo by Marcel Dicke).

several studies call for conservation of dung beetles to protect their ecosystem services (Beynon *et al.*, 2015; Nichols *et al.*, 2008).

2.3.3 Removal of carrion

Cadavers represent more than 1% of organic resources undergoing decomposition in terrestrial ecosystems (Carter *et al.*, 2007). Dead animals provide food to a dynamic community of invertebrates with major roles for insect detritivores such as flies and beetles. The community colonizing a cadaver can comprise hundreds of species (Merritt and De Jong, 2015). Thus, decomposition of carrion is an important ecosystem process. Burying beetles (Coleoptera: Silphidae) have an interesting biology. As soon as a male finds a corpse of a small animal, such as a mouse, it attracts a female and together they quickly bury the corpse to avoid competition with other saprophagous animals. The females lay eggs on the corpse and these develop on food that the adults have pre-digested, and regurgitate to feed the young. A male may collaborate with more than one female and this may be dependent on the degree of competition from e.g. flies. When competition from flies is severe there is a benefit to be gained from rapidly burying the dead animal (Scott, 1998). Although burying beetles can also feed on larger corpses, these resources are usually first exploited by calliphorid flies and too large for burying beetles to handle. Calliphorid flies have a well-developed sense of smell and can perceive a decomposing corpse from large distances through the perception of e.g. dimethyl trisulfide (Brodie *et al.*, 2014; Smith and Wall, 1998). Their arrival at the carrion (Figure 2.4) is the start of a dynamic succession of arthropod species colonizing the cadaver that can easily comprise 100 species from tens of insect families (Farwig *et al.*, 2014; Richards and Goff, 1997). As a result, most of the soft tissues can be decomposed within a few days to weeks, leaving the bones and hairs for longer-lasting decomposition (Pechal *et al.*, 2014). These activities of insects, just like dung burial, result in nutrient cycling and prevent the accumulation of cadavers that may bring risks of pathogens and pests. The early arrival of necrophagous insects importantly enhances the decomposition of the carcass (Pechal *et al.*, 2014). This results in

Figure 2.4. Dead roe deer eaten by calliphorid maggots (photo courtesy Bram Winkelman, Wageningen University, Wageningen, the Netherlands).

the rapid introduction of nitrogen, phosphorous, calcium and magnesium into the soil. The study of the arthropod community on dead human bodies can be exploited to assess the post mortem interval, an important component in forensic research (Greenberg, 1985). For this, the study of community ecology of necrophagous insects is essential, as well as investigating the effects of environmental factors on the development rate of the insects. Swine are often used as substitute for humans in such investigations (Mohr and Tomberlin, 2015), but people can donate their body for such research in the Anthropological Research Facility of the University of Tennessee, Knoxville, TN, USA and the Australian Facility for Taphonomic Experimental Research of the University of Technology, Sydney, Australia.

Necrophagous insects, such as blow flies, may even be used to study the ecology of vertebrates through analysing the local insect fauna for genetic material from their vertebrate resources (Calvignac-Spencer *et al.*, 2013a,b). This may be used to assess the presence of rare species, species diversity and mortality rates. The latter is the result of blow flies exploiting carcasses and usually being one of the first visitors after an animal died (Calvignac-Spencer *et al.*, 2013a).

2.3.4 Pest control

Most potential pests on earth are under natural biological control, exerted by natural enemies including predators, parasitoids and pathogens (Van Lenteren, 2012a). The global economic value of this ecosystem service is estimated to be 400 billion US$ per year (Costanza *et al.*, 1997). For the US economy, the contribution of insects to natural control of pests is estimated to be 4.49 billion US$ per year. In most agricultural systems, humans experience pest problems. As long as humankind practiced agriculture, pests have occurred on their crops and livestock. These have been managed in various ways. The earliest known example of the use of insect predators and parasitoids to protect crops relates to Chinese agriculture around the year 300 AD. Chinese farmers are known to have exploited the presence of ant nests in their orchards to combat insect pests (Van Lenteren, 2012a). Knowledge of the biology and ecology of natural enemies of insect pests has been collected over centuries (Van Lenteren and Godfray, 2005). The first example of classical biological control, i.e. the introduction of an exotic natural enemy to control an exotic pest dates back to 1887, when the cottony cushion scale, *Icerya purchasi,* was successfully controlled by the release of the coccinellid beetle *Rodolia cardinalis* (DeBach, 1964). Since then a total of 2,627 introductions have resulted in satisfactory control against 172 pest species, the majority of which with a long lasting effect (Cock *et al.*, 2016). A relatively recent example is the control of the cassava green mite (*Mononychellus tanajoa*) by predatory mites imported from Latin America, the origin of the cassava green mite (Yaninek *et al.*, 1993). In addition to natural biological control, also augmentative biological control has been initiated and currently more than 200 arthropods are available for biological control. These species are reared in specialized rearing programmes by a range of companies or governmental programmes worldwide and feature in Integrated Pest Management strategies in various crops. For instance, the parasitoid *Aphytis lingnanensis* has been used since 1906 to control scale insects in citrus, the predatory mite *Phytoseiulus persimilis* has been used since 1968 to control two-spotted spider mites in a diversity of crops both in greenhouses and in open fields since 1968 and *Trichogramma* parasitic wasps that attack lepidopteran eggs have been used extensively for decades (Van Lenteren, 2012b). The crop area in which augmentative biological control is used globally is, however, much smaller than the area under natural control (i.e. 0.16 vs 3.5 million km^2) (Van Lenteren, 2012a).

2.3.5 Recreation and commercial fisheries

A large proportion of birds and smaller game are insectivorous. For instance, 89% of American birds are partially or primarily insectivorous, ranging from e.g. swifts to owls (Losey and Vaughan, 2006). Therefore, insects are at the basis of many food chains. While this means that insects provide fundamental ecosystem services, it also means that insects provide important services to humans ranging from recreational fishing and wildlife observation to commercial fish production (Losey and Vaughan, 2006). For instance, bird watching can be dominated by observing insectivorous birds and so without insects only few birds can be watched. Seeing great tits forage for caterpillars in spring or a European honey buzzard feeding on wasps is an enjoyable activity. This is also true for observing woodpeckers searching for wood-boring insects in the forest or young of wild boars searching for soil-borne insects. However, even seeing a buzzard feed on birds or frogs is an event relying on insects as food for the buzzard's prey. Those that enjoy recreational fishing rely on many insectivorous fish, as is clear when you ask what anglers put on the hook, which is often calliphorid larvae. The value of naturally occurring insects in the USA for recreation have been estimated to be ca US$ 50 billion and thus this makes up the majority of the total value of naturally occurring insects in the USA (Losey and Vaughan, 2006).

Apart from recreational fishing, commercial fisheries are more and more aware of the insectivorous nature of the fish they rear. Insects such as house fly (*Musca domestica*) or black soldier fly (*Hermetia illucens*) pupae can be used to different degrees as feedstuff for rainbow trout (St-Hilaire *et al.*, 2007), turbot (Kroeckel *et al.*, 2012), tilapia, sturgeon (Ushakova *et al.*, 2016), salmon and others (Henry *et al.*, 2015; Makkar *et al.*, 2014). It is important to note that each fish species has its own dietary requirements and, therefore, the relative value of individual insect species as fish feed remains to be investigated for individual fish-insect combinations.

Thus, natural ecosystem services that had been disregarded are likely to regain momentum and are likely to stimulate a new insect-producing industry. This is an important observation because aquaculture is one of the fastest-growing food producing sectors (FAO, 2014).

2.4 Conclusions

As a dominant form of animal biomass and life on earth, insects have a wide range of ecological functions in their natural ecosystems, including herbivory, carnivory, and feeding on detritus. Their tremendous species richness is exhibited in a diverse palette of forms and biological characteristics. This diversity translated into natural ecosystem services that are diverse and extensive. Moreover, humankind has exploited insects in many ways, some of which have been highlighted in this chapter while many more could not be highlighted here. Insects are, for instance, also used as medicine (Li *et al.*, 2004; Van Huis, 2003) and the major component of their exoskeleton, chitin, has many applications in agriculture, medicine, food and non-food industries (Tharanathan and Kittur, 2003). Moreover, insects are a source of inspiration to artists active in the visual arts and music (Berenbaum, 1995; Dicke, 2000). Being such an intricate part of life on earth and an invaluable part of ecosystems, it is astonishing that this role still needs to be highlighted to the general public. However, even in the scientific literature ecosystem services of insects still receive limited attention: in the Web of Science database a search on "ecosystem service*" results in 13,165 entries, whereas the combination of "ecosystem service*" and insect* matches with only 701 papers, so only 5% of the total number of papers without insects. Thus, research on ecosystem services of insects deserves thorough stimulation. This includes the ecosystem services of edible insects that are harvested from agricultural systems and that include species involved in pollination, biological control or silk production (Payne and Van Itterbeeck, 2017).

Being a nutritious source of food to a vast range of vertebrates, including amphibians, birds and mammals, it should be no surprise that they have been an important source of nutrition to mankind throughout evolutionary history as well as to our livestock and fish.

References

Berenbaum, M.R., 1995. Bugs in the system. Insects and their impact on human affairs. Addison-Wesely, Reading, MA, USA.

Beynon, S.A., Wainwright, W.A. and Christie, M., 2015. The application of an ecosystem services framework to estimate the economic value of dung beetles to the UK cattle industry. Ecological Entomology 40: 124-135.

Boukraâ, L., 2013. Honey in traditional and modern medicine. CRC Press, Boca Raton, FL, USA, 454 pp.

Bresseel, J. and Constant, J., 2014. Giant sticks from Vietnam and China, with three new taxa including the second longest insect known to date (Phasmatodea, Phasmatidae, Clitumninae, Pharnaciini). European Journal of Taxonomy 104: 1-38.

Brodie, B., Gries, R., Martins, A., VanLaerhoven, S. and Gries, G., 2014. Bimodal cue complex signifies suitable oviposition sites to gravid females of the common green bottle fly. Entomologia Experimentalis et Applicata 153: 114-127.

Brueland, H., 1995. Highest lifetime fecundity. In: Walker, T.J. (ed.) Book of insect records. University of Florida, Gainesville, FL, USA. Available at: http://entomology.ifas.ufl.edu/walker/ufbir.

Calvignac-Spencer, S., Leendertz, F.H., Gilbert, M.T.P. and Schubert, G., 2013a. An invertebrate stomach's view on vertebrate ecology. Bioessays 35: 1004-1013.

Calvignac-Spencer, S., Merkel, K., Kutzner, N., Kuhl, H., Boesch, C., Kappeler, P.M., Metzger, S., Schubert, G. and Leendertz, F.H., 2013b. Carrion fly-derived DNA as a tool for comprehensive and cost-effective assessment of mammalian biodiversity. Molecular Ecology 22: 915-924.

Campana, M.G., Garcia, N.M.R. and Tuross, N., 2015. America's red gold: multiple lineages of cultivated cochineal in Mexico. Ecology and Evolution 5: 607-617.

Carter, D.O., Yellowlees, D. and Tibbett, M., 2007. Cadaver decomposition in terrestrial ecosystems. Naturwissenschaften 94: 12-24.

Cock, M.J.W., Murphy, S.T., Kairo, M.T.K., Thompson, E., Murphy, R.J. and Francis, A.W., 2016. Trends in the classical biological control of insect pests by insects: an update of the BIOCAT database. Biocontrol 61: 349-363.

Costanza, R., d'Arge, R., De Groot, R., Farber, S., Grasso, M., Hannon, B., Limburg, K., Naeem, S., Oneill, R.V., Paruelo, J., Raskin, R.G., Sutton, P. and Van den Belt, M., 1997. The value of the world's ecosystem services and natural capital. Nature 387: 253-260.

Crittenden, A.N., 2011. The importance of honey consumption in human evolution. Food and Foodways 19: 257-273.

Davis, A.L.V., 1996. Seasonal dung beetle activity and dung dispersal in selected South African habitats: implications for pasture improvement in Australia. Agriculture Ecosystems and Environment 58: 157-169.

DeBach, P., 1964. Biological control of insect pests and weeds. Cambridge University Press, Cambridge, UK, 844 pp.

Deblauwe, I., Dupain, J., Nguenang, G.M., Werdenich, D. and Van Elsacker, L., 2003. Insectivory by *Gorilla gorilla gorilla* in southeast Cameroon. International Journal of Primatology 24: 493-502.

Dicke, M., 2000. Insects in western art. American Entomologist 46: 228-236.

Dirzo, R. and Raven, P.H., 2003. Global state of biodiversity and loss. Annual Review of Environment and Resources 28: 137-167.

Food and Agriculture Organisation (FAO), 2014. The state of world fisheries and aquaculture 2014. FAO, Rome, Italy.

Farwig, N., Brandl, R., Siemann, S., Wiener, F. and Muller, J., 2014. Decomposition rate of carrion is dependent on composition not abundance of the assemblages of insect scavengers. Oecologia 175: 1291-1300.

Fostiridou, A., Karapanagiotis, I., Vivdenko, S., Lampakis, D., Mantzouris, D., Achilara, L. and Manoudis, P., 2016. Identification of pigments in Hellenistic and Roman funeral figurines. Archaeometry 58: 453-464.

Gallai, N., Salles, J.M., Settele, J. and Vaissiere, B.E., 2009. Economic valuation of the vulnerability of world agriculture confronted with pollinator decline. Ecological Economics 68: 810-821.

Garibaldi, L.A., Carvalheiro, L.G., Vaissiere, B.E., Gemmill-Herren, B., Hipolito, J., Freitas, B.M., Ngo, H.T., Azzu, N., Saez, A., Astrom, J., An, J.D., Blochtein, B., Buchori, D., Garcia, F.J.C., Da Silva, F.O., Devkota, K., Ribeiro, M.D., Freitas, L., Gaglianone, M.C., Goss, M., Irshad, M., Kasina, M., Pacheco, A.J.S., Kiill, L.H.P., Kwapong, P., Parra, G.N., Pires, C., Pires, V., Rawal, R.S., Rizali, A., Saraiva, A.M., Veldtman, R., Viana, B.F., Witter, S. and Zhang, H., 2016. Mutually beneficial pollinator diversity and crop yield outcomes in small and large farms. Science 351: 388-391.

Garibaldi, L.A., Steffan-Dewenter, I., Winfree, R., Aizen, M.A., Bommarco, R., Cunningham, S.A., Kremen, C., Carvalheiro, L.G., Harder, L.D., Afik, O., Bartomeus, I., Benjamin, F., Boreux, V., Cariveau, D., Chacoff, N.P., Dudenhoffer, J.H., Freitas, B.M., Ghazoul, J., Greenleaf, S., Hipolito, J., Holzschuh, A., Howlett, B., Isaacs, R., Javorek, S.K., Kennedy, C.M., Krewenka, K.M., Krishnan, S., Mandelik, Y., Mayfield, M.M., Motzke, I., Munyuli, T., Nault, B.A., Otieno, M., Petersen, J., Pisanty, G., Potts, S.G., Rader, R., Ricketts, T.H., Rundlof, M., Seymour, C.L., Schuepp, C., Szentgyorgyi, H., Taki, H., Tscharntke, T., Vergara, C.H., Viana, B.F., Wanger, T.C., Westphal, C., Williams, N. and Klein, A.M., 2013. Wild pollinators enhance fruit set of crops regardless of honey bee abundance. Science 339: 1608-1611.

Garratt, M.P.D., Breeze, T.D., Jenner, N., Polce, C., Biesmeijer, J.C. and Potts, S.G., 2014. Avoiding a bad apple: insect pollination enhances fruit quality and economic value. Agriculture Ecosystems and Environment 184: 34-40.

Ghosh, M., Gupta, S. and Kumar, V.S.K., 2015. Studies on the loss of gloss of shellac and polyurethane finishes exposed to uv. Maderas-Ciencia Y Tecnologia 17: 39-44.

Gould, J.L. and Gould, C.G., 1988. The honey bee. Scientific American Library, Armonk, NY, USA.

Greenberg, B., 1985. Forensic entomology: case studies. Bulletin of the Entomology Society of America 31: 25-28.

Gullan, P.J. and Cranston, P.S., 2010. The insects: an outline of entomology. Blackwell Publishing, Hoboken, NJ, USA, 584 pp.

Havsteen, B.H., 2002. The biochemistry and medical significance of the flavonoids. Pharmacology and Therapeutics 96: 67-202.

Henry, M., Gasco, L., Piccolo, G. and Fountoulaki, E., 2015. Review on the use of insects in the diet of farmed fish: past and future. Animal Feed Science and Technology 203: 1-22.

Hu, G., Lim, K.S., Horvitz, N., Clark, S.J., Reynolds, D.R., Sapir, N. and Chapman, J.W., 2016. Mass seasonal bioflows of high-flying insect migrants. Science 354: 1584-1587.

Huigens, M.E., De Jong, P.W., Dicke, M., Van Huis, A., Van Lenteren, J.C. and Smid, H.M., 2006. Muggenzifters en mierenneukers. Insecten onder de loep genomen. Wageningen University, Wageningen, the Netherlands.

Klein, A.-M., Vaissiere, B.E., Cane, J.H., Steffan-Dewenter, I., Cunningham, S.A., Kremen, C. and Tscharntke, T., 2007. Importance of pollinators in changing landscapes for world crops. Proceedings of the Royal Society B-Biological Sciences 274: 303-313.

Kroeckel, S., Harjes, A.G.E., Roth, I., Katz, H., Wuertz, S., Susenbeth, A. and Schulz, C., 2012. When a turbot catches a fly: evaluation of a pre-pupae meal of the black soldier fly (*Hermetia illucens*) as fish meal substitute – growth performance and chitin degradation in juvenile turbot (*Psetta maxima*). Aquaculture 364: 345-352.

Lawrence, B.D., 2014. Processing of *Bombyx mori* silk for biomedical applications. In: Kundu, S.C. (ed.) Silk biomaterials for tissue engineering and regenerative medicine. Woodhead Publishing, Cambridge, UK, pp. 78-99.

Lech, K. and Jarosz, M., 2016. Identification of Polish cochineal (*Porphyrophora polonica* L.) in historical textiles by high-performance liquid chromatography coupled with spectrophotometric and tandem mass spectrometric detection. Analytical and Bioanalytical Chemistry 408: 3349-3358.

Lemelin, R.H., 2013. The management of insects in recreation and tourism. Cambridge University Press, Cambridge, UK, 353 pp.

Lemire, B. and Riello, G., 2008. East and west: textiles and fashion in early modern Europe. Journal of Social History 41: 887-916.

Lev, E., 2003. Traditional healing with animals (zootherapy): medieval to present-day Levantine practice. Journal of Ethnopharmacology 85: 107-118.

Li, W.L., Zheng, H.C., Bukuru, J. and De Kimpe, N., 2004. Natural medicines used in the traditional Chinese medical system for therapy of diabetes mellitus. Journal of Ethnopharmacology 92: 1-21.

Losey, J.E. and Vaughan, M., 2006. The economic value of ecological services provided by insects. Bioscience 56: 311-323.

Makkar, H.P.S., Tran, G., Henze, V. and Ankers, P., 2014. State-of-the-art on use of insects as animal feed. Animal Feed Science and Technology 197: 1-33.

McGovern, P.E., Zhang, J.H., Tang, J.G., Zhang, Z.Q., Hall, G.R., Moreau, R.A., Nunez, A., Butrym, E.D., Richards, M.P., Wang, C.S., Cheng, G.S., Zhao, Z.J. and Wang, C.S., 2004. Fermented beverages of pre- and proto-historic China. Proceedings of the National Academy of Sciences of the United States of America 101: 17593-17598.

McGrew, W.C., 2014. The 'other faunivory' revisited: insectivory in human and non-human primates and the evolution of human diet. Journal of Human Evolution 71: 4-11.

Merritt, R.W. and De Jong, G.D., 2015. Arthropod communities in terrestrial environments. In: Benbow, M.E., Tomberlin, J.K. and Tarone, A.M. (eds.) Carrion ecology, evolution, and their applications. CRC Press, Boca Raton, FL, USA, pp. 65-91.

Mitterboeck, T.F., Fu, J.Z. and Adamowicz, S.J., 2016. Rates and patterns of molecular evolution in freshwater versus terrestrial insects. Genome 59: 968-980.

Mohr, R.M. and Tomberlin, J.K., 2015. Development and validation of a new technique for estimating a minimum *postmortem* interval using adult blow fly (Diptera: Calliphoridae) carcass attendance. International Journal of Legal Medicine 129: 851-859.

Nichols, E., Spector, S., Louzada, J., Larsen, T., Amequita, S., Favila, M.E. and Scarabaeinae Res, N., 2008. Ecological functions and ecosystem services provided by Scarabaeinae dung beetles. Biological Conservation 141: 1461-1474.

Payne, C.L.R. and Van Itterbeeck, J., 2017. Ecosystem services from edible insects in agricultural systems: a review. Insects 8: 24.

Pearnchob, N., Siepmann, J. and Bodmeier, R., 2003. Pharmaceutical applications of shellac: moisture-protective and taste-masking coatings and extended-release matrix tablets. Drug Development and Industrial Pharmacy 29: 925-938.

Pechal, J.L., Benbow, M.E., Crippen, T.L., Tarone, A.M. and Tomberlin, J.K., 2014. Delayed insect access alters carrion decomposition and necrophagous insect community assembly. Ecosphere 5: 21.

Polilov, A., 2015. How small is the smallest? New record and remeasuring of *Scydosella musawasensis* Hall, 1999 (Coleoptera, Ptiliidae), the smallest known free-living insect. ZooKeys 526: 61-64.

Rao, P.V., Krishnan, K.T., Salleh, N. and Gan, S.H., 2016. Biological and therapeutic effects of honey produced by honey bees and stingless bees: a comparative review. Revista Brasileira De Farmacognosia-Brazilian Journal of Pharmacognosy 26: 657-664.

Richards, E.N. and Goff, M.L., 1997. Arthropod succession on exposed carrion in three contrasting tropical habitats on Hawaii island, Hawaii. Journal of Medical Entomology 34: 328-339.

Schoonhoven, L.M., Van Loon, J.J.A. and Dicke, M., 2005. Insect-plant biology. Oxford University Press, Oxford, UK, 400 pp.

Scott, M.P., 1998. The ecology and behavior of burying beetles. Annual Review of Entomology 43: 595-618.

Sheppard, S.K., Henneman, M.L., Memmott, J. and Symondson, W.O.C., 2004. Infiltration by alien predators into invertebrate food webs in Hawaii: a molecular approach. Molecular Ecology 13: 2077-2088.

Smith, K.E. and Wall, R., 1998. Estimates of population density and dispersal in the blowfly *Lucilia sericata* (Diptera: Calliphoridae). Bulletin of Entomological Research 88: 65-73.

St-Hilaire, S., Sheppard, C., Tomberlin, J.K., Irving, S., Newton, L., McGuire, M.A., Mosley, E.E., Hardy, R.W. and Sealey, W., 2007. Fly prepupae as a feedstuff for rainbow trout, *Oncorhynchus mykiss*. Journal of the World Aquaculture Society 38: 59-67.

Stouthamer, R., Breeuwer, J.A.J. and Hurst, G.D.D., 1999. *Wolbachia pipientis*: microbial manipulator of arthropod reproduction. Annual Review of Microbiology 53: 71-102.

Sutton, M.Q., 1995. Archaeological aspects of insect use. Journal of Archaeological Method and Theory 2: 253-298.

Takken, W., 1980. Studies on the role of serum proteins in the ovarian function of the tsetse fly, *Glossina palpalis palpalis*. PhD thesis, Wageningen University, Wageningen, the Netherlands, 87 pp.

Takken, W. and Knols, B.G.J., 2007. Emerging pests and vector-borne diseases in Europe. Ecology and control of vector-borne diseases. Wageningen Academic Publishers, Wageningen, the Netherlands, 498 pp.

Tharanathan, R.N. and Kittur, F.S., 2003. Chitin – the undisputed biomolecule of great potential. Critical Reviews in Food Science and Nutrition 43: 61-87.

Tyndalebiscoe, M. and Vogt, W.G., 1991. Effects of adding exotic dung beetles to native fauna on bush fly breeding in the field. Entomophaga 36: 395-401.

Ushakova, N.A., Bastrakov, A.I., Kozlova, A.A., Ponomarev, S.V., Bakaneva, Y.M., Fedorovykh, Y.V., Zhandalgarova, A.D. and Pavlov, D.S., 2016. Features of the effect of a complex probiotic with *Bacillus* bacteria and the larvae of *Hermetia illucens* biomass on Mozambique tilapia (*Oreochromis mossambicus x O-niloticus*) and Russian sturgeon (*Acipenser gueldenstaedti*) fry. Biology Bulletin 43: 450-456.

Van Huis, A., 2003. Medical and stimulating properties ascribed to arthropods and their products in sub-Saharan Africa. In: Motte-Florac, E. and Thomas, J.M.C. (eds.) Les insectes dans la tradition orale – insects in oral literature and traditions. Peeters Publishers, Leuven, Belgium, pp. 367-382.

Van Huis, A., Dicke, M. and Van Gurp, H., 2014. The insect cookbook – food for a sustainable planet. Columbia University Press, New York, NY, USA.

Van Huis, A., Van Itterbeeck, J., Klunder, H., Mertens., E., Halloran, A., Muir, G. and Vantomme, P., 2013. Edible insects; future prospects for food and feed security. Food and Agriculture Organisation, Rome, Italy.

Van Lenteren, J.C., 2012a. Internet book of biological control. International Organization for Biological Control, Zürich, Switzerland. Available at: http://tinyurl.com/zk3rdrr.

Van Lenteren, J.C., 2012b. The state of commercial augmentative biological control: plenty of natural enemies, but a frustrating lack of uptake. Biocontrol 57: 1-20.

Van Lenteren, J.C. and Godfray, H.C.J., 2005. European science in the enlightenment and the discovery of the insect parasitoid life cycle in the Netherlands and Great Britain. Biological Control 32: 12-24.

Venkataraman, V.V., Kraft, T.S. and Dominy, N.J., 2013. Tree climbing and human evolution. Proceedings of the National Academy of Sciences of the United States of America 110: 1237-1242.

William, D.M., 2001. Largest. In: Walker, T.J. (ed.) Book of insect records. University of Florida, Gainesville, FL, USA. Available at: http://entomology.ifas.ufl.edu/walker/ufbir.

Yang, L.H. and Gratton, C., 2014. Insects as drivers of ecosystem processes. Current Opinion in Insect Science 2: 26-32.

Yaninek, J.S., Onzo, A. and Ojo, J.B., 1993. Continent-wide releases of neotropical phytoseiids against exotic cassava green mite in Africa. Experimental and Applied Acarology 17: 145-160.

Zhang, C.-X., Tang, X.-D. and Cheng, J.-A., 2008. The utilization and industrialization of insect resources in China. Entomological Research 38: 38-47.

Zhang, Z.-Q., 2011. Animal biodiversity: an introduction to higher-level classification and taxonomic richness. Zootaxa 3148: 7-12.

3.

Pachnoda marginata

Environmental impact of insect production

D.G.A.B. Oonincx
Laboratory of Entomology, Wageningen University and Research, P.O. Box 16, 6700 AA Wageningen, the Netherlands; dennisooonincx@gmail.com

Abstract

This chapter deals with the environmental aspects of insect production. First an overview of mitigation measures, which can be used in conventional livestock systems is provided. Then the direct emissions of greenhouse gasses and ammonia from insect production are discussed and compared to conventional production animals. Subsequently, the life cycle assessment, which is a method to quantify environmental impact of a product over its entire production chain, is explained. This is followed by an overview of the outcomes of life cycle assessments conducted for insects and for insect derived products used for food, feed, or fibres. Then the main factor determining environmental impact of animal production systems, feed utilisation, is discussed for conventional production animals and compared with insects. Special attention is given to the utilisation of co-products and waste. Lastly, the advantages and disadvantages of insect production are summarized and directions of further advancements and for further studies are given.

Keywords: insect production, environmental impact, sustainability, greenhouse gas emissions, life cycle assessment

3.1 Introduction

Livestock production systems are associated with a high environmental impact (Herrero *et al.*, 2013, 2015; Steinfeld *et al.*, 2006). Several mitigation strategies have been suggested to decrease this impact. First and foremost, a reduction in the consumption of animal products would decrease environmental impact (Herrero *et al.*, 2015; Smetana *et al.*, 2016). This strategy is especially relevant in developed countries where protein consumption outpaces requirements. Another approach would be to increase the efficiency of livestock production systems. For instance, by selective breeding to increase health, prolong productive lifespan and increase reproductive efficiency, which would limit environmental impact (Herrero *et al.*, 2015; Smetana *et al.*, 2016). Also, optimisation of forage production, for instance by over-sowing of grasses with legumes, or improving feed quality would lower environmental impact (Herrero *et al.*, 2015; Smetana *et al.*, 2016). Furthermore, improved manure management, which can decrease volatilisation and leaching of nitrogenous compounds, is another potential measure (Ndegwa *et al.*, 2008).

A different strategy would be a switch from high impact to lower impact species. For instance substituting ruminants by monogastrics which would decrease methane emissions (Herrero *et al.*, 2015). A similar replacement strategy could be exchanging conventional livestock by insects, because insect production is suggested to have a lower environmental impact than conventional animal production systems (Barwa, 2009; Hardouin *et al.*, 2003; Looy *et al.*, 2013; Premalatha *et al.*, 2011). However, so far there are few published studies that have actually quantified the negative effects of insect farming on the environment. This chapter serves to describe the main findings regarding the environmental impact of insect production and explains the meaning of these findings.

3.2 Direct emissions of greenhouse gases and ammonia

In the last decade, climate change has received much attention. Climate change is driven by anthropogenic greenhouse gases (GHG) emissions (IPCC, 2014). Animal production systems are associated with the generation of large amounts of these GHG (Steinfeld *et al.*, 2006). One can distinguish between direct generation of GHG within the animal production system, and indirect generation of GHG, which occurs elsewhere in the production chain. Direct GHG are due to animal respiration, or the metabolism of associated bacteria, either living inside as a gut symbiont or living on excreted manure. The primary GHG associated with respiration is carbon dioxide (CO_2). The carbon in carbon dioxide stems from feed provided to an animal. This carbon was previously sequestered in plant biomass, and because it is in a closed cycle (air-plant-feed-animal-air) respiration is not considered a net source of carbon dioxide (Steinfeld *et al.*, 2006). However, if carbon in the feed is sequestered in methane, which is 25 times as potent as carbon dioxide, there is a net effect (IPCC, 2007). Ruminants produce large amounts of methane and thereby contribute to climate change via gut symbionts which digest cellulose rich materials such as grass (De Vries and De Boer, 2010). Similarly, insects do not produce methane themselves, but certain insect species have gut symbionts that do, with examples of this found in termites (Isoptera), scarab beetles (Coleoptera) and cockroaches (Blattodea) (Hackstein and Stumm, 1994). These symbionts can utilize substrates with limited direct nutritional value for their host. In this way, they facilitate a more efficient utilisation of substrates. For five species of edible insects the direct GHG production has been quantified and compared to pigs and beef cattle (Oonincx *et al.*, 2010). Three species; house crickets (*Acheta domesticus*), migratory locusts (*Locusta migratoria*), and yellow mealworms (*Tenebrio molitor*) do not produce methane. However, Argentinean cockroaches (*Blaptica dubia*) and sun beetles (*Pachnoda marginata*), do produce methane (1.4 and 4.9 g/kg of mass gain, respectively), which probably is due to the presence of gut symbionts. These amounts are similar to pigs (1.9-4.0 g/kg of mass gain), but far lower than for beef cattle (114 g/kg of mass gain).

Another GHG associated with animal production systems is nitrous oxide. The effect on climate change of one nitrous oxide molecule is 298 times as strong as carbon dioxide (IPCC, 2007). Approximately two thirds of the total anthropogenic nitrous oxide production is due to manure and urine in animal production systems (Steinfeld *et al.*, 2006). The five previously mentioned insect species all produce nitrous oxide. Sun beetle larvae, house crickets and Argentinean cockroaches emit low amounts of nitrous oxide (1.0-5.7 mg/kg of mass gain), whereas this is higher for yellow mealworms (25.5 mg/kg of mass gain) and migratory locusts (59.5 mg/kg of mass gain). Although the differences between species are large, nitrous oxide production from these insects is low compared to pig production systems (106-3,457 mg/kg of mass gain). (Oonincx *et al.*, 2010)

Besides the production of GHG, also ammonia (NH_3) emissions are relevant because these lead to nitrification and acidification of soil. Furthermore, ammonia can be converted to nitrous oxide by soil bacteria (Wrage *et al.*, 2001). Approximately two thirds of all anthropogenic ammonia emissions are due to animal production systems (Steinfeld *et al.*, 2006). Uric acid fermentation in chicken manure and urea fermentation in mammals are the main sources of these emissions (Cole *et al.*, 2005; Fabbri *et al.*, 2007; Lacey, 2002). The ammonia emission of yellow mealworms and sun beetle larvae is very low (1-3 mg/kg of mass gain), for Argentinean cockroaches and migratory locusts these seem higher (36-54 mg/kg of mass gain), and house crickets are associated with the highest ammonia emissions (142 mg/kg of mass gain). These values can be considered low compared to pigs (1,140-1,920 mg/kg of mass gain). Beef cattle produce 2-3 times as much ammonia per kg of body mass compared to pigs.

In conclusion, the direct emissions of GHG and ammonia of insects are lower than for pigs and beef cattle. However, in order to evaluate the environmental impact of insect production, one should also include indirect emissions. Therefore, a more comprehensive method which takes the production chain into account, is required.

3.3 Assessing environmental impact

Life cycle assessment (LCA) is the holistic method to quantify the environmental impact of a product (or service) over its entire production chain; the life cycle of a product (De Vries and De Boer, 2010; Halloran *et al.*, 2016). The basic concept of a LCA is that environmental impact associated with materials or services used during production are allocated to the final product. A first step in LCA methodology is defining the goal of the study; most often this is comparing two products with a similar function, or comparing two production methods (Halloran *et al.*, 2016). For instance, insects are rich in protein and are often seen as a dietary source of protein (Oonincx and De Boer, 2012; Smetana *et al.*, 2015). Benchmarks for insects intended for human consumption can therefore be chicken, pork, beef, or protein rich meat substitutes. Insects intended for animal consumption are better compared to protein sources used in animal feed such as soybean meal or fishmeal (Salomone *et al.*, 2017; Van Zanten *et al.*, 2015).

Subsequently, the production system needs to be defined, taking into account the physical system itself, and the materials and services entering and exiting the system. The starting boundary of the system often contains the entire production process of the entering materials and services. The final boundary is less straightforward. It can be defined as the live insects ready to leave the production facility (Oonincx and De Boer, 2012), as processed insect meal (Salomone *et al.*, 2017; Van Zanten *et al.*, 2015), or as a ready to eat insect derived product (Smetana *et al.*, 2015). This relates to the form and purpose in which a product is used, and leads to the functional unit. For instance, if mealworms are considered as a human protein source, then an appropriate functional unit is grams of edible protein. The definition of this functional unit influences the total impact allocated to a product and allows a direct comparison of environmental impact of different production processes, or different products. Therefore, when comparing products, it is important that the chosen system boundaries and the functional unit are comparable as well. Currently, methods used for the processing of insects for human consumption differ greatly, which impairs accurate quantification of the environmental impact associated with this part of the production chain.

A next step to quantify environmental impact is determining relevant impact categories. The most commonly used at this moment are climate change (emission of greenhouse gases), and the usage of energy (fossil fuels), land, and water. In future studies the addition of other impact categories, for instance eutrophication, ecotoxicity, biodiversity, or ozone depletion, will provide further insight in the environmental impact of insect production (Halloran *et al.*, 2016; Smetana *et al.*, 2016).

As an example of LCA methodology, we can look at the production of mealworms. One of the main contributors to the environmental impact of mealworm production systems is their compound feed. This is made from various ingredients, for instance milled wheat. This wheat is produced in several steps, all of which contribute to the environmental impact of mealworm production. In Figure 3.1. ammonia and nitric acid are used to produce ammonium nitrate, which is subsequently applied as fertilizer on wheat fields. Transport of this fertilizer and the produced wheat is included (Figure 3.1; left side). Wheat processed at Dutch mills can come from several countries, which leads to differences in the GHG emissions associated with the wheat production process. Therefore, a proportional average is calculated (central part of the figure). The same procedure is used for all ingredients of the mealworm feed, which in turn allows the calculation of the total GHG emissions associated with feed production.

In order to perform an LCA all inputs to a production process need to be clearly defined and their environmental impact needs to be known. When several products are derived from a production process, the environmental impact of this process is divided over the main and the co-products. This division can be based on the respective mass of these products. However, the most common way to allocate these impacts is by means of economic allocation; for instance,

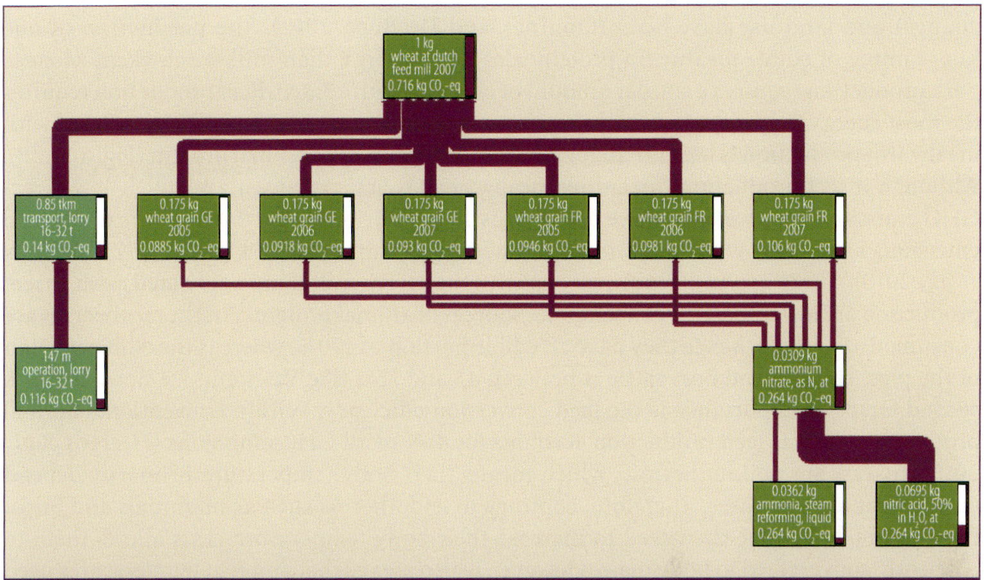

Figure 3.1. A breakdown of factors contributing to the energy use associated with wheat production (adapted from Viets, T., wheat, Ge and FR, at Dutch mill, Ecoinvent database version 2.2).

part of the environmental impact of wheat production is allocated to the co-product straw. Because straw has a lower economic value than wheat bran the environmental impact of one kg of straw is considered to be lower than for one kg of wheat bran.

Within an insect production system, faeces can be seen as a form of waste, but could also be considered a co-product. To which extent insect faeces have a positive economic value depends on the produced species, the utilized feed ingredients, which influence composition and structure, the legislation, availability of similar products and technological measures for processing. In certain cases it might also be possible to utilize the faeces as a substrate in bioreactors to produce energy, in other cases it could be used as crop fertilizer or it might be discarded in landfills (Smetana *et al.*, 2016). Faeces from black soldier fly (*Hermetia illucens*) larvae, produced on food waste, are considered a good fertilizer based on chemical analysis (Choi *et al.*, 2009; Salomone *et al.*, 2017). These faeces were equivalent to artificial fertilizer in a field study on cabbage plants (Choi *et al.*, 2009). Storage of the faeces, or frass, can result in GHG and ammonia emissions (Halloran *et al.*, 2016). However, because insect faeces are often dry, this might be to a lesser extent than with conventional livestock. The utilisation of co-products from insect production systems for other purposes can help to decrease the environmental impact allocated to the insect product itself.

3.4 Insects for human consumption

For yellow mealworms a LCA has been performed and the parameters energy use (an indicator for fossil fuel depletion), global warming potential (GWP; an indicator for climate change) and land use have been (Oonincx and De Boer, 2012). The production of one kilogramme of edible mealworm protein uses more energy than milk or chicken, whereas pork production requires a similar amount of energy (Figure 3.2A). Beef production requires the most energy for producing one kilogramme of edible protein. The GWP associated with mealworm production is lower than for conventional sources of animal protein (Figure 3.2B). Although insect production (including mealworms) is associated with relatively low direct GHG emissions (respiration), these direct emissions contribute only 0.29% to the total GHG emissions associated with mealworm production (Oonincx and De Boer, 2012; Oonincx *et al.*, 2010). There are two main reasons why the GHG emissions associated with insect production are lower than for conventional sources of animal protein. Firstly, mealworms are consumed as a whole, hence they have an edible portion of 100%, whereas the edible portion of the pigs, chicken and beef cattle is between 43 and 56% (De Vries and De Boer, 2010). A second factor of importance is the feed conversion efficiency. Within conventional animal production systems feed production accounts for 45% of all GHG emissions (Herrero *et al.*, 2015). Insects are poikilothermic, which means their body temperature primarily depend on ambient temperatures. Therefore, contrary to endotherms such as mammals and birds, they do not catabolize their feed to increase their body temperature. This allows them to efficiently convert feed to body mass. However, it also means that ambient temperatures need to be regulated more strictly than with endotherms. The latter is an explanation for the fairly high energy use in insect production (Figure 3.2A). Land use associated with mealworm production is lower than for conventional production animals (Figure 3.2C). This low land use is due to their efficient conversion of feed as feed production is the main driver of land use (Oonincx and De Boer, 2012).

Miglietta *et al.* (2015) estimated water use associated with mealworm production, based on data from the previously mentioned study. They expressed the water use per gram of edible protein and compared it with pork, chicken and beef. Mealworms use 23 litres of water, whereas chicken require approximately 34 litres, pigs require 57 litres and beef cattle require 112 litres to produce one gram of protein. Water use is strongly correlated with feed

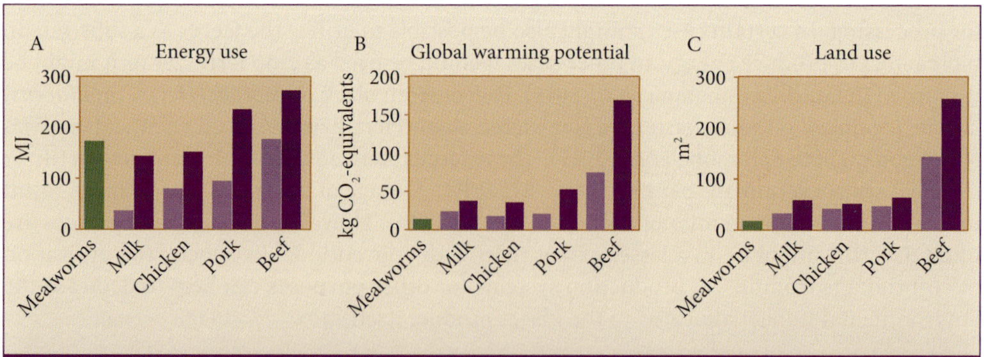

Figure 3.2. (A) Energy use, (B) global warming potential, and (C) land use associated with the production of one kg of edible protein by means of mealworms (green bars; Oonincx and De Boer, 2012) and conventional sources of animal protein (minimum (light) and maximum (dark) values from De Vries and De Boer, 2010).

production (Herrero *et al.*, 2015). Therefore, the previously mentioned high feed conversion efficiency explains the relatively low water use.

3.5 Insects for livestock consumption

Many different insect species have been used as a source of feed, both for companion animals and for production animals (Finke and Oonincx, 2013; Makkar *et al.*, 2014). An example from Togo on small scale use of insects as feed was described by Farina *et al.* (1991); a calabash (a fruit) would be placed in a termite mount and after a few weeks it would be removed and provided to young chickens in order to improve their growth and development. Similarly, there are accounts of termites which are provided with leftovers of husks and straw and subsequently this mixture of provided roughage and termites is used as feed for cattle or poultry (Abasiekong, 1997; Munyuli Bin Mushambanyi and Balezi, 2002). The latter method can be considered an extensive way of insect farming, in which local climatic conditions are used and the yield is optimized by providing more suitable feed than the insects would gather in nature by themselves. As long as the insects are not overharvested, these systems can be considered sustainable because their energy, land, and water use are negligible. This extensive way of farming in which waste is provided to termites would result in higher GHG emissions due to the production of methane by gut symbionts (Hackstein and Stumm, 1994). However, because GHG emissions of feed production are prevented this effect might be limited. These systems are suitable for smallholder farmers to increase their production, but when large amounts of insects are required, more controlled circumstances would be needed. Furthermore, harvesting these insects from nature needs to be done with caution in order to prevent effects on biodiversity.

Potentially more viable approaches to produce large amounts of insects for feed are described by Van Zanten *et al.* (2015) and Salomone *et al.* (2017). The first study describes a system in which poultry manure and food waste is used as feed for house fly larvae, which are subsequently used as pig feed. The second study describes a system in which food waste is used as feed to produce prepupae of the black soldier fly. Benchmarks chosen for both of these studies are soybean meal and fishmeal, two common protein rich ingredients of animal feed. The direct energy use related to the production of one ton of protein via the house fly production system (22,132 MJ) is lower than for fishmeal (44,615 MJ) but higher than for soybean meal (17,118 MJ). However, one tonne of protein from dried black soldier fly prepupae is associated with an energy use of only 15,100 MJ (Salomone *et al.*, 2017). The latter study is based on a pilot plant in Italy, whereas the house fly larvae of Van Zanten *et al.* (2015) are produced in the Netherlands. The difference in energy use between these two studies might be partially explained by a more suitable climate for insect production in Italy compared to the Netherlands. However, it needs to be noted that overly high ambient temperatures require the cooling of insect rearing facilities, which also leads to a high-energy use. Approximately 70% of the energy use during the production of the house fly larvae is due to usage of electricity and gas. Similar to the production of mealworms, this energy is primarily used for maintaining suitable ambient temperatures (Oonincx and De Boer, 2012). The energy consumption for the black soldier fly prepupae is primarily because these are dried, prior to being milled (Salomone *et al.*, 2017). Overall, energy efficiency is one of the main priorities to improve the sustainability of insect production systems.

Energy consumption has a direct relation with global warming potential. The GWP related to the production of one tonne of protein from black soldier fly prepupae (2,100 CO_2-eq) and house fly larvae meal (1,827 kg CO_2-eq) is lower than for fishmeal (3,113 kg CO_2-eq) and slightly higher than for soybean meal (1,764 kg CO_2-eq) if land use changes are excluded (Salomone *et al.*, 2017; Van Zanten *et al.*, 2015). These land use changes pertain to changes in the function of a piece of land leading to GHG emissions, for instance forest which is cut down to convert the land to farm soybeans. This decreases carbon sink capacity and the carbon emissions are attributed to soybean production. Therefore, if land use changes are included, the production of soybean meal is associated with higher GHG emissions (5,901 kg CO_2-eq) than the two insect meals.

The land use for the production of one tonne of protein from black soldier fly prepupae (50 m^2) and house fly larvae meal (76 m^2) are far lower than for soybean meal (3,543 m^2), but higher than for fishmeal (15 m^2) (Van Zanten *et al.*, 2015). These apparently large differences are easily explained; soybean meal production is a land based activity whereas fishmeal production is not. The house fly feed is considered a low value by-product and therefore its allocated land use is low.

The data above pertain to environmental impact, which are attributed directly to a production process. However, the availability of an alternative product can replace the usage of current products. For instance, insect meal can replace soybean meal or fishmeal, thereby preventing their environmental impact. On the other hand feeding insects with food waste decreases the availability of this food waste for other processes. In certain countries, amongst which the Netherlands, food waste is used for biogas production, which in turn is a source of energy. Hence the consequence of using this food waste as insect feed is a decreased energy production. An LCA that takes also takes these consequences into account is called a consequential LCA (cLCA). Based on cLCA calculations the energy use per tonne of house fly meal increases from 22,132 to 86,625 MJ, the GWP increases from 1,827 to 7,430 CO_2-eq, and the land use increases from 66 to 157 m^2 (Van Zanten *et al.*, 2015). This indicates that the context of a cLCA is of importance; the outcome would be very different in a location where food waste is composted instead of used for energy production. Inclusion of these indirect effects therefore makes a large difference in the environmental effects of a production system, and hence provides valuable information for policy makers.

3.6 Other insect products; honey and silk

Besides the usage of insects as a source of food or feed, certain species are used for the production of other insect derived products such as honey and silk.

Honey is probably the best-known insect derived product worldwide. For the production of one kilogramme of honey 1.3-5.6 MJ is used, of which the range is explained by the transport needs due to the distance between production and consumption (Carlsson-Kanyama *et al.*, 2003). In any case, this energy usage is low compared to products with a similar function, such as jam (11-16 MJ/kg), sugar (9.8 MJ/kg) and candy (18 MJ/kg). Although land use and global warming potential were not quantified in that study, the nature of apiculture makes it highly likely that the environmental impact in those categories would be very low. The relatively low environmental impact of this system can be explained by the fact that bees forage on plants in the surroundings and hardly consume feed specifically produced for them. Pollinator services can be considered a co-product or co-service of honey production and could also be considered in LCA's (Crenna *et al.*, 2016).

India and China are the main producers of silk; together they account for 90% of the global production. Environmental impact of silk production in southwest India was quantified by Astudillo *et al.* (2014). The global warming potential, energy use and water use are at least four times higher for silk than for the same amount of cotton, nylon or sheep wool (Astudillo *et al.*, 2014; Vollrath *et al.*, 2013). Although wool production requires approximately 50% more land than silk, the land use of silk production is higher than for cotton and nylon. Hence, these authors concluded silk production to be input intensive. This is due to low production efficiency; only 3.1% of the feed (on a dry matter basis) is converted to silk. Although another 6.3% of the feed mass is converted to pupal mass this is not considered as a co-product because the pupae were not utilized in the study location. It is noteworthy that in northwest India silkworm pupae are used as a source of food (Chakravorty *et al.*, 2013; DeFoliart, 1997), as is the case in several countries in South-East Asia (DeFoliart, 1997; Hanboonsong *et al.*, 2013). This indicates the potential to use this material as a co-product. Generally speaking, increased use of co-products reduces environmental impact of production processes if economic allocation is used. In China co-product management is better optimized than in India, which probably means that the environmental impact of Chinese silk is lower than for Indian silk (Gao, 2011). Even in ancient Chinese sericulture combinations were made in which spent pupae were used as feed for fish and the pond mud was used as fertiliser for the mulberry trees (Vollrath *et al.*, 2013). According to Vollrath *et al.* (2013) co-products such as leaves, dead worms, pupae, and waste silk are mainly used as compost for the mulberry trees, although spent pupae are sometimes sold as fish feed in India.

3.7 Feed utilisation

Feed utilisation is a key driver of resource use and therefore of environmental impact in conventional production systems (Herrero *et al.*, 2013). Furthermore, feed utilisation is the link between land use and animal production systems. These production systems differ between species, climate regions, and level of development within a region. For instance, dairy production systems in Europe and North America use more energy dense ingredients such as grains, whereas in sub-Saharan Africa fibrous products such as grass and stovers are used. Land productivity is lower in the latter region compared to the first, and due to the low energy density of grass and stovers the feed efficiency in sub-Saharan Africa is far lower than in Europe and North-America (Herrero *et al.*, 2013). Although the feed efficiency for dairy production in sub-Saharan Africa is low, marginal grasslands are unsuitable for agricultural crop production. Therefore, from a food security perspective dairy production is an efficient use of available resources in that region (Herrero *et al.*, 2015; Van Zanten *et al.*, 2016). Another aspect to take into account is differences in feed efficiency between species; the production of ruminant meat requires more feed than pork production and poultry production requires less feed than pork (Herrero *et al.*, 2013). However, globally cropland use per unit of protein from ruminant systems is similar to pork and poultry, but in intensive ruminant systems several times as much cropland is used per unit of protein when compared to pork and poultry production (Herrero *et al.*, 2015).

In conventional livestock systems protein efficiency increases when feed with a lower protein content is utilized (Ndegwa *et al.*, 2008). High protein efficiency decreases emissions of nitrous oxide, ammonia and nitrate (Herrero *et al.*, 2015). Although more data are required before definitive conclusions can be drawn, it seems likely that the same applies to insect production systems. Furthermore, it seems that protein efficiency in insects is high compared to their more conventional counterparts (Oonincx *et al.*, 2015b), which at least partially explains their high feed conversion efficiency. Furthermore, growth rates of animals play a role in feed efficiency. Relative growth rates of cattle (0.3% per day) are lower than for pork (3.2%), whereas this can be 7.2% for house crickets, 7.3% for yellow mealworms and even 19.6% for migratory locusts (Herrero *et al.*, 2015; Oonincx *et al.*, 2010). High growth rates mean that a smaller fraction of the energy in the feed is allocated to maintenance metabolism instead of growth (Herrero *et al.*, 2015). This is of lesser concern for insects because they are poikilothermic and hence the energy requirements for maintenance are smaller than for conventional livestock. However, lower growth rates, or increased development times, do lead to a higher energy usage required to create suitable ambient temperatures (Smetana *et al.*, 2016).

When calculated for a production system, feed conversion efficiency should also take into account the replacement of adult breeders. The number of offspring for a cow (~1 per year) is less than for a pig (20-25 per year) whereas a hen can produce more than 100 offspring per year. Hence, the percentage of produced animals required to replace breeder animals is higher for cows than for poultry (Herrero *et al.*, 2015). Many insect species have a far higher reproductive capacity. For instance a house cricket can produce 200-300 offspring in a period of 12 weeks, a black soldier fly female can lay a clutch of more than 1000 eggs within one week after emerging from her pupal case and a super worm female can lay 1,500 eggs in her lifetime (approximately 1 year) (Oonincx, 2015).

In animal production systems, a main driver of environmental impact is feed production. There is a wide variety of insect species suitable as feed or food and the variety in potential dietary ingredients for these species is even wider. This allows for the selection of feed materials with a low environmental impact, for instance the use of co-products from the food industry. Suitability of various co-products has been tested in several insect species (Barry, 2004; Hem *et al.*, 2008; Lalander *et al.*, 2014; Lardé, 1989; Leong *et al.*, 2015; Li *et al.*, 2011; Lundy and Parrella, 2015; Myers *et al.*, 2008; Newton *et al.*, 2005; Oonincx *et al.*, 2015a; Oonincx *et al.*, 2015b; Pastor *et al.*, 2015; Popa and Green, 2012; Van Broekhoven *et al.*, 2015; Yang *et al.*, 2012; Yehuda *et al.*, 2011). Certain species, for instance house crickets, require well-designed diets. Other species, for instance several fly species and mealworms, survive well on a large range of diets. Although, these species survive well, in certain cases development time, or final weight, is affected (Oonincx *et al.*, 2015b; Van Broekhoven *et al.*, 2015). Increased development times result in prolonged production times, which in turn increase energy use during the rearing period if ambient temperatures are insufficient (Halloran *et al.*, 2016). Such trade-offs need to be taken into account when determining which diets are suitable to lower the environmental impact of insect production systems. Furthermore, changes in diet affect the chemical composition of insects, which in turn can have an effect on their suitability as a source of food or feed (Oonincx *et al.*, 2015b; Van Broekhoven *et al.*, 2015).

Besides co-products, the use of waste could be a viable path to follow in the creation of a sustainable insect production sector. Waste reduction could play an important role in a more sustainable food chain (Herrero *et al.*, 2015). Inadequate collection and treatment of waste is considered a large problem regarding GHG emissions, especially in developing countries (Van Huis, 2013). However, waste can also be seen as: 'resources out of place' (Kollikkathara *et al.*, 2009). The latter approach suggests the valorisation of waste streams, for example via insects. The economic value of substrates considered as waste is often negligible, or even negative, as is the case for manure in certain countries (Oonincx *et al.*, 2015a). If the waste is a suitable substrate for an insect species, for instance cattle manure or municipal waste for black soldier fly larvae, this decreases environmental impact and could even have a positive impact, based on an cLCA approach (Smetana *et al.*, 2016).

Other examples of substrates that can be considered 'waste' could be substrates that are contaminated with for instance heavy metals or mycotoxins, making them unsuitable as a source of food or feed in most animal production systems. Recent studies indicate that certain insects are able to utilize substrates containing some of these contaminants (Bosch *et al.*, 2017; Fels-Klerx *et al.*, 2016). In this way nutrients present in these substrates, else destined for destruction, can be upcycled and returned to the food production chain. In most cases these types of solutions would result in environmental benefits. However, exceptions are possible. For instance, the use of manure as feed for flies can result in large proportions of nitrogen being emitted in the form of ammonia or nitrous oxide, which can have a great effect on GWP and acidification potential (Čičková *et al.*, 2012; Morgan and Eby, 1975; Oonincx *et al.*, 2015a).

3.8 Conclusions

The environmental impact of only a handful of insect derived products has been quantified. In general, it seems that the energy use is high, global warming potential is similar and land use is low compared to conventional products. Further development of insect production systems should therefore focus on limiting energy use, for instance by using residual heat from other production systems, or by using heat exchangers. Currently the insect production sector is in its infancy and would benefit greatly by implementing technological innovations. Furthermore, optimizing feed utilisation by using suitable ingredients with a low environmental impact will lead to a more sustainable insect production sector. Increased production volumes are likely to further improve efficiency and thereby lower environmental impact. In the years to come, further data collection on the environmental impact of insect production will provide more insight in sources of environmental impact which can be used to optimize the sustainability of insect derived products. Future studies should include more impact factors in order to provide a better insight in the scope of environmental impact of insect production.

References

Abasiekong, S.F., 1997. Effects of termite culture on crude protein, fat and crude fibre contents of fibrous harvest residues. Bioresource Technology 62: 55-57.

Astudillo, M.F., Thalwitz, G. and Vollrath, F., 2014. Life cycle assessment of Indian silk. Journal of Cleaner Production 81: 158-167.

Barry, T., 2004. Evaluation of the economic, social, and biological feasibility of bioconverting food wastes with the black soldier fly (*Hermetia illucens*). University of North Texas, Denton, TX, USA, 185 pp.

Barwa, E., 2009. Increasing household protein consumption through minilivestock production in Nigeria. Information Manager 9: 10-13.

Bosch, G., Fels-Klerx, H.J.V.d., Rijk, T.C.D. and Oonincx, D.G.A.B., 2017. Aflatoxin B1 tolerance and accumulation in black soldier fly larvae (*Hermetia illucens*) and yellow mealworms (*Tenebrio molitor*). Toxins (Basel) 9(6): E185.

Carlsson-Kanyama, A., Ekström, M.P. and Shanahan, H., 2003. Food and life cycle energy inputs: consequences of diet and ways to increase efficiency. Ecological Economics 44: 293-307.

Chakravorty, J., Ghosh, S. and Meyer-Rochow, V.B., 2013. Comparative survey of entomophagy and entomotherapeutic practices in six tribes of eastern Arunachal Pradesh (India). Journal of Ethnobiology and Ethnomedicine 9: 50.

Choi, Y.-C., Choi, J.-Y., Kim, J.-G., Kim, M.-S., Kim, W.-T., Park, K.-H., Bae, S.-W. and Jeong, G.-S., 2009. Potential usage of food waste as a natural fertilizer after digestion by *Hermetia illucens* (Diptera: Stratiomyidae). International Journal of Industrial Entomology 19: 171-174.

Čičková, H., Pastor, B., Kozánek, M., Martínez-Sánchez, A., Rojo, S. and Takáč, P., 2012. Biodegradation of pig manure by the housefly, *Musca domestica*: a viable ecological strategy for pig manure management. PLoS ONE 7: e32798.

Cole, N.A., Clark, R.N., Todd, R.W., Richardson, C.R., Gueye, A., Greene, L.W. and McBride, K., 2005. Influence of dietary crude protein concentration and source on potential ammonia emissions from beef cattle manure. Journal of Animal Science 83: 722-731.

Crenna, E., Sala, S., Polce, C. and Collina, E., 2016. Pollinators in life cycle assessment: towards a framework for impact assessment. Journal of Cleaner Production 140: 525-536.

De Vries, M. and De Boer, I.J.M., 2010. Comparing environmental impacts for livestock products: a review of life cycle assessments. Livestock Science 128: 1-11.

DeFoliart, G.R., 1997. An overview of the role of edible insects in preserving biodiversity. Ecology of Food and Nutrition 36: 109-132.

Fabbri, C., Valli, L., Guarino, M., Costa, A. and Mazzotta, V., 2007. Ammonia, methane, nitrous oxide and particulate matter emissions from two different buildings for laying hens. Biosystems Engineering 97: 441-455.

Farina, L., Demey, F. and Hardouin, J., 1991. Production de termites pour l'aviculture villageoise au Togo. Tropicultura 9: 181-187.

Fels-Klerx, H.V.d., Camenzuli, L., Lee, M.V.d. and Oonincx, D., 2016. Uptake of cadmium, lead and arsenic by *Tenebrio molitor* and *Hermetia illucens* from contaminated substrates. PLoS ONE 11: e0166186.

Finke, M.D. and Oonincx, D.G.A.B., 2013. Insects as food for insectivores. In: Morales-Ramos, J.A., Rojas M.G. and Shapiro-Ilan, D.I. (eds.) Mass production of beneficial organisms: invertebrates and entomopathogens. Academic Press, London, UK.

Gao, S., 2011. On sustainable development of sericulture industry – an empirical study on Pengan county of Sichuan province. International Journal of Business and Management 6: 216-220.

Hackstein, J.H. and Stumm, C.K., 1994. Methane production in terrestrial arthropods. Proceedings of the National Academy of Sciences of the United States of America 91: 5441-5445.

Halloran, A., Roos, N., Eilenberg, J., Cerutti, A. and Bruun, S., 2016. Life cycle assessment of edible insects for food protein: a review. Agronomy for Sustainable Development 36: 57.

Hanboonsong, Y., Jamjanya, T. and Durst, P.B., 2013. Six-legged livestock: edible insect farming, collection and marketing in Thailand. Food and Agriculture Organization of the United Nations, Bangkok, Thailand.

Hardouin, J., Thys, É., Joiris, V. and Fielding, D., 2003. Mini-livestock breeding with indigenous species in the tropics. Livestock Research for Rural Development 15: 3-4.

Hem, S., Toure, S., Sagbla, C. and Legendre, M., 2008. Bioconversion of palm kernel meal for aquaculture: experiences from the forest region (Republic of Guinea). African Journal of Biotechnology 7: 1192-1198.

Herrero, M., Havlík, P., Valin, H., Notenbaert, A., Rufino, M.C., Thornton, P.K., Blümmel, M., Weiss, F., Grace, D. and Obersteiner, M., 2013. Biomass use, production, feed efficiencies, and greenhouse gas emissions from global livestock systems. Proceedings of the National Academy of Sciences of the USA 110: 20888-20893.

Herrero, M., Wirsenius, S., Henderson, B., Rigolot, C., Thornton, P., Havlík, P., De Boer, I. and Gerber, P.J., 2015. Livestock and the environment: what have we learned in the past decade? Annual Review of Environment and Resources 40: 177-202.

Intergovernmental Panel on Climate Change (IPCC), 2007. Summary for policymakers. IPCC, Cambridge, UK and New York, NY, USA.

Intergovernmental Panel on Climate Change (IPCC), 2014. Climate change 2014: synthesis report. Contribution of working groups I, II and III to the fifth assessment report of the intergovernmental panel on climate change. IPCC, Geneva, Switzerland, 151 pp.

Kollikkathara, N., Feng, H. and Stern, E., 2009. A purview of waste management evolution: special emphasis on USA. Waste Management 29: 974-985.

Lacey, R.E., Redwine, J.S. and Parnell, C.B., 2002. Emission factors for broiler production operations: a stochastic modeling approach. Paper Number: 024212 of the 2002 ASAE Annual International Meeting/CIGR XVth World Congress. Chicago, IL, USA. Available at: http://tinyurl.com/yd8ooapn.

Lalander, C.H., Fidjeland, J., Diener, S., Eriksson, S. and Vinnerås, B., 2014. High waste-to-biomass conversion and efficient *Salmonella* spp. reduction using black soldier fly for waste recycling. Agronomy for Sustainable Development 35: 261-271.

Lardé, G., 1989. Investigation on some factors affecting larval growth in a coffee-pulp bed. Biological Wastes 30: 11-19.

Leong, S.Y., Kutty, S.R.M., Malakahmad, A. and Tan, C.K., 2015. Feasibility study of biodiesel production using lipids of *Hermetia illucens* larva fed with organic waste. Waste Management 47A: 84-90.

Li, Q., Zheng, L., Cai, H., Garza, E., Yu, Z. and Zhou, S., 2011. From organic waste to biodiesel: black soldier fly, *Hermetia illucens*, makes it feasible. Fuel 90: 1545-1548.

Looy, H., Dunkel, F.V. and Wood, J.R., 2013. How then shall we eat? Insect-eating attitudes and sustainable foodways. Agriculture and Human Values 31: 131-141.

Lundy, M. and Parrella, M., 2015. Crickets are not a free lunch: protein capture from scalable organic side-streams via high-density populations of *Acheta domesticus*. PLoS ONE 10: e0118785.

Makkar, H.P., Tran, G., Heuzé, V. and Ankers, P., 2014. State-of-the-art on use of insects as animal feed. Animal Feed Science and Technology 197: 1-33.

Miglietta, P.P., De Leo, F., Ruberti, M. and Massari, S., 2015. Mealworms for food: a water footprint perspective. Water 7: 6190-6203.

Morgan, N. and Eby, H., 1975. Fly protein production from mechanically mixed animal wastes. Israel Journal of Entomology 10: 73-81.

Munyuli Bin Mushambanyi, T. and Balezi, N., 2002. Utilisation des blattes et des termites comme substituts potentiels de la farine de viande dans l'alimentation des poulets de chair au Sud Kivu, République Démocratique, du Congo. Tropicultura 20: 10-16.

Myers, H.M., Tomberlin, J.K., Lambert, B.D. and Kattes, D., 2008. Development of black soldier fly (Diptera: Stratiomyidae) larvae fed dairy manure. Environmental Entomology 37: 11-15.

Ndegwa, P.M., Hristov, A.N., Arogo, J. and Sheffield, R.E., 2008. A review of ammonia emission mitigation techniques for concentrated animal feeding operations. Biosystems Engineering 100: 453-469.

Newton, L., Sheppard, C., Watson, D.W. and Burtle, G., 2005. Using the black soldier fly, *Hermetia illucens*, as a value-added tool for the management of swine manure. Report for Mike Williams, Director of the Animal and Poultry Waste Management Center, North Carolina State University, Raleigh, NC, USA.

Oonincx, D., Van Huis, A. and Van Loon, J., 2015a. Nutrient utilisation by black soldier flies fed with chicken, pig, or cow manure. Journal of Insects as Food and Feed 1: 131-139.

Oonincx, D.G., 2015. Insects as food and feed: nutrient composition and environmental impact. Wageningen University, Wageningen, the Netherlands.

Oonincx, D.G.A.B. and De Boer, I.J.M., 2012. Environmental impact of the production of mealworms as a protein source for humans – a life cycle assessment. PLoS ONE 7: e51145.

Oonincx, D.G.A.B., Van Broekhoven, S., Van Huis, A. and Van Loon, J.J.A., 2015b. Feed conversion, survival and development, and composition of four insect species on diets composed of food by-products. PLoS ONE 10: e0144601.

Oonincx, D.G.A.B., Van Itterbeeck, J., Heetkamp, M.J., Van den Brand, H., Van Loon, J.J. and Van Huis, A., 2010. An exploration on greenhouse gas and ammonia production by insect species suitable for animal or human consumption. PLoS ONE 5: e14445.

Pastor, B., Velasquez, Y., Gobbi, P. and Rojo, S., 2015. Conversion of organic wastes into fly larval biomass: bottlenecks and challenges. Journal of Insects as Food and Feed: 1-16.

Popa, R. and Green, T.R., 2012. Using black soldier fly larvae for processing organic leachates. Journal of Economic Entomology 105: 374-378.

Premalatha, M., Abbasi, T., Abbasi, T. and Abbasi, S.A., 2011. Energy-efficient food production to reduce global warming and ecodegradation: the use of edible insects. Renewable and Sustainable Energy Reviews 15: 4357-4360.

Salomone, R., Saija, G., Mondello, G., Giannetto, A., Fasulo, S. and Savastano, D., 2017. Environmental impact of food waste bioconversion by insects: application of life cycle assessment to process using *Hermetia illucens*. Journal of Cleaner Production 140, part 2: 890-905.

Smetana, S., Mathys, A., Knoch, A. and Heinz, V., 2015. Meat alternatives: life cycle assessment of most known meat substitutes. International Journal of Life Cycle Assessment 20: 1254-1267.

Smetana, S., Palanisamy, M., Mathys, A. and Heinz, V., 2016. Sustainability of insect use for feed and food: life cycle assessment perspective. Journal of Cleaner Production 137: 741-751.

Steinfeld, H., Gerber, P., Wassenaar, T., Castel, V., Rosales, M. and De Haan, C., 2006. Livestock's long shadow. Environmental issues and options, FAO, Rome, Italy.

Van Broekhoven, S., Oonincx, D.G.A.B., Van Huis, A. and Van Loon, J.J.A., 2015. Growth performance and feed conversion efficiency of three edible mealworm species (Coleoptera: Tenebrionidae) on diets composed of organic by-products. Journal of Insect Physiology 73: 1-10.

Van Huis, A., 2013. Potential of insects as food and feed in assuring food security. Annual Review of Entomology 58: 563-583.

Van Zanten, H.H., Mollenhorst, H., Oonincx, D.G., Bikker, P., Meerburg, B.G. and De Boer, I.J., 2015. From environmental nuisance to environmental opportunity: housefly larvae convert waste to livestock feed. Journal of Cleaner Production 102: 362-369.

Van Zanten, H.H.E., Mollenhorst, H., Klootwijk, C.W., Van Middelaar, C.E. and De Boer, I.J.M., 2016. Global food supply: land use efficiency of livestock systems. International Journal of Life Cycle Assessment 21: 747-758.

Vollrath, F., Carter, R., Rajesh, G., Thalwitz, G. and Astudillo, M.F., 2013. Life cycle analysis of cumulative energy demand on sericulture in Karnataka, India. In: Proceedings of the 6th BACSA International Conference: Building Value Chains in Sericulture (BISERICA 2013), 7-12 April 2013, Padua, Italy.

Wrage, N., Velthof, G.L., Van Beusichem, M.L. and Oenema, O., 2001. Role of nitrifier denitrification in the production of nitrous oxide. Soil Biology and Biochemistry 33: 1723-1732.

Yang, S., Li, Q., Zeng, Q., Zhang, J., Yu, Z. and Liu, Z., 2012. Conversion of solid organic wastes into oil via *Boettcherisca peregrine* (Diptera: Sarcophagidae) larvae and optimization of parameters for biodiesel production. PLoS ONE 7: e45940.

Yehuda, B., Marchaim, U., Glatman, L., Drabkin, V., Chizov-Ginzburg, A., Mumcuoglu, K.Y. and Gelman, A., 2011. Bioconversion of poultry and fish waste by *Lucilia Sericata* and *Sarcophaga Carnaria* larvae. Asian Journal of Water, Environment and Pollution 8: 69-75.

Tropical production systems

4.

Imbrasia belina (sun-dried)

Introducing small production systems for edible insects

A. van Huis
Laboratory of Entomology, Wageningen University, P.O. Box 16, 6700 AA Wageningen, the Netherlands; arnold.vanhuis@wur.nl

In tropical countries, harvesting insects from the wild is a normal practice. It is only recently that insects both in the tropical and temperate zones are being reared in large quantities for either human food or animal feed. The last form is domestication, which can be defined as 'a sustained multi-generational relationship in which one group of organisms assumes a significant degree of influence over the reproduction and care of another group to secure a more predictable supply of resources from that second group' (Zeder, 2015). However, there is an intermediate form which is called semi-domestication. The definition of semi-domestication is 'a captive state of a wild animal in which its living conditions and often its rearing are controlled by humans' (Merriam Webster dictionary).

There are several simple examples of semi-domestication (e.g. in central Africa where the local population cut branches with young caterpillars from trees, and transport and place those on the same tree species, but near the village (Malaisse, 1997; pp. 207-208)). Another example is the palm weevil in Latin America given by Dufour (1987), who studied the use of insects as food of the Tukanoan Indians in the Northwest Amazon. He noticed the Tukanoa on hunting and fishing trips, cut palm trees, with the expectation that they would be invaded by the palm weevil (*Rhynchophorus palmarum*). Larvae were then harvested on their return trip two to three months later. DeFoliart (1993) mentioned similar practices of harvesting palm weevils in Latin American countries. The Guajibo Indians in the Amazon even used plastic containers to feed the larvae with parts of sugar cane and palm trees (Cerda *et al.*, 2001).

Van Itterbeeck and Van Huis (2012) described other systems to increase the predictability and availability of edible insects: providing egg-laying sites of reed and grasses for aquatic Hemiptera in lakes of Mexico, and manipulating shifting cultivation and fire regimes to improve forest caterpillar exploitation

For bamboo caterpillars (*Omphisa fuscidentalis*), two methods were proposed for mass production: controlled mating in a nylon cage covering bamboo shoot plot or releasing moths for natural mating into the bamboo shoot plot (Kayikananta, 2000). Bamboo caterpillars were traditionally collected by cutting down entire bamboo clumps to harvest the caterpillars.

Deforestation became a serious problem as whole bamboo forests were cut down to harvest the insects. To prevent this, a sustainable harvesting method of the bamboo caterpillar is now proposed. The first or second internode from the base is examined for the denseness of an internode as this indicates the presence of bamboo caterpillars. The caterpillars are harvested by cutting a rectangular hole at the internodes hosting the bamboo caterpillars. In this way it is not necessary to cut the whole plant. The infested bamboo culms are utilized to make bamboo handicrafts and construction poles, which are even stronger than the non-infested ones (Hanboonsong *et al.*, 2013).

Sleeve nets have also been proposed for silkworm species, such as *Anaphe panda* (Mbahin *et al.*, 2010) and *Gonometa postica* (Ngoka *et al.*, 2008), although it was meant for silk cultivation, both species are also edible (Jongema, 2017). One of the reasons for using sleeve nets is to protect the larvae from parasitoids and predators. The same method was employed to protect the mopane caterpillar *Imbrasia belina* against natural enemies. Even the egg batches proved vulnerable to parasitoid attack (Ghazoul, 2006). Another problem was disease spreading through the captive population. The conclusion was that rearing mopane caterpillars was technically feasible but it was not cost effective.

In Africa, farmers break open small termite nests (*Microtermes* spp.) or parts of termite mounds to provide access to chickens and guinea fowls to feed on the termites (Van Huis, 2017). Farmers also have another method to collect termites as chicken feed. They use a clay container in which they place for example dry stems of sorghum or other cereals, add some water and then put the pot upside down with the opening on a termite gallery. After 3 to 4 weeks the pot has gathered a number of termites and the pot is emptied for the chicks. The method was also used by farmers in Togo, and Farina *et al.* (1991) describes the process in detail.

The Asian weaver ant *Oecophylla smaragdina* is one of the most favoured edible insects in the Lao PDR and Thailand. The queen brood, which is seasonally available, is the target commodity. Collectors leave intentionally a large fraction of the colony undisturbed, in particular the small nests as these yield small amounts of brood (Van Itterbeeck *et al.*, 2014). Removing the queen would mean the end of the colony, but the queen nest is unintentionally not harvested, because the small nest is inconspicuously located near the top of a tree (Van Itterbeeck *et al.*, 2015). The weaver ant is also a biological control agent as predator of pests in tropical crops among which mango orchards. Providing the ants with tuna-based food doubled the yield and can be considered as a way of ant farming with a dual purpose: achieving biological control and obtaining a protein food source (Offenberg and Wiwatwitaya, 2009).

There are two on-line available books which describe collection and rearing techniques for a number of edible insects in Thailand and Lao PDR (Hanboonsong and Durst, 2014; Hanboonsong *et al.*, 2013).

In this section three chapters deal with simple farming systems in tropical countries: rearing cricket species in a sustainable way as human food in tropical countries, rearing palm weevils in Africa using semi-farming techniques, and the rearing of several grasshopper species in India, mainly to be used as feed.

References

Cerda, H., Martinez, R., Briceno, N., Pizzoferrato, L., Manzi, P., Ponzetta, M.T., Marin, O. and Paoletti, M.G., 2001. Palm worm: (*Rhynchophorous palmarum*) traditional food in Amazonas, Venezuela – nutritional composition, small scale production and tourist palatibility. Ecology of Food and Nutrition 40: 13-32.

DeFoliart, G.R., 1993. Hypothesizing about palm weevil and palm rhinoceros beetle larvae as traditional cuisine, tropical waste recycling, and pest and disease control on coconut and other palms – can they be integrated? Principes 37: 1, 42-47.

Dufour, D.L., 1987. Insects as food: a case study from the Northwest Amazon. American Anthropologist 89: 383-397.

Farina, L., Demey, F. and Hardouin, J., 1991. Production de termites pour l'aviculture villageoise au Togo. Tropicultura 9: 181-187.

Ghazoul, J., 2006. Mopani woodlands and the mopane worm: enhancing rural livelihoods and resource sustainability. Final Technical Report, DFID, London, UK.

Hanboonsong, Y. and Durst, P.B., 2014. Edible insects in Lao PDR: building on tradition to enhance food security. RAP Publication 2014/12, Food and Agriculture Organization of the United Nations. Regional Office for Asia and The Pacific, Bangkok, Thailand. Available at: http://www.fao.org/3/a-i3749e.pdf.

Hanboonsong, Y., T. Jamjanya, T. and P.B. Durst, P., 2013. Six-legged livestock: edible insect farming, collection and marketing in Thailand. Food and Agriculture Organization of the United Nations, Regional Office for Asia and the Pacific, Bangkok, Thailand. Available at: http://tinyurl.com/hdq2veu.

Jongema, Y., 2017. List of edible insect species of the world. Wageningen University, Wageningen, the Netherlands. Available at: http://tinyurl.com/zb9u3b9.

Kayikananta, L., 2000. Biological study and rearing techniques on bamboo caterpillar, *Omphisa fuscidentalis* Hampson. In: Puangchit, L, Thaiutsa, B. and Thamnicha, S (eds.) Bamboo 2000 International Symposium, August, 2-4, 2000, Chiang Mai, Thailand, pp. 186-195.

Malaisse, F., 1997. Se nourrir en foret claire africaine: approche ecologique et nutritionnelle. Les Presses Agronomiques de Gembloux, Gembloux, Belgium/Technical Centre for Agricultural and Rural Cooperation, Wageningen, the Netherlands, 384 pp.

Mbahin, N., Raina, S.K., Kioko, E.N. and Mueke, J.M., 2010. Use of sleeve nets to improve survival of the Boisduval silkworm, *Anaphe panda*, in the Kakamega Forest of western Kenya. Journal of Insect Science 10: 1-10.

Ngoka, B.M., Kioko, E.N., Raina, S.K., Mueke, J.M. and Kimbu, D.M., 2008. Semi-captive rearing of the African wild silkmoth *Gonometa postica* (Lepidoptera: Lasiocampidae) on an indigenous and a non-indigenous host plant in Kenya. International Journal of Tropical Insect Science 27: 183-190.

Offenberg, J. and Wiwatwitaya, D., 2009. Sustainable weaver ant (*Oecophylla smaragdina*) farming: harvest yields and effects on worker ant density. Asian Myrmecology 3: 55-62.

Van Huis, A., 2017. Cultural significance of termites in sub-Saharan Africa. Journal of Ethnobiology and Ethnomedicine 13: 8.

Van Itterbeeck, J., Sivongxay, N., Praxaysombath, B. and Van Huis, A., 2014. Indigenous knowledge of the edible weaver ant *Oecophylla smaragdina* Fabricius Hymenoptera: Formicidae from the Vientiane Plain, Lao PDR. Ethnobiology Letters 5: 4-12.

Van Itterbeeck, J., Sivongxay, N., Praxaysombath, B. and Van Huis, A., 2015. Preliminary observations on gravid queen protection in *Oecophylla smaragdina*: evacuation and retinue function. Asian Myrmecology 7: 155-158.

Van Itterbeeck, J. and Van Huis, A., 2012. Environmental manipulation for edible insect procurement: a historical perspective. Journal of Ethnobiology and Ethnomedicine 8: 1-19.

Zeder, M.A., 2015. Core questions in domestication research. Proceedings of the National Academy of Sciences 112: 3191-3198.

5.

Gryllus bimaculatus

Small-scale production of crickets and impact on rural livelihoods

R. Caparros Megido*, É. Haubruge and F. Francis
Entomologie Fonctionnelle et Evolutive, Gembloux Agro-Bio Tech, Université de Liège, Passage des Déportés, 2, 5030 Gembloux, Belgium; r.caparros@ulg.ac.be

Abstract

Small-scale production of edible insects could be a sustainable, cost-effective, and nutritious alternative to traditional livestock in poor rural areas. Among the diversity of edible insect species, crickets are valuable candidates as they reproduce quickly in high numbers, require a reduced-size rearing space and could easily recycle agricultural or industrial by-products, while keeping high feed conversion ratio. This chapter will present an overview of four cricket species, *Acheta domesticus*, *Gryllus bimaculatus*, *Teleogryllus mitratus*, and *Brachytrupes membranaceus* reared at a small-scale level and will describe their corresponding rearing system. Strategies to promote and develop small-scale cricket farming in different countries will be presented throughout a state of play of the past and current projects.

Keywords: field cricket, peasant production, insect farming, entomophagy, edible insects

5.1 Introduction

Edible insects are considered a sustainable sources of animal protein for the future (Belluco *et al.*, 2013; Caparros Megido *et al.*, 2015a; FAO, 2009; Ramos-Elorduy, 1997). In Asia, Africa, South America, and Oceania, the diversity of edible insects is very high (approximately 2,000 species), and these insects are principally collected from the wild or semi-cultivated. In Western countries, entomophagy promoters only rely on the industrial farming of approximately 10 insect species (Caparros Megido *et al.*, 2015b; Van Huis, 2003; Van Huis *et al.*, 2013). Nevertheless, the development of edible insect production, rather than the capture of wild insects, could allow a continuous supply of product, while also having a lower impact on the environment (Van Huis, 2003).

Small-scale insect farming for human food has long been rooted in areas where insects are part of people's culture and historically has supported rural livelihood by facilitating food provision (particularly against seasonal shortage of food) and thus provided additional cash for basic expenditure including food, farm implements, and education (Gahukar, 2012, 2016; Taponen, 2015). The system also offers an opportunity for improvement of local diets, employment, and earned income for landless inhabitants including women who are involved in the gathering, farming, processing, and sale of insects (Agea *et al.*, 2008; Gahukar, 2016; Hope *et al.*, 2009).

For cost-effective production of edible insects, an endemic and nutritionally interesting species must be selected with a high acceptability to human consumers or palatability to animals (Caparros Megido *et al.*, 2015b; Rumpold and Schlüter, 2013a). The insect should be selected based on size, social behaviour (reduced cannibalism), danger to handlers, tendency for pathogen transmission, potential for reproduction and survival, nutritional benefits, and potential and storage merchantability (Schabel, 2010). Finally, when introducing edible insects as a more sustainable production than traditional livestock, it is advantageous to use diets from a local and more sustainable source, such as agriculture or industrial by-products (Van Broekhoven *et al.*, 2015). Among the different species of edible insects, and as you will read in other chapters in this text, crickets have shown particular promise in addressing food and nutrition insecurity (Van Huis, 2013). Though their high nutritional quality, a variety of cricket species are attractive candidates for supporting improved nutritional status in rural communities (Ayieko *et al.*, 2016; A. Halloran *et al.*, unpublished data; Rumpold and Schlüter, 2013b). Moreover, crickets have significantly shorter life cycles than traditional livestock and could be reared with by-products with high feed conversion rates, making them attractive to fight food insecurity and to allow quick return on investment for small farmers (Durst and Hanboonsong, 2014; Gahukar, 2012; Miech *et al.*, 2016; Oonincx *et al.*, 2015).

Three types of business scales for mass production of crickets will be described. The small scale, which revolves around maintaining a few containers, where farmers only raise crickets for their own need and produce a few kilogrammes for surrounding markets (around 50 kg per year or less than 100 hundred boxes); the medium scale where farmers are more like traders with a well-established business but still with a limited marketing coverage (hundreds of kg of crickets per year for surrounding markets), and the large scale where several tons of crickets are produced annually (Fuah *et al.*, 2016). Subsequent paragraphs will describe the most common small-scale farming system per cricket species with a particular focus on the actual sector's stakeholders and established strategies to promote and develop the practice in corresponding countries. All the information is based on secondary data and on a review of literature on crickets.

5.2 Species and characteristics of cricket rearing

5.2.1 House crickets

The house cricket (*Acheta domesticus* (L. 1758)) is a field species native to Southwestern Asia, but has become distributed worldwide (Ghouri, 1961). In Europe, they are commercially bred as food for pets and human consumption (DeFoliart, 1992; Van Huis, 2013). The life cycle of *A. domesticus* ranges from 30 to 49 days at 28 °C and occurs without diapause at 30 °C (Booth and Kiddell, 2007; Durst and Hanboonsong, 2014; Gahukar, 2016).

Typical enclosures for rearing house crickets are concrete tubes or cylinders, plywood or concrete rectangular pens, or plastic boxes. Small farms principally use concrete cylinder pens (80 cm diameter × 50 cm height; Figure 5.1) that are inexpensive and easy to maintain. Producers can manage 20-150 pens with 2-4 kg of crickets per unit. Unfortunately, concrete cylinder pens cannot be moved easily and need considerable space, while plastic drawers deteriorate and must be replaced more frequently. In comparison, plastic drawers (80×180×30 cm) are easier to move and produce 6-8 kg of crickets per unit (Hanboonsong *et al.*, 2013). Furthermore, as plastic drawers can be stacked, less space is required; however, the top drawers have greater mortality due to overheating (Hanboonsong *et al.*, 2013). The use of plywood boxes (120×240× 50 cm) and concrete block pens (120×240× 60 cm) are suitable for medium- to large-scale farms and could produce 20-30 kg of crickets per box or pen (Hanboonsong *et al.*, 2013). Regardless of the rearing structure used, farmers need to design

Figure 5.1. Cylinder pens for cricket breeding (photo by Afton Halloran).

the container to ensure they represent the damp and dark natural habitat of crickets (Fuah *et al.*, 2016) as a means to ensure successful production.

Some basic tenets for mass production of crickets are used regardless of the design. Nets are placed over the enclosures to prevent crickets from escaping and predators from eating the crickets (Gahukar, 2016; Hanboonsong *et al.*, 2013). Plastic bottles or a plate with water and stones are used to provide water. Sticky tape or plastic tablecloths are glue on the inner side of the walls just below the edge to keep the crickets from crawling out of the area. A layer of sandy loam soil with dry grasses (bamboo shoots) is placed on the bottom of the container as hiding places for the crickets. Chicken egg cartons can also be placed in the pens to provide hiding places for the growing crickets. The area where the cricket containers are located is surrounded by a narrow strip of water sometimes containing small fish to prevent access by ants (Gahukar, 2016; Hanboonsong *et al.*, 2013). The production cost for one cycle is approximately two US dollars per kg in concrete cylinder pen and 1.6 US dollars per kg in plastic drawers. Insects could be sold from 3.10 to 4.30 US dollars per kg (Hanboonsong *et al.*, 2013).

Feed is the biggest cost with cricket farming and can account for as much as half of total costs. Traditionally, most farmers use high-protein commercial chicken feed to rear their crickets but new products, specific to crickets have been formulated (Durst and Hanboonsong, 2014). To reduce costs, some farmers utilize fruits and vegetables as feed for crickets when they reach 21 days in development. Vegetables used can vary from cassava and morning glory leaves to water melon as well as fresh pumpkin or squash (Durst and Hanboonsong, 2014; Gahukar, 2016; Hanboonsong *et al.*, 2013). It should be noted, house crickets could have similar or better feed conversion than pigs or chicken when they are reared on well balanced organic side-streams of relatively high-quality (Collavo *et al.*, 2005; Lundy and Parrella, 2015; Oonincx *et al.*, 2015).

The farm gate price for crickets is generally between 3.00 to 5.00 US dollars per kg in Thailand, from 3.70 to 4.90 US dollars per kg in Lao People's Democratic Republic (PDR) and around 2.30 US dollars per kg in Kenya. Bowls of cricket eggs (typically 20 cm in diameter bowls) can be sold for 0.92 to 6.50 US dollars depending on location and availability; in Thailand; this need is driven by the desire to prevent inbreeding problems (Durst and Hanboonsong, 2014; A. Halloran *et al.*, unpublished data).

Eggs are a valuable resource for mass production of crickets. Egg bowls are sold from 1.20 to 1.90 US dollars in Lao PDR and around 2.00 US dollars in Kenya. An average size farm might produce 50 to 60 egg containers per cycle with four to six cricket cycles per year (Durst and Hanboonsong, 2014). Each egg bowl could produce 3 kg of adult crickets with one to three reproductive cycle per generation (Hanboonsong *et al.*, 2013).

Cricket waste is also sold as biofertilizer, earning farmers from 1.09 to 1.67 US dollars for each 20-30 kg bag of fertilizer in Thailand (Durst and Hanboonsong, 2014; A. Halloran *et al.*, unpublished data). Prices of biofertilizer are not available for Kenya and Lao PDR. Actually, prices are similar between countries, and intermediaries receive the greatest profit margin as they purchase insect from farmers at low prices, process them (packing and freezing) and sell them at higher prices (Hanboonsong *et al.*, 2013). In the future, farmers need to process (freezing or cooking) and sell on their own crickets if they want to increase their profit. Nevertheless, some farmers stop production due to not having the necessary business skills to sell at the market (Halloran *et al.*, 2016a). Establishing a cooperative approach for distribution and sale of crickets could facilitate farmer involvement. As an example, in Thailand (province of Mahasalakam), one wholesale cricket trader buys from a cooperative of 200 cricket farmers from surrounding areas at good price and could sell up to 700-800 kg of crickets per day. Farmers could also increase locations where they sold crickets or the way they do it but that could become very time consuming.

Thailand has a long history of rearing house crickets for sale. Initially, local species were reared but now, most farmers have shifted to *A. domesticus* as it is easy to rear and very popular with consumers (Durst and Hanboonsong, 2014). Due to interest in mass production of crickets, most producers are considered medium- to large-scale (Gahukar, 2016; Hanboonsong *et al.*, 2013).

Khon Kaen University has become an epicentre of knowledge and dissemination of cricket farming techniques in Thailand for a number of reasons (Halloran *et al.*, 2016b). Firstly, successful cricket farmers began to encourage and teach peers, resulting in the expansion of cricket farming throughout other regions of the country (Halloran *et al.*, 2016b). Secondly, the establishment of community cooperatives for cricket farmers who disseminate information on technical farming, marketing and business issues, particularly in Northern Thailand, increased exposure and recruitment of new producers (Hanboonsong *et al.*, 2013). Finally, expansion of cricket farming into neighbouring countries, such as Lao PDR, and Vietnam has occurred (Durst and Hanboonsong, 2014; A. Halloran *et al.*, unpublished data; Hanboonsong *et al.*, 2013; Sanewong Na Ayudhaya, 2011). In fact, the introduction of Thai cricket farming technology from 2010-2015 in Lao PDR was assessed with the help of several projects such as 'Sustainable insect farming and harvesting for better nutrition, improved food security and household income' funded by the Food and Agriculture Organization of the United Nations (FAO); and 'Improving rural livelihoods and food security' and 'Insects for nutrition' funded by Veterinarians without Borders (A. Halloran *et al.*, unpublished data). These projects allowed the building of approximately 75 small-scale cricket farming units in the country. In 2012, house cricket farming was introduced to Kenya with the help of a Flying Foods project from the government of the Netherlands. Today, approximately 50 cricket farms can be found in the Great Lake region of Kenya (Homa Bay, Siaya and Kisumu Counties) where the innovation was first introduced. Jaramogi Oginga Odinga University of Science and Technology has been the epicentre of studies on cricket farming and the development of new techniques in Kenya. An experimental cricket farm can also be found at Jomo Kenyatta University of Agriculture and Technology (Ayieko *et al.*, 2016; A. Halloran *et al.*, unpublished data).

5.2.2 Two-spotted cricket

The African or Mediterranean field cricket, or the two-spotted cricket (*Gryllus bimaculatus* De Geer 1773), is a popular food source for insectivorous animals like spiders and reptiles kept as pets or in zoos. Under optimal conditions (27 °C and a 16 h light/ 8 h dark photoperiod), the cricket need approximately 65 days to develop from eggs to last larval instar (Ibler *et al.*, 2009). An additional 10 days are needed for the crickets to reach adulthood (growth of wings)

(Fuah *et al.*, 2016). The reproduction period of females lasts approximately 22 days during which they could possibly lay up to 1,375 eggs. The relative humidity required for egg hatching ranged between 65 to 80% with an approximate temperature of 26 °C (Fuah *et al.*, 2016).

This species seems to be preferred by some farmers as they are larger and easier to take care of than *A. domesticus* (A. Halloran *et al.*, unpublished data). In Thailand and Lao PDR, this species could be sold for a better price than house crickets depending on consumer preferences. Some consumers prefer *G. bimaculatus* due to the bigger size and their perception of this species being 'more natural', while others opt for *A. domesticus* due to the softer texture (Durst and Hanboonsong, 2014; A. Halloran *et al.*, unpublished data). In Kenya, *G. bimaculatus* has also been targeted as a possible cricket species to rear (Ayieko *et al.*, 2016). In Indonesia, *G. bimaculatus*, or *kalung* crickets, are used as feed for birds and other insect-eating animals due to having a very soft body skin (Fuah *et al.*, 2016). Cricket farmers usually use wood, plywood, cardboard, plastic or bamboo to produce box-shaped cricket cages (200×300×50 cm). Much like with *A. domesticus*, cages are covered with nets and surrounded by oil or water to avoid predator attacks (e.g. ants, mice or lizards). Inside cages, cardboard egg cartons and dried banana leaves are used as shelter for crickets. Feed is given *ad libitum* or three times a day and can consist of banana stalk, squash, pumpkin, cabbage leaves, field grass, mustard greens, or fruit peels. In addition, breeders provide chicken feed (20 to 22% of protein content) to increase growth performance. Cassava leaves and stems have also been used to enhance cricket development. Once a week, cages are cleaned and dried banana leaves are replaced without removing crickets and regularly, previous food remains are removed to prevent decay and fungal development. Crickets to be sold are generally harvested at the end of the last nymphal stage (35 to 45 days after hatching), when they are still wingless, and are used as feed for ornamental birds. Crickets used for reproduction remain in cages and, when males begin to stridulate, egg-laying media are placed in a plastic or a plywood container inside cages. Damp sandy media (with a thickness of 3-4 cm) is very effective for enhancing concentrated egg laying and hatching. Bedding consisting of a mixture of soil and sand, cotton and linen could also be used with possible differences in egg incubation time (e.g. eggs placed on cotton hatched faster than in sand). Between each cycle, cages are cleaned with soap and water. With these methods, approximately 30 kg of crickets could be produced with 85 g of cricket eggs (Fuah *et al.*, 2016). Concerning the marketability of crickets in Indonesia, one kg of crickets could be sold for 3.75 US$ and a farmer with an average of 60 boxes could expect an income of approximately 800 US$ without labour charges (Fuah *et al.*, 2015).

5.2.3 Oriental ground cricket

The oriental ground cricket, or field cricket, (*Teleogryllus mitratus* (Walker 1869)) is a native species inhabiting paddy fields and fallow land in Cambodia (Miech *et al.*, 2016). Traditionally, this species harvested from the wild but can also be reared on local weed species (e.g. cassava leaves), agricultural (e.g. rice bran) and food industry (e.g. mung bean sprout production) by-products in facilities. Cassava leaves and *Cleome rutidosperma* Pyrame de Candolle seem to have a great potential as cricket feed among different tested by-products (Caparros Megido *et al.*, 2015c; Miech *et al.*, 2016).

In Cambodia, small-scale cricket farms use 60 L plastic containers to breed this species (Figure 5.2). Medium-scale cricket farms use rectangular brick and cement pens (500×200

Figure 5.2. 60 l plastic insect breeding container in Cambodia: (A) side view and (B) top view.

cm; Figure 5.3) (Caparros Megido *et al.*, 2015c; A. Halloran *et al.*, unpublished data; Miech *et al.*, 2016). As with the previous two systems nets are placed on the pens' top, while a water channel around the base of the pen helps to avoid predators' access to the rearing unit. Egg cartons are placed inside the pen to offer shelters to crickets. These cartons are elevated to avoid the accumulation of food and faeces. Water is provided through a chicken trough filled with gravel; by plates filled with gravel; or by water vaporisation of walls and soil (Caparros Megido *et al.*, 2015c; A. Halloran *et al.*, unpublished data; Miech *et al.*, 2016). Densities of individuals could be from 700 crickets to 1,400 crickets per m^3 (Miech *et al.*, 2016). Plastic bowls (4 per m^3) containing biochar or burned and moistened rice balls must be placed as a

Figure 5.3. Concrete pens for cricket breeding.

substrate for eggs laying. At approximately 30 °C and 70% relative humidity, crickets reach maturity and lay eggs within three months (Caparros Megido *et al.*, 2015c; Miech *et al.*, 2016). One kg of crickets can be sold from 3.75 to 5.00 US dollars depending on the market demand and their taste. Egg cricket bowls and faeces bags (approximately 30 kg) are sold at 2.50 US dollars per bowl or per bag. Knowledge on cricket farming is classically traded with a guarantee of egg bowl purchase from the learner. It should be noted that from 2013 to 2014, a European Union project (Annâdya) developed a field cricket-rearing program in Cambodia. This program was followed by new farmers and project officer and developed a practical guide for future farmers. In 2014, 500 Cambodian households have adopted cricket farming.

5.2.4 The tobacco cricket

The tobacco cricket (*Brachytrupes membranaceus* (Drury, 1770)) is a crop pest native in Africa and is the most consumed cricket in the Kinshasa area in Democratic Republic of Congo (DRC) (A. Halloran *et al.*, unpublished data; Hill, 2008). Traditionally harvested from the wild with other Orthoptera species, rearing assays has been recently undertaken (A. Halloran *et al.*, unpublished data; Nsevolo *et al.*, 2016).

In order to valorise crop and wild plant residues in DRC, the leaves, fruits and/or grains from the following plant species, such as mango, maize, avocado, peanut, cassava, amaranth, cabbage, orange and carrots, are included in the cricket diets. Some diets were exclusively supplemented with leaves from one or several plant species, or in association with peanut waste or maize (A. Halloran *et al.*, unpublished data). As production is very new, the actual cricket farming units are 50 l plastic boxes covered with nets and filled with 30 crickets, egg cartons for shelter and plates filled with gravel and water. Nothing is known on the efficiency of the rearing units or its profitability. Only data on selling prices were available and one kg of fresh adult crickets seems to be sold for 13 US dollars on Kinshasa market. Sometimes young larvae are sold but never eggs or other co-products (A. Halloran *et al.*, unpublished data).

In 2015, the FAO technical cooperation program fully financed a project planning to build 200 insect farms DRC). Moreover, a centre for the promotion of edible insects as well as teaching modules on cricket rearing have been created in partnership with school and universities (A. Halloran *et al.*, unpublished data).

5.3 Conclusions

This chapter section emphasizes cricket farming can have significant implications on rural development and rural economy in low- and middle-income countries. Moreover, crickets have the potential to address food and nutrition insecurity and promote food sovereignty through the promotion of local production and consumption (A. Halloran *et al.*, unpublished data). Effectively, production, processing and marketing of edible insects has the potential to provide an important contribution to food security and nutrition by increasing food access at a household level, income, employment and livelihood opportunities (Durst and Hanboonsong, 2014). Cricket farming could be a profitable business for unemployed people with low education level as it easy to do and it need low investment. The first step to farm cricket in a cost-effective way is to build rearing units with available natural resources (e.g. wood) or low cost materials (e.g. concrete) that recreate the natural environment of cricket (i.e. damp and dark conditions). In these rearing units, crickets must be protected from flying predators by using nets and from walking ones by surrounding boxes with water or oil. Rearing units must be placed in stable environments that are shaded and out of the wind and rain. Inside rearing units, shelters made of cardboard egg cartons or dried leaves must be provided for crickets as they are very fearful animals. Once rearing units are ready, farmers could invest in cricket egg from another producer. To allow an efficient egg hatching and insect development, constant temperature (25 to 30 °C) and relative humidity (60 to 80%) must be provided. Insects must be feed *ad libitum* from neonate nymphs till imago reproduction. Feed cost is generally important as the most common ingredient used is chicken feed but several studies have shown that it could be easily replaced by available by-products, such as fruit and vegetable leaves, or peels. Moreover, different feed producers are developing optimal cricket feed that could increase cricket growth performances (Halloran *et al.*, 2016). At the end of their development, adult crickets will need moist substrates for egg laying such as soil, sand, loam or rice balls. After one or two weeks of reproduction, adults could be harvested for processing (e.g. boiling, freezing or frying) and retail.

The next major challenge to cricket farming will be to increase the support and awareness from stakeholders from governments instead of external project and decrease the feed cost by improving knowledge regarding low-cost feed ingredients that are suitable for optimal cricket development and reproduction (A. Halloran *et al.*, unpublished data). Moreover, standard farm management practices, as presented before, from nursery to cricket harvest are actually needed and technical staff must be consistently trained to allow new skill transfers to farmers on rearing management, nutrition, pest/disease control and farm hygiene (Hanboonsong *et al.*, 2013).

Next challenge will be to propose training courses focused on the cricket quality improvement (e.g. crickets with a standard size), its marketing, the understanding of the effect of seasonal and temperature variations, the inclusion of additional feed sources to decrease cost, and also the learning of basic animal husbandry (Halloran *et al.*, 2016b). Technical staff or extension agent must not only be trained but must also be easily accessible to all farmers. Effectively, even in Thailand, universities and centre of competence are involved in cricket farming training, farmers were not acquainted with the latest knowledge concerning inbreeding, disease and up-to-date farming technology principally due to a low access to the centres of competence or extension agents (Halloran *et al.*, 2016b). Finally, to increase sales urban areas or in other countries, an improvement of the labelling, packaging and the shelf life of crickets must be achieved (Durst and Hanboonsong, 2014).

References

Agea, J.G., Biryomumaisho, B., Buyinza, M. and Nabanoga, G.N., 2008. Commercialization of *Ruspolia nitidula* (nsenene grasshoppers) in Central Uganda. African Journal of Food, Agriculture, Nutrition and Development 8: 319-332.

Ayieko, M.A., Ogola, H.J. and Ayieko, I.A., 2016. Introducing rearing crickets (gryllids) at household levels: adoption, processing and nutritional values. Journal of Insects as Food and Feed 2: 203-211.

Belluco, S., Losasso, C., Maggioletti, M., Alonzi, C.C., Paoletti, M.G. and Ricci, A., 2013. Edible insects in a food safety and nutritional perspective: a critical review. Comprehensive Reviews in Food Science and Food Safety 12: 296-313.

Booth, D.T. and Kiddell, K., 2007. Temperature and the energetics of development in the house cricket (*Acheta domesticus*). Journal of Insect Physiology 53: 950-953.

Caparros Megido, R., Alabi, T., Haubruge, E. and Francis, C., 2015a. Are edible insects really green? Food, Science and Law 3: 98-104.

Caparros Megido, R., Alabi, T., Larreché, S., Alexandra, L., Haubruge, E. and Francis, F., 2015b. Risks and valorization of insects in a food and feed context. Annales de la Société Entomologique de France 51: 215-258.

Caparros Megido, R., Alabi, T., Nieus, C., Blecker, C., Danthine, S., Bogaert, J., Haubruge, E. and Francis, F., 2015c. Optimization of a cheap and residential small-scale production of edible crickets with local by-products as an alternative protein-rich human food source in Ratanakiri Province (Cambodia). Journal of the Science of Food and Agriculture 96: 627-632.

Collavo, A., Glew, R.H., Huang, Y.-S., Chuang, L.-T., Bosse, R. and Paoletti, M.G., 2005. House cricket small-scale farming. In: Paoletti, M.G. (ed.) Ecological implications of minilivestock, ecological implications of minilivestock. potential of insects, rodents, frogs and snails. CRC Press, Enfield, NH, USA, pp. 515-540.

DeFoliart, G.R., 1992. Insects as human food. Crop Protection 11: 395-399.

Durst, P.B. and Hanboonsong, Y., 2014. Small-scale production of edible insects for enhanced food security and rural livelihoods: experience from Thailand and Laos. Journal of Insects as Food and Feed 1: 25-31.

Food and Agriculture Organization of the United Nations (FAO), 2009. How to feed the world in 2050 ? FAO, Rome, Italy. Available at: http://tinyurl.com/5ranufw.

Fuah, A.M., Siregar, H.C.H., Endrawati, Y.C., Astuti, D.A., Khotijah, L., Winarno, H. and Sunyoto, A., 2016. Cricket farming in Indonesia: challenge and opportunity. Lambert Academic Publisher, Saarbrucken, Germany.

Fuah, A.M., Siregar, H.C.H. and Endrawati, Y.C., 2015. Cricket farming for animal protein as profitable business for small farmers in Indonesia. Journal of Agricultural Science and Technology 5: 296-304.

Gahukar, R.T., 2016. Edible insects farming: efficiency and impact on family livelihood, food security, and environment compared with livestock and crops. In: Dossey, A.T., Morales-Ramos, J.A. and Rojas, M.G. (eds.) Insects as sustainable food ingredients. Academic Press, San Diego, CA, USA, pp. 85-111.

Gahukar, R.T., 2012. Entomophagy can support rural livelihood in India. Current Science 103: 10.

Ghouri, A.S.K., 1961. Home and distribution of the house cricket *Acheta domesticus* L. Nature 192: 1000.

Halloran, A., Roos, N., Flore, R. and Hanboonsong, Y., 2016a. The development of the edible cricket industry in Thailand. Journal of Insects as Food and Feed 2: 91-100.

Halloran, A., Roos, N. and Hanboonsong, Y., 2016b. Cricket farming as a livelihood strategy in Thailand. Geographical Journal 183: 112-124.

Hanboonsong, Y., Jamjanya, T. and Durst, P.B., 2013. Six-legged livestock: edible insect farming, collecting and marketing in Thailand. Food and Agriculture Organization of the United Nations, Bangkok, Thailand.

Hill, D.S., 2008. Pests of crops in warmer climates and their control. Springer Science, Amsterdam, the Netherlands.

Hope, R.A., Frost, P.G.H., Gardiner, A. and Ghazoul, J., 2009. Experimental analysis of adoption of domestic mopane worm farming technology in Zimbabwe. Development Southern Africa 26: 29-46.

Ibler, B., Makert, G.R. and Lorenz, M.W., 2009. Larval and adult development and organisation of a systemic rearing of the Mediterranean field cricket (*Gryllus bimaculatus* de Geer, 1773). Zoologische Garten 78: 81-101.

Lundy, M.E. and Parrella, M.P., 2015. Crickets are not a free lunch: protein capture from scalable organic side-streams via high-density populations of *Acheta domesticus*. PLoS ONE 10: e0118785.

Miech, P., Berggren, Å., Lindberg, J.E., Chhay, T., Khieu, B. and Jansson, A., 2016. Growth and survival of reared Cambodian field crickets (*Teleogryllus testaceus*) fed weeds, agricultural and food industry by-products. Journal of Insects as Food and Feed 2: 285-292.

Nsevolo, P., Taofic, A., Caparros, R., Sablon, L., Haubruge, É. and Francis, F., 2016. La biodiversité entomologique comme source d'aliments à Kinshasa (République démocratique du Congo). Annales de la Société Entomologique de France 52: 57-64.

Oonincx, D.G.A.B., Van Broekhoven, S., Van Huis, A. and Van Loon, J.J.A., 2015. Feed conversion, survival and development, and composition of four insect species on diets composed of food by-products. PLoS ONE 10: e0144601.

Ramos-Elorduy, J., 1997. Insects: a sustainable source of food? Ecology of Food Nutrition 36: 247-276.

Rumpold, B.A., Schlüter, O.K., 2013a. Potential and challenges of insects as an innovative source for food and feed production. Innovative Food Science and Emerging Technologies 17: 1-11.

Rumpold, B.A. and Schlüter, O.K., 2013b. Nutritional composition and safety aspects of edible insects. Molecular Nutrition and Food Research 57: 802-823.

Sanewong Na Ayudhaya, A., 2011. Current status of cricket farming (in Thai). News report 25 April 2011. Bureau of Agricultural Commodities Promotion and Management, Bangkok, Thailand.

Schabel, H.G., 2010. Forest insects as food: a global review. In: Durst P.B., Johnson D.V., Leslie R.N. and Shono, K. (eds.) Edible forest insect: human bite back. Proceedings of a Workshop on Asia-Pacific Resources and their Potential for Development. Food and Agriculture Organization of the United Nations, Bangkok, Thailand.

Taponen, I., 2015. Supply chain risk management in entomology farms. Case: high scale production of human food and animal feed. BBA thesis, Helsinki Metropolia University of Applied Sciences, Helsinki, Finland.

Van Broekhoven, S., Oonincx, D.G.A.B., Van Huis, A. and Van Loon, J.J.A., 2015. Growth performance and feed conversion efficiency of three edible mealworm species (Coleoptera: Tenebrionidae) on diets composed of organic by-products. Journal of Insect Physiology 73: 1-10.

Van Huis, A., 2003. Insects as food in sub-Saharan Africa. Insect Science and its Application 23: 163-185.

Van Huis, A., 2013. Potential of insects as food and feed in assuring food security. Annual Review of Entomology 58: 563-583.

Van Huis, A., Van Itterbeeck, J., Klunder, H.C., Mertens, E., Halloran, A., Muir, G. and Vantomme, P., 2013. Edible insects: future prospects for food and feed security. Food and Agriculture Organization of the United Nations, Rome, Italy.

6.

Rhynchophorus ferrugineus

Palm weevil farming contributing to food security in sub-Saharan Africa

F.J. Muafor[1*], A.A. Gnetegha[1], E. Dounias[2], P. Le Gall[3] and P. Levang[4]

[1]Living Forest Trust (LIFT), c/o IRD, BP 1857 Yaoundé, Cameroon; Ministry of Forestry and Wildlife (MINFOF), BP 34430, Yaoundé, Cameroon; [2]French National Research Institute for Sustainable Development (IRD), Center for International Forestry Research (CIFOR), UMR5175 CEFE, 1919 route de Mende, 34293 Montpellier cedex 5, France; [3]Laboratoire Evolution, Génomes, Comportement and Ecologie, CNRS, IRD, Université Paris-Sud, Université Paris-Saclay, 91198 Gif-sur-Yvette, France; [4]Center for International Forestry Research (CIFOR), BP 2008, Yaoundé, Cameroon; French National Research Institute for Sustainable Development (IRD), UMR 220 GRED, 911 avenue Agropolis, BP 64501, 34394 Montpellier Cedex 5, France; muafor1@yahoo.com

Abstract

African palm weevils (*Rhynchophorus* spp., Dryophthoridae) are amongst the most widely consumed species of insects in sub-Saharan Africa, especially in their larval form. Studies have shown that these insects are very rich in essential food nutrients, with protein, carbohydrates, fat, and energy values comparable to that of beef and fish. Moreover, they are rich in micronutrients, and their exploitation is an important source of income for many forest dependent communities in sub-Saharan Africa. However, grubs are usually collected from raffia stems in the wild and their exploitation is strenuous, seasonal and sometimes unsustainable. In an attempt to promote the use of these resources as an alternative source of protein, this paper describes a farming design that could be used by small-scale farmers to produce these edible larvae in plastic boxes at home within a period of 30 days per cycle. Though fractions of raffia stem tissue are used as substrate in this farming system, it is by far less destructive to the raffia ecosystem and more productive than collection in the wild. If disseminated, grub farming could provide an opportunity for a continuous year round production of these nutrient rich insects, while securing their place as an important alternative to protein and income source in Africa.

Keywords: insects, farming, food security, rural communities

6.1 Introduction

Food insecurity on the African continent has worsened in recent years, and the proportion of the malnourished population in sub-Saharan Africa has surpassed 35%, with the greatest problem being accessibility to edible protein sources (Rosegrant *et al.*, 2005). Most frequently, diets in this sub-region are comprised of cereal or root staple crops, with limited sources of proteins and micronutrients. Because of bushmeat depletion and excessively costly substitution by industrial livestock production (Alkemade *et al.*, 2013), proteins are either too expensive, or locally less available and their access is becoming increasingly critical. They are unequally distributed within households or generally not considered as household priority when incomes are insufficient to meet the needs of a high quality diet (Chastre *et al.*, 2007). Protein deficiency and reduced dietary diversity have serious effects on the nutrition and health of both rural and urban populations and account for a number of illness and death in many countries in Africa (Elemo *et al.*, 2011).

Indigenous food resources constitute the bedrock of the diversity in traditional and indigenous food systems of developing country (FAO, 2014). Though a wide range of indigenous foods are neglected, underutilized, and less promoted by agricultural policies in most parts of Africa, they could be fundamental in resolving some of the many problems induced by malnutrition and food insecurity in most rural areas. Moreover, traditional and indigenous foods are less deleterious to the environment and are vital in the preservation of the cultural heritage of local communities (FAO, 2014). In sub-Saharan Africa where more than half of the population is classified as 'poor' (Beegle *et al.*, 2016), the exploitation of indigenous food resources is irrefutable to secure food security. The need to develop non-timber forest products (NTFPs) in particular as an alternative to improve indigenous livelihoods while cutting down human pressure on biodiversity has largely been acknowledged (Arnold and Ruiz Pérez, 2001; Belcher *et al.*, 2005; Ndoye *et al.*, 1997).

Amongst NTFPs, edible insects are quite abundant, and the gathering of insects for food has for long been part of the culture of many human societies in the developing world (Muafor *et al.*, 2012, 2014). Nowadays, palm weevil grubs, caterpillars, termites, grasshoppers and locusts are the most popular in Africa in general and Central Africa in particular (FAO, 1995; Muafor *et al.*, 2015; Stack *et al.*, 2003). The place of the African palm weevil larvae *Rhynchophorus* spp. is one of the most salient as it is consumed in nearly all countries of sub-Saharan Africa. The African palm weevil larvae are extremely rich in essential food nutrients. According to Womeni *et al.* (2012), the consumption of 100 g of these larvae can effectively satisfy the body daily requirements in energy, proteins, fats, glucose, vitamins (thiamine/B1, riboflavin/B2 and pyridoxine/B6), and minerals (K, Ca, Mg, Zn, P and Fe).

In almost all countries where these larvae are eaten, harvesting is done by systematically extracting them from the trunks of raffia in the wild and/or from aging oil palms felled to produce palm wine. Traditionally, collectors used to explore the palms that were naturally infested by the grubs and such exploration required a true ecological expertise. Felling of the weevil infested palms by this indigenous exploitation methods contributed to the sustenance of healthy raffia stands (Dounias, 2003). However, harvesting was seasonal and satisfied only a domestic consumption and, at best, a small-size market trade. In some countries like Cameroon for example, the exploitation of this resource is no longer sustainable as it encourages collectors to clear-cut vast areas of raffia swamps in order to induce weevil infestation in the decaying cut stems. By contrast to past practices, such speculative form of exploitation has become highly destructive (Muafor *et al.*, 2015). In response to these problems and in line with increasing concerns on how to develop mini-enterprises for the production of insect based food and feed in Africa, a team of researchers from the Living Forest Trust (LIFT), the Center for International Forestry Research and the French National Research Institute for Sustainable Development have implemented a farming technique for the production of edible larvae by rural small scale producers in Cameroon. This farming technique, successfully tested in Obout village, is presently fine-tuned and disseminated among smallholders with similar and alternative substrates in selected villages of Cameroon.

6.2 Marketing and socioeconomic potential of grubs in Cameroon

Palm weevil grubs are particularly important economic resources and they provide complementary income to many rural people who depend on their exploitation as main or part-time activity (Dounias, 2004). From the dense humid semi-deciduous forest zone in the East to the highland savannah in the West region, this insect is traded, either cooked or raw, by small-scale roadside vendors. From a recent study (Muafor *et al.*, 2015), in addition to small-scale sales in villages located near harvesting sites, grubs are increasingly traded in larger urban markets. In some markets like Mvog-Mbi and Nkondongo in Yaoundé, or in Bertoua, Abong-Mbang and Ayos a specialised section is dedicated to grub sales (Figure 6.1).

The marketing chain consists of retailers who live in big cities, intermediate suppliers in smaller cities and producers in villages. Each of the producers may sell directly to the retailers or the intermediate suppliers. The intermediate suppliers buy from producers to resell in markets either in local towns or to retailers in the big cities. For instance, market price in a local town like Abong-Mbang is 500 FCFA (USD 1) for a glass containing 25 to 30 individuals of palm weevil grubs. Retailers in bigger cities like Yaoundé and Douala resell the same amount at 1,500 FCFA (USD 3) during seasons of abundance and 2,500 FCFA (USD 5) in periods of scarcity. Brochettes of prepared grubs are equally sold at 100 FCFA (USD 0.2) per

Figure 6.1. Small-scale traders of palm weevil grubs in Abong-Mbang market, Cameroon.

stick of 3 to 4 individuals. Trans-border trade is also important, with sales to neighbouring countries – Gabon, Equatorial Guinea, Nigeria – and even as far as France and Belgium.

In terms of income generated, Dounias (2004) reported that in 2004 the average monthly income for larvae harvesters in rural areas was about 35,500 FCFA (USD 71), for live larvae sold to retailers supplying city markets and 25,000 FCFA (USD 50) for roasted larvae sold in bars and along roadsides. In 2015 figures are much higher with an average income generated by professional grub collectors in Obout village ranging from 90,000 FCFA (USD 180) to 300,000 FCFA (USD 600), representing 30 to 75% of the collectors household income (Muafor *et al.*, 2015). Out of the eight farmers who experimented the proposed farming technique, most were able to complement their income with a regular monthly revenue of 50,000 FCFA (USD 100) to 70,000 FCFA (USD 140) by selling grubs harvested on farm.

6.3 Reproductive material for palm weevil grubs farming

Adult individuals of the African palm weevils are the reproductive animal material, which is used in the grubs farming system. Members of this species belong to the genus *Rhynchophorus* in Dryophthoridae. Adult individuals are characterized by a snout-like projection of their mandibles called a rostrum, while the larvae (or grubs) have relatively large mandibles and are legless. The modified mouthparts in the adults are used for feeding and to make holes in the host plant material where eggs are laid (Tambe *et al.*, 2013). Palm weevils are distributed throughout tropical Africa where they mainly live in oil palm, date palm, raffia palm and coconut palm (Bong *et al.*, 2008; Gries *et al.*, 1994). However, concrete knowledge on the diversity of this genus in Africa is lacking. In Cameroon, two species – notably *Rhynchophorus phoenicis* (Fabricius, 1801) and *Rhynchophorus quadrangulus* Quedenfeldt, 1888 – are frequently cited as edible. *R. quadrangulus* is adapted to highland areas, while *R. phoenicis* is mostly found in humid lowland areas. The life cycle of these two species is similar to that of other *Rhynchophorus* species and is known to undergo complete metamorphosis. In the wild, adult weevils are generally attracted by wounded and sap producing raffia and/or palm species on which they feed, mate and establish for pupation. During this process, fecundated females deposit eggs on the decaying parts of the trunks, which later hatch into young larvae within a period of one week. These young larvae develop within a few weeks into mature larvae that can be harvested.

6.4 Palm weevil grubs farming technique on experimentation in Cameroon

The experimental method implemented by LIFT NGO to promote palm weevil grubs farming in Cameroon consists of rearing adult palm weevils in plastic boxes of 60 cm long, 40 cm wide and 40 cm high. In this system, the plastic boxes are suspended with the help of woody poles at a height of 80 to 100 cm above the ground in a shaded but air circulating environment. Once the farming dispositive is established, the rearing process begins with the selection and coupling of healthy adult palm weevils in small plastic cups for copulation (Figure 6.2). The fecundated beetles are then introduced into larger plastic boxes for eggs production, incubation, hatching and development into edible grubs. Slices of young woody stem tissues of raffia or other palms are introduced as food substrate and/or laying ground (Figure 6.2).

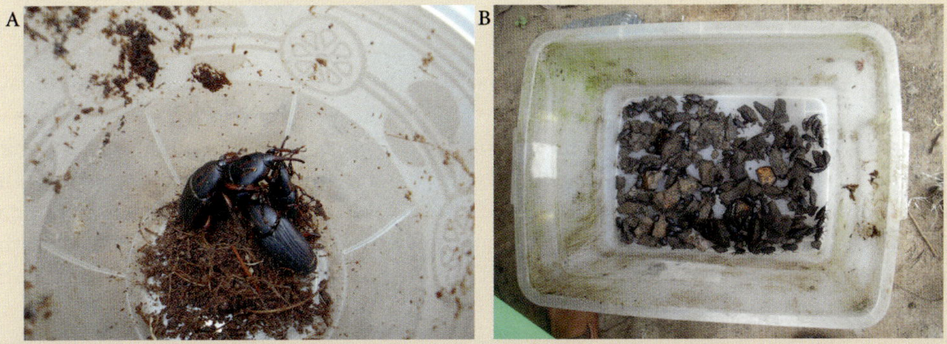

Figure 6.2. (A) Coupling of adults for fecundation and (B) fecundated adults transferred to a larger box for laying and hatching.

Once the eggs are hatched, young grubs are fed by adding pieces of young woody stem tissues from raffia into the boxes containing the young larvae. A handful of fresh stem tissues of raffia is added once a week within the first two weeks, after which the substrate is replaced by hard stem tissues for two additional weeks. With this technique, well fed grubs in suitable farming conditions grow to maturity within a period of 25 to 30 days, when they can be harvested.

6.5 Suitable substrate for rearing the African palm weevil

One of the most challenging aspects in African palm weevils farming is to find the right substrate for adult fecundation, incubation, and development of the young larvae. Various plant materials have been tested as substrate to farm palm weevils in Africa. In Nigeria, Ebenebe and Okpoko (2016) report that it is possible to rear African palm weevils with sugarcane tops and spoilt watermelon. Nonetheless, the authors indicate that growth of the larvae and yields are relatively lower in sugarcane tops than in spoilt watermelon, but are unable to pupate in spoilt watermelon. Monzenga Lokela (2015) similarly studied the possibility of using sugarcane stalks and palm stem tissues in the farming process. In Cameroon, stem tissues from raffia and other palm species as food substrate gave the best results.

Trials conducted by LIFT to test the efficiency of substrate from selected palm species concluded that slices of young woody tissues from *Raphia hookeri* G. Mann & H. Wendl., 1864, *Raphia monbuttorum* Drude 1895 and *Elaeis guineensis* Jacq., 1763 are suitable food substrates for fecundated adults. Moreover, these three materials are equally convenient for egg laying and hatching into young larvae. However, only stem tissues from *R. hookeri* were proven to be suitable food substrate for the development of young grubs after hatching. In all the trials, the quantity of matured grubs harvested in boxes containing *R. monbuttorum* and *E. guineensis* was much lower than the quantity obtained in boxes containing *R. hookeri*. For two or three fecundated female adult weevils that were introduced in each of the boxes containing the different food substrate, the average grubs productivity of boxes containing

R. hookeri was more than twice that of boxes containing *R. monbuttorum* and *E. guineensis* (Figure 6.3).

During the farming trials, it was observed that after the hatching of eggs (about 3 or 4 days after laying), all the boxes were colonised by young larvae, most of which did not survive to maturity in boxes containing *R. monbuttorum* and *E. guineensis*. This might be explained by the fact that the efficiency of the plant tissues on the growth and development of the larvae and the productivity of the farming dispositive depend on the changes that occur within the farming conditions as the substrates decay over time. It was observed that the decomposition of *R. hookeri* was gradual and the medium remained relatively fresh, while

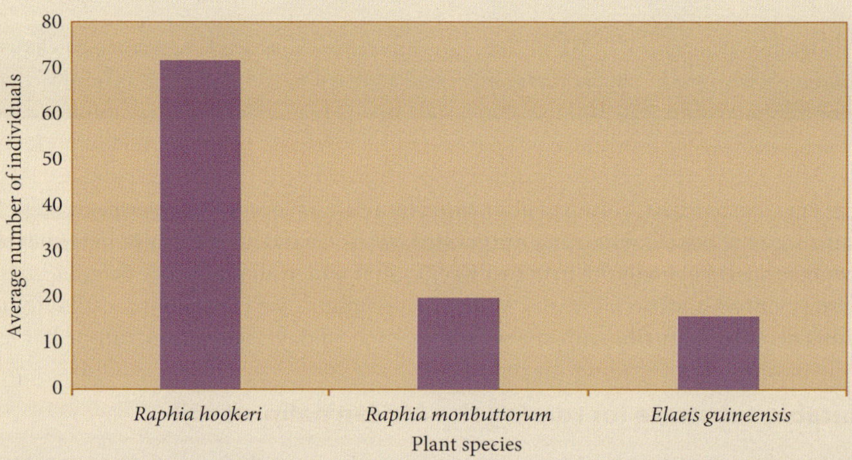

Figure 6.3. Average productivity of grubs in boxes containing different plant tissues as food substrate.

Figure 6.4. Difference in the decomposition and efficiency of substrates from (A) *Raphia hookeri* and (B) *Elaeis guineensis*.

that of *R. monbuttorum* and *E. guineensis* rapidly decomposed, with signs of high level of fermentation, dehydration, and the production of a pungent odour (Figure 6.4).

When composed of *R. monbuttorum* and *E. guineensis*, the medium becomes toxic to the young grubs, most of which die before reaching maturity. This situation is especially frequent when farming procedures are not properly respected.

6.6 Required farming conditions and work schedule

Nevertheless even in farms adopting *R. hookeri*, tissues from matured stems should be harvested as substrate. Some farming conditions need to be respected to ensure the survival and growth of young grubs. Of the many different parameters that must be taken into account for a successful palm weevil farming, humidity of the substrate and temperature in the farming boxes are the most acute. These two parameters are controlled by perforating the farming boxes with six holes of about 1 cm in diameter on the sides and on the base to guarantee a suitable temperature and humidity. In addition to temperature and humidity regulation, the perforations on the sides also ensure the renewal of oxygen in the boxes. The holes at the base are essential for evacuating part of the detritus exudates resulting from the decomposition of the substrate. Periodical cleaning is needed in order to keep the holes clear from any obstacle. The various farming conditions must be controlled during each intervention programmed on a work schedule predetermined in the farming calendar. A typical work schedule for securing good farming conditions is synthetized in Table 6.1.

6.7 Dissemination attempts and performance of small-scale grub farms in Cameroon

The grub farming technique is currently being promoted among small-scale farmers in selected villages in southern Cameroon. Beneficiary villages are selected based on the proximity of a swamp raffia forest and existing exploitation of the forest to harvest palm weevil grubs. In each of the villages, the grub farming is carried out using a Farmer Participatory Research (FPR) approach in which farmers are directly involved in a farming research-based project. Farmers are selected by experts from LIFT, trained on palm weevil farming process and assisted in the establishment and follow-up of individual small-scale pilot grubs farms (Figure 6.5).

Since the last quarter of 2015, a total of 25 small-scale farmers and a cooperative have been trained and assisted in the establishment of grub farms. In the different farms, experts from LIFT collaborate with the farmers to test different farming parameters. Apart from humidity and temperature, two other parameters are being tested through the FPR approach: the first one is the number of fecundated females that should be introduced in a box to assure an optimum yield; the second one is the suitable food substrate for the development of young larvae, by comparing the efficiency of stem tissues from the two raffia species (*R. hookeri* and *R. monbuttorum*) and oil palm (*E. guineensis*).

The PFR approach allows for data collection while continuously adjusting the technique based on remarks and propositions formulated in return by the farmers. It efficiently combines farmer training with on-going improvement of the farming technique inspired from their

Table 6.1. Work schedule and farming conditions for a successful palm weevil grubs farm.

Status	Farming material	Substrate	Conditions	Farming day	Activities
Week 1					
fecundation	small plastic containers with lids	young woody stem tissue of *Raphia hookeri*	two adult couples per container	1	• perforate small plastic containers on the lid • put adult couples in small plastic containers
			fecundation best at 20-25 °C	2	• observe if females have started laying eggs
incubation	large plastic storage containers	young woody stem tissue of *R. hookeri*	incubation is best at 24-25 °C	3	• perforate large plastic storage containers on the sides and at the base • introduce 1 kg of substrate into the container • transfer eggs and fecundated females into the containers (3 females per container)
Week 2 to 4					
growth and development of larvae	large plastic storage containers	hard woody stem tissue of *R. hookeri*	humidity of raffia tissues should be conserved	3 to 6	• control perforations at the base of the containers and reopen if closed
				7	• add 0.5 kg of fresh substrate
				8 to 13	• control perforations at the base of the containers and reopen if closed
			growth is best at 25-30 °C	14	• add 0.5 kg of fresh substrate
				15 to 21	• control perforations at the base of the containers and reopen if closed
			humidity of raffia tissues should be conserved	22	• add 0.5 kg of fresh substrate
				23 to 29	• control perforations at the base of the containers and reopen if closed
				30	• harvesting of grubs

own initial knowledge and 'learning by doing' experience (Freeman, 2001). Selected farmers will be trained to become expert farmers on the grub farming technique. These expert farmers will serve as models to disseminate the grub farming technique to the neighbouring farmers. Production data collected during two experimental farming cycles in 10 small-scale pilot farms and an average of 3 boxes per farm show that a total of 3,956 grubs were produced from 33 boxes that necessitated 12 raffia stems as substrate (Table 6.2).

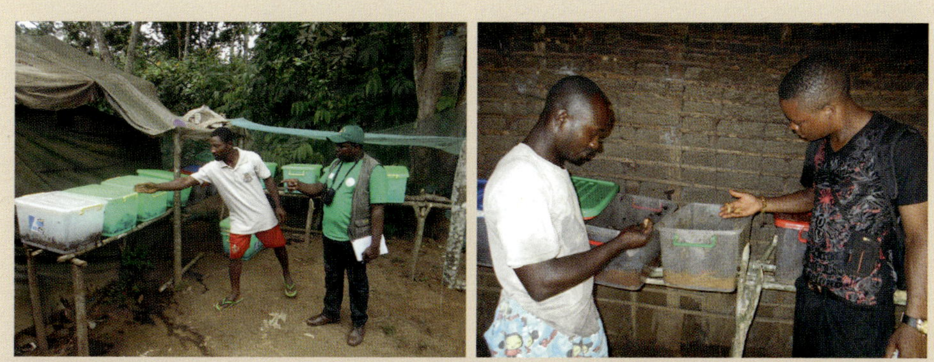

Figure 6.5. Individual small-scale pilot grub farms promoted by the Living Forest Trust in the Nyong Basin area.

Table 6.2. Grubs produced by small-scale farmers in Obout village in Cameroon.

Codes of individual farmers	Number of farming boxes	Number of raffia stems exploited for all the boxes	Number of grubs produced in 1st trial	Number of grubs produced in 2nd trial	Total
Farmer 1	3	1	191	161	352
Farmer 2	3	1	158	168	326
Farmer 3	3	1	253	192	445
Farmer 4	3	1	148	160	308
Farmer 5	3	1	230	135	365
Farmer 6	3	1	201	abandoned	201
Farmer 7	3	1	273	118	391
Farmer 8	6	2	590	512	1,102
Farmer 9	3	1	159	168	327
Farmer 10	3	1	139	abandoned	139
Total	33	12	2,342	1,614	3,956

The advantage of this farming technique is that the productivity of a single stem of raffia that is used as substrate in the farming system is at least four times higher than that in the wild. Generally, raffia stems in the wild hardly produce more than 60 individuals. Moreover, it allows farmers to produce grubs at any period of the year for household consumption and/or trade (Figure 6.6).

Figure 6.6. Grubs harvested from small-scale grub farms in Obout village in Cameroon.

However, the dispositive must be closely supervised in order to ensure good yields. Some farmers who could not provide the required attention to the farming dispositive abandoned the experiment after the first trial, while others only obtained low yields during both trials for reasons that remain unexplained.

6.8 What future for insect based enterprises in sub-Saharan Africa?

Insect farming is quite new in sub-Saharan Africa, in spite of the fact that insects have for long been part of human food in this sub-region. Traditionally, forest dependent people have been gathering insects from the wild for family consumption and trade. However, with increasing concerns over the need to optimize the potential of insect resources as an alternative in the development of food and feed, procurement from the wild is reaching its limits and insect farming is alternatively developing in different corners of the world. Although there is no legislation forbidding the consumption of insects or the use of processed insect products in human and/or animal nutrition in sub-Saharan Africa, no significant effort has been made by authorities in charge of agriculture and even private investors to develop enterprises that promote the processing of insect resources into innovative products. In an area where entomophagy is culturally well established, this growing economic opportunity needs to be elicited to alleviate endemic poverty and malnutrition while promoting highly valued food resources.

According to Van Huis *et al.* (2013), the establishment of insect-based enterprises and the processing of insects into street foods and animal feed could easily be achieved in low-income countries where it is relatively easy to bring insects to the market, and where demand for edible insects is high. However, this can only be achieved if people are trained in insect farming and if the processing of insects into quality products is secured. Some organisations have started investing in the promotion of insect-based enterprises in sub-Saharan Africa and the local farmers are gradually being educated on the economic potential of the edible-insect trade sector. One rare example of a pioneer organisation that promotes insect-based enterprises in Cameroon and Central Africa at large is LIFT that is developing and training farmers on palm weevil grubs farming. Other promising initiatives to develop insect-based industry in sub-Saharan Africa are those of ASPIRE that has similarly introduced palm weevil farming in Ghana and the efforts of FASOPro in the processing and trade of edible insects in Burkina Faso. Apart from grubs farming, other types of insects like mealworms and crickets are being introduced for farming in sub-Saharan Africa. The increasing interest in the development of insect-based enterprises and products is a strong incentive to pursue research exploration on the controlled rearing of other edible insects so far picked out from the wild.

6.9 Conclusions

The development of palm weevil farming is an important gateway for the development of insect-based small-scale enterprises in sub-Saharan Africa. Palm weevil grubs (Figure 6.7) are largely accepted as food in most of the countries in the sub-region. The technique of domestication promoted by LIFT is relatively simple and easy to implement. The enthusiasm of small-scale farmers in the experimental grub farming in Cameroon demonstrates the high potential of edible-insect farming as an economic activity. With an existing market for grubs and other species of insects, this sub-region presents a huge potential for the development of insect-based mini-enterprises and processed food and feed products. However, the region is still fledgling in the domain of growing insect economy and more efforts need to be deployed jointly by authorities, entrepreneurs, and researchers to transform traditional gathering practices into a more profitable and specialised insect-based economic sector. The effective

Figure 6.7. *Rhynchophorus phoenicis* larvae.

development of this sector in sub-Saharan Africa might be geared by a number of factors, amongst which:

- enhancement of research on farming procedures and conditions of grubs and other edible insect species;
- putting in place of suitable mechanisms that facilitate access to funds for research and development of insect product-based enterprises;
- active involvement of all development stakeholders (government, NGOs, donors, private investors, local communities) in the development of this new economic sector;
- transfer of technologies and training of small-scale farmers;
- development of a legal framework for securing product quality;
- promotion of entomophagy among non-insect consumers and the use of insect-based products as food and feed.

Acknowledgements

We would like to thank the C2D-PAR program of the Ministry of Scientific Research and Innovation (MINRESI) for providing substantial funds to promote insect farming and insect-based enterprises in Cameroon. Great thanks to the Center for International Forestry Research (CIFOR), the managers of the CGIAR CRP-FTA fund and particularly Dr Peter Cronkleton for supporting our palm weevil grubs farming project. We are equally grateful to the French National Research Institute for Sustainable Development (IRD), as well as the RAPHIA Project of IRD for providing supplementary funds for the training of small-scale farmers. Special thanks to Michel Owona Tseungui for his decisive contribution during the experimental farming phase.

References

Alkemade, R., Reid, R.S., Van den Berg, M., De Leeuw, J. and Jeuken, M., 2013. Assessing the impacts of livestock production on biodiversity in rangeland ecosystems. Proceedings of the National Academy of Sciences 110: 20900-20905.

Arnold, J.E.M. and Ruiz Pérez, M., 2001. Can non-timber forest products match tropical forest conservation and development objectives? Ecological Economics 39: 437-447.

Beegle, K., Christiaensen, L., Dabalen, A. and Gaddis, I., 2016. Poverty in a rising Africa: overview. World Bank, Washington, DC, USA.

Belcher, B., Ruiz Pérez, M. and Achdiawan, R., 2005. Global patterns and trends in the use and management of commercial NTFPs: implications for livelihoods and conservation. World Development 9: 1435-1452.

Bong, C.J., Er, C., Yiu, P. and Rajan, A., 2008. Growth performance of the red-stripe weevil *Rhynchophorus schach* Oliv. (Insecta: Coleoptera: Curculionidae) on meridic diets. American Journal of Agricultural and Biological Sciences 3: 403-409.

Chastre, D., Duffield, A., Kindness, H., LeJeune, S. and Taylor, A., 2007. The minimum cost of a healthy diet: findings from piloting a new methodology in four study locations. Save the Children, London, UK.

Dounias, E., 2003. L'exploitation méconnue d'une ressource connue: la collecte des larves comestibles de charançons dans les palmiers raphias au sud Cameroun. In: Motte-Florac, E. and Thomas, J.M.C. (eds.) Les insectes dans la tradition orale. Peeters Publishing, Paris, France, pp. 205-226.

Dounias, E., 2004. Edible weevil larvae: a pest for palm trees but a delicacy for city-dwellers. In: López, C. and Shanley, P. (eds.) Riches of the forest: for health, life and spirit in Africa. CIFOR-DFID-EC, Bogor, Indonesia, pp. 9-12.

Ebenebe, C.I. and Okpoko, V.O., 2016. Preliminary studies on alternative substrate for multiplication of African palm weevil under captive management. Journal of Insects as Food and Feed 2: 171-177.

Elemo, B.O., Elemo, G.N., Makinde, M.A. and Erukainure, O.L., 2011. Chemical evaluation of African palm weevil, *Rhychophorus phoenicis*, larvae as a food source. Journal of Insect Science 11: 146.

Food and Agriculture Organisation (FAO), 1995. Non-wood forest products for rural income and sustainable forestry. Non-wood forest products 7. FAO, Rome, Italy.

Food and Agriculture Organisation (FAO), 2014. Promotion of underutilized indigenous food resources for food security and nutrition in Asia and the pacific. FAO Office for Asia and the Pacific, Bangkok, Thailand, 200 pp.

Freeman, D., 2001. Teacher learning and student learning in TESOL. TESOL 35: 608-609.

Gries, G., Gries, R., Perez, A.L., Gonzalez, L.M., Pierce Jr., H.D., Oehlschlager, A.C., Rhainds, M., Zebeyou, M. and Kouame, B., 1994. Ethyl proprionate: synergistic kairomone for African palm weevil, *Rhynchophorus phoenicis* L. (Coleoptera: Curculionidae). Journal of Chemical Ecology 20: 889-897.

Monzenga Lokela, J.C., 2015. Ecologie appliquée de *Rhynchophorus phoenicis* Fabricius (Dryophthoridae: Coleoptera): phénologie et optimisation des conditions d'élevage à Kisangani. PhD thesis, Université Catholique de Louvain, Louvain-la-Neuve, Belgium, 201 pp. Available at: http://hdl.handle.net/2078.1/157580.

Muafor, F.J., Levang, P. and Le Gall. P., 2014. A crispy delicacy: Augosoma beetle as alternative source of protein in East Cameroon. International Journal of Biodiversity: 214071.

Muafor, F.J., Levang, P. and Le Gall, P., 2012. Making a living out of forest insects: beetles as income source in Southwest Cameroon. International Forestry Review 14: 314-325.

Muafor, F.J., Gnetegha, A.A., Le Gall, P. and Levang, P., 2015. Exploitation, trade and farming of palm weevil grubs in Cameroon. CIFOR Working Paper 178, Bogor, Indonesia, 32 pp.

Ndoye, O., Ruiz-Perez, M. and Eyebe, A., 1997. The markets of non-timber forest products in the humid forest zone of Cameroon. ODI Network Paper 22c, Overseas Development Institute, London, UK.

Rosegrant, M.W., Ringler, C., Msangi, S., Cline, S.A. and Sulser, T.B., 2005. International Model for Policy Analysis of Agricultural Commodities and Trade (IMPACTWATER): model description. International Food Policy Research Institute, Washington, DC, USA.

Stack, J., Dorward, A., Gondo, T., Frost, P., Taylor, F. and Kurebgaseka, N., 2003. Mopane worm utilisation and rural livelihoods in southern Africa. CIFOR Livelihood Conference, 19-23 May 2003, Bonn, Germany. Available at: http://tinyurl.com/y73t3by9.

Tambe, T.J., Riolo, P., Okolle, N.J., Isidoro, N., Fanciulli, P.P. and Dallai, R., 2013. Sexual size differences and color polymorphism of *Rhynchophorus phoenicis* in the southwest region of Cameroon. Bulletin of Insectology 66: 153-159.

Van Huis, A., Van Itterbeeck, J., Klunder, H., Mertens, E., Halloran, A., Muir, G. and Vantomme, P., 2013. Edible insects: future prospects for food and feed security. Food and Agriculture Organization of the United Nations (FAO), Rome, Italy, FAO Forestry Paper no. 171, 187 pp. Available at: http://www.fao.org/docrep/018/i3253e/i3253e.pdf.

Womeni, H.M., Tiencheu, B., Linder, M., Nabayo, C.M.E., Tenyang, N., Mbiapo, T.F., Villeneuve, P., Fanni, J. and Parmentier, M., 2012. Nutritional value and effect of cooking, drying and storage process on some functional properties of *Rhynchophorus phoenicis*. International Journal of Life Science and Pharma Research 2: 203-219.

7.

Locusta migratoria

Farming of grasshoppers for feed and food

P. Haldar[1*] and C. Malakar[2]

[1]Visva-Bharati University, Santiniketan 731235, West-Bengal, India; [2]Suri Vidyasagar College, Suri, Birbhum 731101, West-Bengal, India; phaldar5@yahoo.co.in

Abstract

Grasshoppers feed on plant material. They also act as prey, notably for birds and mammals. However, they are also a recognized food utilized for centuries by various human cultures throughout the world. Farming of grasshoppers can play a vital role in fulfilling the acute demand for animal protein. Although a large number of species belong to family Acrididae, only very few species are suitable for farming. Short-horned grasshoppers (acridids) pass from 4 to 7 nymph stages before reaching the adult stage. To allow for optimal biomass production of grasshoppers, temperature should range from 30 to 35 °C, relative humidity from 70 to 80%, light intensity from 500±25 lux, and photoperiod from 14:10 L:D. For farming, adults are field-collected and put in an insectary for acclimatizing. Then, they are placed in specially designed cages with fresh food. Sterilized sand in plastic cups is used for collecting the eggs. Soon after hatching the nymphs are transferred to rearing jars. After they have developed to third instars, they are placed in different compartments and allowed to mature to adults. After approximately five days in these compartments, the adult grasshoppers can be harvested. The freshly harvested grasshoppers (live) can be used for poultry feed. Whole dried grasshoppers can be used as animal feed or for human food. They can also be crushed into a powder for further processing. Grasshopper powder could be utilized to obtain the following products: (1) grasshopper powder itself (for preparation of animal feed); (2) lipids; (3) protein isolates/concentrates; and/or (4) chitin. Some grasshoppers contain omega-3 fatty acid lipids. Data from nutrition studies indicate that grasshoppers (as oxyameal) are superior to the conventional protein supplements such as fishmeal/soymeal.

Keywords: primary consumer, biomass, omega-3 fatty acid, oxyameal, protein supplements

7.1 Introduction

In recent years, interest in using insects for human consumption and for animal feed has increased substantially. However, in order to satisfy the growing demand, harvesting from nature is no longer adequate and insects need to be farmed. 'Farming' means producing and harvesting the organisms and their products.

Taylor (1975) first proposed the concept of actual insect rearing for human consumption. Kok (1983) discussed the potential of insect farming and remarked that such farms can be utilised in a spaceship or domed megalopolis. In this context detailed information of farming on insects such as mealworms, crickets and silkworm are found in Van Huis *et al.* (2013). Dossey *et al.* (2016) provided detailed information about production, processing, and food applications of insects.

Farming of grasshoppers, mainly the acridids (short horned grasshoppers), plays a vital role in fulfilling the acute demand for animal protein. Haldar and his research scholars in the Department of Zoology, Visva-Bharati University, India, have been conducting research on insect farming since 1992. Research has focused on several aspects of various short horned grasshopper species of India aiming to establish a grasshopper farm. In addition to rearing grasshoppers, farming of grasshoppers plays a vital role in producing nonconventional food and feed. Haldar and Nandi (1997) for the first time highlighted the feasibility and importance of grasshopper farming. This text serves as a vital resource for understanding the methods utilised for grasshopper farming.

In establishing any grasshopper farm, knowledge on nutritive value of grasshopper is of primary importance. Information on the quantity and the quality of nutrients, anti-nutrients, fatty acids, amino acids, vitamins, minerals and calorie content in grasshoppers is given by Uvarov (1966), Ramos Elorduy *et al.* (2007, 2012), Anand *et al.* (2008), Das and Mandal (2013), Ganguly *et al.* (2013), Chakrovorty *et al.* (2014), Ghosh *et al.* (2014), and Paul *et al.* (2016a).

7.2 Species of acridids

Reports on experiments on rearing of many species of acridids (short horned grasshoppers) are as follows:

> *Aiolopus thalassinus* (Sanjayan and Murugan, 1987); *Anabrus simplex* (DeFoliart, 1989); *Aularches punctatus* (Katiyar, 1955); *Bryophyta debilis* (Luong-Skovman, 2001); *Chorthippus biguttulus* (Kohler *et al.*, 1987); *Chorthippus brunneus* (Kelly-Stebbings and Hewitt, 1972); *Chorthippus parallelus* (Kelly-Stebbings and Hewitt, 1972); *C. parallelus* (Kohler *et al.*, 1987; Paul *et al.*, 2016b); *Diabolocatantops pinguis* (Vedham, 1994); *Eyprepocnemis alacris alacris* (Muralirangan and Ananthakrishnan, 1977); *Eyprepocnemis plorans meridionalis* (Jago, 1963); *Gesonula punctifrons* (Sanjayan and Murugan, 1987); *Gomphocerus rufus* (Kohler *et al.*, 1987); *Hieroglyphus nigrorepletus* (Riffat and Wagan, 2007); *Melanoplus bivittatus* (Bailey and Mukerjee, 1976; Henry, 1985; Nayar, 1964); *Melanoplus differentialis* (Barnes, 1963; Henry, 1985; Pfadt, 1949); *M. femurrubrum* (Bailey and Mukerjee, 1976); *M. mexicanus mexicanus* (Parker, 1930);

Melanoplus sanguinipes (Hinks and Olfert, 1999); *Myrmeleotettix maculates* (Kelly-Stebbings and Hewitt, 1972); *Oamnula pellucid* (Parker, 1930); *Oedaleus senegalensis* (Cheke, 1990); *Orthacris maindroni* (Ananthakrishnan *et al.*, 1985); *Oxya nitidula* (Fanny *et al.*, 1999); *Oxya velox* (Anonymous, 1988, 1992); *Parahieroglyphus bilineatus* (Katiyar, 1956).

But there are only a few experiments on farming of grasshopper species such as: *Acrida exaltata* (Das, 2002; Nath and Haldar, 1992); *Hieroglyphus banian* (Das *et al.*, 2002); *Oxya fuscovittata* (Das, 2002; Ganguly *et al.*, 2013, 2014; Haldar and Nandi, 1997; Malakar *et al.*, 2009); *Spathosternum prasiniferum prasiniferum* (Anand, 2008); and *Oxya hyla hyla* (Das *et al.*, 2012; Ghosh *et al.*, 2014, 2016).

7.3 Biology

Grasshoppers are usually found in shrubs, herbs, all sorts of vegetation, and agricultural fields. Some show phototaxism and/or diapauses. They are almost cosmopolitan in distribution except in some parts of the polar regions and some oceanic islands. They have slender, elongated body and exhibit perfect bilateral symmetry. The length of these insects ranges from 15-70 mm. Their colour may be yellow, yellowish-green, leafy green, brown or grey (Kirby, 1914).

Grasshoppers are mainly phytophagous and depend largely on vegetation for food. Some show cannibalism. They pass through 4 to 7 nymph stages before reaching the adult stages. Total number of eggs laid by a single female may be as high as 299 (Bhowmik, 1986). The warmer months are most favourable for population growth. Grasshoppers also act as prey, notably for birds and small mammals and considered as a key taxon in the food chain. Some grasshoppers are proposed as ecological indicators of ecosystem health (Latchininsky *et al.*, 2011; Saha and Haldar, 2009).

7.4 Rearing conditions

Ali (1982) observed that with the increase in temperature from 20 to 37 °C the egg-laying capabilities increased for *A. exaltata*. Numerous biotic and abiotic factors (mainly temperature, precipitation, and food quality) affect the fecundity and the survival of these insects (Joern and Gaines, 1990). *A. exaltata* shows a higher growth rate at the third instar at 23 to 35 °C (Nath and Haldar, 1993). Temperature is a very important factor when rearing grasshoppers (Hinks and Erlandson, 1994). The reproductive potential in grasshoppers is highly dependent on temperature (Blanford and Thomas, 2000; Whitman, 1988). Ganguly *et al.* (2010) reported that a temperature of 35±2 °C and a photoperiod of 12:12 L:D are the optimum values for mass rearing *O. hyla hyla*. Das *et al.* (2012) reared *O. hyla hyla* and *S. prasiniferum prasiniferum* under laboratory conditions and obtained the highest biomass production using a temperature of 30 to 35 °C, a relative humidity of 70 to 80%, a light intensity of 500±25 lux, and a photoperiod of 14:10 L:D. During their rearing it was also observed that if the specimen in cages gets direct sunlight, especially in the morning, their growth and development improves.

7.5 Diet

Most of the grasshoppers are gramnivores. McFarlane and Thorsteinson (1980) reported that mixed plant diets are superior for *M. bivittatus*, for promoting higher survival, larger adults and higher growth indices. Chapman and Joern (1990) provided detailed information on food selection of the grasshopper and remarked that depending upon the food availability/shortage, some acridids exhibit polyphagous behaviours. For *A. exaltata*, Bermuda grass *Cynodon dactylon* is a better diet for nymphs than for adults (Nath and Haldar, 1992). The nutritional requirements of grasshoppers influence dietary selection (Behmer and Joern, 1993). Fanny *et al.* (1999) found that Guinea grass *Panicum maximum* gives the shortest nymphal period of *O. nitidula*. Das *et al.* (2012) reported that *C. dactylon, Dactyloctenium aegyptium*, Egyptian crowfoot grass *D. aegyptium* (family Poaceae) and California grass *Brachiaria mutica* are most suitable foods for *O. hyla hyla* to produce a higher annual biomass. The grasshopper species *Phlaeoba panteli* can live on dry grasses in insectary condition. Feeding on sorghum (*Sorghum halepense*) gave the shortest nymph duration for *O. fuscovittata* and *S. prasiniferum prasiniferum* but wheat (*Triticum aestivum*) gave the shortest result for the grasshopper species *Oedaleus abruptus* (Ganguly *et al.*, 2013). Some grasshoppers prefer grasses from Poaceae and Cyperaceae family (Das, 2014). Le Gall and Behmer (2014) found significant effects of protein and carbohydrate on mass gain and development time in *M. differentialis*. Grasshoppers can also be reared on semi-synthetic diet (Sharaby *et al.*, 2010; Uvarov, 1966).

7.6 Rearing equipment

For farming of grasshoppers, the following equipments are essential: (1) sweep net of conventional design (to collect grasshoppers); (2) rearing cages of various sizes with nylon net gauge on a wooden or metal frame (wood is better for grasshoppers because it does not conduct temperature); (3) glass/plastic cups (200 ml/300 ml) (for hatching eggs); (4) enamel tray (10×10×4.5 cm^3 for collecting egg pods along with eggs); (5) electric bulb (source of light); (6) sieve (for egg pod collection from the sand); (7) water filled conical flask (300 ml) to keep the grass blade (food) fresh and also to maintain relative humidity in cages; (8) transparent plastic jars (25×10 cm diameter) (to rear nymphs up to third instar stage); (9) environmental/rearing compartment (to control the temperature, humidity, light intensity and air circulation); (10) dial hygrometer (for measurement of relative humidity in rearing cages/jars/chambers); (11) lux meter (for measurement of light intensity in the environmental rearing compartment); (12) cold room (for storing egg pods in adverse condition); (13) hot air oven (to sterilize the sand and to dry the dead grasshoppers); and (14) blow torch (to disinfect medium and large size equipment).

7.7 Reproduction unit

During the initial stage of grasshopper farming, the desired species are collected from nearby agricultural fields or grasslands with a sweep net and transferred to rearing cages. Shoots of fresh grasses dipped in water filled conical flask (300 ml) are used to provide food in the rearing units (Figure 7.1). The water in the conical flask not only keeps the food fresh but also helps in maintaining humidity in the rearing units. After acclimatization in the insectaria condition, mature grasshoppers start copulation (Figure 7.2). Copulation takes place from

Figure 7.1. Supply of fresh food in rearing unit.

Figure 7.2. Copulating grasshoppers inside cage.

few minutes to more than three hours. In the cages, after mating, females oviposit egg pods inside the sterilised sand of the small enamel trays or glass or plastic cups. The females lay their eggs in egg-pods at a depth of 3 to 5 cm. inside the sand, depending on the size of the species (Uvarov, 1966). Multivoltine species are suitable for farming as the univoltine species have a longer diapause stage. Multivoltine species yield higher biomass in comparison to the univoltine species (Anand, 2008).

7.8 Production unit

For further rearing, the egg pods in the enamel trays or glass or plastic cups are to be collected by sieving; it prevents the eggs from being damaged (Figure 7.3). Then the egg pods need to be transferred to another set of glass or plastic cups (200 ml). In the cups, egg pods should be placed at a depth of 2.5-5 cm in the sand. Egg-pods are kept 3-5 cm apart depending on the species. The mouths of the cups are to be covered with fine nylon net, using rubber bands. Covered nylon net of cups is to be kept free from the sand surface for easy movement of the newly hatched nymphs. More than one egg pods can be placed in one cup, based on the pod size. Slifer (1948) described the method for breaking diapause in the egg of *M. differentialis*. Regularly, a few drops of water are to be added in the cups to keep the sand moist. Cups with egg pods in sand are to be placed in the environmental compartment (Figure 7.4). The alternate moistening and drying of sand is essential for proper hatching of the eggs. Regular observation is required to notice the hatching inside the cups. Newly hatched nymphs (i.e. first instars hatchlings) need to be transferred immediately to a nymph rearing jar (Figure 7.5 and 7.6). For this, covered nylon nets of cups are to be unwrapped only after transferring the cups in the nymph rearing jars. A fine nylon net is spread over previous day food (grass leaves) along with the nymphs inside the rearing jar. Fresh grass leaves are placed over it. Nymphs move from old leaves to the fresh food through the meshes. Now they can be easily

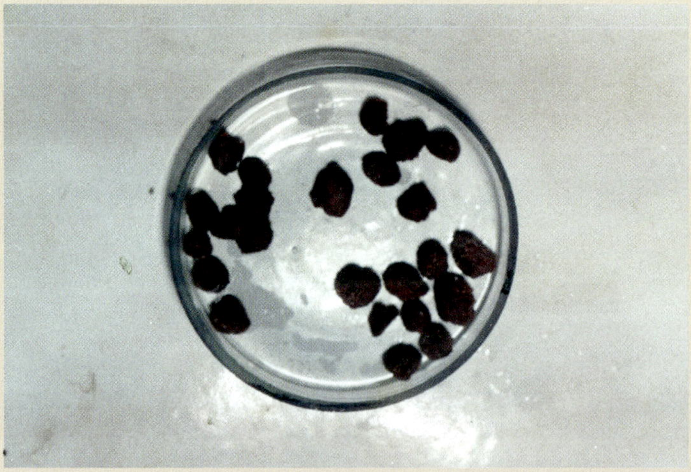

Figure 7.3. Egg pods collected in Petri dish by sieving.

Figure 7.4. Egg pods in cups for hatching inside environmental compartment.

Figure 7.5. Hatchlings are coming outside, above sand surface, by making tunnels.

transferred along with the nylon net from the old jar to the new jar without any injury of the nymphs. Hatchlings are reared up to third instars stage in the rearing jar. More developed nymphs and adult grasshoppers can be transferred gently by holding them on either side of the thorax from above with the fingers. A regular supply of fresh grass leaf as food needs

Figure 7.6. Hatchlings of *Oxya fucovittata* along with ensheathing membrane.

to be provided. From fourth instars stage the individuals are to be shifted in adult rearing compartment for further development (Das, 2014) (Figure 7.7). After five days of the final moult, the insects are harvested and at least 10-15 pairs are carefully transferred to the adult rearing compartment for their mating and oviposition. A sex-ratio of 3:1 is ideal for mass culture. This adult stock serves as a colony for future generations (Das, 2014). No effects of

Figure 7.7. Advance nymphs in right and adults in left rearing compartment.

inbreeding have been observed during various experiments on rearing of several species of grasshoppers in our laboratory during the last 30 years.

7.9 Hygiene and sanitation

Experiments on rearing of several grasshopper species (*A. exaltata, A. thalassinus tamulus, Eyprepocnemis resea, G. punctifrons, Heteracris pulcher, H. nigrorepletus, O. abruptus, O. fuscovittata, O. hyla hyla, O. velox, S. prasiniferum prasiniferum, P. panteli, Trilophidia annulata*) in our laboratory revealed that they are very resistant to infection. Fungal outbreaks occurred only in a few instances. A mixture of 2% formalin and 0.05% detergent is effective in disinfecting the cages and preventing fungal outbreaks. It is best to keep the humidity low and the enclosure needs to be dry rather than wet. Sanitation in the cages is achieved by regular cleaning of uneaten food and faecal matter and maintenance of proper humidity. Application of lime dust in combination with bleaching powder in and around rearing houses and premises improve hygiene in the environment. Clearing of wooden, plastic and polystyrene equipment with washing soda and metallic equipment with a blow torch give satisfactory disinfection results.

7.10 Processing

The freshly harvested grasshoppers (live) can be used for poultry feed. Whole dried grasshoppers can be used as animal feed or for human food. They can also be made into powder for further processing. Grasshopper powder can be utilized to obtain the following products: (1) grasshopper powder itself (for preparation of animal feed); (2) lipids; (3) protein isolates/concentrates; and/or (4) chitin. Commercially drying and grinding can be achieved in more efficient and cost effective ways to ensure minimum damage to nutritional components. The powder is stored in sealed aluminium foil bag at 4 °C for use in future (Das, 2012).

The grasshopper powder can be defatted using a screw press, centrifugation, or solvent extraction (Tzompa-Sosa *et al.*, 2014). The defatted powder can be used to prepare protein isolates and concentrates. Chitin can be removed using acid caustic treatments as described by Finke (2007). Alternatively grasshopper powder (non-defatted) can be used to produce oxyameal, as it is blended with many other ingredients effectively, without compromising the texture or structural integrity of the product. A study (Anand *et al.*, 2008) on nutritive potential of different acridid species, especially on *Oxya* spp. revealed that acridids are superior to the conventional protein supplements, such as soymeal, and fishmeal due to the large amount of calories (4.5-7 kcal/gm), high protein content (60-66%), crude fat (6-7.5%), carbohydrates (3.6-7.5%) and minerals. The production cost of oxyameal (a product of Oxya-farm) is cheaper than fishmeal in the Indian market. Protein extracted from grasshopper may be used to manufacture high protein tablets for treatment of protein deficiency in humans (Anand, 2008). Acridid faecal matter is rich in nitrogen content. So, the production of the large amount of faecal matter during rearing of grasshoppers could be utilised as bio fertilizer (Das *et al.*, 2010).

Grasshoppers are not only a rich source of protein, but some species also contain up to 22% lipids on a dry weight basis (Paul *et al.*, 2016b; Rumpold and Schlüter 2013). Some grasshoppers contain omega-3 fatty acids (Das and Mandal, 2013; Paul *et al.*, 2017; Yang *et al.*, 2006). Consumption of omega-3 fatty acids is linked with various health promoting effects in the human body, such as obesity, metabolic syndrome, and cardiovascular diseases (Lorenta-Cebrián *et al.*, 2013). Therefore the lipids can be successfully harvested as one of the important by-products from grasshopper farming (Paul *et al.*, 2017). Lipid extracted from acridid tissue may be used to produce biodiesel. An attempt was made to produce biodiesel from the lipid extracted from *Oxya fuscovitata* in the Department of Zoology, Visva Bharati University, in collaboration with Professor Mihir Kumar Purkait, Indian Institute of Technology, Gauhati, India, in the year 2010. The lipid was extracted, filtered and dehumidified. Thereafter, trans-esterification was performed to get biodiesel. Various fuel properties of produced biodiesel (like calorific value, cetane number, pour point, flash point, density and kinematic viscosity) were determined and compared with the commercially available diesel. Both the quantity and the quality of biodiesel were satisfactorily obtained from this species.

7.11 Conclusions

Grasshoppers are a potential source for food and feed since they contain a reasonable level of protein, essential amino acids, fat especially unsaturated fatty acids, vitamins, micronutrients, minerals and anti-nutrients. Farming of suitable grasshopper species with modern techniques would increase their availability and subsequent commercial value. It would also decrease the pressure exerted on the conventional sources of proteins from livestock industries.

References

Ali, S., 1982. Effect of temperature and humidity on the development and fertility-fecundity of *Acrida exaltata*. Proceedings of the Indian Academy of Sciences Animal Sciences 3: 267-273.

Anand, H., 2008. Evaluation of the feed value of acridids and its utilization in formulating economic poultry feed. PhD-thesis, Visva Bharati University, Bolpur, India.

Anand, H., Das, S., Ganguly, A. and Haldar, P., 2008. Biomass production of Acridids as possible animal feed supplement. Journal of Environment and Sociobiology 5: 181-190.

Ananthakrishnan, T.N., Dhileepan, K. and Padmanabhan, B., 1985. Behavioural responses in terms of feeding and reproduction in some grasshoppers (Orthoptera: Insecta). Proceedings of the Indian Academy of Sciences Animal Sciences 94: 443-461.

Anonymous, 1988. Thailand and Nepal. Insect Food Newsletter 1: 5.

Anonymous, 1992. Sky prawns and other dishes: the food and future. Expat World 4: 1.

Blanford, S. and Thomas, M.B., 2000. Thermal behaviour of two acridid species: effects of habitat and season on body temperature and the potential impact on biocontrol with pathogens. Environmental Entomology 29: 1060-1069.

Barnes, O.L., 1963. Food-plant tests with the differential grasshopper. Journal of Economic Entomology 56: 396-399.

Bailey, C.G. and Mukerjee, M.K., 1976. Consumption and utilization of various host plants by *Melanoplus bivittatus* (Say) and *M. femurrubrum* (DeGeer) (Orthoptera: Acrididae). Canadian Journal of Zoology 54: 1044-1050.

Behmer, S. and Joern, A., 1993. Diet choice by a grass-feeding grasshopper based on the need for a limiting nutrient. Functional Ecology 7: 522-527.

Bhowmik, H.K., 1986. Grasshopper fauna of West Bengal. Zoological Survey of India, Calcutta, India, 180 pp.

Chakravorty, J., Ghosh, S., Jung, C. and Meyer-Rochow, V.B., 2014. Nutritional composition of *Chondacris rosea* and *Brachytrupes orientalis*. Journal of Asia-Pacific Entomology 17: 407-415.

Cheke, R.A., 1990. A migrant pest in the Sahel – the Senegalese grasshopper, *Oedaleus senegalensis*. Philosophical Transactions of the Royal Society B 328: 539-553.

Chapman, R.F. and Joern, A., 1990. Biology of grasshoppers. John Wiley and Sons, Hoboken, NJ, USA.

Das, A., 2002. Strategies to yield high biomass in Acridid: Orthoptera. PhD thesis, Visva Bhatati University, Bolpur, India.

Das, A., 2014. Insects used as food with emphasis to acridid culture, biomass production and scope of application. International Journal of Zoology Research 4: 82-87.

Das, A., Das, S. and Haldar, P. 2002. Effect of food plants on the growth rate and survivability of *Hieroglyphus banian* (Fabricius) (Orthoptera: Acridoidea), a major paddy pest in India. Applied Entomology and Zoology 37: 207-212.

Das, M., 2012. Acridid resource management for poultry development. PhD thesis, Visva Bharati University, Bolpur, India.

Das, M., Ganguly, A. and Haldar, P., 2010. Nutrient analysis of grasshopper manure for soil fertility enhancement. American-Eurasian Journal of Agricultural and Environmental Sciences 7: 671-675.

Das, M., Ganguly, A. and Haldar, P., 2012. Determination of optimum temperature and photoperiod for mass production of *Oxya hyla hyla* (Serville). Turkish Journal of Zoology 36: 329-339.

Das, M. and Mandal, S., 2013. Assessment of nutritional quality and anti-nutrient composition of two edible grasshoppers (Orthoptera: Acrididae) – a search for new food alternative. International Journal of Pharmacy and Pharmaceutical Sciences 3: 31-48.

DeFoliart, G.R., 1989. The human use of insects as food and as animal feed. Bulletin of the Entomological Society of America 35: 22-35.

Dossey, A.T., Morales-Ramos, J.A. and Rojas, M.G., 2016. Insects as sustainable food ingredients: production, processing and food applications. Academic press, Amsterdam, the Netherlands.

Finke, M.D., 2007. An estimate of chitin in raw whole insects. Zoo Biology 26: 105-115.

Fanny, P., Ravikumar, T., Muralirangan, M.C. and Sanjayan, K.P., 1999. Influence of host plants on the duration of post-embryonic development and food utilisation of *Oxya nitidula* (Walker) (Orthoptera: Acrididae). Journal of Orthoptera Research 8: 119-173.

Ganguly, A., Chakravorty, R., Sarkar, A. and Haldar, P., 2010. Johnson grass (*Sorghum halepense* (L.) Pers.): a potential food plant for attaining higher grasshopper biomass in acridid farms. Philippine Agricultural Scientist 93: 329-336.

Ganguly, A., Chakravorty, R., Das, M., Gupta, M., Mandal, D.K. and Haldar, P., 2013. A preliminary study on the estimation of nutrients and anti-nutrients in *Oedaleus abruptus* (Thunberg) (Orthoptera: Acrididae). International Journal of Nutrition and Metabolism 5: 60-65.

Ganguly, A., Chakravorty, R., Sarkar, A., Mandal, D.K., Haldar, P. and Ramos-Elorduy, J., 2014. A preliminary study on *Oxya fuscovittata* (Marschall) as an alternative nutrient supplement in the diets of *Poecilia sphenops* (Valenciennes). PLoS ONE 91: 111-848.

Ghosh, S., Haldar, P. and Mandal, D.K., 2014. Suitable food plants for mass rearing of the short-horn grasshopper *Oxya hyla hyla* (Orthoptera: Acrididae). European Journal of Entomology 111: 448-452.

Ghosh, S., Haldar, P. and Mandal, D.K., 2016. Evaluation of nutrient quality of a short horned grasshopper, *Oxya hyla hyla* Serville (Orthoptera: Acrididae) in search of new protein source. Journal of Entomology and Zoology Studies 4: 193-197.

Haldar, P. and Nandi, N.C.,1997. Feasibility of grasshopper farming in Bengal: a review on laboratory based study. Journal of Interacademicia 1: 160-161.

Henry, J.R., 1985. *Melanoplus spp.* In: Singh, P. and Moore, R.F. (eds.) Handbook of insect rearing. Elsevier Science Publisher, Amsterdam, the Netherlands, pp. 451-464.

Hinks, C.F. and Erlandson, M.A., 1994. Rearing grasshoppers and locusts: review, rationale and update. Journal of Orthoptera Research 3: 1-10.

Hinks, C.F. and Olfert, O., 1999. Growth and survival of early-instar grasshoppers on selected perennial grasses, with observations on plant constituents which may influence performance. Journal of Orthoptera Research 8: 237-242.

Jago, N.D., 1963. Some observations on the life cycle of *Eyprepocnemis plorans meridionalis* Uv. 1921, with a key for separation of nymphs at any instars. Proceedings of the Royal Entomological Society of London 38: 113-124.

Joern, A. and Gaines, S.B., 1990. Population dynamics and regulation in grasshoppers. In: Chapman, R.F. and Joern, A. (eds.) Biology of grasshoppers. John Wiley & Sons, New York, NY, USA, pp. 415-428.

Katiyar, K.N., 1955. The life history and ecology of the northern spotted grasshopper, *Aularches punctatus* Drury (Orthoptera: Acrididae). Agra University Journal of Research, Science 4: 395-414.

Katiyar, K.N., 1956. The life history and ecology of the short horned grasshopper, *Parahieroglyphus bilineatus* Bolivar (Orth.: Acrididae). Agra University Journal of Research, Science 5: 179-192.

Kelly-Stebbings, A. and Hewitt, G.M., 1972. The laboratory breeding of British Gomphocerine grasshoppers (Acrididae: Orthoptera). Acrida 1: 233-245.

Kirby, W.F., 1914. The fauna of British India, including Ceylon and Burma. Orthoptera (Acridiidae*)*. Taylor and Francis, London, UK, 276 pp.

Kohler, G., Brodhun, H.P. and Schaller, G., 1987. Ecological energetics of central-European grasshoppers (Orthoptera, Acrididae). Oecologia 74: 112-121.

Kok, R., 1983. The production of insects for human food. Canadian Institute of Food Science and Technology Journal 16: 5-18.

Latchininsky, A., Sword, G.A., Sergeev, M.G., Cigliano, M. and Lecoq, M., 2011. Locusts and grasshoppers: behavior, ecology and biogeography. Psyche: 578327.

Le Gall, M. and Behmer, S.T., 2014. Effects of protein and carbohydrate on an insect herbivore: the vista from a fitness landscape. Integrative and Comparative Biology 54: 1-13.

Lorente-Cebrián, S., Costa, A.G.V., Navas-Carretero, S., Zabala, M., Martínez, J.A., Moreno-Aliaga, M.J., 2013. Role of omega-3 fatty acids in obesity, metabolic syndrome, and cardiovascular diseases: a review of the evidence. Journal of Physiology and Biochemistry 69: 633-651.

Luong-Skovman, M.H., 2001. Some behavioral and biological traits under rearing conditions of *Bryophyma debilis* (Karsch) (Orthoptera, Acrididae, Cyrtacanthacridinae), a grasshopper from Burkina Faso. Journal of Orthoptera Research 10: 165-169.

Malakar, C., Ganguly, A. and Haldar, P., 2009. Influence of cadmium on growth, survival and clutch size of a common indian short horned grasshopper, *Oxya fuscovittata.* American-Eurasian Journal of Toxicological Sciences 1: 32-36.

McFarlane, J.H. and Thorsteinson, A.J., 1980. Development survival of the two striped grasshopper *Melanoplus bivittatus* (Say) (Orthoptera: Acrididae) on various single and multiple plant diets. Acrida 9: 63-76.

Muralirangan, M.C. and Ananthakrishnan, T.N., 1977. Biological studies on *Eyprepocnemis alacris alacris* (Serv.) (Orthoptera: Acrididae) growth in terms of weight, length and eye stripe number. Entomon 2: 193-200.

Nath, S. and Haldar, P., 1992. Studies on the growth rate of *Acrida exaltata* (Walker) (Orthoptera: Acrididae). Proceedings of the Zoological Society of India 45A: 389-393.

Nath, S. and Haldar, P., 1993. Effects of food on the reproductive potential of a common Indian grasshopper. Environment and Ecology (Kalyani) 11: 450-452.

Nayar, J.K., 1964. The nutritional requirements of grasshoppers: I. Rearing of the grasshopper, *Melanoplus Bivittatus* (Say), on a completely defined synthetic diet and some effects of different concentrations of B-vitamin mixture, linoleic acid, and β-carotene. Canadian Journal of Zoology 42: 11-22.

Parker, J.R., 1930. Some effects of temperature and moisture upon *Melanoplus mexicanus* Saussure and *Oammula pellucida* Scudder (Orth.). Bulletin Montana Agricultural Experiment Station 223: 1.

Paul, A., Frederich, M., Uyttenbroeck, R., Hatt, S., Malik, P., Lebecque, S., Hamaidia, M., Miazek, K., Goffin, D., Willems, L., Deleu, M., Fauconnier, M., Richel, A., Pauw, E., Blecker, C., Monty, A., Francis, F., Haubruge, E. and Danthine, S., 2016a. Grasshopper as a food source? A review. Biotechnologie, Agronomie, Société et Environnement 20: 337-352.

Paul, A., Frederich, M., Uyttenbroeck, R., Malik, P., Filocco, S., Richel, A., Heuskin, S., Alabi, T., Megido, R.C., Franck, T. and Bindelle, J., 2016b. Nutritional composition and rearing potential of the meadow grasshopper (*Chorthippus parallelus* Zetterstedt). Journal of Asia-Pacific Entomology 19: 1111-1116.

Paul, A., Frederich, M., Megido, R.C., Alabi, T., Malik, P., Uyttenbroeck, R., Francis, F., Blecker, C., Haubruge, E., Lognay, G. and Danthine, S., 2017. Insect fatty acids: a comparison of lipids from three Orthopterans and *Tenebrio molitor* L. larvae. Journal of Asia-Pacific Entomology 20: 337-340.

Pfadt, R.E., 1949. Range grasshoppers as an economic factor in the production of livestock. Wyoming Range Management 7: 1-7.

Ramos-Elorduy, J., Medeiros Costa Neto, E., Pino, J.M. and Zetina, D.H., 2007. Knowledge about useful entomofauna in the county of La Purisima Palmar de Bravo, Puebla State, Mexico. Biotemas 20: 121-134.

Ramos-Elorduy Blasquez, J., Pino Moreno, J.M. and Martinez Camacho, V.H., 2012. Could grasshoppers be a nutritive meal? Food and Nutrition Sciences 3: 164-175.

Riffat, S. and Wagan, M.S., 2007. Life history and economic importance of *Hieroglyphus nigrorepletus* Bolivar (Hemiacridinae:Acrididae: Orthoptera) from Pakistan. International Journal of Entomology 4: 379-386.

Rumpold, B.A. and Schlüter, O.K., 2013. Nutritional composition and safety aspects of edible insects. Molecular Nutrition and Food Research 57: 802-823.

Saha, H.K. and Haldar, P., 2009. Acridids as indicators of disturbance in dry deciduous forest of West Bengal in India. Biodiversity and Conservation 18: 2343-2350.

Sanjayan, K.P. and Murugan, K., 1987. Nutritional influence on the growth and reproduction in two species of acridids (Orthoptera: Insecta). Proceedings of the Indian Academy of Sciences Animal Sciences 96: 229-237.

Sharaby, A., Montaser, S.A., Mahmoud, Y.A. and Ibrahim, S.A., 2010. The possibility of rearing the grasshopper *Heteracris littoralis* (R.) on semisynthetic diets. Journal of Agriculture and Food Technology 1: 1-7.

Slifer, E.H., 1948. A simplified procedure for breaking diapause in grasshopper eggs. Science 107: 152.

Taylor, R.L., 1975. Butterflies in my stomach: insects in human nutrition. Santa Barbara, USA, Woodbridge Press Publishing Company.

Tzompa-Sosa, D.A., Yi, L., Van Valenberg, H.J.F., Van Boekel, M.A.J.S. and Lakemond, C.M.M., 2014. Insect lipid profile: aqueous versus organic solvent-based extraction methods. Food Research International 62: 1087-1094.

Uvarov, B., 1966. Grasshoppers and locusts. Cambridge University Press, London, UK, pp. 85.

Van Huis, A., Van Itterbeeck, J., Klunder, H., Mertens, E., Halloran, A., Muir, G. and Vantomme, P., 2013. Edible insects: future prospects for food and feed security. Food and Agriculture Organization of the United Nations (FAO), Rome, Italy, FAO Forestry Paper no. 171, 187 pp. Available at: http://www.fao.org/docrep/018/i3253e/i3253e.pdf.

Vedham, K., 1994. Bioecology and energetics of *Diabolocatantops pinguis* (Walker) (Orthoptera: Acrididae). PhD thesis, University of Madras, Tamil Nadu, India.

Whitman, D.W., 1988. The function and evolution of thermoregulation in the grasshopper *Taeniopoda eques*. Journal of Animal Ecology 57: 369-383.

Yang, L.F. and Siriamornpun, S. and Li, D., 2006. Polyunsaturated fatty acid content of edible insects in Thailand. Journal of Food Lipids 3: 277-285.

Production technology and management

8.

Alphitobus diaperinus

Insect production and facility design

R. Kok
McGill University, Department of Bioresource Engineering, Macdonald Campus, 21111 Lakeshore, Ste-Anne-de-Bellevue, QC H9X 3V9, Canada; robert.kok@mcgill.ca

Abstract

The industrial farming of insects has the potential to significantly impact a number of issues related to food and feed production, organic waste treatment, the conversion of low-grade materials to more valuable ones, the production of new antibiotics and other interesting biochemicals, etc. 'Entotechnology' is the term used to refer to the extensive array of techniques and methods used and being developed by this rapidly evolving industry. The intent in this chapter is to familiarize the reader with some aspects of entotechnology and to help him/her address the following questions: 'What should I think about before I decide to start farming insects?', 'Where and how should I start?' and 'How should I proceed?'. The following subjects are dealt with: organisms and feeds, process types, reactor types, environmental requirements, heat and mass transfer, materials handling, process control, air conditioning, waste management, cleaning, strain breeding, culture maintenance, population synchronicity, and safety. This chapter is oriented to the mass production of insects by industrial methods, meaning that a plant will minimally have the capacity to output 6,000 tonnes live weight per year, which is equivalent to one truck load per day. World-scale plants are expected to have capacities of one million tonnes or more per year of live weight.

Keywords: entotechnology, industrial, bio-conversion

8.1 Introduction

The overall objective of mass producing insects is to convert low-cost inputs into high-value outputs on an industrial scale in a manner that is both profitable and environmentally benign. In this context 'mass production' refers to industrial plants which will minimally have the capacity to output 6,000 tonnes live weight per year, which is equivalent to one truck load per day. World-scale plants are expected to have capacities of one million tonnes or more per year of live weight (>30 kg/s). Typically, the bulk of the inputs will be organic by-products from agri-based industries and the target outputs will be protein, fat, chitin, and other materials isolated from the insects. In some cases the objective may simply be to reduce the volume and mass of a feed material that has negative value, i.e. as in waste treatment.

The intent in this chapter is to discuss the major aspects of farming insects so as to give the reader an overall impression of what is required to start up an insect production facility and what major issues need to be considered in its conception, design, construction, and operation. Although the focus of the chapter is large-scale production as defined above, many of the issues discussed are also relevant to smaller-scale facilities which are operated to produce relatively modest quantities of insects for biological control, specialty goods, medical and pharmaceutical uses, etc. However, the financial aspects of operating a smaller-scale facility will usually be quite different from industrial-scale production. This is largely due to the market prices of the outputs, which will be much higher for organisms produced in relatively small amounts for very specific applications. Mass production, in order to be competitive with the more conventional outputs of intensive animal rearing, will have to generate insect live weight at a cost of about 1.00 $US/kg (2017). This is comparable to prices for low-quality beef, hog and chicken which are, as of this writing (April 2017, North America), approximately 1.25, 1.00 and 1.13 $US/kg live weight, respectively.

In this chapter only the production (farming) of insects is dealt with; the subsequent conversion of the insects into protein isolate, fat, chitin and other bio-materials is not discussed. Due to space limitations this is a general discussion only; no detailed calculations or design directives are presented. The objective here is more to help the reader formulate answers to the questions: 'What should I think about before I decide to start producing insects?', 'Where should I start?' and 'How should I proceed?'

8.2 General considerations

The farming of insects is, in essence, no different from the production of other animals in confinement. Cows, goats, chickens, rabbits, mink, ostriches, salmon, horses, etc. all require housing, feed, water, and oxygen while their waste products must be removed and environmental conditions must be maintained so that they can grow, reproduce, and end up in a final state that is useful to the farmer. Ultimately, they must be slaughtered and dissembled into components and outputs that have commercial value. If a production process is well thought out and well designed, there will be a number of valuable outputs resulting and a minimum of material that has negative value and thereby presents a disposal problem. In farming insects the main opportunities lie in their various abilities to break down and digest unusual feedstocks and to require less inputs than competing animals or other organisms to achieve similar biochemical conversions. They may also be able to produce outputs that are difficult to obtain from other organisms or via synthetic pathways. Overall, as for any type of farming, the viability of insect production depends on the financial performance of the enterprise which may, in some cases, be affected by the capacity of the species of interest to thrive in very high population densities and at relative low water activity. The latter factor is a thermodynamic measure related to the availability of water, with some insects being able to function at fairly low water activities, thus limiting the number and variety of organisms that are able to compete for their feed. One significant aspect of insect farming is that, until now, animal rights legislation is not applicable in most jurisdictions. Commensurate with this, there has been very little societal discussion on the acceptability of mass producing and slaughtering of insects. A drawback is that insects are still regarded with suspicion in occidental countries and are only allowed to be used as animal feed or human food under limited circumstances.

One way in which insects are quite different from other farmed animals is the number of species available. In nature several million species of insects exist and at least several thousand are known to be edible by humans. Undoubtedly, many more species will be found to be edible, or potentially edible through breeding, so that it should be possible to closely match a species/strain with any specific target feed.

For example, if the proposed feed is chicken manure or a mixture of manures, then one would select all known insect species that can live on this type of feedstock (either in raw or modified form) and investigate the suitability and potential of each. This might result in the selection of several species of flies or manure beetles (e.g. *Hermetia illucens*, *Oryctes boas*, etc.) which would then each need to be investigated in more detail. On the other hand, if the proposed feed is mostly cellulosic in nature (forestry waste, corn stalks, straw, etc., perhaps augmented with nitrogen and nutraceuticals) a different group of organisms (such as termites, drugstore beetles) are likely to be found suitable, whereas omnivorous species such as cockroaches might be found to be more appropriate for mining domestic garbage for nutrient content. Overall, for any proposed farming process the organism/feed pair is a central precept. As for any novel undertaking, before it can be established whether a specific organism/feed pair can be successfully (profitably) used in a farming operation, the entire production sequence for that specific pair must be examined. For this, minimally, the following is required: detailed entomological knowledge of the organism, its life cycle, pathology, performance on the target feed, tolerance of high population density, and the capacity to be bred into different strains. Characterization of the feed, especially its variability, and how that will influence the quality of the final product is also needed. Another important factor to be considered at the very beginning is the cost of the feed and how that is likely to be influenced by a major demand arising from its utilization. The choice of organism/feed combination will strongly influence the process type that will be selected, the materials handling methods, and the equipment such as the reactors and separators that can be used. In the end, the success of any proposed organism/feed pair will be influenced by many factors including political forces and market conditions. It is therefore difficult to foresee the success for any pair with any accuracy, but by means of careful initial design and consideration of available information the likelihood of business success can be maximized. It should be noted that in many cases adequate information may not be available for a proposed organism/feed pair (e.g. conversion rates and growth rates as functions of temperature and water activity, egg-laying rates, pupal weight loss during metamorphosis, variability between individuals, sensitivity to parasites, variability in the feed). Generally, It will be desirable to obtain such information through research before proceeding with further process development, and then identify further information needs as process development proceeds. Throughout the procedure of organism/ feed choice and subsequent process development it is prudent to continuously monitor the world patent literature to avoid unpleasant surprises in the future. A general diagram of a typical production scheme is presented in Figure 8.1.

8.3 Organism environmental requirements at various life stages

Besides the provision of an adequate food supply, a large number of other factors will affect the development, growth, reproduction and overall viability of an insect population. Hence, in order to make an insect farm (also referred to here by the industrial term 'plant') productive, a number of process variables need to be controlled for the process to remain stable and the population to grow, as well as for the feed to remain available to the target organism. Although it is always desirable to 'optimize' a process, generally this is not possible at first and often it is a sufficient challenge for the operator to simply start up the production process and keep it running stably for some time while learning about the intricacies of the particular organism/feed combination chosen. Some of the social insects (e.g. termites, ants) live as semi-integrated, non-unitary entities and establishing and maintaining a stable, productive colony

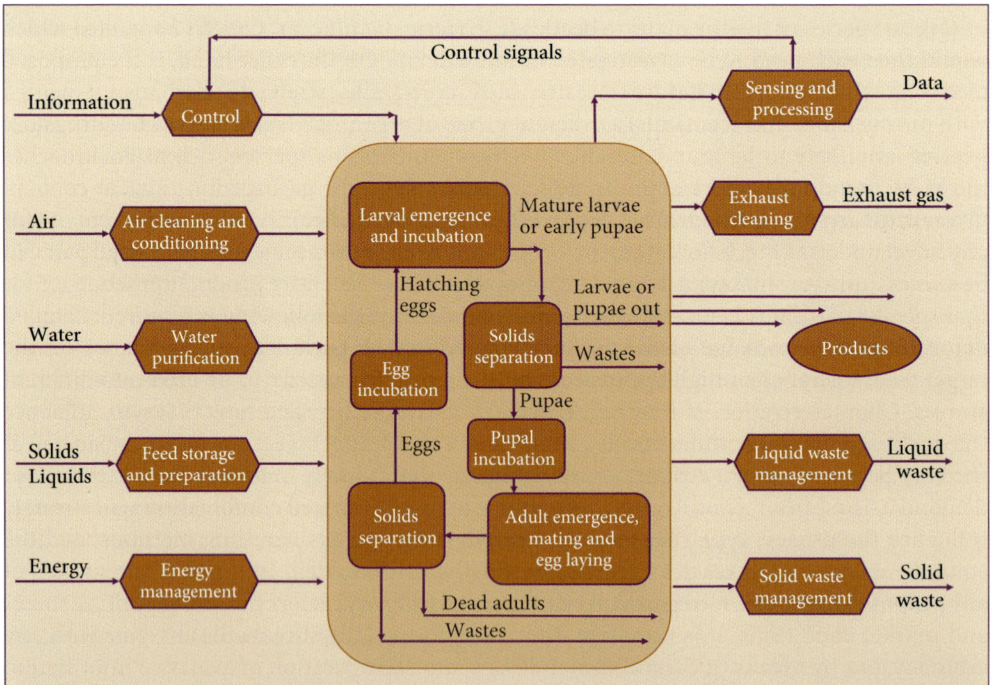

Figure 8.1. A general production scheme for holometabolous insects.

of them can be particularly challenging. But even the commonly-grown Black Soldier Fly (*H. illucens*) is influenced by social structure and this must always be taken into account in the design and operation of its environment. Thus, it is recommended that the potential insect farmer first learn about the target insect (and the chosen feed), and then establish a very small farm (up to 10 g live weight/day) to familiarize him/herself with the conditions that allow persistent and stable operation and production for some time before seriously considering moving on to larger scale production, process optimization, etc. One thing to keep in mind is that, because so much of process control theory and technology was developed in the fields of electrical, mechanical and chemical engineering, it is often assumed that there is a 'best' value for each variable and that it is 'best' to maintain that variable stably at that one value. Of course, this is usually the case if one is pulling purified silicon from a melt or welding two steel parts together. However, insects are not pure elements or compounds or metals; they are biological organisms that frequently have rather intractable qualities. Consequently, although environmental conditions and procedures may have been rigorously maintained at steady state, an insect population may undergo a sudden and dramatic change for no readily discernible reason, often to the consternation of the insect farmer. Unfortunately, it may be difficult to ascertain the cause(s) of such a phenomenon. Sometimes it may be ascribed to an infection or a chemical disturbance, but it may also be due to a less traceable factor such as 'colony fatigue' which can be brought on by growth conditions being kept the same for too long. Thus, during preliminary work with a small trial farm, if the process goes unstable, it may be fruitful to try varying the environmental conditions on a regular basis (perhaps in imitation of daily or seasonal temperature and light fluctuations) or to 'shock' the colony by applying a sudden, drastic change in one of the main variables. Overall, this work is best carried out before a major commitment is made to a particular organism/feed pair and production process.

One of the most influential process variables in insect farming, and often one of the easiest to control, is the temperature of the environment in which the animals are reared. As pointed out above, to successfully propagate an insect in captivity one needs to have detailed information available about that organism, and its response to temperature must be well known. Thus, the 'best temperatures' to maintain for egg incubation, the emergence of immatures and their growth, mating, egg laying, etc. should all be found out through research of the selected organism growing on the particular feed(s) of interest. Evidently, being the simpler approach, the literature should be consulted before any experimental work is undertaken. Once 'best values' have been determined it should also be taken into account that even better performance can, perhaps, be obtained by varying the temperature in accordance with a certain pattern during one or more life stages (e.g. pupation might be brought on by lowering the temperature near the end of the final larval stage and then emergence might be increased by increasing it at a certain point) and it may then be desirable to build that functionality into the production process, at the cost of increased structural and operational complexity. Temperature control is often based on the temperature as measured by sensors such as thermistors or thermocouples. But, in reality, that often means it is controlled only at the points at which these sensors are located, with very little assurance that the temperature is the same elsewhere. This is especially problematic when insects are present at high densities (kg of organism/m^3 reactor space) and are in an active stage, metabolizing food at a high rate and therefore giving off a lot of heat (W/m^3 reactor space). This metabolic heat needs to be removed by cooling, but it is not necessarily straightforward to ensure that such cooling occurs uniformly. Hence, unless the process and reactors are designed appropriately, some individuals may be subjected to quite a different time-temperature history than some others, so that they will diverge in their development (which is undesirable). Also, if cooling is extremely non-uniform, some of the organisms may die due to thermal exposure (e.g. this can happen to senior fly larvae that are highly concentrated in their feed and growing at a maximum rate). Temperature control and maintaining the temperature distribution uniform is not as problematic with the less active stages such as eggs and pupae. These issues are discussed in greater detail below.

The temperature variable was used here to illustrate three things: (1) the 'best value', or the most appropriate time-temperature history, for each life stage of the insects should be determined; (2) there should be a way to control that variable at that value, or force it through that particular history; and (3) the variable should be controlled as uniformly as possible for all individuals so as to minimize divergence in the population. The same rules hold for a substantial number of other variables, many of the important ones being related to the composition of the gas phase in the growth environment. The partial pressures of carbon dioxide, oxygen, water, various noxious gases such as carbon monoxide and nitrogen oxides, and any number of volatile organics may all be influenced by the insect itself, the feed material and how it ages, other organisms that are growing incidentally on the feed, as well as the fresh air that is drawn into the system. Thus, they are all dynamic process variables that may require continuous, active control. A large number of other variables are equally important to system and organism performance, but often are less dynamic. Intermediate in this respect are variables such as liquid water availability and radiation (visible light as well as IR and UV), sound, and mechanical vibration. Thus, these may all need to be controlled, and can perhaps be used to manipulate the organism, but not necessarily on a minute-to-minute basis (e.g. irradiation frequency, intensity and duration to influence breeding behaviour). A third class of environmental variables is even less dynamic and includes factors related to the physical conformation of the growth facility. Features such as wall height and smoothness to prevent larvae from climbing up, 3D spacing for flying or jumping adults, and surface texture for egg laying are typical of this class. Generally, physical conformation requirements are quite different for the various life stages of an insect. Obviously, the production facility should be constructed of materials that are not toxic to the organism, either in their original form or as any compounds derived from them through a degradation process (e.g. through galvanic corrosion of stainless steel).

Overall, each organism and each strain thereof will have specific response curves to a number of feed inputs such as the energy and protein source, water, and micronutrients like minerals and vitamins. It will also be sensitive to conditions such as high or low temperature, the presence of some chemical agents, a certain pH range, disease organisms, etc. And it will be tolerant of, or even stimulated by, some other organisms, as well as a host of other factors. The purpose of system design is to define a physical conformation and operating scheme that will satisfy both the input and environmental requirements of the specific organism. For some variables this will require active control. For good system performance, all relevant variables must be addressed simultaneously, the objective being to assure long term system stability while maximizing performance and profitability. As mentioned, to assure long-term stability, it may be necessary to vary some of the controlled variables on a short-term basis. To make system design and operating/control scheme specification possible the requirements, sensitivities, tolerances, and performance of the organism should obviously be known in detail. Thus, organism response curves to various conditions should be obtained before plant design is started. Due to the cost and effort involved, this is best done in the laboratory. An obvious extension of this activity is to improve the organism's performance, reduce its sensitivity, and widen its tolerance range, through a breeding program.

8.4 Heat and mass transfer considerations

In almost all industrial-scale insect production schemes the organism is confined in a cage or on a tray with walls of some type. Depending on the insect, it may live in or on a bed of substrate that consists either of a fairly dry particulate solid or a wet slurry (e.g. mealworms, fly larvae, silkworms), or near a source of substrate (e.g. crickets, house flies). Oxygen is then usually supplied with an air stream that is moved over or through the substrate bed or through the cage by mechanical means. (In this chapter the production of organisms in an aqueous phase, e.g. mosquito larvae, is not dealt with.) Simultaneously, the air stream is used to carry away part of the heat of metabolism as well as the carbon dioxide and other gaseous products such as carbon monoxide, nitrogen oxides, and various organics like ketones, aldehydes, steroids, etc. Chemical components may also be intentionally added to the air stream to inhibit disease organisms, reduce competing and contaminating species, or to manipulate the target organism in some way. Depending on the substrate material and the condition of the air supplied, water may be removed or added to the substrate by the air flowing over or through it. Overall, the cage or tray, the substrate, the growing insects and the moving air stream constitute a complex simultaneous heat and mass-transfer situation in which various forms of heat transfer as well as gaseous diffusion, natural convection and forced convection all play a role. Moreover, the substrate bed may be subject to considerable stirring and mixing as the organisms move about on it or through it. Especially when larvae are grown in high density this can be a very significant aspect of the heat and mass transfer; in some cases the larvae are so dynamic that the substrate-organism mix can take on the appearance of a slow-boiling, high-viscosity liquid, with substantial material turnover from the top to the bottom of the bed occurring. In most industrial processes that exist today the insects are grown in batches so that the substrate-organism mix changes in character as the organism develops and the heat and mass-transfer situation changes with it. Furthermore, many production processes are of the 'fed-batch' type (see below), further complicating the situation

Many insects develop in three or four distinct stages, depending on whether they undergo incomplete or complete metamorphosis. For instance, grasshoppers, crickets, termites, cockroaches, praying mantises, lice and true bugs develop through three stages: egg, nymph and adult whereas beetles, butterflies, flies, ants, bees and wasps develop through four: egg, larva, pupa and adult. For production purposes these developmental stages may be further differentiated into sub-stages such as several larval instar stages, a pre-pupal stage, a pupal stage, an early adult stage, a mature adult stage, etc. In an industrial setting these stages will often be dealt with separately since their requirements may be quite different. Keeping the stages well separated in the production facility will also inhibit the transmission of diseases. With respect to heat transfer considerations, the most significant aspect of the developmental sequence is that in some stages the organism is relative quiescent, whereas in others it is extremely dynamic, the metabolic rate (W/kg body mass) varying with the degree of activity. The most challenging heat transfer situation occurs when the organism metabolic rate is elevated while, at the same time, its density on or in the substrate is high.

The microbial degradation of the substrate can also make a major contribution to the total heat produced. To what extent this will occur depends on the nature of the substrate as well its preparation, the environmental conditions, and the process type. For example, if the substrate is an easily-digested organic material (e.g. manure) that is not well sterilized before use and present in a bed as a wet slurry, microbial activity may account for a substantial fraction of total heat production. In such cases a considerable portion of the substrate may be lost to microbial metabolism and much excess heat and carbon dioxide, etc. may be given off. On the other hand, if the substrate is more difficult to digest (e.g. cellulosic), is sterilized before use, and is present at relatively low water activity, microbes are much less likely to make a major contribution to heat production.

8.4.1 Heat transfer – eggs and pupae

Insect eggs may be incubated at a fairly high density (kg egg/m^3), often mixed with some cushioning material or finely-ground substrate for the emerging immatures to feed on. Some eggs can be incubated in warm water. Because their metabolic rate is fairly low, heat production by eggs is quite limited and it is not difficult to achieve accurate, uniform control of the temperature throughout the egg mass. Of course, because insect eggs are small, beds of egg-substrate mix tend to be quite small too, even in larger production facilities, and that also makes it simpler to regulate heat flow in and out of the bed and achieve good temperature control. The situation for pupal incubation is quite similar to that of eggs, except that pupae are larger than eggs so that the quantity of material to deal with in a given production process is larger. Like eggs, most pupae can also be incubated at a fairly high density (kg pupae/m^3) without it being complicated to achieve accurate, uniform control of the temperature throughout the mass. Some insects pass through a 'pre-pupal' stage during which they are not yet immobile although they have stopped their intense larval movement and feeding behaviour. At this point their metabolic rate is dropping but can still be significant and it is probably worthwhile to incubate the organism at this stage at an intermediate density and in a separate location to accommodate the specific combination of metabolic rate and the organism's need to move around or simply flex in place while completing its transition to pupa.

8.4.2 Heat transfer – larvae

As pointed out above, a major heat transfer challenge occurs when rapidly metabolizing (and growing) larvae are present at a high density in a bed of substrate, resulting in an elevated volumetric heat production rate (W/m^3 bed). This is especially significant for larvae in their final instar stages, which are close to their final weight, but still rapidly gaining weight and accumulating fat in preparation for pupation. At this stage, seeing that the larvae have been growing in the substrate for some time, microbial activity may also be making a significant contribution to the overall heat production regardless of what precautions were taken in terms of substrate pasteurization, etc. Thus, a substantial amount of heat will continuously need to be removed from the bed and several approaches are commonly employed for this, generally corresponding to the various mechanisms of heat transfer recognized. Accordingly, cooling of a larval bed can be achieved via convective, conductive and radiative heat transfer. Of these, convective cooling is the simplest to achieve and also the most commonly used, with both sensible and latent heat loss usually playing a part. The simplest arrangement is to force air to flow across the surface of the bed, during which the air will increase in temperature (sensible heat gain) and, as well, in moisture content (latent heat gain). Of course, this will result in the surface cooling more than the deeper part of the bed, tending to cause temperature stratification in the vertical direction (assuming the beds are in horizontal trays, the usual arrangement). At the same time, if the air flow is in one direction across the bed surface it will cool one end more than the other, leading to stratification in the horizontal direction. Such effects may, however, be mitigated to some degree by the bed contents being re-mixed in all directions through the dynamic activity of the larvae who may actually enhance this result by actively following a temperature gradient in the bed. Regardless of any temperature stratification inside the bed, under the circumstance described all the heat must be lost through the bed's surface, and the maximum allowable depth of the bed is limited by this. That maximum depth will be strongly affected, for a specific heat production rate (W/m^2 bed), by the minimum and maximum temperatures at which the larvae feel comfortable, the solids/slurry mixing the larvae are able to achieve, and the heat and mass transfer coefficients that can be attained at the bed surface. The latter will depend largely on surface roughness and air velocity across the surface of the bed. As mentioned above, this heat and mass-transfer situation is quite complex, even for the simple physical arrangement of air being blown across a flat tray, and it is not a straightforward exercise to design a system that will maintain reasonably uniform temperature and moisture distributions throughout the beds in the trays. The main conclusion here is that the beds should not be 'too thick' (deep). However, the shallower the bed, the more surface area (trays) will be needed for a given bed volume, with space between them all for air flow, and the more complicated the physical arrangement will become.

A different approach to cooling a larval bed, still based on convection and involving both sensible and latent heat loss, is to blow air through the bed rather than just over its surface. This can be achieved by, e.g. installing a porous bottom in the tray and feeding conditioned air via a plenum below it. A similar approach is to put porous tubes inside the bed and feed the air through those. Of course, either of these approaches or both can be combined with surface aeration. Through-bed aeration has the major advantage that the bed can be thicker than with surface aeration only, allowing for deeper but fewer trays, albeit at the cost of somewhat greater mechanical intricacy. And it offers the possibility of reduced temperature and humidity stratification in the bed. Of course, one can only pass so much air through the substrate-organism mixture before the bed and the larvae will become disturbed. With other words, the allowable superficial velocity of the air is quite limited (m^3 air/m^2 surface area.s or, simply, m/s), and this is especially so if the substrate is a slurry which may start to bubble up or foam when air is passed through it. The through-bed approach also has the disadvantage that the pressure drop across the bed can be significant, so that the energy cost for blowing the air through the bed will be larger than for removal of heat by means of surface aeration.

Yet another way to remove heat from a larval bed is via conduction. For example, finned tubes may be put into the bed and a coolant pumped through them. Or, the tray floor and/or walls may be made ribbed and hollow and thus function as cooling tubes. Or, if the tray is made of metal, cool air can be blown against the floor and walls on the outside. A major advantage of conductive cooling is that, since only sensible heat is removed, the mass transfer situation inside the bed is not much affected, except when the walls are cooled to such a degree that water will start to condense on them. Disadvantages are that the physical facility becomes more complicated, especially if the coolant is a liquid, and ribbed or finned structures that have been in extended contact with a larval bed could be rather difficult to clean. As well, because effective solid-solid heat transfer (finned tube to substrate/organism mix) and conductive heat transfer inside a particulate bed are difficult to achieve, temperature stratification inside the bed may occur more readily than when convective cooling is used. Although radiative cooling is possible it doesn't seem practical for the larval growth situation. In animal production radiative heat transfer is more often used for heating than for cooling and there may be a useful role for it in the incubation of eggs and pupae, as well as in environmental control for adults.

8.4.3 Heat transfer – adults

Most adult insects of present industrial interest are quite mobile. Some of the species walk, others jump, and yet others fly. Adult crickets are able to locomote by all three means, flies can both walk and fly with some species mating only in flight, and some beetles are mostly content to remain on a surface and walk although they may be capable of flight, if given the opportunity. The details of adult motility and behaviour depends a lot on the specifics of the species/strain. For instance, most silk moths have been bred to not fly anymore. The wide variety of motility behaviours of adult insects means that adult facilities must always be designed and operated specifically for the particular species/strain being produced. This is also due to adult insects often engaging in social behaviours, detailed mating rituals, and displaying particular egg-laying comportments. Thus, for flies a full 3D environment of minimal dimensions (cages) must be provided, with provisions for social behaviour, feeding and watering, as well as for egg laying. Crickets are quite similar to flies in their requirements of the dimensionality of their environment, whereas some other organisms such as roaches and termites are intermediate in this respect, requiring perhaps 2.5D. For many beetles a largely 2D environment (trays with a layer of feed on them, perhaps 2.1D) may suffice in which the adults crawl around on or in the bed of substrate and in which they lay their eggs. Although the adult insects may, individually, have a substantial metabolic rate (W/kg body mass), in no case are they so densely packed in their environment that a significant specific heat production rate will occur (W/m^3 facility or W/m^2 bed). Thus, in adult facilities the problem is usually not so much one of heat removal, but of adequate air conditioning. In many ways, this is similar to dealing with the quiescent or dormant stages like eggs and pupae. It is, however, a more complicated situation because the adults are motile and their behaviour can be strongly influenced by minor variations in temperature, both in space and time, local air velocity and gas phase composition. Both humidity and volatile organics in the air stream are of major significance to the latter aspect. As well, like other animals, they can be strongly influenced by radiation, including frequencies in the IR, UV and microwave ranges, with many organisms having an absolute requirement for some visible light. Generally, in adult facilities convection-mediated sensible heat exchange is the most important, with radiative transfer sometimes playing a role. Very often, latent heat exchange will also occur in such convective systems, closely linked to the mass transfer situation.

8.4.4 Mass transfer – eggs and pupae

Here, only convective gas flow and gas diffusion will be discussed; bulk mass flow and movement of solids will be dealt with under 'Materials handling'. As discussed above, at various stages of the insects' lives, the organisms may require either cooling or air conditioning, both often accomplished by means of blowing a stream of air over or through a bed or a cage of some type. This air stream will then carry away (or supply) sensible heat, carbon dioxide, organics, etc. But, frequently, it will also carry away latent heat in the form of gaseous water that has evaporated from the substrate bed or the organisms, and sometimes this will be unintentional. Thus, when dormant eggs or pupae are aerated to supply oxygen and remove carbon dioxide and various organics generated during development, water may also be lost from the organism. This may have three effects: (1) simple loss of water from the organism; (2) due to evaporation of water, cooling well below the dry-bulb temperature of the air flow; and (3) loss of energy from the organism to supply the water's heat of evaporation. As a result, the eggs or pupae may dry out, experience a temperature below optimal for development, and suffer a loss of energy which would otherwise be available for their transformation. Obviously, all three can lead to lowered levels of emergence. How intense these effects are will depend on a number of factors, including: water diffusion through the shell of the eggs or the skin of the pupae, water movement through the bed of substrate, the mass transfer coefficient at the interface between the medium and the air stream, and the difference in water activity between the air and the organism-substrate mix. To aerate the quiescent stages a good approach is therefore to keep the water activity of the air stream close to that of the organism/substrate mix so that the driving force for mass transfer of water is minimized. This is often done by keeping the relative humidity of the aeration stream fairly high. Also, because only small quantities of gaseous products need to be removed from the quiescent stages and there is no need for major cooling, the velocity of the aeration stream coming in contact with the organism-substrate mix can be kept modest so that the mass transfer coefficient will be kept low.

8.4.5 Mass transfer – adults

Adults can be sensitive to the relative humidity of the gas phase (closely related to the water activity) as well as to the presence or absence of a variety of other gaseous components. Thus, in some species mating and oviposition are triggered or inhibited by certain conditions or patterns of time-varying humidity (as well as temperature and visible light), with some insects preferring low-humidity and/or drying conditions as adults and other a higher relative humidity and a non-drying situation. This effect depends almost entirely on the species and the specific strain under consideration and this is why it is best to develop detailed performance information on a proposed organism/feed pair early in a project. Although most adult insects are well protected against dehydration by their exoskeleton, similar mass transfer considerations apply to it as to eggs and pupae. There must be an air flow through the adult living space to remove carbon dioxide and volatiles while the rate of water loss from the adults will depend on the difference in water activity between the air and the insect's internals, as well as on the mass transfer coefficient achieved between the air and the adult body's surface, the latter being strongly influenced by the air velocity. Under certain conditions an individual may be subject to excessive water loss, causing local cooling so that an individual may experience a temperature that is quite different from the dry-bulb temperature in the facility. As discussed for the other stages, water loss implies energy loss since it takes metabolic energy to evaporate the water. Because adults are usually kept at a modest volumetric density (kg adult/m^3 facility) overall heat, water and volatiles removal is usually not a problem, so that the overall aeration rate through the adult space (m^3 air/m^3 space.s) can be kept relatively low. However, under such circumstances it can be quite challenging to ensure uniform conditions throughout the space and it can be a major design problem to avoid major non-uniformities in the fluid flow field of the adult cages. The non-uniformities can manifest themselves as phenomena such as dead spaces, by-pass streams and vortices in which conditions may be quite different from the average so that adults which find themselves in such places will be subjected to sub-optimal conditions for mating and egg-laying. The easiest approach to avoid this issue is to ensure laminar gas flow throughout the adult space, this approach being particularly applicable to flying and hopping insects and even those living on the surface of a bed. If the adults live inside a bed of substrate it can be attempted to achieve an even, low-velocity flow through the bed to ensure uniformity of conditions. Generally, it requires careful planning and system design to achieve laminar flow and uniform conditions.

8.4.6 Mass transfer – larvae

For most industrial organisms it is during larval growth that the highest conversion rate occurs (kg substrate/kg insect.s), together with the highest metabolic rate (W/kg organism). At the same time, it often is also when the organism lives at a high density (kg organism/m^3 reactor space) so that the requirement for heat removal is at its maximum at this stage (W/m^3 reactor space). As discussed above, the heat of metabolism, both of the insects as well as other organisms such as composting bacteria, can be removed by conduction and/or convection, with convective heat transfer occurring through two major mechanisms: sensible and latent heat loss. The latter depends on the evaporation of water from the substrate/larvae mix, thus requiring mass transfer. If cooling is mostly achieved by blowing air across the surface of a bed, the water must evaporate from that surface, the rate of evaporation depending on the difference in water activities at the surface and in the air stream, as well as on the mass transfer coefficient between the two. As water evaporates from the surface, taking heat with it, it is replaced with water from the rest of the bed, moving up from the bed's body to the surface. This water movement through the bed may depend on a number of mechanisms: (1) if the bed is a wet slurry there may be bulk flow of water, probably influenced by surface tension; (2) if the bed is porous or semi-porous there may be diffusive mass transfer through it; and (3) if the bed is stirred by the organisms' movement (or by the process operator) there may be a bulk flow of wet solids from the deeper part of the bed to the surface. Under all circumstances, heat removal via the bed surface will cause the surface to be cooler than the rest of the bed so that larvae can be exposed to very different circumstances. As well, the water activity at the surface will be lower than throughout the rest of the bed, and great care must be taken to avoid excessive drying at the bed's surface and creating a crust. Thus, in designing a system, there is a balance to be maintained between how much cooling can be safely achieved via the surface and ensuring that conditions at the bed surface remain within acceptable limits of temperature and water activity. This limits the allowable thickness of the bed that can be cooled in this way.

As discussed in the section on heat transfer, an alternative is to blow air through the bed rather than just across the surface. In this case latent heat will be removed by evaporation of water throughout the volume of the bed, the balance between sensible heat and latent heat removal depending on the temperature difference between the air stream and the bed, the overall heat transfer coefficient achieved, the water activity potential between the air stream and the bed, and the overall mass transfer coefficient achieved. Generally, it is easier to attain uniform conditions throughout the bed with this approach, albeit at a cost. As pointed out above, through-bed aeration requires a more intricate physical arrangement which makes it more expensive to build, operate, clean and maintain. Because of pressure losses, the energy requirement will also be higher than for surface aeration. As well, material close to the aeration tubes or panels may dry out and thus become less habitable to the larvae, or even form into crusts that are impenetrable. Also, when the bed material is not very porous, e.g. a wet slurry, foaming at the surface may occur or, in extreme cases, the bed contents may be lifted out of its enclosure. Through-bed aeration is most suitable for beds of relatively porous material and allows for tighter control of temperature and temperature distribution than surface aeration. For instance, it is possible to reduce sensible heat removal to nil and to cool entirely by latent heat removal throughout the bed volume, thus achieving a more even temperature distribution. How desirable and/or necessary this is depends on the specifics of the situation.

8.4.7 Simultaneous heat and mass transfer

Heat and mass transfer are basic to the functioning of any insect mass production system because several major aspects of the insect environment are impacted by their metabolism and movement. This is true for all the growth stages. Therefore, organism needs and tolerances for temperature and concentrations of various substances should be determined early in the planning stages so that required process operating conditions can be defined and system heat and mass transfer performance specifications can be set at the onset of actual system design. Energy and mass flows will then be key to the functioning of the process and will depend on various mechanisms of heat and mass transfer operating simultaneously. And, as for most biological production systems, there will be an intricate and complex interplay between these. Dealing at the same time with substrate and water supply, organism development, competing organisms growing on the substrate, removal of metabolic products, temperature control, and heat removal by means of several mechanisms, all while trying to maintain uniform conditions, can be a design challenge for which sufficient information is often not available. Preliminary design can serve as a guide to determine what further data needs to be developed.

The simple situation of larvae growing in a shallow bed of substrate is shown in Figure 8.2. Here, air flows over the surface of the bed, coming in at temperature T1, humidity level H1, oxygen content O1 and carbon dioxide content CD1. The oxygen will need to transfer throughout the bed to the organisms while carbon dioxide and other gases need to transfer from the bed to the air stream. At the same time, sensible heat is transferred to the air, warming it up, while latent heat is transferred from the bed to the air stream via the evaporation of water. Typically, T2, H2, and CD2 will be higher than T1, H2, and CD1 while O2 will be lower than O1. It depends on the organism/substrate combination whether other gases will be given off and removed.

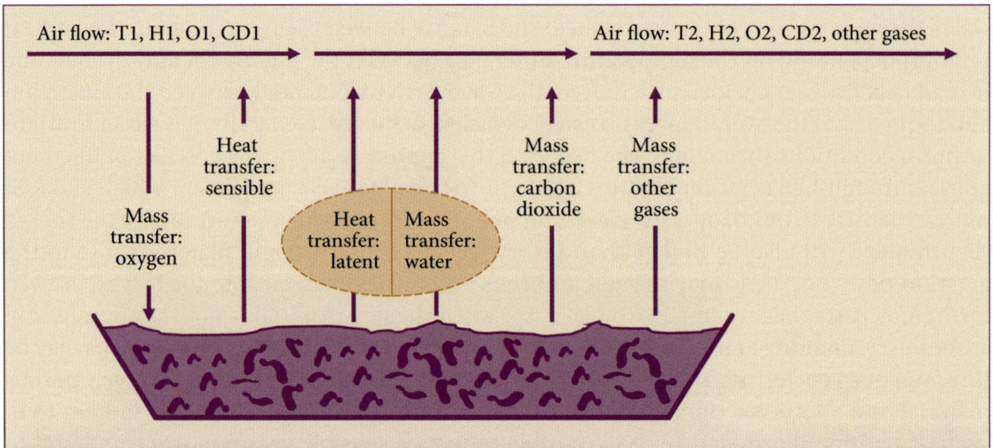

Figure 8.2. Simultaneous heat and mass transfer in a larval bed (T = temperature; H = humidity level; O = oxygen content; CD = carbon dioxide content).

8.5 Process types

The simplest type of process is 'straight batch'. In this, all the ingredients except for gases that are absorbed or vented, are added to the reactor at the same time and stay there for the duration of the process. In a small straight-batch process one might, for instance, add some mealworm eggs to a tray of substrate, wait until the larvae have grown to the desired size, separate the tray contents, and then put the larvae in a birdfeeder while the rest of the tray contents are added to the compost heap. Alternately, after the larvae are harvested, part of the 'colony' can be retained in the tray, allowed to pupate, transform into adults, and mate and lay eggs. If the eggs are then separated from the remaining solids the batch cycle can be repeated almost *ad infinitum* (cyclical batch process). The straight-batch approach is suitable for rearing small quantities and is appropriate for educational and investigative situations. The need for equipment and resources is minimal, but productivity per unit volume of 'reactor space' will also be quite limited. Generally, the organism concentration is low (kg organism/m^3 reactor space) and heat and mass transfer happen via natural convection, or only require that a fan be pointed at the surface of the bed. Sometimes the organism metabolic activity will rise steeply at the end of the growth period and oxygen depletion and overheating can occur. A common approach to gather more information about a given species is to raise them in many small batches under slightly different circumstances. A simple scheme for cyclical straight-batch production is illustrated in Figure 8.3. The author kept a colony of confused flour beetles (several million individuals) alive and healthy for several years in this way.

Straight-batch operation generally leads to a low volumetric conversion rate (kg substrate converted/m^3 reactor space.s) and a high cost of production ($/kg organism produced). To improve system performance the 'fed-batch' process can be used, often combined with movement of the organisms from one location to another. In practice this means that eggs will first be added to a small tray of substrate and then, as the immatures grow, substrate will

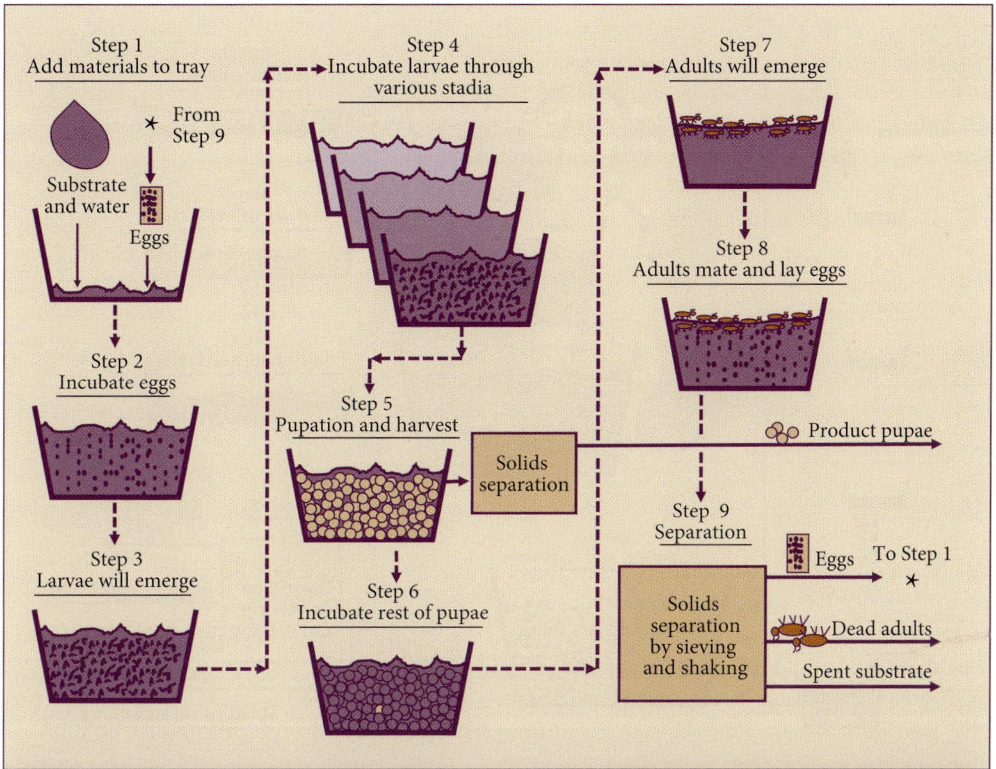

Figure 8.3. A simple scheme for cyclical straight-batch production.

be added regularly while the batch of substrate/organism mix may also be moved to larger quarters several times during larval development. In this case, one aspect of straight-batch operation is preserved in that the organisms in the batch are all similar in age, although the substrate material and wastes are not. Depending on the species and system sophistication, different organism stages may be housed separately and treated slightly differently. Another advantage of the fed-batch over the straight-batch approach is that substrate is not as available to competing organisms such as composting bacteria. That is because, instead of being put out all at once, the substrate is supplied as needed by the target organism and thus is less subject to spoilage. A smaller fraction of the substrate will, therefore, be consumed by non-target organisms, less metabolic heat will be given off, less carbon dioxide will be produced, etc. A cyclical fed-batch system for holometabolous insects is illustrated in Figure 8.4.

To further improve system productivity and cost performance a number of batches can be grown in parallel, but not necessarily all in phase. This is known as the 'parallel-batch' approach in which the individual reactors are operated in the fed-batch mode. Typically, every day a group of small trays are inoculated with eggs. These will then be parallel, i.e. not mixed with one another, and in-phase, and their contents will have the same age structure. The groups of batches will, however, be out-of-phase by a given interval period, e.g. one

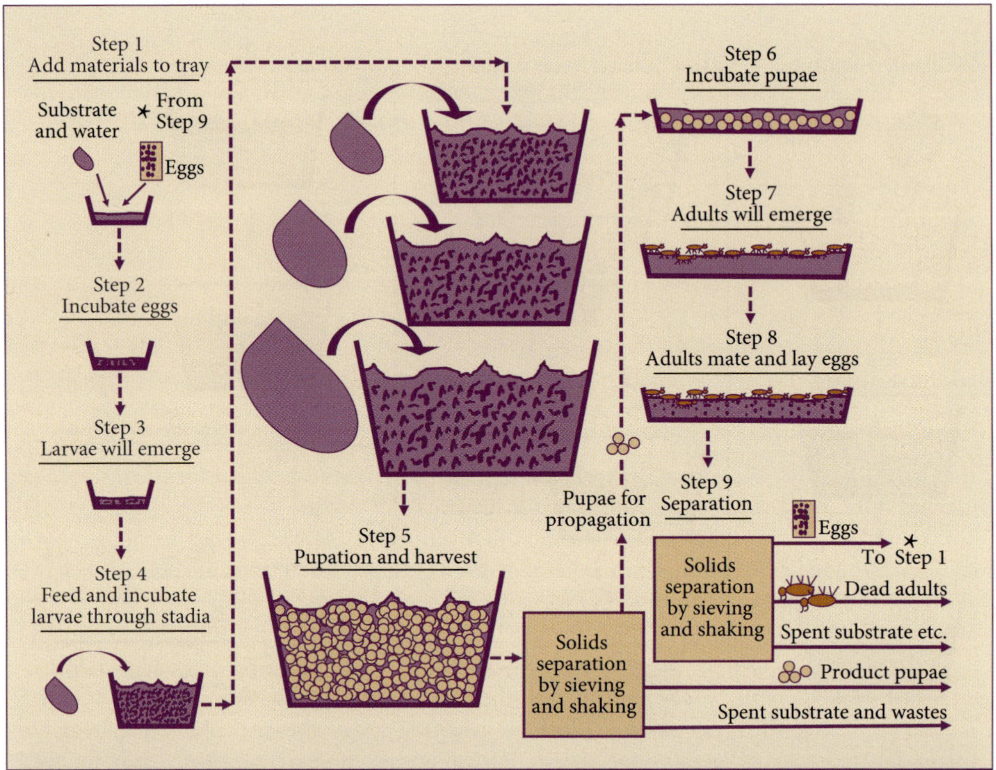

Figure 8.4. A cyclical fed-batch system for holometabolous insects.

day, resulting in a semi-continuous flow of product. One advantage of this approach is that production and handling equipment is used more intensively so that its cost can be written off over a larger amount of product. Also, personnel can be made more productive and the approach lends itself to process scale-up, albeit at the cost of much greater structural and operational complexity and, therefore, investment. Typically, in this type of process, egg incubation, the rearing and incubation of immatures, and adult maintenance will occur in separate spaces. The flow of the larval incubation sequence for a parallel, fed-batch system is shown in Figure 8.5 on day 20 after start-up for the following conditions: the larval incubation period is 20 days, the larvae are grown in trays, a new group of trays is started each day, the larvae are fed each day, and the tray contents are moved into larger trays four times during the incubation period. Thus, on any given day there will be twenty different groups of trays present in the larval incubation facility. Twenty days after start-up it will be day 20 (harvest day) for group #1, day 17 for group #4, and day 9 for group #12. Thirty days after start-up it will then be day 20 (harvest day) for group #11 and day 1 for group #30, etc.

Continuous industrial production tends to lead to high rates of equipment utilization and cost-effective operation, but is usually applicable only to large-scale operations. With respect to the food industry, the production of homogenized milk and deep-fried snacks like potato

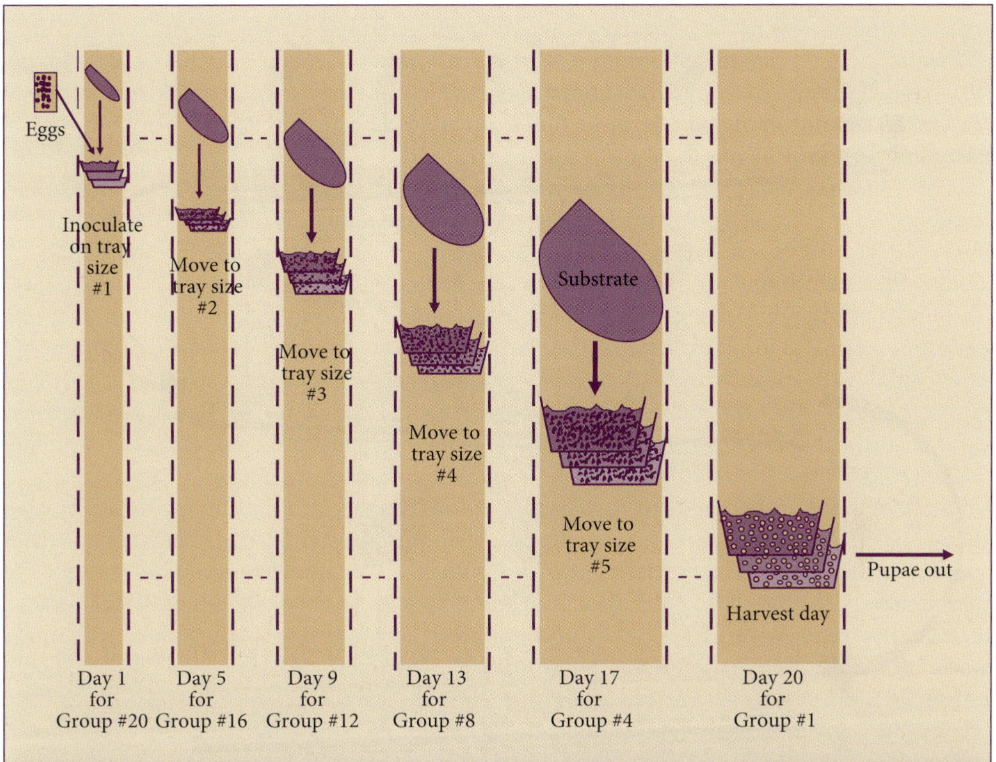

Figure 8.5. The larval-growth part of a parallel fed-batch system.

and corn chips come to mind. In these, daily production volumes are large and the investment in the plant high. Typically, the plants are operated in 'plug-flow' mode, with material passing through the reactor (processing chain) not being mixed with older or newer material, so that it remains age-segregated at all points. Thus, it is quite similar to the parallel-batch process except that the batches are infinitely small and there are infinitely many of them moving through the system continuously. Although true continuous production has not been explored very much for other types of animals that are commonly farmed, e.g. fish, birds and mammals, because of its potential for a high conversion rate per unit investment (kg substrate converted/$ invested.s) it may be a cost-effective approach for insects. The latter are much smaller in size than any of the aforementioned types of animals and are difficult to deal with individually. At the same time, they are much larger than microorganisms such as yeast cells which are commonly cultured continuously. At this point it is not clear at all whether it is feasible to grow insects, and which species, in this manner, and how that should be done. One of the great advantages of plug-flow mode continuous operation is that automatic control is very compatible with it. This means that, often, large production volumes can be achieved at a fairly low cost per unit. The larval growth part of a plug-flow continuous production scheme is illustrated in Figure 8.6.

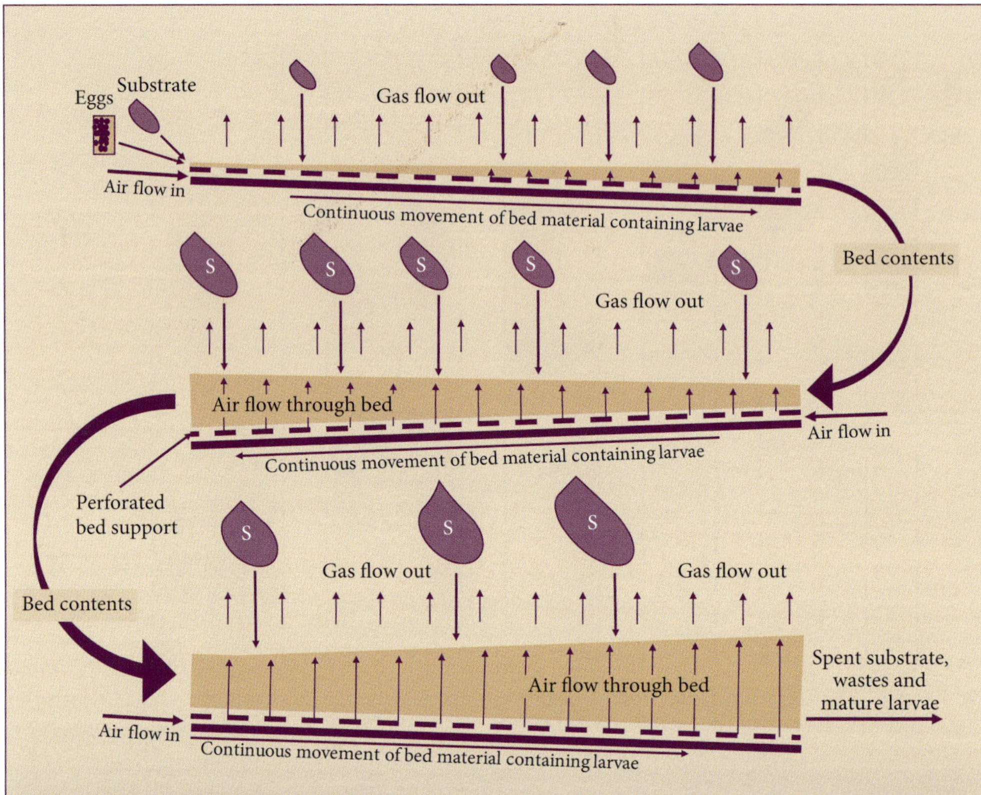

Figure 8.6. Larval-growth part of a plug-flow continuous production scheme.

An entirely different approach to continuous production is to have all the substrate-organism material age-mixed in one location, rather than keep it age-segregated as in the 'plug-flow' mode described above. This type of system also tends to come about naturally. Thus, coffee husks might be discarded in a heap and become infested with flies, resulting in continuous, but entirely unsynchronized production of eggs, larvae, pupae and adults. A similar situation occurs with compost beetles in manure heaps. Climatic cycling may impose some degree of synchronization on this type of system. Organism synchronization is dealt with in in a separate section below. Although they are easy to set up and operate 'heap systems' (like a compost heap; one simply adds and subtracts material on a regular basis), it is usually difficult to obtain a consistent flow of product from them, e.g. larvae of a certain age. In some cases this may not be important. For instance, if the fly larvae and pupae of various ages are simply harvested by chickens which simultaneously turn over the mixture, all is well. Also, if the intent is merely to reduce the substrate volume and its volatile organic content this approach can be quite acceptable and inexpensive. It then becomes a form of insect-mediated composting; to facilitate the process, insect eggs may be added to the incoming material. But if the intent is, for example, to harvest mature larvae for consumption the age-mixed approach is practical only if labour costs are low or if the separation method can be automated. It remains to be seen what role age-mixed, continuous systems will play in industrial production schemes for insects.

Many variants of the basic process types described above are possible. In some, only several nutrients will be added during a batch cycle, but not others. Or, part of the population may be removed all at once from a mixed-age system on a regular basis. Or, some stages of a process can be operated batchwise, and the rest continuously, etc. As more industrial processes are developed a great variety of them will undoubtedly come online, with some of them proving successful and others less so.

8.6 Reactor types

In this context the term 'reactor' is used to refer to any container, cage, tray, tube, tank or vessel together with associated equipment in which the organism grows, develops, metamorphoses, and reproduces. Thus, for eggs that are developing while residing in a tray filled with substrate, the tray is the reactor. For adult crickets confined to a cage in which they breed and lay eggs, the cage is the reactor. For larvae in a continuous plug-flow process living in a layer of feed on a moving belt, the enclosure and the belt are the reactor. The entire production process is thus seen, described, and designed as happening in a series of reactors, each having inflows and outflows, and materials being moved between them, with organism feeding as well as heat and mass transfer all happening inside them.

Hence, reactors are central to any process and must correspond to the process type and the needs of the organism. They must be made of materials that are not toxic to the organism, nor impart toxicity to them. For batch processes they can either be disposable or re-usable. In the latter case they should be easy to clean. In a cyclical process they should be resistant to frequent handling and have a low cost per cycle. For example, if beetle larvae are grown in batches at relatively low density in deep beds of feed, the trays could be made of cardboard, a durable plastic, or stainless steel. A cardboard tray would then be used for only one cycle, a plastic one for numerous cycles, and the stainless steel one for a large number of cycles. Any of these might suffice for operating the process; which is the more favourable will depend on the costs of acquisition and inventory, handling, cleaning, disposal, etc.

Many of today's processes are based on the cyclical parallel-batch approach in which at least one or two stages reside in trays of substrate. As soon as the process reaches an industrial scale it becomes too cumbersome to deal with individual trays and they are usually placed in and dealt with via service frames. In this way feed, water, oxygen and carbon dioxide supply and removal as well as heat exchange and monitoring and control services are all mediated via frames which interface with the larger system. The reactor will then consist of a large space containing a number of frames each holding a number of trays. Typically, the trays will be charged and loaded outside the reactor and then loaded onto the frames. In some cases the frames remain inside the larger reactor space at all times which makes it more difficult to clean them, but easier to keep them connected to the larger system; in other cases the frames are loaded up with trays outside the larger reactor space and then re-inserted, this making it easier to clean them, but more complicated to re-connect them to the system. Obviously, there is a myriad of possible arrangements, all with advantages and disadvantages, which are often a function of the species being considered. It is evident that the chosen species, the process type, the operating sequence, and the reactor configuration are closely related, with many combinations and possibilities presently being investigated and tried out throughout the industry. A typical parallel batch reactor is illustrated in Figure 8.7.

8.7 Materials handling

Materials handling refers to all aspects of moving, stocking, mixing, and separating substrate and wastes as well the handling and transport of live organisms. In the types of processes under consideration the materials will mostly be complex mixtures of particulate solids, sometimes mixed with water to form a thick slurry. It includes the receiving of feed components and the shipping of products and by-products at the loading dock, the mixing of feed ingredients, the distribution of feed to the reactors, the separation of live organisms from spent substrate, the segregation of by-products into different classes, etc. The operations that are used in a process will depend very much on the specifics of the situation. Thus, if crickets are raised on a wet, particulate medium the materials handling operations employed will be quite different from those in a mealworm facility in which the feed will more likely be finely-divided, dry, and powdery.

Methods commonly employed for the operations mentioned above are: mechanical and pneumatic conveying, mixing and stirring of solids, sieving and shaking, and bagging and de-bagging. Special machines are available for all these, often used in combination with less specialized equipment such as fork-lift trucks, hand-trucks, pumps, blowers, cyclones, etc.

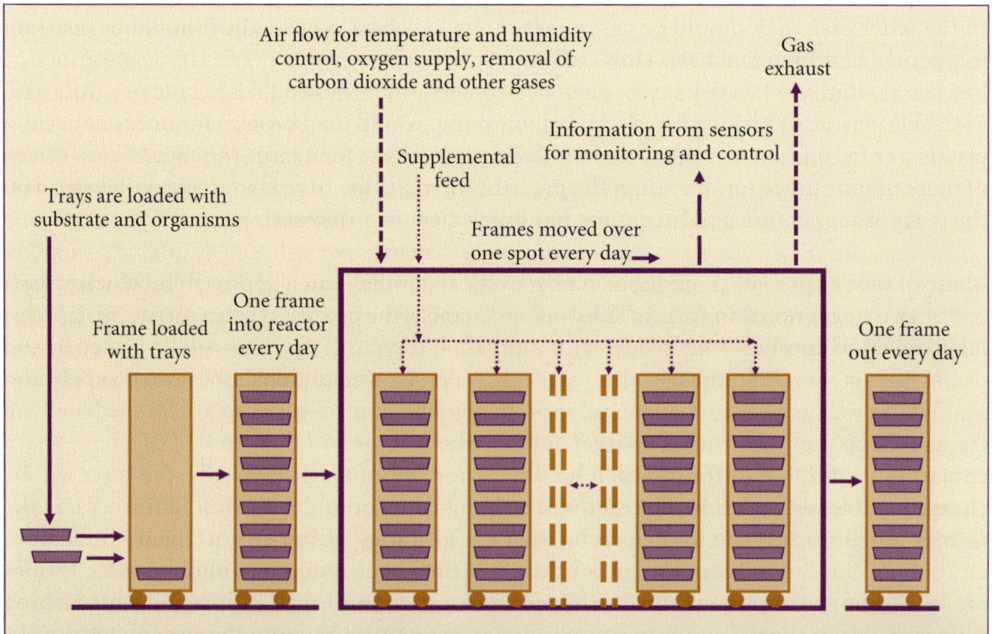

Figure 8.7. An air-cooled parallel-batch reactor with moving service frames.

Generally speaking, the materials handling system assembles the solids input streams (and semi-solids) for the reactors and transports them hence, and then deals with the solids output streams, often by fractionating their components and sending some of those on to the next stage reactors. Gaseous streams are dealt with similarly by the heating, ventilation, and air-conditioning (HVAC) system.

When designing an industrial process and a plant in which to run it, after deciding on an organism/feed pair, the process type, and the reactor types for the different stages, the next set of questions must relate to how materials will flow through the process, where they are to be stored, combined, and separated and how all this is to be accomplished in an efficient and cost-effective manner. Thus, materials handling has a major impact on plant design and plant layout and should be considered in detail early in the design process, with the specifics depending almost entirely on the earlier choices. The materials handling system will occupy a substantial part of the plant's floor space and can represent a major fraction of the capital cost. Evidently, if the cost of the materials handling system and its operation turns out to be very high, the earlier choices about organism/feed pair, etc. should be re-considered. A very general flow diagram for a materials handling system is presented in Figure 8.8, together with an overall diagram of an HVAC system and the input section of a water supply system. The latter two are dealt with further below.

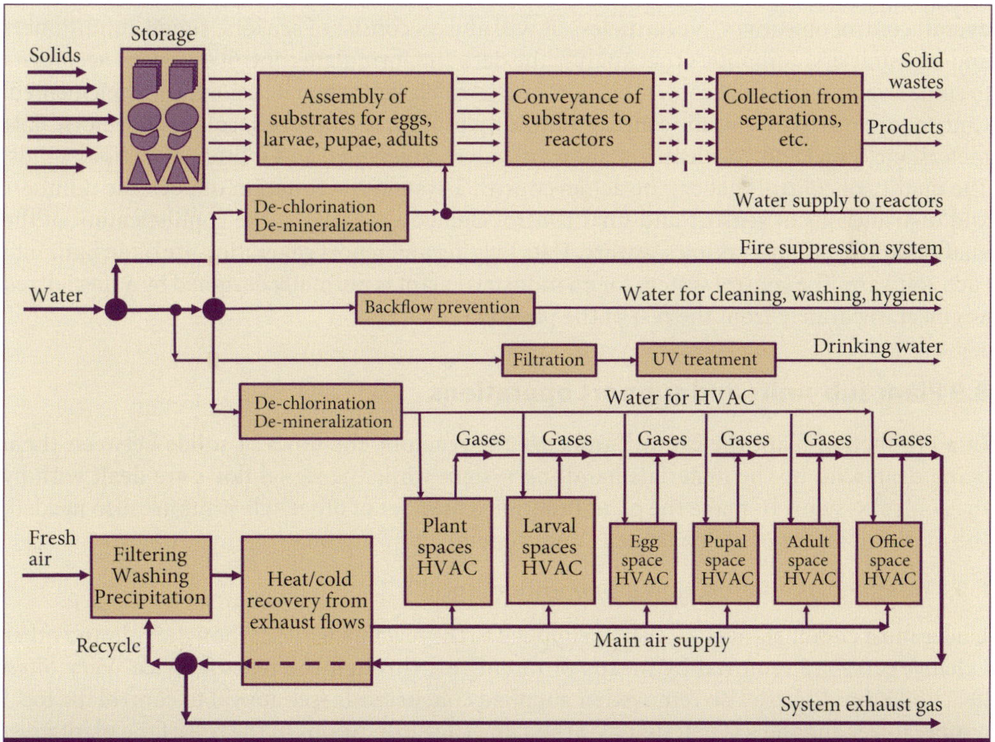

Figure 8.8. Materials handling, heating, ventilation, and air-conditioning (HVAC) and water supply systems.

8.8 Process control

Much of the process control in an insect-based conversion facility will be oriented to assure the accuracy, constancy, and uniformity of environmental conditions. Thus, typical target variables will be temperature, water activity, and gas phase composition. These are then controlled through the manipulation of variables such as the rate of application or removal of thermal energy, water spray intensity, air flow rate, etc. In an industrial process the feed conversion rate per unit volume of reactor will be high (kg substrate/m^3.s), meaning that conditions can change very quickly. This is especially so for the gas phase composition seeing that the amount of gaseous mass in the system present is very small, due to its low density. The control system must, therefore, address two kinds of objectives at the same time: (1) maintain conditions within critical limits so assure organism survival; and (2) keep conditions as stable as possible and as close as possible to optimal values. Generally, it is more difficult to implement good quality control for a batch process than for a continuous one since during the former the values of variables such as the rate of carbon dioxide production, the rate of heat generation, etc., will vary much more. One reason to keep conditions as uniform as possible is to keep population stratification and de-synchronization to a minimum. The quality of control achieved by a system can significantly impact these. This is discussed further in Section 8.10.

The specifics of a control system will depend on the species, the process, the reactors, and the overall control objectives. Nevertheless, it will always consist of sensors, signal conditioners and transmitters, computers that analyse the data and formulate control decisions according to some strategy, and final control elements with which the system decisions are implemented. Common sensors are thermocouples, humidity probes, level probes, colour detectors, flow meters, etc. Final control elements often are valves, solenoids, and motors of various kinds. The quality of control that can be achieved with a system is a function of both the numbers and distributions of sensors and final control elements, as well as the sophistication of the analysis and decision-making software. Data logging and report generation are often built into such software. The control system for an industrial plant is normally designed by a specialized engineer, separately from the rest of the plant.

8.9 Plant sub-units and support operations

In a production plant the central units are the reactors, the flows of solids between them being dealt with by the materials handling system while the signal flows are dealt with by the control system. To make the plant function a number of other sub-units are also needed, the most important of which is the HVAC system.

8.9.1 Heating, ventilation, and air-conditioning

Under most circumstances, oxygen is supplied to the reactors while carbon dioxide and other exhaust gases are removed by means of forcing air through the reactor space. Very often heat and humidity are also removed or supplied concurrently (see above) to control the local temperatures and the water activity. At the same time, conditions in the rest of the plant must be maintained as well at acceptable levels of temperature, humidity, dust, gas composition, air velocity, etc. while the specifications for these may vary considerably between regions inside the plant. For instance, the requirements for loading dock conditions can be quite different from those for the control room. All in all, the HVAC system must be able to simultaneously handle varying loads in the reactors and service a number of other regions of the plant while external climatic conditions may vary considerably. Whether that is all accomplished by a single, integrated system or with a number of semi-autonomous, smaller units depends on the plant's designer. The HVAC system's functioning will be under the supervision of the plant's overall control system and a substantial fraction of the total control effort will be dedicated to regulating various aspects of the gas phase. In many ways, the HVAC system is to the gas phase what the materials handling system is to solids. A general flow diagram for the HVAC system is presented in Figure 8.8.

8.9.2 Management of input and waste streams

Insects are biological entities and are as much subject to poisoning and disease as any other. The extensive use of insecticides, especially in the agricultural sector, can pose a severe danger to any production plant. For instance, if a cloud of pesticide spray or dust enters the air intake of a plant it could spread throughout within minutes and strongly affect the entire colony. Thus, all air taken in should be thoroughly cleaned before being allowed into the plant. Traditional techniques such as water washing and electrostatic precipitation can be combined with HEPA filtering, carbon filtering, etc. to achieve any degree of air cleanliness. To further reduce contamination, the plant can be operated at positive pressure and the loading docks can be equipped with air curtains. Also, the plant may be divided into zones for which different contamination rules apply.

As important as it is to keep contaminants from entering the plant, it is equally important to clean up the output streams so as to prevent contamination of the external environment. The exhaust gas stream from an insect production facility will normally contain a substantial amount of carbon dioxide. It may also contain volatile nitrogen compounds such as ammonia, sulphur compounds that may be malodorous and any number of organics, some of which may be biologically active molecules like pheromones. As well, it often contains fine dust and small insect parts that may present inhalation hazards. Hence, although there is obviously a cost involved, plants should be designed and operated in an eco-responsible manner, so that all waste streams are properly cleaned up and rendered safe before discharge. This principle should be applied to all gaseous, liquid, slurry, and solids output streams. As mentioned below, liquid and solid waste management should probably be carried out in ancillary units, well away from the production facility. The processing of gaseous waste streams is normally done locally.

8.9.3 Plant and equipment cleaning

Of the various operations required to re-use reactors and reactor components, cleaning is often the most complicated. Some organisms will leave faeces and other excretions like slime trails on the walls and these, especially when dried out, can be difficult to remove. In certain cases mechanical removal by scrubbing, sometimes done manually, is required. In this regard, the cleaning of adult cages can be particularly challenging (e.g. for house flies). Overall, all surfaces that have been in contact with the insects and/or the feed should undergo deep cleaning between uses. For mobile equipment this is best done at a special cleaning station. For stationary equipment in-place cleaning methods developed for the food industry, such as high-pressure spraying, may be applicable. In order to avoid the establishment of subsidiary colonies in the plant (no matter what, there always will be escapees) it is important to prevent the formation of dust, substrate and moisture deposits throughout. This can be partly accomplished with the HVAC system, but not entirely. As well, the entire plant must be thoroughly cleaned on a regular basis. That also prevents competing organisms from establishing themselves and lowers the rate of infection. An emergency 'total cleaning' procedure should also be established and practiced (like fire drills) in case the production unit is subject to large-scale infection.

8.9.4 Support services

In order to have a plant run smoothly and its personnel stay healthy, considerable resources must be devoted to a number of support systems. The electrical system and the backup electrical system are the most important since almost all other systems are powered via them. For instance, if the ventilation to a larval incubator stops it takes only several minutes (if that much) for the organisms to overheat and/or run out of oxygen. A simple approach is to have a backup generator; a better approach is to have battery backup to power the most essential systems for a short time while the backup generator is brought online because this eliminates major power fluctuations; a complete approach is to have a number of such backup units in parallel. With respect to the cost-benefit of electrical backup it is worthwhile to consider how much can be lost during thirty minutes without electrical power while the single backup generator is being coaxed into action by personnel working in the dark.

Other essential and critical systems are related to the water supply and fire suppression. Depending on the location, water may be acquired from a municipal supply or be pumped from a well. In the latter case it may need to be filtered and sterilized by means of chlorine, ozone, UV, etc. In both cases part of the water supply may need to be de-chlorinated, de-ionized or distilled, depending on the tolerance of the organism. Although many fire suppression systems are water-based, if one or more feed components are dry powders that have the capacity to explode (especially when present as dense dust during handling operations), other systems may be needed to assure a safe environment. An overall flow diagram of the input section of a plant's water supply is shown in Figure 8.8.

Personnel need offices and meeting rooms, washrooms with change and shower facilities, and a lunchroom. The water supply for personnel should come through a separate connection from the mains to reduce the risk of backflow contamination from the laboratories and production units.

8.9.5 Breeding, culture maintenance, fresh inoculum supply, waste management

Throughout the history of agriculture people have started off with wild strains and bred them to become today's crops and domesticated animals. This was done to improve yield, performance under various conditions (including robustness and stability), uniformity and synchronicity (discussed further below) as well as handling characteristics, all of which contribute to a higher return on labour/investment. Undoubtedly, the mass production of insects will follow the same path, with specific strains of organisms being bred by different means to maximize their protein content, fat composition, performance on various feeds, etc. Thus, specific strains of mealworms will probably be developed to, e.g. grow well on corn starch and have a high omega-3 content. Some of this is already being done but the spectrum of possibilities has hardly been touched, one enticing aspect of insect culture being the sheer number of species and wild strains that are readily available for further development. The application of modern methods in breeding, including artificial insemination and biotechnological procedures, will likely lead to exciting results. Nevertheless, considering the possible harmful consequences of such efforts, however remote, it is necessary that they be carried out in a properly equipped containment facility so that novel organisms or strains cannot exit until they are proven, within reasonable limits, to not be hazardous. It is, therefore, advisable that the breeding program be carried out by qualified personnel and in a separate building to which access is limited, well away from the main production unit. Strain breeding is an essential aspect of the mass culture of insects and can impact process performance quite significantly through a number of variables. Thus, it cannot be ignored, although it can be outsourced to specialized suppliers who would then breed and supply strains to the industrial producer.

Once one develops a particular strain of an organism, it needs to be maintained. How complicated that is depends on its inherent stability. Whereas in wild populations there is usually a wide distribution of any given characteristic, e.g. colour, thickness of the exoskeleton, resistance to disease, etc., in a cultured production strain it is desirable to have very narrow distributions. Thus, all individuals should have similar developmental response surfaces, i.e. under identical circumstances they will have closely related ontogenies and develop at the same rate, have the same body composition, etc. In that vein, to facilitate harvesting and marketing, all the Granny Smith apples in the orchard should have the same colour and be ripe at the same time, all Holstein cows in the herd should have similar lactation curves, all soya beans should be ready to leave their pods on the same day, etc. Of course, there are some drawbacks to such strain uniformity. For instance, all the plants in the field will be equally non-resistant to the same virus or to a lack of rain so that the entire harvest can be lost due to one factor. For this reason, in agrarian societies the use of cultured strains is often avoided, the people preferring to live with a high likelihood of a moderate harvest to running the risk of no harvest at all. Cultured strains have a natural tendency to drift and for the distributions of their characteristics to widen so that, without guidance, a strain will revert to a wild form. This is why, for instance, responsible dog breeders cull a substantial number of their pups, so as to keep the breed within specifications. In an industrial setting these same two issues must be addressed by strain maintenance: (1) to head off the danger of rapid and massive colony infection by disease organisms; and (2) to offset the natural tendency of the distributions of the strain's characteristics to widen. To reduce the risk of contamination, strain maintenance should be carried out away from breeding and from production, also preferably in a separate building to which access is limited. For each strain a number of small sub-colonies should be maintained separately, and each of these should be monitored and guided consistently to stay within the strain specifications. Although there is a cost associated with strain maintenance, it provides considerable protection against dramatic loss and also allows for more rapid system recovery should the main production unit become infected and needs to be put through a 'total cleaning' procedure.

Under most circumstances a production process will be operated in a cyclical manner, with the eggs for the next generation being derived from the adults of the previous one.

Usually, however, there is a limit to the number of cycles before the strain starts to drift and the population becomes overly desynchronized. Therefore, it is advisable to regularly interrupt the egg-to-adult cycle and to re-start it from a fresh inoculum supplied by the culture maintenance unit, with a very narrow age distribution. The flows between strain development, strain maintenance, and production are illustrated in Figure 8.9. As shown, the ancillary units could also supply new strains and fresh inoculum to other production facilities. Thus, breeding and culture maintenance need not necessarily be done at all production locations. Waste management (liquids and solids) is, however, normally done at the production location and, to mitigate the risk of contamination, it is usually desirable to also house these activities in separate, ancillary units. As for all other animal husbandry undertakings, it will be a challenge to minimize the cost of disposal from insect production facilities and to maximize the value of the 'wastes'. Composting may be one of the most viable methods to deal with the mixture of left-over feed, insect bodies and parts, faeces, off-spec product, etc. Hyper-thermal composting has the advantage that none of the insects in any form survive the treatment. It is likely that standard, industrial, liquid waste treatment and exhaust cleaning methods can be adapted to give satisfactory performance in an insect production setting, but this needs to be confirmed. A desirable arrangement of the main production unit of the plant and the ancillary units, as well as the flows between them, is presented in Figure 8.9.

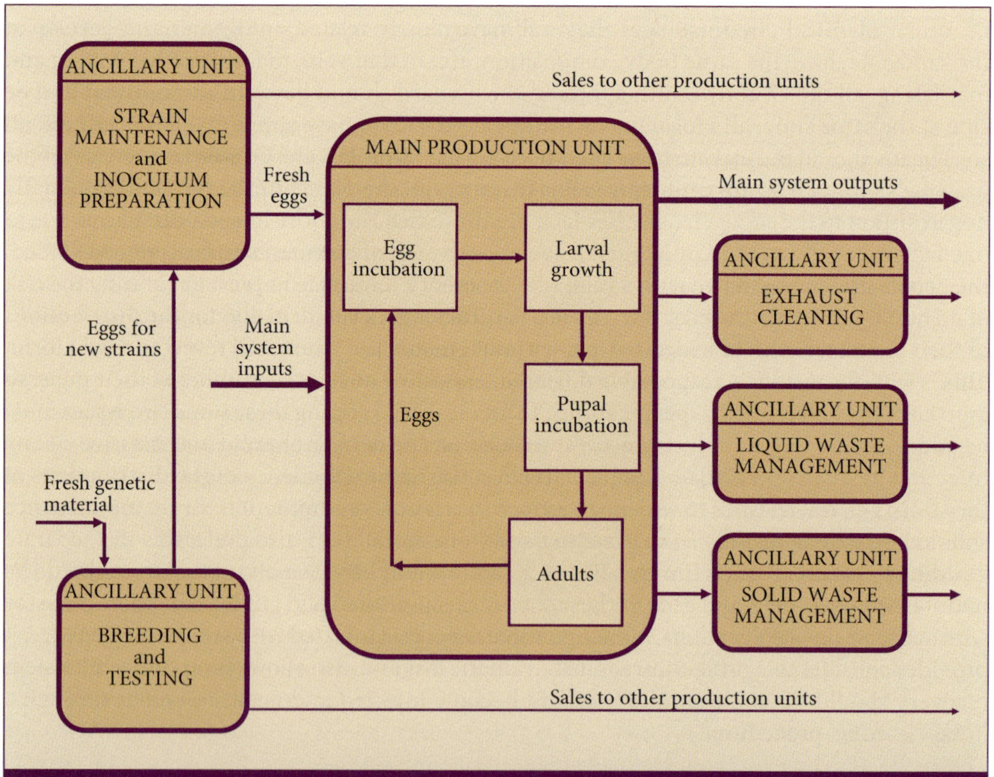

Figure 8.9. Arrangement of plant units and flows between them.

8.10 Other issues

8.10.1 Synchronization and its maintenance

As mentioned earlier, to maximize productivity it is important to have the population synchronized as much as possible. Ideally, larvae in the same batch should emerge from their eggs at the same time, moult from stadia to stadia in step and, in the case of holometabolous insects, start their transformation to pupae at the same time. As well, adults should emerge and start and stop their oviposition together. Population synchronization is here examined from the point of view of what factors cause and influence desynchronization in a process, there being two main aspects: static and dynamic. The term 'static' here refers to nothing changing in the makeup or behaviour in the organism whereas the term 'dynamic' refers to changes in the organism's response to external conditions through adaptive or evolutionary mechanisms. Such mechanisms can act either within the organism itself or in any of its symbionts or parasites.

Main factors that contribute to desynchronization under static conditions (starting with eggs) are: (1) the age distribution of the eggs; (2) the development distribution of the individual organisms, the latter being due to two subsidiary factors: (3) innate differences between individuals in their developmental responses to given conditions; and (4) differences in their treatment, i.e. the conditions under which they are grown. After a certain period of incubation following inoculation with eggs, the individual immatures development distribution will then be the convolution of the treatment distribution and the individual organism response distribution. At the same time, the immatures population development distribution will be the convolution of the initial egg age distribution and the individual immatures development distribution. This is illustrated in Figure 8.10.

The same principle applies to the other stages of organism development so that, depending on how stratified the treatment conditions are and how wide the various distributions are, eggs produced by the adults to form the next generation may have a much wider age distribution than those initially used for inoculation. In a cyclical process (Figure 8.9) this effect then becomes cumulative so that, if unchecked, the population can become less and less synchronized, as happens naturally in a continuous 'heap system'.

Each of the factors contributing to desynchronization can be addressed in several ways. The age distribution of the eggs used for inoculation can be kept fairly narrow by occasionally interrupting the adult-egg cycle and re-starting it with fresh inoculum that has been carefully age-synchronized in the inoculum preparation step. At the same time, during the direct adult-egg cycling in the main production unit (Figure 8.9) the egg age distribution can be narrowed through frequent and complete harvesting of any eggs present. This, in effect, is the equivalent of taking 'thin slices' from the wider distribution. As well, to reduce natural spread, very early and very late eggs can be rejected. Of course, there are costs of extra effort and egg loss associated with each of these approaches.

The innate differences between individuals can be reduced by strain breeding and producing inoculum eggs consistently under standard conditions. Differences between treatments

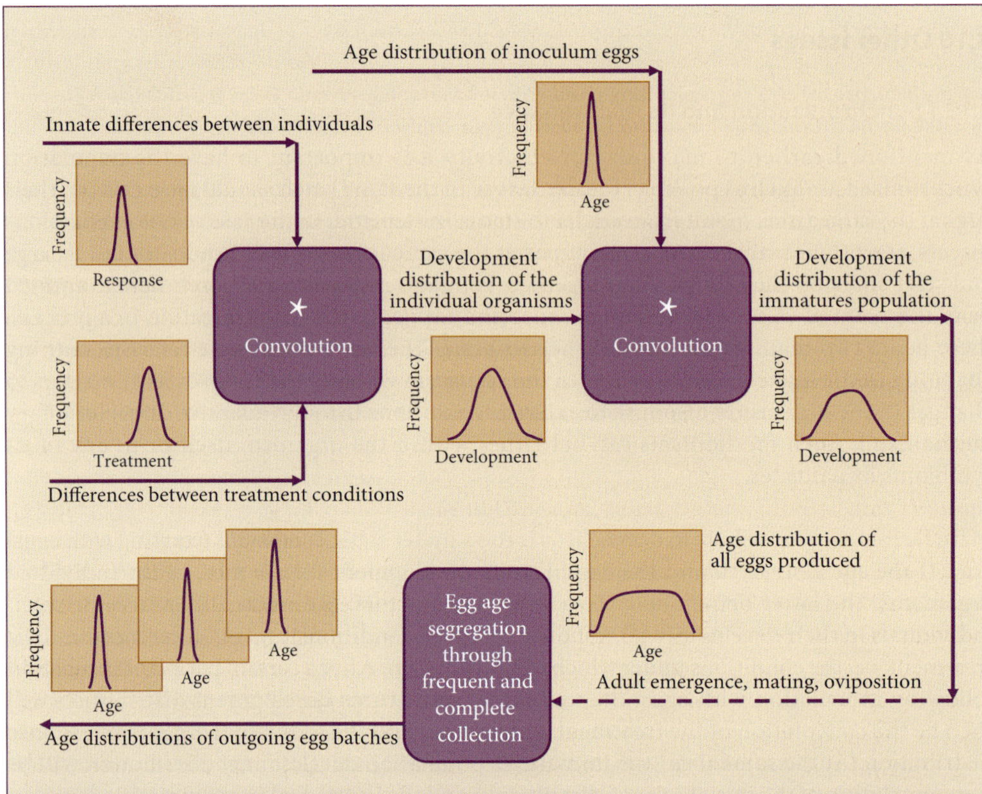

Figure 8.10. Factors influencing desynchronization under static conditions.

will, however, often occur due to process conditions. For instance, if larvae are raised in trays and the temperature throughout the feed layer is not uniform then some larvae may develop faster than others and the development distribution of the individual organisms, and thereby of the population, will be widened. This means that even if the inoculum egg age distribution is extremely narrow and, as well, the organism has been bred to be very uniform in its response to processing conditions, the population may still become desynchronized as a result of stratification inside the reactors due to uneven heat and mass transfer. Hence, ensuring the uniformity of organism handling and conditions in the reactors is important to maintain synchronicity and maximizing productivity. As pointed out above, there are some disadvantages to uniformity; any system whose members are narrowly distributed will be less stable, robust, and resistant than one whose members are more diversified. Because of this, typically, cultured systems require more management and protection than wild systems.

The second aspect of desynchronization is related to the dynamic nature of biological structures. Regardless of how carefully one handles a strain, there is always a tendency for the distribution of any one characteristic to widen. Thermodynamics predicts it and experience bears it out. Unless that effect is desired, this needs to be countered through strain maintenance. This is another reason why an occasional re-start of the production cycle with fresh inoculum is a good management practice. Basically, it is a management practice that will limit desynchronization.

8.10.2 Safety aspects

There are two safety aspects briefly considered here: worker safety and public safety. Apart from the usual dangers present in an industrial setting, the entotechnology employee can face several extra hazards. Besides the normal products of metabolism such as carbon dioxide and methane, insects can give off a wide variety of by-products that can be inhalation hazards. These can be present as gases, aerosols, or very small particles such as scales or setae. Some such materials may be poisonous, allergenic, or even carcinogenic. Some of the effects may surface only after longer-term exposure. The health status of employees should be regularly monitored if they are in frequent contact with large quantities of insects. The use of robotics in industrial plants can minimize chronic employee exposure.

Because most insects will be raised in confinement public safety issues arising from the animals themselves is probably not a major concern. However, the possibility and likely impacts of a massive organism escape both within and from the production plant should be considered. If the adult insects can fly, should the plant envelope fail, e.g. as the result of an earthquake, a cloud containing several tonnes of insects could land on a nearby community and cause panic even if they are harmless as such. This type of impact can be mitigated by minimizing connections (and traffic) between sections of the plant so that an incident in one section will not strongly affect the others. Passive methods such as the installation of netting and double doors between sections, and active methods such as optical escapee detection and automated removal, can all help reduce problems. At the same time, they can also be effective in reducing the nuisance potential of escapees to neighbours and in reducing disease transmission within the plant. To deal with safety situations, an emergency response team of employees should be trained and an overall response plan worked out together with local emergency response authorities.

More formidable public safety concerns arise from the ecological impact of escapees and the effects of the plant exhaust on the surrounding environment and communities. If the insect is foreign to the locale but able to adapt and thrive in it, the ecological consequences may be extensive and significant. And, unless the exhaust is properly treated and cleaned, the area around the plant, or downwind from it, may be dangerous to live in. Health hazards may arise from either acute, or long-term exposure. Overall, before a species is chosen for massive production on an industrial scale an extensive safety audit should be carried out to assess the risks.

8.11 Conclusions

Entotechnology is still in its formative stages and the industry built upon it, although rapidly expanding, is yet in its infancy. So far, worldwide, only a number of pilot plants and several small production units have been built and brought online. As of this date (April 2017) there are not yet any large-scale factories in operation. But there are some in the planning stages and the potential is tremendous. Until now, apart from traditional insects like the honey bee and the silkworm, only a small number of species are grown in any quantity, mostly the cricket, the black soldier fly and the house fly, and several species of mealworm. Many other insects are, however, attracting interest because of their special abilities (e.g. break down cellulose) and other attributes. Termites, cockroaches, weevils, and many types of degradation beetles are prime candidates, with each species presenting a special challenge for its domestication and mass production. As the field of entotechnology develops, insects will be used as the agents in a wide variety of conversion processes, become the source of many new bio-materials and other bio-products, and will also be used as bio-robots in mining, waste separation, construction, etc. After all, nature provides for a way to grow a complete bio-robot for the price of an old newspaper; now what we have to do is learn how to program it and interface it with our command structure. How well the products and capabilities of entotechnology will be accepted by society remains, of course, a question for the future to answer.

Acknowledgements

I want to thank all my friends who have stimulated my interest in entotechnology and its development over the years. We have dreamed together and worked together to create new possibilities. It has been an exciting journey so far and that journey is far from over. I hope we can keep on dreaming and creating and maybe, in the future, be together when a robot ant serves us a roasted giant cockroach smothered in termite gravy.

9.

Gryllodes sigillatus

Breeding and maintaining high-quality insects

K. Jensen[1,2], T.N. Kristensen[1,3], L.-H.L. Heckmann[4] and J.G. Sørensen[3*]

[1]Department of Chemistry and Bioscience, Section for Biology and Environmental Science, Aalborg University, Fredrik Bajers Vej 7H, 9220 Aalborg, Denmark; [2]Department of Bioscience, Section for Terrestrial Ecology, Aarhus University, Vejlsøvej 25, 8600 Silkeborg, Denmark; [3]Department of Bioscience, Section for Genetics, Ecology and Evolution, Aarhus University, Ny Munkegade 116, Building 1540, 8000 Aarhus C, Denmark; [4]Life Science Division, Danish Technological Institute, Kongsvang Allé 29, 8000 Aarhus C, Denmark; jesper.soerensen@bios.au.dk

Abstract

Insects have a large potential for sustainably enhancing global food and feed production, and commercial insect production is a rising industry of high economic value. Insects suitable for production typically have fast growth, short generation time, efficient nutrient utilization, high reproductive potential, and thrive at high density. Insects may cost-efficiently convert agricultural and industrial food by-products into valuable protein once the technology is fine-tuned. However, since insect mass production is a new industry, the technology needed to efficiently farm these animals is still in a starting phase. Here, we discuss the challenges and precautions that need to be considered when breeding and maintaining high-quality insect populations for food and feed. This involves techniques typically used in domestic animal breeding programs including maintaining genetically healthy populations by allowing large population sizes, artificially selecting animals for breeding that possess valuable production traits, and allowing natural selection to operate and thereby maintain high fitness and facilitate adaptation to the production environment. We furthermore emphasize the importance of adjusting the rearing environment, especially density and temperature, as well as optimizing dietary nutrient availability and composition, to maximize production. In particular, the ratio of protein to non-protein energy has high impact on e.g. body size and composition as well as development time. Optimal temperature and dietary nutrient composition both vary with species, life stage and sex depending on density and disease risk, and different temperatures or nutrient compositions maximize different life-history traits. Thus, it is possible to favour specific traits by adjusting temperature and diet, and balance the cost of the feed against production efficiency and outcome. Importantly, valuable qualities such as efficient protein utilization must be maintained within populations, and the fastest production conditions might not be most optimal as physiological and adaptive utilization efficiency may be compromized.

Keywords: artificial and natural selection, dietary nutrient composition, effective population size, protein utilization efficiency, rearing temperature

9.1 Insects as a new means to enhance food security

Insects are increasingly promoted as an alternative, environmentally sustainable means to produce protein for animal feed and human consumption, which may help maintaining food security as the requirement for food and especially protein rises in an increasingly populated world (Van Huis, 2013). Using insects for protein production may be a feasible way to fully exploit and recycle organic by-products from industry and agriculture, as insects may efficiently ingest and assimilate nutrients from diffuse, already exploited resources, and concentrate them in their body tissue (Van Huis *et al.*, 2013). Moreover, insects can be maintained at high density and have short development times compared to traditional production animals, requiring little space for high-throughput production. However, several factors still need to be explored and optimized before efficient high-throughput production systems are established for producing high-quality insects at a low cost.

9.2 Current conditions and challenges for insect mass-rearing and production

In comparison to other terrestrial production animals, it is far more feasible to construct a production environment for insects that ensures species-specific biological stimulation throughout the entire life cycle. Insect production can be split into two major parts: (1) breeding; and (2) biomass production. The breeding part relates to the adult individuals and the process of facilitating optimal reproduction to populate production trays for juvenile growth and thereby biomass production. Breeding also serves to found future generations of the breeding stock. Population genetic concepts (e.g. genetic variation, inbreeding, drift, selection) should be applied here. The biomass production part relates to the juvenile life stages (either larvae or nymphs) that build up biomass during a period of weeks to a few months depending on the species and rearing conditions. Considering environmental effects (e.g. temperature and diet) and environment by genotype interactions are especially important for the biomass production. From an industrial perspective, there are currently only a few insect species where protocols and facilities have been developed allowing production at tonnage level (Makkar *et al.*, 2014; Van Huis, 2013). These includes: (1) the tenebrionid beetles *Tenebrio molitor* (common mealworm; Oonincx *et al.*, 2010; Van Broekhoven *et al.*, 2015; Zielińska *et al.*, 2015), *Alphitobius diaperinus* (lesser mealworm; Van Broekhoven *et al.*, 2015) and *Zophobas morio* (giant mealworm; Van Broekhoven *et al.*, 2015); (2) crickets, e.g. *Acheta domesticus* (house cricket; Oonincx *et al.*, 2010) and *Gryllodes sigillatus* (banded or decorated cricket; Zielińska *et al.*, 2015), (3) the locusts *Locusta migratoria* (migratory locust; Oonincx *et al.*, 2010) and *Schistocerca gregaria* (desert locust; Zielińska *et al.*, 2015); (4) cockroaches, e.g. *Periplaneta americana* (American cockroach; Sánchez-Muros *et al.*, 2014) and *Blaptica dubia* (dubia cockroach; Oonincx *et al.*, 2010); (5) the flies *Hermetia illucens* (black soldier fly; Sheppard *et al.*, 2002; Surendra *et al.*, 2016; Tomberlin *et al.*, 2002) and *Musca domestica* (house fly; Hussein *et al.*, 2017); and (6) lepidoptera, e.g. *Bombyx mori* (silkworm; Sánchez-Muros *et al.*, 2014). However, many more species will likely follow in the future. In addition, a number of insect species are mass-reared for use in biological control (Vreysen *et al.*, 2007).

The biology of these species is understood to a level that facilitates domestication, but still a lot is unknown and the true potential of these and other insect species remain to be fully unfolded. For instance, this includes:

- development of diets to meet species- and life stage-specific nutritional needs leading to maximal reproduction or biomass yield/turnover;
- optimization of feeding frequency and feed load in the production trays to facilitate the best feed conversion efficiency and reduce impact of internal competition amongst juveniles;
- optimization of reproductive output through ensuring that mainly gravid females inoculate production trays (e.g. by selection of production populations with a skewed sex ratio towards more female individuals per male);
- optimization of juvenile density to enable optimum biological temperature in the production trays and hence maximum output during the biomass production cycle, e.g. by differentiating ambient temperature in the production area according to the developmental stage of the juveniles;
- increasing production output by increasing the number (height) of trays per square meter in the production facilities (which will again challenge the biological temperature experienced in the individual trays); and
- reduction of disease risk (e.g. facilitated through application of probiotics in the diets), which will inevitably increase with increased production amount and with insect density.

The development of efficient and sustainable insect production for food and feed needs a common understanding and direction. This need has for example been recognized by the International Platform of Insects for Food and Feed (http://ipiff.org). Currently, European insect producers can deliver products adhering to relevant EU regulations regarding food and feed grade quality. However, there are clearly many challenges to overcome before insect production reaches a more mature and industrialized level, and the interaction between multiple disciplines is needed to facilitate this transition (Smith and Barnes, 2015); e.g. integration of biology and engineering by implementation of automation and digital visual monitoring to overcome the physical workload and economic costs of handling future factories with more than e.g. 100,000 production trays stacked in racks of several meters in height.

9.3 Quality production animals

Maintaining genetic quality of mass-reared insects is a key component of commercial production of insects for food and feed. However, mass- or laboratory-rearing have potential effects on both short- and long-term performance via environmental effects of the rearing facility, effects of diet, loss of genetic variation, laboratory adaptation and inbreeding depression with following potential changes to life history traits, tolerance and robustness. To evaluate the consequences, to allow optimization through artificial selection (breeding), and to potentially counter any unwanted changes, there is a need to gain knowledge of the basic biology of the organism in question, including the basic population genetics and the effects of temperature, diet, and density. Research on insect thermal biology, physiological and morphological changes of individuals within generations (phenotypic plasticity) and the potential to change traits through artificial selection has revealed knowledge of relevance for applied insect mass-rearing, including maintaining or further improving quality in this production (Sørensen *et al.*, 2012).

9.3.1 What characterizes a high-quality production animal?

For efficient production, important characteristics of production animals are fast growth and maturation and high survival to harvest or reproduction. In addition, large body mass is preferred to increase biomass production, as well as a high reproductive capacity to ensure offspring for the next generation (Honěk, 1993). Furthermore, it is an advantage if the animals have a high content of desired or valuable nutrients, e.g. particular fatty acids or minerals, or a high relative protein content which increases the value of the produced biomass. Importantly, high-quality production animals should have high feed, especially protein, utilization efficiency to convert their feed into body growth and proteins. Since space is also an important limiting factor in production, it is desirable if animals can be maintained at high density without disease outbreaks or destructive aggressive encounters among individuals. In order to maintain healthy populations, conditions should be optimized to maximize performance across generations. This includes establishing breeding programs that reduce inbreeding and genetic drift to ensure presence of the required genetic diversity for maximal performance and enable evolutionary changes of the populations. Additionally, environmental factors such as temperature and diet should be optimized to match the requirements at different life-history stages. Although much research is required to establish optimal conditions for individual species and strains, some general principles and patterns may apply to insect production at broad.

9.3.2 Possible discrepancies between the outcomes of artificial and natural selection

As in other domesticated animals, artificially selecting for traits favourable for biomass production and survival under domestication is likely to produce domestic populations that are adapted to conditions under domestication and production, which likely differ in some aspects from those environments that populations were adapted to in nature. This could be caused by shifts in the genetic regulation of a number of physiological trade-offs that further control life-history related trade-offs. For example, if there is always ample feed under production conditions, animals may evolve high consumption capacity and fast growth, but this could be at the expense of low starvation tolerance or feed conversion efficiency. Optimal conditions under production might therefore reduce adaptations to conditions in the wild, which might have negative consequences. If relaxed selection under production conditions also leads to higher sensitivity to environmental challenges, including lower disease resistance or feed utilization efficiency, then populations unintentionally selected for higher consumption and reduced feed utilization efficiency would theoretically be selected for under *ad libitum* feeding conditions.

9.3.3 Density and costs to disease resistance

Under production conditions, insects are typically maintained at high density, which increases disease risk. In response, immune defences are upregulated to increase disease resistance (Barnes and Siva-Jothy, 2000; Wilson *et al.*, 2001, 2002), which in turn is shown to incur costs and trade-off with other life-history traits such as growth and reproduction (Bascuñán-García *et al.*, 2010; Schwenke *et al.*, 2016). Therefore, high density not only increases the risk of disease outbreaks, but may also change individual resource allocation towards defence at the cost of biomass production. In insects, the mechanisms of disease resistance include the formation of a thicker and less permeable cuticle, and the production and maintenance of phenoloxidase enzymes in the haemolymph, which requires protein and energy (Lochmiller and Deerenberg, 2000). This may cause smaller individuals with thicker cuticles and a lower efficiency of feed utilization for nutrient production. At the same time, maintaining a natural defence to resist potential disease risks in the population is important to prevent outbreaks and to avoid the necessity of applying pesticides to control bacteria and fungi. Although pesticides should be avoided to the extent possible to ensure a trace-free final product, regular cleaning the insect facilities with detergents is undoubtedly essential for preventing disease and maintaining healthy populations. Furthermore, providing a substrate for physical structure is essential to utilize volume for space within enclosures and increase distance between individuals to reduce pathogen transmission and aggressive encounters. Similar hygiene precautions are standard for domestic animals traditionally kept for production purposes. Developing a new industry based on insects for food and feed will also demand the development of antibiotics and the education of veterinarians with the required skills.

9.4 Genetic aspects of mass production

Mass-rearing of insects for food and feed is still in its startup phase in most parts of the world, and genetic issues currently do not have major focus (Sørensen *et al.*, 2012). We anticipate this will change and advocate producers of insects to adapt the extensive knowledge generated in basic research disciplines such as population and quantitative genetics and statistics. This knowledge has been utilized successfully in applied sciences such as animal and plant breeding for many decades, and we suggest that it can be successfully used in breeding of insects as well. Breeding for traits of economic importance in domestic animals has led to tremendous genetic progress. For example, the same genetic principles applied to breeding programs that optimize growth rate and egg, milk and wool production as well as general health in domesticated animals, can also be applied to the breeding of more disease resistant insects with higher protein content or feed conversion efficiency. Animal breeding is based on the principles of identifying genetically superior animals with desired traits and high breeding values, and using these to produce the next generation. Over generations, such selection will change allele frequencies within populations, and thus the general expression of the preferred traits.

If insect breeders take advantage of the experience and knowledge that already exists from breeding programs on traditional production animals, the capacity of their production would most likely increase compared to that of a randomly mating population. This also includes realizing the challenges and trade-offs that are faced by traditional animal breeding programs. For example, to maximize the gain of alleles that promote a desired trait within a population, only a small number of animals possessing this trait should be used in breeding, which increases the intensity of selection for the trait (Falconer and Mackay, 1996). This in contrast causes a small effective population size (N_e), which increases the rate of genetic drift and inbreeding (see section on 'Population genetics for animal breeding'). Another challenge in selective breeding is the fact that some traits are genetically correlated, either due to close genetic linkage or to genetically determined physiological correlations and trade-offs. This implies that selecting for one trait indirectly selects for the other trait as well, and it can be very difficult or impossible to separate the two traits. For example, cattle breeders are challenged by the fact that fertility and disease traits are often unfavourably genetically and phenotypically correlated with milk production and other production traits of economic interest (Rauw *et al.*, 1998). High-producing dairy cattle thus typically suffer from mastitis (udder infection), have increased mortality, and ca. 30% of the population is affected by infertility and reproductive disorders (Erb and Martin, 1980).

In this section, we aim to give a brief overview of some of the genetic opportunities and challenges associated with breeding in domestic populations. For a more thorough understanding we refer to texts dedicated to these topics (Falconer and Mackay, 1996; Lynch and Walsh, 1998).

9.4.1 From gene to phenotype

The phenotype of an individual is affected by its genotype and environmental factors, and by possible interactions between genotype and the environment (Figure 9.1). To rank animals according to their genetic suitability for production, it is important to be able to account for factors such as herd or strain, year, and season. The more pedigree and performance information is available, the more accurately an individual's genetic potential and value for breeding can be estimated. To correct for systematic environmental effects, so-called best linear unbiased prediction models are often used (Henderson, 1973). These statistical models utilize all available information from the animal itself and from genetically related animals. Breeding values are normally estimated for a number of economically important traits. Each of these traits is weighted based on the proposed economic importance and often collected in a total merit index on which selection of animals can be based (Kinghorn *et al.*, 2000).

Keeping track of pedigrees and estimating breeding values for individual animals is challenging from a practical point of view, and feasible alternatives for insect breeding exist. These include group selection where selection is based on characteristics of a group of individuals (Wade *et al.*, 2010), and phenotypic selection where those animals with specific characteristics are used as parents for the next generation. Such breeding procedures do not demand keeping track of pedigrees, and despite not being as efficient as selection based on pedigree information, it can lead to genetic progress (Bubliy and Loeschcke, 2005). Novel tools in molecular biology and bioinformatics also allow for genomic selection, which is a new approach for improving quantitative traits in large animal breeding populations that uses whole genome molecular markers (high-density markers and high-throughput genotyping). Genomic prediction combines marker data with phenotypic and pedigree data (when available) in an attempt to increase the accuracy of the prediction of breeding and genotypic values (Goddard and Hayes, 2007). These approaches are currently used in many domestic animal species and are expected to lead to higher genetic progress compared to traditional methods (Schaeffer, 2006). This is due to more accurate predictions of breeding values, higher intensities of selection and shorter generation intervals (Pryce and Daetwyler, 2012). The cost of genotyping individuals at thousands of genetic loci has decreased. For non-model insects of relevance for consumption, genomic selection could be routinely used to efficiently select for traits of selection within the next decade. This technique can also be used to ensure that levels of inbreeding are kept under control (Pryce and Daetwyler, 2012).

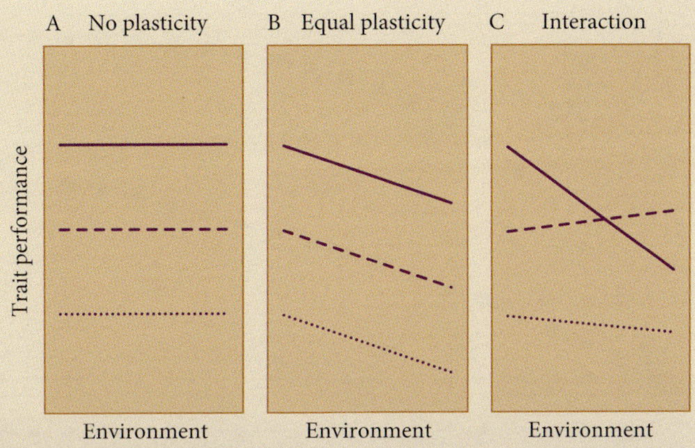

Figure 9.1. The relationship between an environmental variable and the performance of a given phenotypic trait in three different genotypes. Panel A shows no plasticity, i.e. the trait value is independent of the environmental factor, even if different genotypes show different trait values. Panel B shows equal plasticity of each genotype, i.e. the trait value depends on the environmental factor, but each genotype responds in an identical manner. Panel C shows a more complex response where each genotype has its own degree of plasticity in response to the environment. This complex response depicts a biologically realistic response that calls for genotype- and trait-specific investigations.

9.4.2 Population genetics for animal breeding

Population genetic theory predicts an association between population sizes, levels of inbreeding, and random genetic drift. Compared to larger populations, smaller populations are expected to experience: (1) greater genetic drift leading to loss of (primarily rare) alleles thus lowering levels of genetic variation within populations; and (2) higher rates and levels of inbreeding (Figure 9.2), potentially causing inbreeding depression (i.e. a decrease in fitness due to increased homozygosity). Genetic drift is a random process leading to changes in allele frequencies owing to chance events that determine which alleles are carried over to the next generation. Genetic drift influences the likelihood that populations will maintain the same genetic constitution across generations, and the impact of drift on this process is greater in smaller populations. Inbreeding is defined as the mating between two individuals in a population that are more closely related to each other than two randomly selected individuals from the population. Inbreeding increases the level of homozygosity in the population and thereby the expression of recessive deleterious alleles that are typically rare and hidden in heterozygotes in large outbreeding populations (Kristensen and Sørensen, 2005).

A central concept in population genetics for the quantification of levels of inbreeding and genetic drift is the effective population size (N_e) (Figure 9.2). The effective population size is the size of an idealized population that would experience the same amount of inbreeding and genetic drift as the population of interest. In popular terms, N_e is the number of individuals

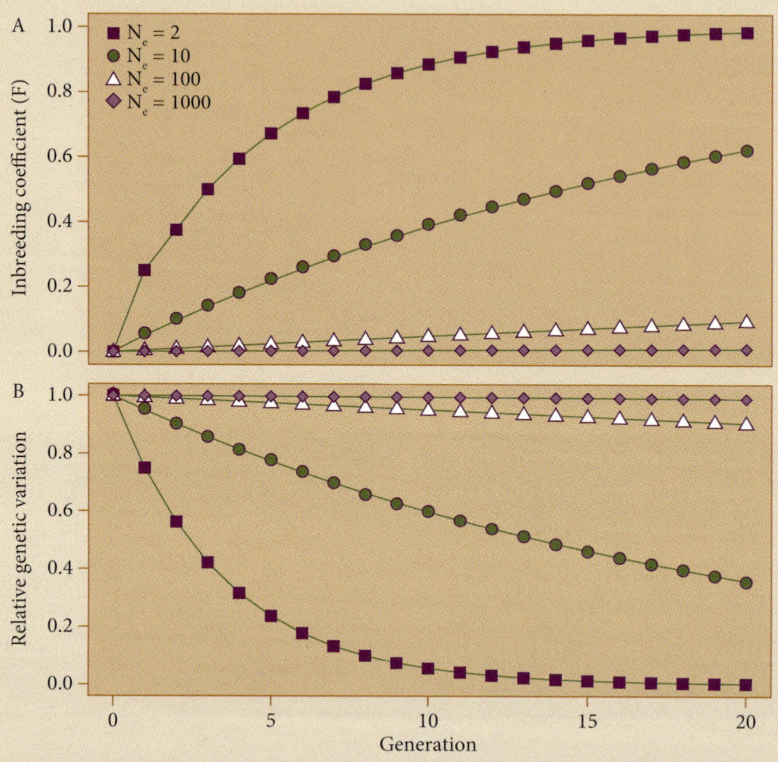

Figure 9.2. The theoretical relationship between (A) inbreeding coefficient and (B) loss of genetic variation across generations given different constant effective population sizes (N_e). At very low N_e, the increase in inbreeding and loss of genetic variation is progressing at a high rate, however, at high effective population sizes, detrimental effects do accumulate to noticeable levels across generations. As production animals can be expected to be maintained for many generations, it is important to consider the balance between adaptation of the breeding stocks and the accumulation of detrimental genetic effects.

in a population that contribute offspring to the next generation. The ratio between the census size, N (the total number of individuals in a population) and N_e (N/N_e) is typically around 0.2 (Frankham, 1995; Palstra and Fraser, 2012), but this ratio is highly species- and environment-specific (Nunney, 1995). The N/N_e ratio can be lowered for a multitude of reasons including a low number of founders establishing a population, general low or highly fluctuating population size, skewed sex ratios or assortative mating. Quantification of N_e can be laborious, but good estimates can be obtained from molecular data or from knowledge about the number of individuals founding a population and the number of females and males reproducing. Expected loss of genetic variation per generation and the increase in inbreeding per generation, typically estimated by the inbreeding coefficient F (which is the probability that an individual has received the same allele from its mother and father descending from a common ancestor), is the inverse of twice the effective population size ($1/[2 \times N_e]$). Numerous assumptions should be fulfilled to justify the use of this formula. However, the estimates of levels of inbreeding and genetic drift obtained from the formula are simple and useful practical tools to judge the 'genetic health' of populations. Larger effective population sizes are needed to avoid the negative impacts of inbreeding on fitness traits and long-term evolutionary potential. The exact required number is debated in the literature, but current estimates suggest that N_e of a few hundred are needed to avoid negative impacts of inbreeding and at least one thousand to maintain enough genetic variation to allow for evolutionary adaptation (Frankham *et al.*, 2014; Jamieson and Allendorf, 2012).

Based on this short introduction to the concepts of inbreeding and genetic drift it becomes clear that in sexually reproducing species and from a genetic perspective, larger N_e is better (Figure 9.2). At the same time, a highly adapted breeding strain with robust survival and growth but little genetic variation might facilitate the most constant and high-yielding production, as each individual is performing maximally under the given conditions with little variation between individuals. In the case of the introduction of disease or other changes in the production environment, however, the potential for evolution and adaptation to the new circumstance is limited if N_e is low. This take home message is relevant in diverse biological disciplines, including conservation and evolutionary biology and animal and plant breeding, and also applies in the emerging field of producing insects for food and feed. We propose that tools and lessons from population-, quantitative- and evolutionary genetics are important for this field and that those lessons should be actively used. In the following, we propose a more detailed discussion of the effects of inbreeding and how inbreeding and drift can be effectively managed.

9.4.3 Managing inbreeding and genetic drift

There is evidence in the literature that inbreeding causes inbreeding depression in all sexually reproducing species. This is especially manifested in traits closely associated with fitness, and often inbred individuals have decreased disease resistance and reproductive output and increased developmental time compared to less inbred individuals (Bechsgaard *et al.*, 2013; Kristensen and Sørensen, 2005). Also morphological traits in recently domesticated animal populations often express inbreeding depression in the form of reduced daily weight gain and body size (Nagy *et al.*, 2013), and with more morphological defects observed in inbred compared to outbred individuals (Mansfield and Land, 2002; Olsson *et al.*, 1996). Finally and importantly, inbreeding has been shown to have negative impact on the quality of food products (Gjerde *et al.*, 1983; Mc Parland *et al.*, 2008; Neira *et al.*, 2016). In conclusion, there is strong evidence that inbreeding has a negative impact on traits that will be important for breeding cultures of insects for food and feed. Thus, we advocate that emphasis is put on reducing rates of inbreeding in insect breeding programs.

Strong genetic drift will reduce genetic variation and thus the potential for improving performance in selected traits through artificial selection. Responses to natural and artificial selection partly depend on standing genetic variation, and loss of genetic variation will over time lead to reduced responses. In studies from domestic animals, model organisms, and natural populations, this claim is supported although there is some recent controversy regarding the association between the speed of evolutionary change and the amount of genetic variation (Hoffmann *et al.*, 2017; Kristensen *et al.*, 2015; Wood *et al.*, 2016). As for other domestic populations, breeding programs can be designed for mass-rearing of insects to be used for food and feed. For these programs to successfully change the mean and variance of traits of interest, genetic variation is a prerequisite, and maintenance of genetic variation through reduction of genetic drift is therefore essential.

9.4.4 Genetic management

Inbreeding rates and levels of genetic drift can be controlled and kept low in several ways, and to do so maintaining a high N_e is of key importance. First, the number of individuals founding a population is important. Second, keeping the population size high with approximately equal sex ratio across generations is another mean to reduce inbreeding and genetic drift. Third, crosses between independent populations will increase N_e and reduce genetic drift. This latter approach is sometimes used in conservation genetics to rescue genetically depauperate populations, and similarly crossbreeding between different breeds of livestock and plants is typically practiced to reintroduce genetic diversity (for recommendations regarding genetic management of populations see Frankham *et al.*, 2010; Meuwissen, 2009). In cases where it is not known whether populations suffer from inbreeding depression and / or low levels of genetic variation, experiments can be performed to investigate whether traits of production relevance suffer from inbreeding depression (e.g. by crossing different populations), and molecular or quantitative genetic approaches can be used to assess levels of genetic variation and N_e in the populations. Thus, numerous tools are available that can be used to secure control of inbreeding and genetic drift in farming of insects used for food and feed.

9.4.5 Low ability to change some production traits through selection

It is well known in evolutionary and quantitative genetics that some traits are easy to change through selection whereas it is more difficult for others (Hoffmann *et al.*, 2016; Mousseau and Roff, 1987). Generally, traits closely associated with fitness have lower heritabilities compared to morphological traits (Hoffmann, 2000; Mousseau and Roff, 1987). A commonly held view within the animal breeding world is that all traits can be readily changed through selection if sufficient techniques and selection intensities are applied (Hill and Zhang, 2004; Walsh and Blows, 2009). However, more and more studies point to low additive genetic variance and heritability in key traits also in livestock (Hoffmann *et al.*, 2016). Thus, in relation to the large-scale production of insects for food and feed, it is important to be aware of these limitations in the ability of some traits to change through selection. Reasons for evolutionary constraints include low heritability due to little overall genetic variation, unfavourable genetic correlations, or stem from periods of intense directional selection resulting in fixation of favoured alleles. Low realized heritabilities might also reflect changing environmental conditions and a low genetic correlation for the same trait across environments. Finally, a loss of allelic variation due to DNA decay will constrain evolutionary responses. The DNA decay hypothesis proposes that if selection on a trait is weak, mutations will accumulate, potentially resulting in loss of function of genes, leading to reduced functional genetic variance (for more details see Hoffmann *et al.*, 2017).

9.4.6 Importance of maintaining natural and sexual selection

To maintain high fitness potential within insect breeding populations, it is advised to promote conditions that do not eliminate natural selection forces. Even if we could identify and include all traits contributing to production performance, it would be impossible at the same time to account for fitness related traits during a selection event or breeding scheme. For example, if individuals are allowed to mate under conditions where females have access to several males, this may help maintain high genetic quality within the populations, as the fittest males most likely attain most matings (Whitlock and Agrawal, 2009). It is unlikely that the insect breeder would be able to make as precise an evaluation of male quality as the female insect can. Since sexual display is known to be condition dependent, this would theoretically speed up adaptation to production conditions as the best adapted males that are able to allocate resources to the most physiologically demanding sexual display will sire a greater number of offspring. In addition, the strongest males typically prefer mating with the largest, most fecund females. Free mating conditions including male-male competition and female choice might therefore be an important constituent in healthy animal breeding that maintains genetic quality and increases adaptation to breeding conditions (Hunt *et al.*, 2004). For example, males that invest heavily in an unrequired disease defence or have poor utilization of a limited nutrient have fewer resources to allocate to sexual display and are not favoured in sexual selection. Therefore, it might be advantageous to establish a mating environment where male-male competition and female response to male sexual advertisement facilitate that the fittest genes are selected for the next generation. Condition dependence of the sexual display is shown to be sensitive not only to temperature and parasite pressure, but also to the nutritional composition of the available diet (Rantala *et al.*, 2003; Rapkin *et al.*, 2017), and there is high potential for sexual selection as a mechanism to facilitate adaptation.

9.5 Optimal temperature and thermal adaptations

Temperature affects all aspects of ectothermic life (Clarke, 2014). Thus, insect metabolism, population, and individual growth rates are generally positively related to temperature, while longevity and adult body size are negatively related (Booth and Kiddell, 2007). Especially small ectotherms will have body temperatures that follow the temperature of the surroundings closely, even though some insects have some scope for manipulating body temperatures via wing ventilation or flight muscle heating (Harrison *et al.*, 2012). Furthermore, under artificial conditions there will be limited opportunity for manipulating body temperatures through behavioural choice, as production facilities usually will provide limited habitat heterogeneity. This limitation makes the selection of temperatures a key parameter, which can be manipulated with potential impact on production. Insects are, as mentioned above, affected by temperature on a range of inter-related traits (Angilletta, 2009; Clarke, 2003; Hoffmann *et al.*, 2003). For each trait, the relationship between temperature and performance is given by a thermal performance curve, which for most insects is similar to what is shown in Figure 9.3. Typically, the insect-thermal performance curve is skewed towards low temperatures with a steep decline as temperatures raises above optimal temperature (the peak of the curve).

Optimal temperatures might differ for different traits and/or life stages. Furthermore, there are expected to be trade-offs between important production traits like development time and size – and thus potentially between population growth rate and biomass production. This means that animals for breeding could easily have different optimal temperatures than animals for production (biomass). Production and growth might be higher at higher temperatures, whereas low temperatures (at least during juvenile development) is expected to lead to larger size and

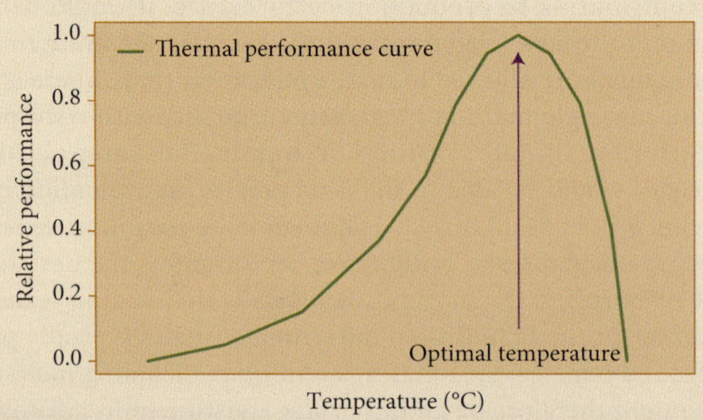

Figure 9.3. A theoretical thermal performance curve, with relative performance (y-axis) against temperature (x-axis). Both stress tolerance and life history traits (thermal tolerance, viability, development time, fecundity, activity) generally follow this scheme. However, curves are not expected to be identical across traits or life stages. Thus, what constitutes an optimal temperature for one trait or life stage might constitute a suboptimal temperature for another trait or life stage, making selection of a single 'optimal' temperature impossible.

thus larger fecundity, but might also mean that laboratory breeding stocks will spend fewer generations in the production facility leading to a decreased genetic erosion through genetic drift and greater genetic robustness against biotic and abiotic challenges (see earlier sections). The optimal temperature for a given trait is not necessarily a fixed value. Insects have potential for thermal plasticity, a phenomenon where the same genotype produces different phenotypes (on some level or trait) depending on the environmental conditions of the organism (Debat and David, 2002). While thermal plasticity mainly has been investigated with regards to extreme thermal tolerance (Angilletta, 2009; Hoffmann *et al.*, 2003), the phenomenon is widespread and has evolved in many forms and for many types of characters in insects (e.g. physiological, morphological, and behavioural) (reviewed in Chown and Terblanche, 2007; West-Eberhard, 2003; Whitman and Ananthakrishnan, 2009). Plasticity might thus affect a range of relevant production traits, and can be both beneficial and maladaptive. Overall, the types of responses can be divided into: (1) beneficial acclimation, where animals will perform best at the temperature they are acclimated to; (2) colder is better, where animals will perform best when acclimated to low temperatures; or (3) warmer is better, where animals will perform best after acclimation to high temperatures (Huey *et al.*, 1999). Most types of thermal plasticity (thermal acclimation) are expected to be advantageous. However, it is not well understood how different traits will respond to acclimation, and there appears to be a high degree of trait specificity. Knowledge of thermal plasticity has the potential to be developed into a valuable tool for manipulating and optimizing insect quality in different thermal environments. Thus, knowledge of the biology of an animal in question and the important traits and how they respond is needed to evaluate effects of temperature and acclimation.

The vast majority of research on thermal biology in insects is performed at constant temperatures (Angilletta, 2009; Hoffmann *et al.*, 2003), and similarly, most insect mass-rearing facilities use constant temperatures as this is expected to support predictable, homogenous production environments. In nature, terrestrial ectotherms are rarely exposed to constant thermal conditions. While research comparing stable and thermally variable conditions such as daily fluctuations indicate that variable temperatures are more stressful than constant temperatures for some traits, the outcome of exposure to fluctuating temperatures seems to depend on the amplitude, predictability and periodicity of fluctuations (Manenti *et al.*, 2014). Thus, variable temperatures might also induce beneficial responses by activation of molecular mechanisms that increase overall performance under unstable thermal conditions (Sørensen *et al.*, 2016). For example, wild caught winter acclimatized *Drosophila subobscura* had superior heat and cold tolerance compared to their laboratory conspecifics maintained at constant temperature (Sørensen *et al.*, 2015). Thus, heterogeneous rearing conditions might create more thermally robust animals, but likely at the expense of optimal physiological adjustment and performance compared to rearing at one stable temperature where many of the activated mechanisms are not required (on both a short and long time perspective), and fewer resources are therefore available for growth. These relationships need to be more thoroughly investigated to draw well-informed decisions for mass-rearing, as different rearing/maintenance conditions might also directly affect growth and body composition.

Protocols for mass-rearing have been developed for several insect species used for biological control, where predators of pest species or sterilized conspecifics (sterile insect technique) are produced and released in the treated habitats. A major challenge of such efforts is to identify what constitutes high quality and to then produce such high-quality insects (Sørensen *et al.*, 2012; Terblanche and Chown, 2007). Based on results from *Drosophila melanogaster* (Kristensen *et al.*, 2008), Chidawanyika and Terblanche (2011) showed that thermal acclimation during rearing could be used to improve field performance after release, at least temporarily. Thus, the thermal conditions applied during rearing directly predicted performance under similar field conditions. While field fitness or field performance is not directly a concern for mass-rearing of insects for food and feed, this example shows that relatively short-term thermal pre-treatments could be a feasible avenue for manipulation of beneficial traits, and that evolutionary and physiological knowledge from basic evolutionary and eco-physiological research are directly applicable for applied insect mass-rearing in an inexpensive fashion based on cross-disciplinary approaches and application of relatively basic techniques.

9.6 Optimal diet in production

Overall, the nutritional requirements of insects are comparable to those in other production animals. Important goals are to optimize the balance between survival (product amount), time until harvest (development time), product quality (nutritional value), and reproduction (offspring for the next generation) to maximize the total economic value of the production. However, the optimal diet might be more economically costly than a less optimal diet, and economic costs must be considered. The optimal diet depends on the species, the life stage and sex, and contains all the nutrients required at the specific life stage in the right amounts and proportions. Nutrients are required for metabolic energy and as building blocks for growth and maintenance of body structure and in the functioning of physiological processes. To maximize performance, insects need to consume required nutrients not only in the right amounts but also in the right proportions (Figure 9.4). The nutritional blend that matches requirements and maximizes fitness changes over time, and is affected by nutritional history, life stage and sex, and by environmental conditions such as temperature and immune challenge. Juvenile development and early adult fecundity and reproduction are the life-history traits that are most important in production insects since individuals are harvested either as late juveniles or early adults.

9.6.1 Essential diet components

In addition to macronutrients, a range of micronutrients including vitamins, salts, and minerals, essential lipids and sterols (e.g. linoleic acid, oleic acid and cholesterol), and 10 specific amino acids (methionine, tryptophan, arginine, lysine, histidine, phenylalanine, isoleucine, threonine, leucine, and valine) are essential in the diet. Comprehensive information on insect nutritional requirements under rearing conditions is given in Cohen (2015). Diets can be successfully manufactured from organic by-products (Oonincx *et al.*, 2015; Van Broekhoven *et al.*, 2015), which implies that although the basic ingredients are known, each ingredient may contain a range of unknown constituents, including micronutrients, phagostimulants and growth factors that may influence consumption and physiological resource allocation and affect performance and life-history trade-offs.

9.6.2 Macronutrients

The macronutrients contain energy and are required in the greatest quantities. They are divided into three major groups – proteins, carbohydrates, and lipids – based on their biochemical

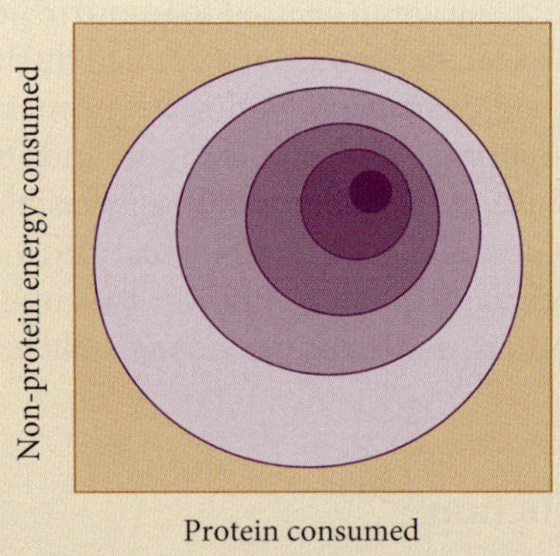

Figure 9.4. Hypothetical performance landscape showing the composition of protein vs non-protein energy (from carbohydrates and lipids) where performance is maximized (dark purple), and the gradual decline in performance as one or both constituents are eaten in suboptimal amounts. Whereas amino acids from protein are required as building blocks for body proteins and therefore growth, carbohydrates and lipids are valuable mainly as energy sources. Empirical nutritional performance landscapes are available in e.g. Simpson and Raubenheimer (2012), Simpson *et al.* (2015), Jensen *et al.* (2015), and Rapkin *et al.* (2017).

composition. Whereas carbohydrates and lipids are mainly utilized for metabolic energy, proteins are composed of amino acids, which are required as building blocks in synthesis of new proteins for tissue growth. Proteins contain nitrogen, which is often a limiting element in natural environments, and herbivorous and omnivorous insects have adapted by utilizing protein at high efficiency particularly under protein-poor conditions. This in turn involves little adaptation for handling protein excesses, and energetic requirements should therefore be covered from carbohydrates and lipids to support stable development and longer lifespan.

9.6.3 Macronutrient balance and life-history trade-offs

Different performance parameters have different nutritional requirements (Bunning *et al.*, 2015; Jensen *et al.*, 2015; Lee *et al.*, 2008a; Rapkin *et al.*, 2017; Simpson and Raubenheimer, 2012). Protein consumption increases growth and development and thereby shortens development time to the adult stage. Furthermore, immature insects fed a relatively protein-rich diet typically attain larger adult body size (Han and Dingemanse, 2017), and females with relatively high protein consumption have greater reproductive rates. In contrast, high protein consumption shortens lifespan, and the highest lifetime reproductive output is attained at an intermediate consumption of protein relative to carbohydrate and lipid. Survival to the adult stage also appears to be highest at intermediate macronutrient ratios where growth rate is not maximized. Individuals that have developed on a protein-rich diet are lean but large and rich in protein, whereas individuals that have developed on a protein-poor but energy-rich diet are rich in lipids and can endure longer periods of starvation (Lee and Jang, 2014; Rho and Lee, 2014).

9.6.4 Protein utilization efficiency

Several insects including caterpillars, grasshoppers and *T. molitor* are shown to utilize ingested protein less efficiently for body growth once a certain requirement has been obtained (Lee *et al.*, 2002; Raubenheimer and Simpson, 1997; Rho and Lee, 2014; Simpson *et al.*, 2002). Therefore, whereas insects typically grow faster when given a high proportion of protein in their diet, the protein-economy of the production facility may be less optimal. In contrast, lipids are stored in direct proportion to ingested energy from carbohydrates and lipids. Lipid-rich insects can therefore be efficiently produced by increasing the ratio of lipids and carbohydrates relative to protein in the diet, whereas insects reared on protein-rich diets are lean. The most efficient production of protein body mass while keeping lipid content low is attained by adding protein at the highest possible ratio where it is still utilized efficiently for tissue growth while carbohydrates and lipids are used for energy. For *T. molitor*, the optimal ratio lies around a 1:1, while for cockroaches a protein to carbohydrate 1:2 appears optimal for growth (Raubenheimer and Jones, 2006; Rho and Lee, 2016).

9.6.5 Nutrient self-selection from complementary feed sources

Premixed diets prevent selective consumption of specific nutrients if properly mixed, and life-history traits can thereby be manipulated through the diet. In contrast, dietary choice between nutritionally complementary feed sources allow individuals to compose an optimally balanced diet maximizing their fitness by optimizing juvenile growth rate and maximizing lifetime reproductive output (Lee *et al.*, 2008a; Simpson and Raubenheimer, 2012; Simpson *et al.*, 2015). Since dietary requirements change over time, individuals with a choice have the chance to continuously target the nutrient composition which maximizes their performance, and individuals given choice between nutritionally complementary feeds can therefore reach higher performance than individuals restricted to a single 'optimal' diet (Bunning *et al.*, 2016; Senior *et al.*, 2015).

9.6.6 Nutrient balance and reproduction

When given dietary choice, *D. melanogaster* females maximize lifetime reproduction by balancing protein and carbohydrate intake, but their reproductive rate is maximal if they are restricted to a more protein-rich diet at the cost of shorter lifespan (Jensen *et al.*, 2015; Lee *et al.*, 2008a). Early or short-term fecundity can thus be improved by restricting reproducing females to a more protein-biased diet than they would self-select. In contrast, adult males are typically most attractive if they are given a carbohydrate-rich diet after eclosing to adults, since their sexual displays are typically energy-demanding and often is directly proportional to the male's energetic resources (lipid reserves). Recent evidence suggests dietary preference may be linked between the two sexes so neither sex reaches its nutritional optimum when given dietary choice (Jensen *et al.*, 2015; Maklakov *et al.*, 2008; Reddiex *et al.*, 2013). Males also require protein for sperm production however, and their dietary choice might balance attractiveness with other traits such as viability (Bunning *et al.*, 2015). Juvenile protein availability also affects adult reproduction as females have the highest reproductive potential at larger size (Honěk, 1993) and typically prefer large males (Hunt *et al.*, 2009).

9.6.7 Nutritional effects on immunocompetence

There are many likely diseases in insect production due to high rearing density (Eilenberg *et al.*, 2015), and poor nutrition leads to increased parasitism and disease risk (DeGrandi-Hoffman and Chen, 2015). Furthermore, mounting an immune response is costly, even if there is no disease (Moret and Schmid-Hempel, 2000; Schwenke *et al.*, 2016). Insect nutritional immunology is a rising field in insect science (Ponton *et al.*, 2013), and diets, which are optimal for growth and survival, are not necessarily optimal for immune functioning (Cotter *et al.*, 2011). In honey bees (*Apis mellifera*), a diverse diet including pollen from a range of plant species was found to promote higher immunocompetence levels (Alaux *et al.*, 2010). A superior immune response is found in caterpillars fed a higher proportion of protein in their diet, and infected individuals selectively consume more protein when given choice (Lee *et al.*, 2006; Povey *et al.*, 2009, 2013). Furthermore, the quality of different protein sources is shown to affect caterpillar immunocompetence by promoting the immune response unevenly (Lee *et al.*, 2008b), and also insect genotype interacts with dietary composition on immune functioning (Unckless *et al.*, 2015a). The immune defence can also be enhanced by including antioxidants in the diet (Ojala *et al.*, 2005).

9.6.8 Adaptation to rearing diet

Omnivorous and herbivorous insects have evolved to efficiently utilize protein that typically limits growth and reproduction in nature. Since protein is relatively expensive, the ability to efficiently utilize protein is one reason why insects should be used in sustainable production of protein for food and feed from organic by-products (Van Huis *et al.*, 2013). Therefore, maximizing growth and reproduction by adding protein to the diet should be balanced against the economic costs of protein, with high protein utilization efficiency maintained within populations. Recent studies with *D. melanogaster* show the amount of food consumed (Garlapow *et al.*, 2015) and the preference of protein-to-carbohydrate balance (Reddiex *et al.*, 2013), are heritable. Furthermore, there is significant genetic variation in the nutritional status of individuals exposed to the same diet (Unckless *et al.*, 2015b), and evidence for a preference for the ancestral diet as well as higher performance on the ancestral diet after many generations of breeding on nutritionally specific diets (Warbrick-Smith *et al.*, 2009). Caterpillars selected on a carbohydrate-rich diet over generations evolved less efficient utilization of ingested energy for lipid storage (Warbrick-Smith *et al.*, 2006). If selected on a protein-rich diet, omnivorous insects such as crickets potentially evolve a more carnivore-like physiology involving respiration of protein for energy instead of efficiently incorporating protein into tissue growth. Keeping populations at a protein-limited stage might therefore be important for maintaining a high protein utilization efficiency within populations.

9.6.9 Toxins

No toxic elements should be in the diet as these will reduce the quality of the insects both by reducing their fitness but more importantly by compromising food safety as the insects may be toxic to the consumer. For example, *S. gregaria* and other orthoperans can include deterrent plant components in their diet when competition for food is high, and in turn become toxic to their predators (Despland, 2005; Sword *et al.*, 2000). Insect species not possessing the biochemical capabilities to synthese toxic compounds, which some species produce for deterrence against predators, should be exploited (Pasteels *et al.*, 1983).

9.6.10 Diet-temperature interactions

As both diet and temperature affect insect life histories, the two factors are likely to interact, although such interactions are not extensively studied. However, some knowledge on the interactions between diet and temperature does exist (Clissold and Simpson, 2015; Clissold *et al.*, 2013; Davidowitz and Nijhout, 2004; Davidowitz *et al.*, 2003; Mirth and Riddiford, 2007). There is a general relationship between temperature and size leading to larger adult insects at lower temperature, and diet appears to play a critical role in modulating this relationship. In caterpillars of *Manduca sexta*, poor diet quality was found to even reverse the temperature-size rule, so individuals reared on a poor quality host plant become larger at high than at low temperature (Diamond and Kingsolver, 2010). In a nutritionally explicit study in *Spodoptera exigua* caterpillars, the ratio of protein to carbohydrate was varied, and the temperature-size rule was not reversed on any diet, but higher temperatures had a significantly positive effect on growth rate when the diet was nutritionally balanced, while individuals on imbalanced diets had more similar growth rate across temperatures (Lee and Roh, 2010). In another study on *Spodoptera litura*, pupal mass was highly affected by temperature following the temperature-size rule in caterpillars reared on a diet with a high ratio of carbohydrate to protein, but was much less affected by temperature in caterpillars reared on more balanced diets (Lee *et al.*, 2015). Together, these studies show that a nutritionally balanced diet buffers the effect of temperature. Furthermore, nitrogen digestion efficiency in *S. exigua* is lower at high temperature, which involves lower protein utilization efficiency and therefore higher protein requirements to maximize growth at high temperature (Lemoine and Shantz, 2016). Thus, a high rearing temperature implies that a smaller proportion of the protein in the feed is converted to insect body protein.

Physiological studies in lizards have shown that different nutritional processes have different thermal optima (Angilletta, 2009), and similar results start appearing in insects leading to trade-offs between different nutritional strategies regulated by dietary conditions. Studies in grasshoppers show these insects will attempt to maximize their nutritional outcome behaviourally by locating microhabitats where they can thermoregulate optimally and maximize nutritional outcomes according to the nutritional value of their diet (Clissold and Simpson, 2015; Clissold *et al.*, 2013). When feed is abundant, migratory locusts actively select a high temperature for digestion, maximizing growth at a cost of lower nutrient utilization efficiency (Miller *et al.*, 2009). In contrast, when restricted to feed of poor nutritional quality, the locusts select a lower temperature that maximizes the utilization of ingested nutrients instead of maximizing growth rate (Coggan *et al.*, 2011). While a high temperature maximizes growth and insect biomass production, a lower temperature thus maximizes the conversion efficiency of feed into insect biomass. Feed quality and temperature may also interact on immune functioning and disease resistance, involving effects of density (Shikano and Cory, 2015; Triggs and Knell, 2012), although more studies are required to clearly characterize such interactions.

9.7 Conclusions

Domesticated insects have a large potential for contributing to future global food and feed requirements (Van Huis, 2013; Van Huis *et al.*, 2013). So far, protocols for a handful of species have successfully been established, and these can be considered as models as many more species might have equal or superior potential for becoming production animals of the 21st century. So far, as has been the case for mass-rearing of insects for pest management, the focus of breeders has been on practical matters such as developing artificial diets and optimizing rearing schemes. However, as a next step and to better expose the full potential, we propose theory from animal breeding, genetics, physiology, and evolutionary biology need to be applied to this developing field. With respect to quality in insect mass-rearing for food and feed, two related aspects are especially noteworthy. The first is to maintain a genetically sound breeding stock while exploiting the genetic potential of a given species. The second is to utilize both genetic and environmental conditions for optimizing the production. As life history, tolerance and physiological traits are expected to be related through genetic and evolutionary trade-offs, a detailed knowledge of desired traits and their constraints is needed.

Acknowledgements

K.J., T.N.K. and J.G.S. acknowledge funding from The Danish Council for Independent Research Technology and Production Sciences (DFF – 4184-00248, Optimizing performance of biological control agents used in Integrated Pest Management).

References

Alaux, C., Ducloz, F., Crauser, D. and Le Conte, Y., 2010. Diet effects on honeybee immunocompetence. Biology Letters 6: 562-565.

Angilletta, M.J., 2009. Thermal adaptation: a theoretical and empirical synthesis. Oxford University Press, Oxford, UK.

Barnes, A.I. and Siva-Jothy, M.T., 2000. Density-dependent prophylaxis in the mealworm beetle *Tenebrio molitor* L. (Coleoptera: Tenebrionidae): cuticular melanization is an indicator of investment in immunity. Proceedings of the Royal Society of London B: Biological Sciences 267: 177-182.

Bascuñán-García, A.P., Lara, C. and Córdoba-Aguilar, A., 2010. Immune investment impairs growth, female reproduction and survival in the house cricket, *Acheta domesticus*. Journal of Insect Physiology 56: 204-211.

Bechsgaard, J.S., Hoffmann, A.A., Sgró, C., Loeschcke, V., Bilde, T. and Kristensen, T.N., 2013. A comparison of inbreeding depression in tropical and widespread *Drosophila* species. PLoS ONE 8: e51176.

Booth, D.T. and Kiddell, K., 2007. Temperature and the energetics of development in the house cricket (*Acheta domesticus*). Journal of Insect Physiology 53: 950-953.

Bubliy, O.A. and Loeschcke, V., 2005. Correlated responses to selection for stress resistance and longevity in a laboratory population of *Drosophila melanogaster*. Journal of Evolutionary Biology 18: 789-803.

Bunning, H., Bassett, L., Clowser, C., Rapkin, J., Jensen, K., House, C.M., Archer, C.R. and Hunt, J., 2016. Dietary choice for a balanced nutrient intake increases the mean and reduces the variance in the reproductive performance of male and female cockroaches. Ecology and Evolution 6: 4711-4730.

Bunning, H., Rapkin, J., Belcher, L., Archer, C.R., Jensen, K. and Hunt, J., 2015. Protein and carbohydrate intake influence sperm number and fertility in male cockroaches, but not sperm viability. Proceedings of the Royal Society of London B: Biological Sciences 282: 20142144.

Chidawanyika, F. and Terblanche, J.S., 2011. Costs and benefits of thermal acclimation for codling moth, *Cydia pomonella* (Lepidoptera: Tortricidae): implications for pest control and the sterile insect release programme. Evolutionary Applications 4: 534-544.

Chown, S.L. and Terblanche, J.S., 2007. Physiological diversity in insects: ecological and evolutionary contexts. Advances in Insect Physiology 33: 50-152.

Clarke, A., 2003. Costs and consequences of evolutionary temperature adaptation. Trends in Ecology and Evolution 18: 573-581.

Clarke, A., 2014. The thermal limits to life on earth. International Journal of Astrobiology 13: 141-154.

Clissold, F.J., Coggan, N. and Simpson, S.J., 2013. Insect herbivores can choose microclimates to achieve nutritional homeostasis. Journal of Experimental Biology 216: 2089-2096.

Clissold, F.J. and Simpson, S.J., 2015. Temperature, food quality and life history traits of herbivorous insects. Current Opinion in Insect Science 11: 63-70.

Coggan, N., Clissold, F.J. and Simpson, S.J., 2011. Locusts use dynamic thermoregulatory behaviour to optimize nutritional outcomes. Proceedings of the Royal Society of London B: Biological Sciences 278: 2745-2752.

Cohen, A.C., 2015. Insect diets: science and technology. CRC Press, Boca Raton, FL, USA.

Cotter, S.C., Simpson, S.J., Raubenheimer, D. and Wilson, K., 2011. Macronutrient balance mediates trade-offs betweenimmune function and life history traits. Functional Ecology 25: 186-198.

Davidowitz, G., D'Amico, L.J. and Nijhout, H.F., 2003. Critical weight in the development of insect body size. Evolution and Development 5: 188-197.

Davidowitz, G. and Nijhout, H.F., 2004. The physiological basis of reaction norms: the interaction among growth rate, the duration of growth and body size. Integrative and Comparative Biology 44: 443-449.

Debat, V. and David, J.R., 2002. Analysing phenotypic variation: when old-fashioned means up-to-date. Journal of Biosciences 27: 191-193.

DeGrandi-Hoffman, G. and Chen, Y., 2015. Nutrition, immunity and viral infections in honey bees. Current Opinion in Insect Science 10: 170-176.

Despland, E., 2005. Diet breadth and anti-predator strategies in desert locusts and other Orthopterans. Journal of Orthoptera Research 14: 227-233.

Diamond, S.E. and Kingsolver, J.G., 2010. Environmental dependence of thermal reaction norms: host plant quality can reverse the temperature-size rule. American Naturalist 175: 1-10.

Eilenberg, J., Vlak, J.M., Nielsen-LeRoux, C., Cappellozza, S. and Jensen, A.B., 2015. Diseases in insects produced for food and feed. Journal of Insects as Food and Feed 1: 87-102.

Erb, H.N. and Martin, S.W., 1980. Interrelationships between production and reproduction diseases in Holstein cows. Journal of Dairy Science 63: 1911-1917.

Falconer, D.S. and Mackay, T.F.C., 1996. Introduction to quantitative genetics. Longman, Burnt Mill, UK.

Frankham, R., 1995. Effective population size/adult population size ratios in wildlife: a review. Genetics Research 66: 95-107.

Frankham, R., Ballou, J.D., Briscoe, D.A. and McInnes, K.H., 2010. Introduction to conservation genetics. Cambridge University Press, Cambridge, UK.

Frankham, R., Bradshaw, C.J.A. and Brook, B.W., 2014. Genetics in conservation management: revised recommendations for the 50/500 rules, red list criteria and population viability analyses. Biological Conservation 170: 56-63.

Garlapow, M.E., Huang, W., Yarboro, M.T., Peterson, K.R. and Mackay, T.F.C., 2015. Quantitative genetics of food intake in *Drosophila melanogaster*. PLoS ONE 10: e0138129.

Gjerde, B., Gunnes, K. and Gjedrem, T., 1983. Effect of inbreeding on survival and growth in rainbow trout. Aquaculture 34: 327-332.

Goddard, M.E. and Hayes, B.J., 2007. Genomic selection. Journal of Animal Breeding and Genetics 124: 323-330.

Han, C.S. and Dingemanse, N.J., 2017. You are what you eat: diet shapes body composition, personality and behavioural stability. BMC Evolutionary Biology 17: 8.

Harrison, J.F., Woods, H.A. and Roberts, S.P., 2012. Ecological and environmental physiology of insects. Oxford University Press, Oxford, UK.

Henderson, C.R., 1973. Sire evaluation and genetic trends. In: Proceedings of the Animal Breeding and Genetics Symposium in Honour of Dr. Jay L. Lush. American Society for Animal Science – American Dairy Science Association – Poultry Science Assosiation, Champaign, IL, USA, pp. 10-41.

Hill, W.G. and Zhang, X.S., 2004. Genetic variation within and among animal populations. In: Simm, G., Villanueva, B., Sinclair, K.D. and Townsend, S. (eds.) Farm animal genetic resources. Nottingham University Press, Nottingham, UK, pp. 67-84.

Hoffmann, A.A., 2000. Laboratory and field heritabilities: some lessons from *Drosophila*. In: Mousseau, T.A., Sinervo, B. and Endler, J.A. (eds.) Adaptive genetic variation in the wild. Oxford University Press, New York, NY, USA.

Hoffmann, A.A., Merilä, J. and Kristensen, T.N., 2016. Heritability and evolvability of fitness and nonfitness traits: lessons from livestock. Evolution 70: 1770-1779.

Hoffmann, A.A., Sgró, C.M. and Kristensen, T.N., 2017. Revisiting adaptive potential, population size, and conservation. Trends in Ecology and Evolution 32: 506-517.

Hoffmann, A.A., Sørensen, J.G. and Loeschcke, V., 2003. Adaptation of *Drosophila* to temperature extremes: bringing together quantitative and molecular approaches. Journal of Thermal Biology 28: 175-216.

Honěk, A., 1993. Intraspecific variation in body size and fecundity in insects: a general relationship. Oikos 66: 483-492.

Huey, R.B., Berrigan, D., Gilchrist, G.W. and Herron, J.C., 1999. Testing the adaptive significance of acclimation: a strong inference approach. American Zoologist 39: 323-336.

Hunt, J., Breuker, C.J., Sadowski, J.A. and Moore, A.J., 2009. Male-male competition, female mate choice and their interaction: determining total sexual selection. Journal of Evolutionary Biology 22: 13-26.

Hunt, J., Bussière, L.F., Jennions, M.D. and Brooks, R., 2004. What is genetic quality? Trends in Ecology and Evolution 19: 329-333.

Hussein, M., Pillai, V.V., Goddard, J.M., Park, H.G., Kothapalli, K.S., Ross, D.A., Ketterings, Q.M., Brenna, J.T., Milstein, M.B., Marquis, H., Johnson, P.A., Nyrop, J.P. and Selvaraj, V., 2017. Sustainable production of housefly (*Musca domestica*) larvae as a protein-rich feed ingredient by utilizing cattle manure. PLoS ONE 12: e0171708.

Jamieson, I.G. and Allendorf, F.W., 2012. How does the 50/500 rule apply to MVPs? Trends in Ecology and Evolution 27: 578-584.

Jensen, K., McClure, C., Priest, N.K. and Hunt, J., 2015. Sex-specific effects of protein and carbohydrate intake on reproduction but not lifespan in *Drosophila melanogaster*. Aging Cell 14: 605-615.

Kinghorn, B.P., Van der Werf, J.H.J. and Ryan, M., 2000. Animal breeding – use of new technologies. The Post Graduate Foundation in Veterinarian Science of the University of Sydney, Sydney, Australia.

Kristensen, T.N., Hoffmann, A.A., Overgaard, J., Sørensen, J.G., Hallas, R. and Loeschcke, V., 2008. Costs and benefits of cold acclimation in field-released *Drosophila*. Proceedings of the National Academy of Sciences of the USA 105: 216-221.

Kristensen, T.N., Hoffmann, A.A., Pertoldi, C. and Stronen, A.V., 2015. What can livestock breeders learn from conservation genetics and *vice versa*? Frontiers in Genetics 6: 38.

Kristensen, T.N. and Sørensen, A.C., 2005. Inbreeding – lessons from animal breeding, evolutionary biology and conservation genetics. Animal Science 80: 121-133.

Lee, K.P., Behmer, S.T., Simpson, S.J. and Raubenheimer, D., 2002. A geometric analysis of nutrient regulation in the generalist caterpillar *Spodoptera littoralis* (Boisduval). Journal of Insect Physiology 48: 655-665.

Lee, K.P., Cory, J.S., Wilson, K., Raubenheimer, D. and Simpson, S.J., 2006. Flexible diet choice offsets protein costs of pathogen resistance in a caterpillar. Proceedings of the Royal Society of London B: Biological Sciences 273: 823-829.

Lee, K.P. and Jang, T., 2014. Exploring the nutritional basis of starvation resistance in *Drosophila melanogaster*. Functional Ecology 28: 1144-1155.

Lee, K.P., Jang, T., Ravzanaadii, N. and Rho, M.S., 2015. Macronutrient balance modulates the temperature-size rule in an ectotherm. American Naturalist 186: 212-222.

Lee, K.P. and Roh, C., 2010. Temperature-by-nutrient interactions affecting growth rate in an insect ectotherm. Entomologia Experimentalis et Applicata 136: 151-163.

Lee, K.P., Simpson, S.J., Clissold, F.J., Brooks, R., Ballard, J.W.O., Taylor, P.W., Soran, N. and Raubenheimer, D., 2008a. Lifespan and reproduction in *Drosophila*: new insights from nutritional geometry. Proceedings of the National Academy of Sciences of the USA 105: 2498-2503.

Lee, K.P., Simpson, S.J. and Wilson, K., 2008b. Dietary protein-quality influences melanization and immune function in an insect. Functional Ecology 22: 1052-1061.

Lemoine, N.P. and Shantz, A.A., 2016. Increased temperature causes protein limitation by reducing the efficiency of nitrogen digestion in the ectothermic herbivore *Spodoptera exigua*. Physiological Entomology 41: 143-151.

Lochmiller, R.L. and Deerenberg, C., 2000. Trade-offs in evolutionary immunology: just what is the cost of immunity? Oikos 88: 87-98.

Lynch, M. and Walsh, B., 1998. Genetics and analysis of quantitative traits. Sinauer Associates, Sunderland, MA, USA.

Makkar, H.P.S., Tran, G., Heuzé, V. and Ankers, P., 2014. State-of-the-art on use of insects as animal feed. Animal Feed Science and Technology 197: 1-33.

Maklakov, A.A., Simpson, S.J., Zajitschek, F., Hall, M.D., Dessmann, J., Clissold, F., Raubenheimer, D., Bonduriansky, R. and Brooks, R.C., 2008. Sex-specific fitness effects of nutrient intake on reproduction and lifespan. Current Biology 18: 1062-1066.

Manenti, T., Sørensen, J.G., Moghadam, N.N. and Loeschcke, V., 2014. Predictability rather than amplitude of temperature fluctuations determines stress resistance in a natural population of *Drosophila simulans*. Journal of Evolutionary Biology 27: 2113-2122.

Mansfield, K.G. and Land, E.D., 2002. Cryptorchidism in Florida panthers: prevalence, features, and influence of genetic restoration. Journal of Wildlife Diseases 38: 693-698.

Mc Parland, S., Kearney, J.F., MacHugh, D.E. and Berry, D.P., 2008. Inbreeding effects on postweaning production traits, conformation, and calving performance in Irish beef cattle. Journal of Animal Science 86: 3338-3347.

Meuwissen, T., 2009. Genetic management of small populations: a review. Acta Agriculturae Scandinavica, Section A – Animal Science 59: 71-79.

Miller, G.A., Clissold, F.J., Mayntz, D. and Simpson, S.J., 2009. Speed over efficiency: locusts select body temperatures that favour growth rate over efficient nutrient utilization. Proceedings of the Royal Society of London B: Biological Sciences 76: 3581-3589.

Mirth, C.K. and Riddiford, L.M., 2007. Size assessment and growth control: how adult size is determined in insects. Bioessays 29: 344-355.

Moret, Y. and Schmid-Hempel, P., 2000. Survival for immunity: the price of immune system activation for bumblebee workers. Science 290: 1166-1168.

Mousseau, T.A. and Roff, D.A., 1987. Natural selection and the heritability of fitness components. Heredity 59: 181-197.

Nagy, I., Gyovai, P., Radnai, I., Kiszlinger, H.N., Farkas, J. and Szendrő, Z., 2013. Genetic parameters, genetic trends and inbreeding depression of growth and carcass traits in Pannon terminal line rabbits. Archiv Tierzucht 18: 191-199.

Neira, R., García, X., Lhorente, J.P., Filp, M., Yáñez, J.M. and Cascante, A.M., 2016. Evaluation of the growth and carcass quality of diallel crosses of four strains of Nile tilapia (*Oerochromis niloticus*). Aquaculture 451: 213-222.

Nunney, L., 1995. Measuring the ratio of effective population size to adult numbers using genetic and ecological data. Evolution 49: 389-392.

Ojala, K., Julkunen-Tiitto, R., Lindström, L. and Mappes, J., 2005. Diet affects the immune defence and life-history traits of an Arctiid moth *Parasemia plantaginis*. Evolutionary Ecology Research 7: 1153-1170.

Olsson, M., Gullberg, A. and Tegelström, H., 1996. Malformed offspring, sibling matings, and selection against inbreeding in the sand lizard (*Lacerta agilis*). Journal of Evolutionary Biology 9: 229-242.

Oonincx, D.G.A.B., Van Broekhoven, S., Van Huis, A. and Van Loon, J.A., 2015. Feed conversion, survival and development, and composition of four insect species on diets composed of food by-products. PLoS ONE 10: e0144601.

Oonincx, D.G.A.B., Van Itterbeeck, J., Heetkamp, M.J.W., Van den Brand, H., Van Loon, J.J.A. and Van Huis, A., 2010. An exploration on greenhouse gas and ammonia production by insect species suitable for animal or human consumption. PLoS ONE 5: e14445.

Palstra, F.P. and Fraser, D.J., 2012. Effective/census population size ratio estimation: a compendium and appraisal. Ecology and Evolution 2: 2357-2365.

Pasteels, J.M., Grégoire, J.-C. and Rowell-Rahier, M., 1983. The chemical ecology of defence in arthropods. Annual Review of Entomology 28: 263-289.

Ponton, F., Wilson, K., Holmes, A.J., Cotter, S.C., Raubenheimer, D. and Simpson, S.J., 2013. Integrating nutrition and immunology: a new frontier. Journal of Insect Physiology 59: 130-137.

Povey, S., Cotter, S.C., Simpson, S.J., Lee, K.P. and Wilson, K., 2009. Can the protein costs of bacterial resistance be offset by altered feeding behaviour? Journal of Animal Ecology 78: 437-446.

Povey, S., Cotter, S.C., Simpson, S.J. and Wilson, K., 2013. Dynamics of macronutrient self-medication and illnessinduced anorexia in virally infected insects. Journal of Animal Ecology 83: 245-255.

Pryce, J.E. and Daetwyler, H.D., 2012. Designing dairy cattle breeding schemes under genomic selection: a review of international research. Animal Production Science 52: 107-114.

Rantala, M.J., Kortet, R., Kotiaho, J.S., Vainikka, A. and Suhonen, J., 2003. Condition dependence of pheromones and immune function in the grain beetle *Tenebrio molitor*. Functional Ecology 17: 534-540.

Rapkin, J., Jensen, K., House, C.M., Sakaluk, S., Sakaluk, J. and Hunt, J., 2017. The complex interplay between specific macronutrient intake, cuticular hydrocarbon expression, and mating success in male decorated crickets. Journal of Evolutionary Biology 30: 711-727.

Raubenheimer, D. and Jones, S.A., 2006. Nutritional imbalance in an extreme generalist omnivore: tolerance and recovery through complementary food selection. Animal Behaviour 71: 1253-1262.

Raubenheimer, D. and Simpson, S.J., 1997. Integrative models of nutrient balancing: application to insects and vertebrates. Nutrition Research Reviews 10: 151-179.

Rauw, W.M., Kanisb, E., Noordhuizen-Stassen, E.N. and Grommers, F.J., 1998. Undesirable side effects of selection for high production efficiency in farm animals: a review. Livestock Production Science 56: 15-33.

Reddiex, A.J., Gosden, T.P., Bonduriansky, R. and Chenoweth, S.F., 2013. Sex-specific fitness consequences of nutrient intake and the evolvability of diet preferences. American Naturalist 182: 91-102.

Rho, M.S. and Lee, K.P., 2014. Geometric analysis of nutrient balancing in the mealworm beetle, *Tenebrio molitor* L. (Coleoptera: Tenebrionidae). Journal of Insect Physiology 71: 37-45.

Rho, M.S. and Lee, K.P., 2016. Balanced intake of protein and carbohydrate maximizes lifetime reproductive success in the mealworm beetle, *Tenebrio molitor* (Coleoptera: Tenebrionidae). Journal of Insect Physiology 91-92: 93-99.

Sánchez-Muros, M.-J., Barroso, F.G. and Manzano-Agugliaro, F., 2014. Insect meal as renewable source of food for animal feeding: a review. Journal of Cleaner Production 65: 16-27.

Schaeffer, L.R., 2006. Strategy for applying genome-wide selection in dairy cattle. Journal of Animal Breeding and Genetics 123: 218-223.

Schwenke, R.A., Lazzaro, B.P. and Wolfner, M.F., 2016. Reproduction – immunity trade-offs in insects. Annual Review of Entomology 61: 239-256.

Senior, A.M., Nakagawa, S., Lihoreau, M., Simpson, S.J. and Raubenheimer, D., 2015. An overlooked consequence of dietary mixing: a varied diet reduces interindividual variance in fitness. American Naturalist 186: 649-660.

Sheppard, D.C., Tomberlin, J.K., Joyce, J.A., Kiser, B.C. and Sumner, S.M., 2002. Rearing methods for the black soldier fly (Diptera: Stratiomyidae). Journal of Medical Entomology 39: 695-698.

Shikano, I. and Cory, J.S., 2015. Impact of environmental variation on host performance differs with pathogen identity: implications for host-pathogen interactions in a changing climate. Scientific Reports 5: 15351.

Simpson, S.J., Clissold, F.J., Lihoreau, M., Ponton, F., Wilder, S.M. and Raubenheimer, D., 2015. Recent advances in the integrative nutrition of arthropods. Annual Review of Entomology 60: 293-311.

Simpson, S.J. and Raubenheimer, D., 2012. The nature of nutrition – a unifying framework from animal adaptation to human obesity. Princeton University Press, Princeton, NJ, USA.

Simpson, S.J., Raubenheimer, D., Behmer, S.T., Whitworth, A. and Wright, G.A., 2002. A comparison of nutritional regulation in solitarious- and gregarious-phase nymphs of the desert locust *Schistocerca gregaria*. Journal of Experimental Biology 205: 121-129.

Smith, R. and Barnes, E., 2015. PROteINSECT: insects as sustainable sources of protein. Report produced by Minerva Health and Care Communications Ltd on behalf of PROteINSECT. Available at: http://tinyurl.com/yctqc376.

Sørensen, J.G., Addison, M.F. and Terblanche, J.S., 2012. Mass-rearing of insects for pest management: challenges, synergies and advances from evolutionary physiology. Crop Protection 38: 87-94.

Sørensen, J.G., Kristensen, T.N., Loeschcke, V. and Schou, M.F., 2015. No trade-off between high and low temperature tolerance in a winter acclimatized Danish *Drosophila subobscura* population. Journal of Insect Physiology 77: 9-14.

Sørensen, J.G., Schou, M.F., Kristensen, T.N. and Loeschcke, V., 2016. Thermal fluctuations affect the transcriptome through mechanisms independent of average temperature. Scientific Reports 6: 30975.

Surendra, K.C., Olivier, R., Tomberlin, J.K., Jha, R. and Khanal, S.K., 2016. Bioconversion of organic wastes into biodiesel and animal feed via insect farming. Renewable Energy 98: 197-202.

Sword, G.A., Simpson, S.J., El Hadi, O.T.M. and Wilps, H., 2000. Density-dependent aposematism in the desert locust. Proceedings of the Royal Society of London B: Biological Sciences 267: 63-68.

Terblanche, J.S. and Chown, S.L., 2007. Factory flies are not equal to wild flies. Science 317: 1678-1678.

Tomberlin, J.K., Sheppard, D.C. and Joyce, J.A., 2002. Selected life-history traits of black soldier flies (Diptera: Stratiomyidae) reared on three artificial diets. Annals of the Entomological Society of America 95: 379-386.

Triggs, A. and Knell, R.J., 2012. Interactions between environmental variables determine immunity in the Indian meal moth *Plodia interpunctella*. Journal of Animal Ecology 81: 386-394.

Unckless, R.L., Rottschaefer, S.M. and Lazzaro, B.P., 2015a. The complex contributions of genetics and nutrition to immunity in *Drosophila melanogaster*. PLoS Genetics 11: e1005030.

Unckless, R.L., Rottschaefer, S.M. and Lazzaro, B.P., 2015b. A genome-wide association study for nutritional indices in *Drosophila*. Genes Genomes Genetics 5: 417-425.

Van Broekhoven, S., Oonincx, D.G.A.B., Van Huis, A. and Van Loon, J.A., 2015. Growth performance and feed conversion efficiency of three edible mealworm species (Coleoptera: Tenebrionidae) on diets composed of organic by-products. Journal of Insect Physiology 73: 1-10.

Van Huis, A., 2013. Potential of insects as food and feed in assuring food security. Annual Review of Entomology 58: 563-583.

Van Huis, A., Van Itterbeeck, J., Klunder, H., Mertens, E., Halloran, A., Muir, G. and Vantomme, P., 2013. Edible insects: future prospects for food and feed security. Food and Agriculture Organization of the United Nations (FAO), Rome, Italy, FAO Forestry Paper no. 171, 187 pp. Available at: http://www.fao.org/docrep/018/i3253e/i3253e.pdf.

Vreysen, M.J.B., Robinson, A.S. and Hendrichs, J., 2007. Area-wide control of insect pests: from research to field implementation. Springer, Dordrecht, the Netherlands.

Wade, M.J., Bijma, P., Ellen, E.D. and Muir, W., 2010. Group selection and social evolution in domesticated animals. Evolutionary Applications 3: 453-465.

Walsh, B. and Blows, M.W., 2009. Abundant genetic variation + strong selection = multivariate genetic constraints: a geometric view of adaptation. Annual Review of Ecology, Evolution, and Systematics 40: 41-59.

Warbrick-Smith, J., Behmer, S.T., Lee, K.P., Raubenheimer, D. and Simpson, S.J., 2006. Evolving resistance to obesity in an insect. Proceedings of the National Academy of Sciences of the USA 103: 14045-14049.

Warbrick-Smith, J., Raubenheimer, D., Simpson, S.J. and Behmer, S.T., 2009. Three hundred and fifty generations of extreme food specialisation: testing predictions of nutritional ecology. Entomologia Experimentalis et Applicata 132: 65-75.

West-Eberhard, M.J., 2003. Developmental plasticity and evolution. Oxford University Press, New York, NY, USA.

Whitlock, M.C. and Agrawal, A.F., 2009. Purging the genome with sexual selection: reducing mutation load through selection on males. Evolution 63: 569-582.

Whitman, D. and Ananthakrishnan, T.N., 2009. Phenotypic plasticity of insects: mechanisms and consequences. Science Publishers Inc., Enfield, USA, pp. 1-894.

Wilson, K., Cotter, S.C., Reeson, A.F. and Pell, J.K., 2001. Melanism and disease resistance in insects. Ecology Letters 4: 637-649.

Wilson, K., Thomas, M.B., Blanford, S., Doggett, M., Simpson, S.J. and Moore, S.L., 2002. Coping with crowds: density-dependent disease resistance in desert locusts. Proceedings of the National Academy of Sciences of the USA 99: 5471-5475.

Wood, J.L.A., Yates, M.C. and Fraser, D.J., 2016. Are heritability and selection related to population size in nature? Meta-analysis and conservation implications. Evolutionary Applications 9: 640-657.

Zielińska, E., Baraniak, B., Karaś, M., Rybczyńska, K. and Jakubczyk, A., 2015. Selected species of edible insects as a source of nutrient composition. Food Research International 77: 460-466.

10.

Zophobas morio

Natural enemies in insect production systems

J. Eilenberg[1*], S.N. Gasque[1] and V.I.D. Ros[2]
[1]Department of Plant and Environmental Sciences, University of Copenhagen, Thorvaldsensvej 40, 1871 Frederiksberg C., Denmark; [2]Laboratory of Virology, Wageningen University & Research, Droevendaalsesteeg 1, 6708 PB Wageningen, the Netherlands; jei@plen.ku.dk

Abstract

Natural enemies (including vertebrate and invertebrate predators, arthropod parasitoids and microorganisms) can be very harmful in insect production systems. We provide an overview of main biological characteristics of significant groups of natural enemies, for example concerning microorganisms and their way to infect insects. Such biological traits will, together with the characteristics of the production systems, be of major importance when analysing which natural enemies can be a problem in a specific production system. Two examples of host insects and their natural enemies are detailed. The first example deals with the red palm weevil, *Rhynchophorus ferrugineus*, and its many different natural enemies. The second example deals with house crickets *Acheta domesticus*, and its viral pathogens. Finally, we give recommendations on how to manage natural enemies in production systems.

Keywords: natural enemies, insect pathogens, predators, prevention of insect diseases

10.1 Characteristics of natural enemies in insect production systems

Insects in production systems will suffer from natural enemies, as is also found for traditional husbandry like poultry, cattle, pigs and sheep. Invasions by such natural enemies must be prevented or managed. Natural enemies of insects include predators and parasites. Predators feed directly on other animals and thereby kill their prey. Predators are generalists, since they mostly feed on several or even multiple animal host species. Vertebrate and invertebrate predators have different tactics when locating and attacking their prey. Vertebrate predators simply overwhelm their insect prey, since they are relatively bigger than their prey. Vertebrate predators may easily switch to different prey species – when put in a new environment they can hunt for a new kind of prey (Hajek, 2004). Invertebrate predators are smaller in size compared to their prey and thereby they cannot in all cases overwhelm the prey in the same way as vertebrate predators. Even though invertebrate predators are mostly generalists, they will prey on a limited number of prey species and many do not easily switch to a new prey species (Hajek, 2004). Invertebrate predators often rely on chemical stimuli for locating their prey. In addition they may use vision or move around in the area of their last meal. Once the prey is located, invertebrate predators can use a wide range of methods to capture it; they can use traps, attack prey in groups, or use an ambush or sit-and-wait strategy (i.e. wait in hiding for their prey and make a surprise attack) (Hajek, 2004).

Parasites are organisms that live at the expense of a host. A true host provides the conditions the parasite needs to reproduce and complete its life cycle (Gullan and Cranston, 2013). A parasite can live within a host (endoparasite) or on the exterior of the host (ectoparasite). Some parasites will live at the expense of one host (direct lifecycle) and others at the expense of multiple hosts. Parasites can be obligate parasites – where they are dependent on their parasitic lifestyle, or they can be facultative parasites – where they can alternate a parasitic with a free-living lifestyle, depending on the circumstances. Parasites visible to the naked eye are called macroparasites, in contrast to microparasites. One group of macroparasites are the so-called parasitoids. Parasitoids live and develop within a single host and finally kill it, and are mainly to be found in the insect orders Hymenoptera and Diptera (Gordh and Headrick, 2011). Parasitoids rarely parasitize the adult stage (Gullan and Cranston 2013), so the vulnerable stages are all the immature stages: eggs, nymphs/larvae and pupae. Phoretic mites can be seen as yet another group of parasites. These ectoparasites do not infect their hosts, and often they are seen as organisms passively using an insect host for transportation. However, some phoretic mites may lower host fitness significantly (great loss of fecundity or high mortality) (see the section on palm weevils).

Microbial parasites (or microparasites) include insect pathogens (viruses, bacteria, fungi, microsporidia and nematodes) that infect their host and either kill the host or at least significantly lower its fitness (Kaya and Vega, 2012). These different groups of insect pathogens only share the feature that they are 'small'. With respect to any other characteristic (infection process, life cycle, etc.) these groups can differ significantly, and also within each group significantly different life styles may be found. For example, insect pathogens may differ in their capacity to survive outside the host, making them obligate, facultative, or opportunistic pathogens (Kaya and Vega, 2012). Some overlap between the two latter groups exists. It may be hard to determine whether a microorganism present in or on a dead insect was the pathogen causing the disease and death of the insect, or whether it was merely a saprotroph benefiting from the available food in the insect cadaver. Insects harbour a wide array of microbiota, and also insects in production systems (e.g. the house cricket *Acheta domestica* and the mealworm *Tenebrio molitor*) contain a diversity of bacterial and fungal microorganisms (Garofalo *et al.*, 2017), which are not natural enemies but are often beneficial to the host. New species of insect pathogens are described every year, and even when they were not found as a natural enemy on one of the insect species in production, especially generalists can still be pathogenic to insects in production. One recent example is the new species *Beauveria medogensis* isolated from soil in Tibet (Imoulan *et al.*, 2016). Pathogenicity of this fungus was shown in the laboratory to the mealworm *T. molitor.*

Insect pathogenic viruses are obligate pathogens and are entirely dependent on their host to complete their replication cycle, while several insect pathogenic fungi and bacteria are facultative pathogens that can survive and even produce spores outside the host. Insect pathogenic nematodes are normally not in themselves seen as being microorganisms due to their (at least for some species and life stages) relatively large size. However, due to their infection biology (via body openings) and due to their association with certain symbiotic bacteria (genera *Xenorhabdus* and *Photorhabdus*), which are needed for some important insect pathogenic nematode genera (*Steinernema and Heterorhabditis*) to fulfil their life cycle, insect pathogenic nematodes are mostly considered insect pathogens. The mealworm *T. molitor* is often used as a model insect to understand infection biology and immune responses in host-pathogen relationships. In a recent study (Binda-Rosetti *et al.*, 2016) it was shown that both live nematodes and live bacterial symbionts could depress the antimicrobial response of the host.

Not all natural enemies known from insects in their natural habitat will be of importance in production facilities. First, parasites with a complicated life cycle may not be able to complete such a life cycle within insects in production. One example concerns *Formica* ants and the lancet liver fluke *Dicrocoelium dendriticum* (Manga-Gonzalez *et al.*, 2001). The fluke needs three hosts to fulfil its life cycle: a land snail, an ant and a grass or ant-eating vertebrate. This life cycle has no chance to be completed when *Formica* ants are taken into production. Second, some natural enemies exist only in some parts of the geographical area of the host's occurrence. Such natural enemies may not be a problem when the host is to be produced elsewhere. However, microorganisms may be dispersed without being noticed, for example within hosts or within substrate or food used for production. Rapidly, such insect diseases may spread and cause huge problems in areas where they were not known to occur previously (see section about viruses in crickets).

Insect production systems may offer different opportunities for insect natural enemies to invade and establish (Eilenberg and Jensen, in press). In general, an open production facility with a high level of contact with the environment enhances the risk of natural enemies entering the facility. Natural enemies can arrive via soil by their own movement (insect predators), via air (microorganisms, birds, parasitoids), or via food and substrate (microorganisms), and they can circulate within the production system for extended periods. Some microorganism may even be present in a latent phase without being detected before an epidemic develops rapidly. Closed production systems (closed buildings and/or closed production rooms within a building) limit the entrance of natural enemies to some extent, although entry via air or food/substrate is still a possibility. Another important factor determining the possibilities of natural enemies' success is the insect production method. There are different types of production methods including straight-batch systems, cyclical fed-batch systems, parallel fed-batch systems and plug-flow continuous production (Kok, 2017). They have different characteristics which may ease or hamper transmission of unwanted microorganisms and small invertebrates within the production stock. In general, continuous production systems are more vulnerable than production system in batches. In the latter case, problems in one batch can more easily be managed without a need to stop the entire production chain.

Several insect species are currently reared in large-scale production systems and the number of species in production will surely increase over the coming years. Since each species is threatened by a range of natural enemies, the number of natural enemies which may pose a problem is high. In Table 10.1 we have compiled some general biological characteristics of the main groups of natural enemies, although we acknowledge that the table will not give the full picture. A major challenge is the diagnosis of insect pathogens among production insects (Eilenberg *et al.*, 2015). Specific expertise is needed and while established insect production systems, like for honey bees or silkworms, have regional services and specialized laboratories for diagnosing diseases, this is not yet the case for other insects in production systems. Concerning insects produced for food and feed we are currently in a phase where such diagnostic services are gradually being developed. Companies do observe problems in their production systems, as documented in a survey performed in 2014 (Eilenberg *et al.*, 2015), but different insect species in production did not suffer from such natural enemies to the same extent. For example, while crickets were subjected to several diseases noted by companies, the black soldier fly *Hermetia illucens* seemed almost unaffected by insect diseases (Eilenberg *et al.*, 2015). Another issue is to which extent insect pathogens may pose a threat to humans or vertebrate husbandry animals eating insects or insect products. We do not treat the aspect of food safety in this chapter, but an excellent recent publication (EFSA, 2015) reviewed the current knowledge on this topic.

In this chapter, we have chosen to exemplify the challenges in insect production systems by selecting two species reared in production facilities and by describing in detail their natural enemies, their biology and their effects on the production insects. Together these two examples give an impression of the challenges encountered and foreseen in insect production systems.

Our first example is the palm weevil *Rhynchophorus ferrugineus*: this is an edible insect, which has been taken into production as human food recently (Hanboonsong *et al.*, 2013). Due to its status as pest insect of global importance in palms, many different types of natural enemies are known. Yet there is a need to apply this biological knowledge into managing this very broad range of natural enemies in palm weevil production systems. Known natural enemies include vertebrate and invertebrate predators, insect parasitoids, insect pathogenic viruses, bacteria, fungi, and nematodes. We outline the diversity of biological characteristics of these natural enemies that are (or will potentially become) important to palm weevil production.

Table 10.1. Characteristics of selected main taxa of natural enemies of insects. Generalists: infecting or predating on hosts from taxonomically unrelated taxa (for example different insect orders). Specialists: infecting or predating either just one host species or several taxonomically closely related host species.

Natural enemy taxa	Mode of action	Host specialization
Predators:		
vertebrates (birds, reptiles, mammals) or invertebrates (insects, spiders and mites)	attacking prey from outside and killing them	generalists
Parasitoids:		
insects mainly from the orders Hymenoptera and Diptera, including important superfamilies Chalcidoidea and Ichneumonidea	depositing eggs in host; larvae exploit the host, which will finally die	specialists
Nematodes:		
from phylum Nematoda, important genera are *Steinernema* and *Heterorhabditis*	entering insect host via body openings, interaction with bacteria	generalists
Viruses:		
from different virus families, important families include *Baculoviridae* and *Parvoviridae* (densoviruses)	oral uptake by insect host, infection via gut epithel	specialists
Bacteria:		
from different bacterial families, important genera are *Bacillus*, *Serratia* and *Pseudomonas*	oral uptake by insect host, infection via gut epithel	generalists or specialists
Fungi:		
from different fungal phyla, most important are species from Ascomycota and Entomophthoromycota	infection through host cuticle or orally, growth in hemolymph	generalists or specialists
Microsporidia:		
unicellular pathogens, a sister group to fungi, an important genus is *Nosema*	oral uptake by insect host, infection via gut epithel	specialists

Our second example is complementary both with respect to insect species and natural enemies. Here, we provide information about crickets (including *Acheta domesticus*, *Gryllus bimaculatus* and other species) and the diversity of one type of natural enemies, namely cricket pathogenic viruses. Crickets have been produced for decades especially as pet food. More recently the production of crickets for human food has increased rapidly (Halloran *et al.*, 2016). We focus on the biology of viruses infecting crickets, their transmission and their effect on cricket production systems. Many types of viruses are known from crickets and several have proven problematic or even disastrous in cricket production systems. Detailing the diversity and infection biology of these viruses provides us with an overview of the challenges with just one important group of pathogens.

10.2 Diversity of natural enemies of the palm weevil *Rhynchophorus ferrugineus*

Of the approximately 2,000 edible insects (Jongema, 2017; Van Huis *et al.*, 2013), beetles (Coleoptera) are the most consumed insects worldwide, by accounting for 31% of the consumed insects (Gahukar, 2016). Weevils is a beetle taxon that comprises a high number of consumed species, including species in the genus *Rhynchophorus* (palm weevils). Larvae of *Rhynchophorus* palm weevils, the so-called palm worms, have been eaten for centuries. Larvae are a high fat and protein source (Van Huis *et al.*, 2013) and contain high amounts of thiamine, riboflavin, zinc and iron (Gullan and Cranston, 2013). Palm weevils have great potential for mass production, and by-products as palm weevil's faeces can be sold as fertilizer (Hanboonsong *et al.*, 2013).

Palm weevils are not only a prime insect food source (Van Huis *et al.*, 2013), but also a major pest on a great range of different palm trees. The red palm weevil, *R. ferrugineus*, is the most important and harmful pest on the coconut palm *Cocos nucifera* (Faleiro, 2006b; Lever, 1969) and is considered the most destructive pest of palms worldwide (Vives 2011). The South American palm weevil, *Rhynchophorus palmarum*, causes the same damage as *R. ferrugineus*, is also a pest to coconut and date palm trees, but is in addition a primary vector of the red ring nematode *Rhadinaphelenchus cocophilus* (Miguens *et al.*, 2011; Moura *et al.*, 1993). This is of special importance because the nematode causes red ring disease, which is lethal to palm trees (Moura *et al.*, 1993; http://www.cabi.org/isc). In the search for a suitable biological control agent to suppress the palm weevil, extensive research has been performed on its natural enemies and many natural enemy species are thus known and have been studied from a biological control perspective. This acquired knowledge can be used to assess the risk of these enemies to semi-cultivation and mass rearing of palm weevils.

The production method for palm weevils may influence the risk of natural enemies to enter. In Venezuela (Van Huis *et al.*, 2013) and Thailand (Hanboonsong *et al.*, 2013) the larvae of palm weevils are commonly farmed in plastic containers. Farming of palm weevils can either be performed on palm mash in plastic containers or on cut palm trunks placed on the ground (Hanboonsong *et al.*, 2013). Both practices are low cost enterprises and can be very profitable for the farmers. Adult beetles from the wild or beetles that have been reared in containers will mate and oviposit their eggs on the mash. Natural enemies (invertebrates and microorganisms) from the wild may invade the containers. In the trunk rearing the farming is not concealed, so natural enemies can attack the weevils as easily as in the wild. In Table 10.2 we have compiled data on natural enemies documented in field populations of *R. ferrugineus*. As seen in the table, some bird species are known as predators on palm weevils. We assume that many more predatory bird species can act as natural enemies of palm weevil, provided they can access the production facilities. Closed production facilities will effectively prevent birds and other vertebrate predators to enter; so only in open production systems birds may be an important threat. Next on the list are insect predators and insect parasitoids. One predator (the black earwig *Chelisoches morio*) was specifically mentioned as a natural enemy (Table 10.2), but as was the case with birds, we assume that many more generalist insect predators can act as enemies, at least against small larvae. Two parasitoids are known, and it has yet to be proven whether more parasitoid species can be found. Invertebrate predators

Table 10.2. Natural enemies of the red palm weevil (*Rhynchophorus ferrugineus*).

Taxonomic position	Genus	Species	Predator (pred.) or parasite (para.)	Attacked host stage[1]	Location of record	References
Vertebrata						
Aves, Cuculidae	*Centropus*	*sinensis*	pred.	u	India	Faleiro (2006b)
Aves, Corvidae	*Dendrocitta*	*vagabunda parvula*	pred.	a	India	Krishnakumar and Sudha (2002, as cited in Basheer and Thomas, 2012)
Insecta						
Dermaptera, Carcinophoridae	*Anisolabis*	*maritima*	pred.	e	Egypt	Sharaby *et al.* (2011)
Dermaptera, Chelisochidae	*Chelisoches*	*morio*	pred.	e, l	unknown	Abraham and Kurian (1973); http://www.cabi.org/isc
Diptera, Sarcophagidae	*Sarcophaga*	*fuscicauda*	para.		India (Kerala)	Lever (1969)
Hymenoptera, Scoliidae	*Scolia*	*erratica*	para.	l	Indonesia (Java), Singapore	Lever (1969); Peter (1989, as cited in Murphy and Briscoe, 1999); http://www.cabi.org/isc
Arachnida						
Acari, Trachyuropodidae	*Aegyptu*	*alhassa*	para.	e, p, a	Saudi Arabia	Al-Dhafar and Al-Qahtani (2012)
	Aegyptus	*rhynchophorus*	para.	p, a	Egypt	Hassan *et al.* (2011); Ali Taha *et al.* (2014)
	Aegyptus	*zaheri*	para.	p, a	Egypt	Hassan *et al.* (2011); Ali Taha *et al.* (2014)
Uropodidae	*Centroupoda*	*almerodai*	para.	p, a	Italy, Maleysia	Mazza *et al.* (2011); Dilipkumar *et al.* (2015)
	Fascuoropod	*marginata*	para.	p, a	Egypt	Hassan *et al.* (2011); Ali Taha *et al.* (2014)
Laelapidae	*Hypoaspis*	*sardoa*	para.	p, a	Egypt	El-Sharabasy (2010); http://www.cabi.org/isc
	Hypoaspis	sp.	para.		India (Tamil Nadu)	Peter (1989, as cited in Murphy and Briscoe, 1999)
Eviphididae	*Iphidosoma*	sp.	para.	all	Egypt	El-Sharabasy (2010)

>>>

Table 10.2. Continued.

Taxonomic position	Genus	Species	Predator (pred.) or parasite (para.)	Attacked host stage[1]	Location of record	References
Arachnida (continued)						
Parasitidae	*Parasitis*	*zaheri*	para.	l, p	Egypt	El-Sharabasy (2010)
Podapolipidae	*Rhynchopolipus*	*rhynchophori*	para.	d	Egypt	Lever (1969); Abdullah (2009)
	Rhynchopolipus	*swiftae*	para.	a, l	Indonesia (N. Sulawesi)	Husband and OConnor (1999)
Nematoda						
Secernentea, Heterorhabditidae	*Heterorhabditis*	*bacteriophora*	para.	l, p, a	Turkey	Atakan *et al.* (2009)
	Heterorhabditis	*indica*	para.	l	India, Saudi Arabia, Egypt, United Arab Emirates	Abbas *et al.* (2001); Faleiro (2006a); http://www.cabi.org/isc
Entaphelenchidae	*Praecocilenchus*	*ferruginophorus*	para.	l, a	India (Kerala)	Rao and Reddy (1980); http://www.cabi.org/isc
Steinernematidae	*Steinernema*	*abbasi*	para.	l, a	Egypt	Abbas *et al.* (2001)
	Steinernema	*carpocapsae*	para.	l, a	India	Fazul Rahaman *et al.* (2000); Vives (2011); http://www.cabi.org/isc
	Steinernema	*riobravis*	para.	l, a	North America	http://www.cabi.org/isc
	Steinernema	sp.	para.	a	Saudi Arabia	Saleh *et al.* (2011)
Fungi						
	Aspergillus	sp.	para.	l, p, a	Italy (Sicily)	Torta *et al.* (2009); Francesca *et al.* (2014)
	Beauveria	*bassiana*	para.	l, p, a	Egypt, Italy (Sicily and Lazio), Spain (SE)	Sewify *et al.* (2009); Torta *et al.* (2009); Dembilio *et al.* (2010); Güerri-Agulló *et al.* (2011); Francardi *et al.* (2012)
	Cladosporium	sp.	para.	a	Italy (Sicily)	Francesca *et al.* (2014)
	Fusarium	sp.	para.	all	Italy (Sicily)	Torta *et al.* (2009)
	Metarhizium	*anisopliae*	para.	a	Egypt, Italy (Sicily)	Merghem (2011); Francardi *et al.* (2012)
	Metarhizium	*pingshaense*	para.	a	Vietnam	Cito *et al.* (2014)

>>>

Table 10.2. Continued.

Taxonomic position	Genus	Species	Predator (pred.) or parasite (para.)	Attacked host stage[1]	Location of record	References
Fungi (continued						
	Metarhizium	sp.	para.	a	Italy (Sicily)	Torta *et al.* (2009)
	Penicillium	sp.	para.	l, p, a	Italy (Sicily)	Torta *et al.* (2009); Francesca *et al.* (2014)
	Trichothecium	sp.	para.	l, a	Italy (Sicily)	Torta *et al.* (2009)
Bacteria						
Bacilli, Bacillaceae	*Bacillus*	*amyloliquefaciens*	para.	a	Italy (Sicily)	Francesca *et al.* (2014)
	Bacillus	*cereus*	para.	a	Italy (Sicily)	Francesca *et al.* (2014)
	Bacillus	*laterosporus*	para.	l, p	Egypt	Salama *et al.* (2004)
	Bacillus	*licheniformis*	para.	a	Italy (Sicily)	Francesca *et al.* (2014)
	Bacillus	*megaterium*	para.	l, p, a	Egypt, Italy (Sicily)	Salama *et al.* (2004); Francesca *et al.* (2014)
	Bacillus	*pumilus*	para.	a	Italy (Sicily)	Francesca *et al.* (2014)
	Bacillus	*subtilis*	para.	a	Italy (Sicily)	Francesca *et al.* (2014)
	Bacillus	*thuringiensis*			Egypt	Alfazariy *et al.* (2003)
	Lysinibacillus	*fusiformis*	para.	a	Italy (Sicily)	Francesca *et al.* (2014)
	Lysinibacillus	*sphaericus*	para.	l, p, a	Egypt, Italy (Sicily)	Alfazairy *et al.* (2003)); Salama *et al.* (2004); Francesca *et al.* (2014)
Gammaproteobacteria, Pseudomonadaceae	*Pseudomonas*	*aeruginosa*	para.	l	India (Kerala)	Banerjee and Dangar (1995)
Enterobacteriaceae	*Serratia*	*marcescens*	para.	l	India (Kerala)	Banerjee and Dangar (1995); Pu and Hou (2016)
Virus						
dsRNA	cytoplasmic polyhedrosis virus (CPV)		para.	all	India (Kerala), Egypt	Gopinadhan *et al.* (1990); Alfazairy *et al.* (2003); El-Minshawy *et al.* (2005)

[1] a = adult; all = all stages; d = different stages; e = eggs; l = larvae; p = pupae; u = unknown.

and parasitoids may need special attention and action (for example blocking possibilities for them to enter) to prevent them from posing problems.

We list some phoretic mites as parasitic natural enemies here. Phoretic mites attach to beetles to be transported over larger distances – a form of symbiosis where one partner benefits without negatively affecting the other (commensalism). Although the effect of such mites on insects is mostly considered to be neutral, heavy loads of phoretic mites might have a negative influence. When populations of mites reach numbers in the hundreds or the thousands, as described by Dilipkumar *et al.* (2015), there might be an antagonistic influence. The phoretic mite *Centroupoda almerodai* of *R. ferrugineus* might be considered as a parasite for two reasons. Firstly, infested beetles have their lifespan reduced by up to one third compared to non-infested beetles (under laboratory conditions). Secondly, deutonymphs of *C. almerodai* fed with weevil body parts moulted into the adult stage (Mazza *et al.*, 2011), which might indicate that these mites not only use the beetles for transportation, but also as a protein source (Mazza *et al.*, 2011). Thus, we consider it relevant to list phoretic mites posing these threats as natural enemies of palm weevils needing attention.

Several species of nematodes (genera *Steinernema*, *Heterorhabditis* and *Praecocilenchus*) have been found to be natural enemies of palm weevils. Both larvae and adults of palm weevils can become infected by nematodes. Nematodes can cause mortalities as high as 66.7% among palm weevil larvae in the field and up to 100% for palm weevils reared in the laboratory (Abbas *et al.*, 2001). Nematode species *Steinernema carpocapsae* and *Heterorhabditis bacteriophora* can actively move and find their host, for example *R. ferrugineus* (Santhi *et al.*, 2015). Due to the moist environment needed for palm weevil larvae production, nematodes should be considered as a threat in palm weevil production systems.

The fungi known naturally to infect palm weevils are diverse. Table 10.2 includes examples of the highly virulent generalist insect pathogenic fungal genera *Metarhizium* and *Beauveria*, which can infect larvae, pupae and adults of palm weevils. It is important that fungi of these genera are detected rapidly in production systems, since they can develop epidemics in many insect species (Eilenberg *et al.*, 2015). Experimental studies have proven that *Beauveria bassiana* also has sublethal effects on *R. ferrugineus*, namely feeding performance, which can lead to growth retardation (Husain *et al.*, 2015).

Furthermore, the list of fungi includes genera like *Penicillium* and *Fusarium*, where it can be hypothesized that fungal species within these genera are facultative pathogens, which only will pose a problem if the insect production stock is weakened and/or the exposure to spores is extremely high. In cases where high mortality of palm weevils in production occurs and high amount of fungal in or on the cadavers can be found, a laboratory screening of pathogenicity and virulence against the palm weevils should be done to evaluate if the fungus (or fungi) present are pathogens. Nevertheless, the species included in the Table 10.2 is from natural infections and the list of fungal genera found on *R. ferrugineus* in Table 10.2 can act as a checklist of the diversity of genera that producers should be aware of.

In the list of bacteria in Table 10.2, obligate and highly virulent insect pathogens and more opportunistic or facultative insect pathogens can be found. For example, well-known insect pathogenic bacteria species (like *Bacillus thuringiensis* and *Serratia marcescens*) have been found on *R. ferrugineus*. Also, as seen in Table 10.2, there is documentation of the presence of opportunistic or facultative pathogenic bacteria, (like *Pseudomonas aeruginosa* and various species from the genus *Bacillus* like *Bacillus cereus*). The diagnosis of insect pathogenic bacteria is sometimes difficult and also, it can be hard to prove that an insect cadaver filled with bacteria was actually killed by these bacteria (Eilenberg *et al.*, 2015). We based our inclusion of several of the bacteria in the table on the assumption that the authors of the articles referred to did carefully check that such bacteria did infect the palm weevils and were not a secondary overgrowth of saprotrophs after the insect died from other causes. However, it should be tested experimentally, if the opportunistic or facultative pathogenic bacteria can infect solely weakened palm weevils.

Finally, a single insect pathogenic virus, a cypovirus (formerly known as cytoplasmic polyhedrosis virus), can infect palm weevils (Table 10.2; Alfazairy *et al.*, 2003; El-Minshawy *et al.*, 2005; Gopinadhan *et al.*, 1990). Insect pathogenic viruses are obligate pathogens and possess the ability, given the right conditions, to rapidly develop from a latent, endemic phase to an epidemic phase in the insect host population. A question is, whether or not more viruses will be found in palm weevils over the coming years. When comparing with the diversity of viruses found in crickets (next section) it seems fair to assume that more virus species infecting palm weevils will be found in the years to come. Therefore, despite the limited knowledge of viruses infecting palm weevils, they remain a threat in the production, and producers should know how to detect, control or mitigate viral diseases.

10.3 Insect viruses infecting crickets

Although several cricket species are being consumed as food, only two cricket species, *A. domesticus* (the European house cricket) and *G. bimaculatus* (the African or Mediterranean field cricket, or the two-spotted cricket), are farmed economically (Van Huis *et al.*, 2013). These species are reared in Europe and North America for the pet-food industry (for insectivorous vertebrates) and as fish bait and in Southeast Asia (mainly Thailand), primarily using small-scale farming, for human consumption (Halloran *et al.*, 2016; Szelei *et al.*, 2011; Van Huis *et al.*, 2013). In Europe and North America, cricket-rearing facilities range from small (100-750 m^2) to very large facilities (total areas over 10,000 m^2) (Szelei *et al.*, 2011). Additional species include the *Gryllus assimilis* (Jamaican field cricket; Szelei *et al.*, 2011), *Gryllodus sigillatus* (tropical or Indian house cricket) (Van Huis *et al.*, 2013) and *Gryllus locorojo* (Weissman *et al.*, 2012). *G. sigillatus* is currently sold in both USA and European pet food stores and is easier to culture than *A. domesticus* (Weissman *et al.*, 2012).

A. domesticus is a good candidate for large-scale rearing for human consumption, considering its nutritional value, taste, and ease of rearing (Van Huis *et al.*, 2013). However, during the last decade disease outbreaks seriously threaten mass rearing of *A. domesticus* in the cricket pet food industry. Especially the A. domesticus densovirus (AdDV) caused serious problems in commercial rearing facilities in the North America and Europe (Szelei *et al.*, 2011; Weissman *et al.*, 2012). Additionally, cricket iridovirus (CrIV) caused mortality in commercial cricket colonies (Just and Essbauer, 2001; Kleespies *et al.*, 1999). Here, we describe these and other viruses that potentially threaten cricket mass rearing (Table 10.3). Besides viruses, crickets are

Table 10.3. Insect viruses infecting crickets.

Virus name	Virus family (genome)	Cricket host species	Susceptible stages	Pathobiology	References
Acheta domesticus densovirus (AdDV)[1]	*Parvoviridae* (ssDNA)	*A. domesticus* *Gryllodus sigillatus* *Gryllus locorojo*	last nymphal stage, young adults	retarded grow, anorexia, lethargy followed by paralysation, finally death	Meynardier *et al.* (1977); Styer and Hamm (1991); Szelei *et al.* (2011); Weissman *et al.* (2012)
Cricket iridovirus (CrIV)[2]	*Iridoviridae* (dsDNA)	*A. domesticus* *Gryllus campestris* *Gryllus bimaculatus* *Gryllus assimilis* *Gryllus texensis*	nymphs and adults	reduced fecundity and life span, apathy and disorientation, death	Kleespies *et al.* (1999); Just and Essbauer (2001); Jakob *et al.* (2002); Adamo *et al.* (2014)
Cricket paralysis virus (CrPV)	*Dicistroviridae* (ssRNA)	*Teleogryllus oceanicus* *Teleogryllus commodus* *A. domesticus*	early- to mid-instar nymphs	paralysis followed by death	Reinganum *et al.* (1970, 1981); Morrissey and Edwards (1977); Scotti *et al.* (1981)
G. bimaculatus nudivirus (GbNV)	*Nudiviridae* (dsDNA)	*G. campestris* *G. bimaculatus* *T. oceanicus* *T. commodus*	nymphs (lethal), adults (chronic)	reduced grow, uncoordinated and lethargic behaviour, finally death	Huger (1985); Wang and Jehle (2009)
A. domesticus volvovirus (AdVVV)	Unassigned (ssDNA)	*A. domesticus* *G. assimilis*	unknown	unknown	Pham *et al.* (2013a,b)
A. domesticus mini ambidensovirus (AdMADV)	*Parvoviridae* (ssDNA)	*A. domesticus*	unknown	unknown	Pham *et al.* (2013c)

[1] Official species name *Orthopteran ambidensovirus 1* (www.ictvonline.org).
[2] Official species name *Invertebrate iridescent virus 6* (IIV-6) (www.ictvonline.org).

indeed also subjected to infections from other insect pathogens, for example fungi (Eilenberg *et al.*, 2015).

10.3.1 Acheta domesticus densovirus

Acheta domesticus densovirus (AdDV; official species name *Orthopteran ambidensovirus 1*; International Committee on Taxonomy of Viruses (ICVT; www.ictvonline.org)) frequently occurs in rearing facilities of *A. domesticus* (Eilenberg *et al.*, 2015; Szelei *et al.*, 2011; Weissman *et al.*, 2012). Densoviruses are small (25 nm), isometric, non-enveloped viruses that have a linear, single-stranded DNA of 4-6 kb (Tijssen *et al.*, 2016). Densoviruses form the subfamily *Densovirinae* of the family *Parvoviridae* and infect invertebrate species. An infection with AdDV causes severe mortality in crickets, predominantly in the last larval stage and in young adults (Szelei *et al.*, 2011). Growth of infected individuals is retarded and individuals become paralyzed for several days, after which they die (Szelei *et al.*, 2011). First symptoms include anorexia and lethargy, followed by flaccidity (Styer and Hamm, 1991; Tijssen *et al.*, 2016).

Cricket rearing facilities have been affected by large-scale AdDNV outbreaks leading to collapses of the colonies. The first report traces back to an outbreak in a commercial Swiss rearing facility in 1977 (Meynadier *et al.*, 1977) where the virus spread rapidly and caused high mortality rates. In 1988, an AdDV outbreak was reported in southern USA (Styer and Hamm, 1991) with mortalities ranging from 20-80% per production box. The Dutch company Kreca experienced a sudden crash of their *A. domesticus* population in 2000, when 50% of their crickets died within 8-12 hours due to an AdDV outbreak (Van Huis *et al.*, 2013). Since 2009, a range of outbreaks have been reported in facilities spread across North America and in Europe, resulting in huge losses in production and an acute crisis in the pet food industry (Szelei *et al.*, 2011; Van Huis *et al.* 2013; Weissmann *et al.*, 2012).

Although AdDV is specific to *A. domesticus*, Szelei *et al.* (2011) showed that other species that are often reared by cricket producers, including the beetles *T. molitor* and *Zophobas morio*, are potential vectors of AdDV that could mechanically transmit the virus via the body surface or the gut (faeces). Even so, experiments showed that the virus may be present in air filters in rearing facilities, suggesting that the virus can be airborne and easily transmitted between rearing rooms (Szelei *et al.*, 2011). The virus can survive on the cuticle of crickets for months and is also present in soil isolates (Weissman *et al.*, 2012). These findings might explain why sanitation measures often appeared unsuccessful. Nowadays, it appears difficult to rear healthy *A. domesticus* (Szelei *et al.*, 2011). Although attempts have been undertaken to select tolerant crickets, an AdDNV-resistant *A. domesticus* strain has not been identified yet (Szelei *et al.*, 2011). The worldwide transport of crickets has played a major role in the spread of AdDV. Weissman *et al.* (2012) tested alternative cricket species received from growers for presence of AdDV, and found that *G. sigillatus* and *G. locorojo* were also heavily infected with AdDNV, while *G. bimaculatus* and *G. assimilis* were found more resistant to AdDNV (Szelei *et al.*, 2011; Weissman *et al.*, 2012). Replacement of *A. domesticus* by *G. assimilis* or *G. bimaculatus* is an option to reduce problems due to AdDV (Szelei *et al.*, 2011), although other viruses might cause problems (see below).

10.3.2 Cricket iridovirus

Cricket iridovirus (CrIV), was detected in commercial colonies of crickets in the Netherlands in 1996 (*Gryllus campestris* and *A. domesticus*; company Kreca) (Kleespies *et al.*, 1999) and in Germany (*G. bimaculatus*) (Just and Essbauer, 2001). CrIV caused mortality and greatly reduced fecundity and life span in nymphs and adults and infected individuals showed apparent behaviour including apathy and disorientation (Just and Essbauer, 2001; Kleespies *et al.*, 1999).

Iridoviruses (family *Iridoviridae*) are large icosahedral cytoplasmic viruses with a linear, double-stranded DNA genome and can be divided into five genera, infecting either vertebrates or invertebrates (genera *Chloriridovirus* and *Iridovirus*) (Kleespies *et al.*, 1999; Williams *et al.*, 2005). A major characteristic of iridoviruses is their large icosahedral capsid and their replication and assembly in the cytoplasm (Jakob *et al.*, 2002). Members of the two genera infecting invertebrates are known as invertebrate iridescent viruses (IIVs) and *Invertebrate iridescent virus 6* (IIV-6, or the recognized synonym Chilo iridescent virus, CIV) is the type species of the *Iridovirus* genus (www.ictvonline.org). IIV-6 infects species of the order Orthoptera, Lepidoptera, Diptera (mosquitoes), Coleoptera and Hymenoptera (Williams, 2008). Comparative genomic analysis showed that CrIV is a strain of IIV-6 (Jakob *et al.*, 2002). An overt IIV infection is often recognizable by the blue iridescent colour of infected insects, which is due to the massive proliferation of virus particles and their paracrystalline arrangement in the host cell's cytoplasm, mainly in fat body cells (Just and Essbauer, 2001; Williams *et al.*, 2005). In natural host populations, the prevalence of overt IIV infections is usually very low, and covert IIV infections are common (Williams *et al.*, 2005).

There are a few reports of iridoviruses isolated from orthopteran species, including mole crickets (*Scapteriscus* spp.) and the crickets *A. domesticus*, *G. bimaculatus*, *G. campestris* and *Gryllus texensis* (Adamo *et al.*, 2014; Boucias *et al.*, 1987; Fowler, 1989; Just and Essbauer, 2001; Kleespies *et al.*, 1999; Williams, 2008). Transmission studies have shown that CrIV (IIV-6) has a broad orthopteran host range (Jakob *et al.*, 2002). CrIV can be transmitted perorally, by feeding high concentrations, to orthopteran species, including *A. domesticus*, *G. campestris*, *G. bimaculatus*, *G. assimilis* and to some locust and cockroach species (Jakob *et al.*, 2002; Kleespies *et al.*, 1999). Dipping of cricket nymphs in concentrated virus suspensions also resulted in infection (Kleespies *et al.*, 1999). However, the exact route of infections is still unclear in many host-IIV systems (Williams *et al.*, 2005). *Per os* infections requires ingestion of very large doses of particles, and most likely cannibalism, predation of infected individuals or transmission by parasitic wasps or nematodes are additional routes of virus transfer. In addition, sexual transmission can occur, as was shown in a recent study by Adamo *et al.* (2014), who detected CrIV in their laboratory colony of *G. texensis* and showed that the virus could be transmitted during mating. Moreover, behavioural changes were observed: infected males were quicker to court females than uninfected males and sickness behaviour, the adaptive reduction of feeding and sexual behaviour that is observed once the immune system is activated, was absent in infected crickets (Adamo *et al.*, 2014; Han *et al.*, 2015).

10.3.3 Cricket paralysis virus

Cricket paralysis virus (CrPV) is known to cause sudden, large-scale mortality in laboratory populations of field-collected crickets (Scotti *et al.*, 1981). CrPV was first isolated from two species of field crickets, *Teleogryllus oceanicus* and *Teleogryllus commodus*, and causes rapid and fatal paralysis (Reinganum *et al.*, 1970). The virus has also been detected in *A. domesticus* (Morrissey and Edwards, 1977; Scotti *et al.*, 1981), although no reports describing CrPV outbreaks in commercial mass rearing of crickets are known (Maciel-Vergara and Ros, 2017). However, since the virus may occur as a covert infection, it could be a thread in mass rearing, where high densities may trigger outbreaks of covert infections.

Dicistroviruses (family *Dicistroviridae*) are small icosahedral positive-strand RNA viruses infecting arthropods, primarily insects. The family name is derived from the two open reading frames (ORFs), or cistrons, encoded in the viral genome (Bonning, 2009; Bonning and Miller, 2010). Three genera are currently known, *Cripavirus* (named after Cricket paralysis virus), *Aparavirus* and *Triatovirus* (www.ictvonline.org). The type species of *Cripavirus* is CrPV, which has a broad host range infecting species of the orders Diptera, Hemiptera, Hymenoptera, Lepidoptera and Orthoptera (Bonning and Miller, 2010; Scotti *et al.*, 1981). CrPV infects gut tissues, fat body, tracheae and muscle tissue, and might show increased neurotropism, resulting in paralysis (Bonning, 2009). The family *Dicistroviridae* includes some additional viruses that paralyze their host and these viruses are often highly pathogenic, whereas other dicistroviruses cause no obvious symptoms. Unapparent, persistent infections may become more pathogenic in response to stress or secondary infections, leading to epizootics and massive die-offs (Bonning and Miller, 2010).

Reinganum *et al.* (1970) first noticed CrPV in mass-reared early instars of *T. oceanicus*, where it spread rapidly and resulted in a loss of 95% of the crickets. Affected *T. oceanicus* and *T. commodus* crickets became uncoordinated due to a progressive immobilization of their hind legs, followed by death. Early- to mid-instar nymphs were the most susceptible to the virus. Some healthy adults were found to contain low virus levels without showing symptoms. CrPV has a high prevalence in *T. commodus* field populations, where CrPV was present in 43% of the 232 sites sampled (Reinganum *et al.*, 1981). CrPV is transmitted horizontally *per os*, through virus particles in the faeces (Bonning, 2009, Bonning and Miller, 2010) or by cannibalism (Scotti *et al.*, 1981). Reinganum *et al.* (1970) demonstrated that CrPV is not transmitted transovarially (within the egg), by decontaminating *Teleogryllus* spp. eggs using solutions of sodium hypochlorite and formalin, which successfully prevented vertical transmission. However, in natural or laboratory populations, vertical transmission might occur by surface contamination of eggs (Scotti *et al.*, 1981).

CrPV was detected in the central nerve system of *A. domesticus* (Morrissey and Edwards, 1977) and in dead *A. domesticus* crickets from a disease outbreak at a cricket farm in Georgia (Scotti *et al.*, 1981). Additional reports describing CrPV outbreaks in commercial mass rearing of crickets are so far unknown (Maciel-Vergara and Ros, 2017). No epizootics due to CrPV have been recorded in natural cricket populations. However, apparently healthy insects, collected in the field and reared in the laboratory, may suddenly succumb to a viral infection (Scotti *et al.*, 1981).

10.3.4 Gryllus bimaculatus nudivirus

Gryllus bimaculatus nudivirus (GbNV) was originally isolated in laboratory colonies of *G. campestris* in Southern Germany (Huger, 1985). Nudiviruses (family *Nudiviridae*) comprise a highly diverse group of viruses with enveloped, rod-shaped nucleocapsids containing a large, circular double-stranded DNA genome (www.ictvonline.org; Williams *et al.*, 2017). Currently, two genera are known, *Alphanudivirus* and *Betanudivirus*. Nudiviruses are found in phylogenetically highly diverse arthropod hosts, including crustaceans and insects in the orders Coleoptera, Diptera, Lepidoptera, Orthoptera, Hymenoptera (as endogenous viral elements) and possibly a range of additional orders (Williams *et al.*, 2017). Surprisingly little is known of their infection cycle in either cells or natural hosts (Wang and Jehle, 2009).

GbNV is a virus of nymphs and adult field crickets. Apart from *G. campestris* the virus also infects *G. bimaculatus*, *T. oceanicus* and *T. commodus* (Huger, 1985). GbNV causes a lethal infection in larvae, but chronic disease in adults (Wang and Jehle, 2009). Infection occurs mainly *per os* during nymphal development, by cannibalistic feeding on diseased or dead specimens (Huger, 1985). Infected crickets have a reduced size, show uncoordinated and lethargic behaviour until they finally die. In the final stage of disease, crickets are often swollen, containing a large amount of viscous and milky opalescent haemolymph. GbNV perorally invades hosts through faecal contaminated foods at nesting and mating sites (Huger, 1985).

10.3.5 Current and new viruses in crickets

While AdDV and CrIV have so far caused serious losses in commercial cricket rearing, other viruses including CrPV and GbNV form potential threats even though major outbreaks or epizootics have not been reported yet for these viruses. Furthermore, recently additional viruses (not recognized yet by the ICTV) have been detected in crickets. While screening North American *A. domesticus* samples from mass cricket die-offs, a circular single-stranded DNA virus was detected (named Acheta domesticus volvovirus (AdVVV); Pham *et al.*, 2013a) and a virus resembling ambisense densoviruses, but with a shorter genome (named Acheta domesticus mini ambidensovirus (AdMADV); Pham *et al.*, 2013c). AdVVV was subsequently detected in *G. assimilis* (Pham *et al.*, 2013b). No further details have been described for these viruses. The risk of new outbreaks increases, due to further intensification of cricket mass rearing, the world-wide trade in crickets and the broad orthopteran host range of viruses from different taxa. Most insect viruses infect insects predominantly *per os*, and since cannibalism and consumption of faeces and dead conspecifics is frequent among crickets, viruses may easily spread within mass rearing facilities. In addition, viruses might be present in a covert form in natural cricket populations and these can be vertically transmitted from mother to offspring. New virus species are continuously being discovered (Maciel-Vergara and Ros, 2017) and the chance of additional pathogen emergence in insect rearing is high. Furthermore, mass rearing, accompanied by stress factors including high density, might trigger the conversion of covert into overt, symptomatic infections (Cory and Myers, 2003; Maciel-Vergara and Ros, 2017).

10.4 Recommendations

Some general advises about diagnostics, prevention and management is given below (based on Eilenberg and Jensen, in press, modified).

The physical appearance of the production facilities is essential. Invasion by major predators (birds, mammals like rats) can be prevented with relatively little effort. Closed, clean facilities with tight doors, closed trays or incubators will be of great value. Smaller predators and other enemies (predatory insects, parasitoids) are more challenging to manage. Again, the advice is to keep production facilities well separated from the environment: nylon mesh, sticky traps and other equipment is very useful. It is even more challenging to keep insect pathogens out. An often used prevention method for especially viruses is to implement a procedure, where the insect stock rearing facilities and the virus production (in insects) facilities are maintained in separate units and by that avoid dispersal of virus by humans or air from virus production rooms into insect rearing rooms.

Rearing conditions (temperature, moisture, food, light, density, etc.) should be optimal for maintaining healthy insects of all instars and therefore maintaining a healthy production stock. Healthy insects will better resist natural enemies and also, a healthy production stock may have a better innate immune system to combat diseases. It is important to evaluate if the production method is appropriate to prevent and/or manage insect diseases. A choice can include if production should take place in batches or in one continuous production system. Also, it is important to ensure that production takes place in separate units to avoid that all production is lost if a natural enemy gets access to one unit.

Frequent observations of the production facilities are needed. Personnel to inspect should be trained and should possess good skills in entomology and microbiology so they are able to identify and report malfunctions immediately. Any insect looking aberrant (discoloured, less active or sluggish) should be discharged and brought away from production facilities, since they could harbour a parasitoid or an insect disease.

Covert virus infections can be present and may cause sudden, disastrous epidemics and immediate action is needed. The decision can be that the entire production stock should be destroyed, irrespective of a temporary loss in income. After careful cleaning and a re-evaluation of production procedures, a new virus-free insect stock can be introduced.

Genetic variation in the production stock should be considered. There are two different perspectives here. A homogenous production stock will make it easier to define the best production environment. On the other hand, inbreeding may occur in these cases and monocultures are more vulnerable to disease outbreaks. To diversify the gene pool in the production stock, insects from the wild may be used. There is, however, a risk to accompany the addition of wild insects, namely the risk of bringing in new parasitoids and new diseases. A way to avoid this is to keep newly collected wild insects in quarantine for a period of time.

Substrates (plant material, soil, etc.) brought from outside should be checked before introduction. Especially potential generalist insect pathogens may pose a problem in slurry and other wastes, and a check should include screening for such microorganisms, typically fungi or bacteria, but also viruses. At the same time, it is important to remove all organic wastes immediately: plant material, soil, dead insects, etc.

Assistance from specialized laboratories may be needed to diagnose for insect diseases or to determine predators and parasitoids. Such experts should be entomologists and insect pathologists. International networking is crucial, since no scientists have competences and methods to diagnose and determine all natural enemies. While diagnosis of well-known pathogens may sometimes be quick, identification of new and unknown pathogens will take longer.

10.5 Conclusions

Insects produced for food and feed meet severe challenges by housing natural enemies: predators, parasitoids, and microorganisms (viruses, fungi, bacteria and more). One specific insect species can be a target for a range of natural enemies. Some of these are known, while more are expected to be revealed when production of insects increases both in quantities of insects produced and in geographical distribution of production and processing facilities. New methods, like next generation sequencing, may reveal the presence of hitherto unknown insect culprits responsible for the disease being manifested. Furthermore, a better coverage of services to diagnose for natural enemies and provide advises on prevention and management of these natural enemies in insects produced for food and feed is needed.

References

Abbas, M.S.T., Saleh, M.M.E. and Akil, A.M., 2001. Laboratory and field evaluation of the pathogenicity of entomopathogenic nematodes to the red palm weevil, *Rhynchophorus ferrugineus* (Oliv.) (Col.: Curculionidae). Journal of Pest Sciences 74: 167-168.

Abdullah, M.A.R., 2009. Biological control of the red palm weevil, *Rhynchophorus ferrugineus* (Olivier) (Coleoptera: Curculionidae) by the parasitoid mite, *Rhynchopolipus rhynchophori* (Ewing) (Acarina: Podapolipidae). Journal of the Egyptian Society of Parasitology 39: 679-686.

Abraham, V.A. and Kurian, C., 1973. *Chelisoches morio* F. (Forficulidae: Dermaptera), a predator on eggs and early instar grubs of the red palm weevil *Rhynchophorus ferrugineus* F. (Curculionidae: Coleoptera). Journal of Plant Crops 1: 147-152.

Adamo, S.A., Kovalko, I., Easy, R.H. and Stoltz, D., 2014. A viral aphrodisiac in the cricket *Gryllus texensis*. Journal of Experimental Biology 217: 1970-1976.

Al-Dhafar, Z.M. and Al-Qahtani, A.M., 2012. Mites associated with the red palm weevil, *Rhynchophorus ferrugineus* Olivier in Saudi Arabia with a description of a new species. Acarines 6: 3-6.

Alfazairy, A.A., Hendi, R., El-Minshawy, A.M. and Karam, H.H., 2003. Entomopathogenic agents isolated from 19 coleopteran insect pests in Egypt. Egyptian Journal of Biological Pest Control 13: 125.

Ali Taha, H., Hassan, M.F., Allam, F. and Reham, A.M., 2014. Studies on some parasitic and predaceous mites associated with the red palm weevil, *Rhynchophorus ferrugineus* Olivier (Coleoptera: Curculionidae). In: Proceedings of the 5th International Date Palm Conference, March 16-18, 2014, Abu Dhabi, United Arabic Emirates. Available at: http://tinyurl.com/ybjqzw4h.

Atakan, E., Elekçioglu, H., Gözel, U., Günes, Ç. and Yüksel, O., 2009. First report of *Heterorhabditis bacteriophora* (Poinar, 1975) (Nematoda: Heterorhabditidae) isolated from the red palm weevil, *Rhyncophorus ferrugineus* (Oliver, 1970) (Coleoptera: Curculionidae) in Turkey. Bulletin OEPP 39: 189-193.

Banerjee, A. and Dangar, T.K., 1995. *Pseudomonas aeruginosa* a facultative pathogen of red palm weevil *Rhynchophorus ferrugineus*. World Journal of Microbiology and Biotechnology 11: 618-620.

Basheer, P.M. and Thomas, S.K., 2012. Indian Treepie *Dendrocitta vagabunda parvula* (Latham, 1790) (Passeriformes: Corvidae) as a natural enemy of the pests of coconut and areca palm plantations. Journal of Biopesticides 5: 205-208.

Binda-Rosetti, S., Mastore, M., Protasoni, M. and Brivio, M.F., 2016. Effects of an entomopathogen nematode on the immune response of the insect red palm weevil: focus on host antimicrobial response. Journal of Invertebrate Pathology 133: 110-119.

Bonning, B.C. and Miller, W.A., 2010. Dicistroviruses. Annual Review of Entomology 55: 129-150.

Bonning, B.C., 2009. The *Dicistroviridae*: an emerging family of invertebrate viruses. Virologica Sinica 24: 415-427.

Boucias, D.G., Maruniak, J.E. and Pendland, J.C., 1987. Characterization of an iridovirus isolated from the southern mole cricket, *Scapteriscus vicinus*. Journal of Invertebrate Pathology 50: 238-245.

Cito, A., Mazza, G., Strangi, A., Benvenuti, C., Barzanti, G.P., Dreassi, E., Turchetti, T., Francardi, V. and Roversi, P.F., 2014. Characterization and comparison of *Metarhizium* strains isolated from *Rhynchophorus ferrugineus*. FEMS Microbiology Letters 355: 108-115.

Cory, J.S. and Myers, J.H., 2003. The ecology and evolution of insect baculoviruses. Annual Review of Ecology, Evolution, and Systematics 34: 239-272.

Dembilio, Ó., Quesada-Moraga, E., Santiago-Álvarez, C. and Jacas, J.A., 2010. Potential of an indigenous strain of the entomopathogenic fungus *Beauveria bassiana* as a biological control agent against the red palm weevil, *Rhynchophorus ferrugineus*. Journal of Invertebrate Pathology 104: 214-221.

Dilipkumar, M., Ahadiyat, A., Mašán, P. and Chuah, T.S., 2015. Mites (Acari) associated with *Rhynchophorus ferrugineus* (Coleoptera: Curculionidae) in Malaysia, with a revised list of the mites found on this weevil. Journal of Asia-Pacific Entomology 18: 169-174.

Eilenberg, J. and Jensen, A.B., in press. Prevention and management of diseases in terrestrial invertebrates. In: Hajek, A.E. and Shapiro-Ilan, D. (eds.) Ecology of invertebrate diseases. John Wiley, Hoboken, NJ, USA.

Eilenberg, J., Vlak, J., Nielsen-LeRoux, C., Cappellozza, S. and Jensen, A.B., 2015. Diseases in insects produced for food and feed. Journal of Insects as Food and Feed 1: 87-102.

El-Minshawy, A.M., Hendi, R.A. and Gadelhak, G.G., 2005. Viability of stored polyhedrosis virus of the red palm weevil, *Rhynchophorus ferrugineus* (Olivier) (Coleoptera: Curculionidae). In: FAO/IAEA International Conference on Area-Wide Control of Insect Pests: integrating the sterile insect and related nuclear and other techniques, May 22-26, 2017, Vienna, Austria, pp. 241-242. Available at: http://tinyurl.com/yb57fd75.

El-Sharabasy, H.M., 2010. A survey of mite species with the red palm weevil, *Rhyncho-phorus ferrugineus* (Olivier) in Egypt. Egyptian Journal of Pest Control 20: 67-70.

European Food Safety Authority (EFSA) Scientific Committee, 2015. Scientific opinion on a risk profile related to production and consumption of insects as food and feed. EFSA Journal 13(4257): 1-60.

Faleiro, J.R., 2006a. A review of the issues and management of the red palm weevil *Rhynchophorus ferrugineus* (Coleoptera: Rhynchophoridae) in coconut and date palm during the last one hundred years. International Journal of Tropical Insect Science 26: 135-154.

Faleiro, J.R., 2006b. Insight into the management of red palm weevil *Rhynchophorus ferrugineus* Olivier: based on experiences on coconut in India and date palm in Saudi Arabia. In: Fundación Agrolimed (ed.) I Jornada Internacional sobre el Picudo Rojo de las Palmeras, Fundación Agrolimed, Valencia, Spain, pp. 35-57.

Fazul Rahaman, P., Sharma, S.B. and Wightman, J.A., 2000. A review of insect-parasitic nematodes research in India: 1927-1997. International Journal of Pest Management 46: 19-28.

Fowler, H.G., 1989. Natural microbial control of cricket populations (Orthoptera: Gryllotalpidae: *Scapteriscus borellii*): regulation of populations aggregated in time and space. Revista Brasileira de Biologia 49: 1039-1051.

Francardi, V., Benvenuti, C., Roversi, P.F., Rumine, P. and Barzanti, G., 2012. Entomopathogenicity of *Beauveria bassiana* (Bals.) Vuill. and *Metarhizium anisopliae* (Metsch.) Sorokin isolated from different sources in the control of *Rhynchophorus ferrugineus* (Olivier) (Coleoptera Curculionidae). Redia Journal of Zoology 95: 49-55.

Francesca, N., Alfonzo, A., Lo Verde, G., Settanni, L., Sinacori, M., Lucido, P. and Moschetti, G., 2014. Biological activity of *Bacillus* spp. evaluated on eggs and larvae of red palm weevil *Rhynchophorus ferrugineus*. Annals of Microbiology 65: 477-485.

Gahukar, R.T., 2016. Edible insects farming: efficiency and impact on family livelihood, food security, and environment compared with livestock and crops. In: Dossey, A.T., Morales-Ramos, J.A. and Rojas, M.G. (eds.) Insects as sustainable food ingredients. Elsevier Inc., Amsterdam, the Netherlands, pp. 85-111.

Garofalo, C., Osimani, A., Milanovic, V., Taccari. T., Cardinali, F., Aquilanto, L., Riolo, P., Rushioni, S., Isidoro, N. and Clementi, F., 2017. The microbiota of marketed processed edible insects as revealed by high-throughput sequencing. Food Microbiology 62: 15-22.

Gopinadhan, P.B., Mohandas, N. and Nair, K.P.V., 1990. Cytoplasmic polyhedrosis virus infecting red palm weevil of coconut. Current Science India 59: 577-580.

Gordh, G. and Headrick, D., 2011. A dictionary of entomology (2nd Ed.). CABI, Wallingford, UK.

Güerri-Agulló, B., López-Follana, R., Asensio, L., Barranco, P. and Lopez-Llorca, L.V., 2011. Use of a solid formulation of *Beauveria bassiana* for biocontrol of the red palm weevil (*Rhynchophorus ferrugineus*) (Coleoptera: Dryophthoridae) under field conditions in SE Spain. Florida Entomologist 94: 737-747.

Gullan, P.J. and Cranston, P.S., 2013. The insects: an outline of entomology (4th Ed.). Wiley-Blackwell, Chichester, UK.

Hajek, A., 2004. Natural enemies – an introduction to biological control. Cambridge University Press, Cambridge, UK.

Halloran, A., Roos, N., Flore, R. and Hanboonsong, Y., 2016. The development of the edible cricket industry in Thailand. Journal of Insects as Food and Feed 2: 91-100.

Han, Y., Van Oers, M.M., Van Houte, S. and Ros, V.I.D., 2015. Virus-induced behavioural changes in insects. In: Mehlhorn, H. (ed.) Host manipulations by parasites and viruses. Parasitology Research Monographs. Springer International Publishing, Cham, Switzerland, pp. 149-174.

Hanboonsong, Y., Jamjanya, T. and Durst, P.B., 2013. Six-legged livestock: edible insect farming, collecting and marketing in Thailand. FAO, Regional Office for Asia and the Pacific, Bangkok, Thailand.

Hassan, M.F., Nasr, A.K., Allam, S.F., Taha, H.A. and Mahmoud, R.A., 2011. Biodiversity and seasonal fluctuation of mite families associated with the red palm weevil, *Rhynchophorus ferrugineus* Oliver (Coleoptera: Curculionidae) in Egypt. Egyptian Journal of Biological Pest Control 21: 317-323.

Huger, A.M., 1985. A new virus disease of crickets (Orthoptera: Gryllidae) causing macronucleosis of fat body. Journal of Invertebrate Pathology 45: 108-111.

Husain, A., Rizwan-ul-Haq, Al-Ayedh, H., Ahmed, S. and Al-Jabr, A.M., 2015. Effect of *Beauveria bassiana* on feeding performance and antioxidant defense of red palm weevil, *Rhynchophorus ferrugineus*. BioControl 60: 849-859.

Husband, R.W. and OConnor, B.M., 1999. Two new ectoparasitic mites (Acari: Podapolipidae) of *Rhynchophorus* spp. (Coleoptera: Curculionidae) from Indonesia, Malaysia, the Philippines and West Africa, International Journal of Acarology 25: 101-110.

Imoulan, A., Wu, H.J., Lu., W.L.L., Li, B.B., Yang, R.H., Wang, W.J., Wang, X.L., Kirk, P. and Yao, Y.J., 2016. *Beauveria medogensis* sp. nov., a new fungus of the entomopathogenic genus from China. Journal of Invertebrate Pathology 139: 74-81.

Jakob, N., Kleespies, R., Tidona, C., Müller, K., Gelderblom, H. and Darai, G., 2002. Comparative analysis of the genome and host range characteristics of two insect iridoviruses: Chilo iridescent virus and a cricket iridovirus isolate. Journal of General Virology 83: 463-470.

Jongema, Y., 2017. List of edible insects of the world. Wageningen University, Wageningen, the Netherlands. Available at: http://tinyurl.com/mestm6p.

Just, F.T. and Essbauer, S.S., 2001. Characterization of an iridescent virus isolated from *Gryllus bimaculatus* (Orthoptera: Gryllidae). Journal of Invertebrate Pathology 77: 51-61.

Kaya, H.K. and Vega, F.E., 2012. Scope and basic principles of insect pathology. In: Vega, F.E. and Kaya, H.K. (eds.) Insect pathology (2nd Ed.). Academic Press, Cambridge, MA, USA, pp. 1-12.

Kleespies, R., Tidona, C. and Darai, G., 1999. Characterization of a new iridovirus isolated from crickets and investigations on the host range. Journal of Invertebrate Pathology 73: 84-90.

Kok, R., 2017. Insect production and facility design. In: Van Huis, A. and Tomberlin, J.K. (eds.) Insects as food and feed: from production to consumption. Wageningen Academic Publishers, Wageningen, the Netherlands, pp. 142-172.

Krishnakumar, R. and Sudha, G., 2002. Indian treepie (*Dendrocitta vagabunda parvula* Whistler and Kinnear) (Corvidae) – a predatory bird of red palm weevil (*Rhynchophorus ferrugineus* Oliv.). Insect Environment 8: 133.

Lever, R.J.A.W., 1969. Invertebrate pests. Pests of the coconut palm. FAO Plant Production and Protection Series. FAO, Rome, Italy.

Maciel-Vergara, G. and Ros, V.I.D., 2017. Viruses of insects reared for food and feed. Journal of Invertebrate Pathology 147: 60-75.

Manga-González, M.Y., González-Lanza, C., Cabanas, E. and Campo, R., 2001. Contributions to and review of dicrocoeliosis, with special reference to the intermediate hosts of *Dicrocoelium dendriticum*. Parasitology 123: 91-114.

Mazza, G., Cini, A., Cervo, R. and Longo, S., 2011. Just phoresy? Reduced lifespan in red palm weevils *Rhynchophorus ferrugineus* (Coleoptera: Curculionidae) infested by the mite *Centrouropoda almerodai* (Uroactiniinae: Uropodina). Italian Journal of Zoology 78: 101-105.

Merghem, A., 2011. Susceptibility of the red palm weevil, *Rhynchophorus ferrugineus* (Olivier) to the green muscardine fungus, *Metarhizium anisopliae* (Metsch.) in the laboratory and in palm tree orchards. Egyptian Journal of Biological Pest Control 21: 179-183.

Meynardier, G., Matz, G., Veyrunes, J.-C. and Bres, N., 1977. Virus de type densonucléose chez les orthoptères. Annales de la Société Entomologique de France 13: 487-493.

Miguens, F.C., Sacremento de Magalhães, J.A., Melo de Amorim, L., Goebel, V.R., Le Coustour, N., Lummerzheim, M., Moura, J.I.L. and Costa, R.M., 2011. Mass trapping and biological control of *Rhynchophorus palmarum* L.: a hypothesis based on morphological evidences. EntomoBrasilis 4: 49-55.

Morrissey, R. and Edwards, J., 1977. Virus-like particles in peripheral nerve of the house cricket. Journal of Invertebrate Pathology 29: 247-249.

Moura, J.I.L., Mariau, D. and Delabie, J.H.C., 1993. Eficiência de *Paratheresia menezesi* Townsend (Diptera: Tachnidae) no controle biológico naturel de *Rhynchophorus palmarum* (L.) (Coleoptera: Curculionidae). Oléagineux 48: 219-223.

Murphy, S.T. and Briscoe, B.R., 1999. The red palm weevil as an alien invasive: biology and prospects for biological control as a component of IPM. Biocontrol 20: 35-45.

Peter, C., 1989. A note on the mites associated with the red palm weevil, *Rhynchophorus ferrugineus* Oliv. in Tamil Nadu. Journal of Insect Science 2: 160-161.

Pham, H.T., Bergoin, M. and Tijssen, P., 2013a. Acheta domesticus volvovirus, a novel single-stranded circular DNA virus of the house cricket. Genome Announcements 1: e00079-00013.

Pham, H.T., Iwao, H., Bergoin, M. and Tijssen, P., 2013b. New volvovirus isolates from *Acheta domesticus* (Japan) and *Gryllus assimilis* (United States). Genome Announcements 1: e00328-00313

Pham, H.T., Yu, Q., Bergoin, M. and Tijssen, P., 2013c. A novel ambisense densovirus, *Acheta domesticus* mini ambidensovirus, from crickets. Genome Announcements 1: e00914-00913.

Pu, Y.C. and Hou, Y.M., 2016. Isolation and identification of bacterial strains with insecticidal activities from *Rhynchophorus ferrugineus* Oliver (Coleoptera: Curculionidae). Journal of Applied Entomology 140: 617-626.

Rao, P.N. and Reddy, Y.N., 1980. Description of a new nematode *Praecocilenchus ferruginophorus* n. sp. from weevil pests (Coleoptera) of coconut palms in South India. Rivista di Parasitologia 16: 93-98.

Reinganum, C., Gagen, S.J., Sexton, S.B. and Vellacott, H.P., 1981. A survey for pathogens of the black field cricket, *Teleogryllus commodus*, in the Western district of Victoria, Australia. Journal of Invertebrate Pathology 38: 153-160.

Reinganum, C., O'Loughlin, G. and Hogan, T., 1970. A nonoccluded virus of the field crickets *Teleogryllus oceanicus* and *T. commodus* (Orthoptera: Gryllidae). Journal of Invertebrate Pathology 16: 214-220.

Salama, H.S., Foda, M.S., El-Bendary, M.A. and Abdel-Razek, A., 2004. Infection of red palm weevil *Rhynchophorus ferrugineus*, by spore-forming bacilli indigenous to its natural habitat in Egypt. Journal of Pesticide Science 77: 27-31.

Saleh, M.M.E., Alheji, M.A., Alkhazal, M.H., Alferdan, H. and Darwish, A., 2011. Evaluation of *Steinernema* sp. SA a native isolate from Saudi Arabia for controlling adults of the red palm weevil. *Rhynchophorus ferrugineus* (Oliver). Egyptian Journal of Biological Pest Control 21: 277-282.

Santhi, V.S., Salamem L., Nakache, Y., Koltai, H., Soroker, V. and Glazer, I., 2015. Attraction of entomopathogenic nematodes *Steinermena carpocapsae* and *Heterorhabditis bacteriophora* to the red palm weevil (*Rhynchoporus ferrugineus*). Biological Control 83: 75-81.

Scotti, P.D., Longworth, J.F., Plus, N., Croizier, G. and Reinganum, C., 1981. The biology and ecology of strains of an insect small RNA virus complex. Advances in Virus Research 26: 117-143.

Sewify, G.H., Belal, M.H. and Al-Awash, S.A., 2009. Use of the entomopathogenic fungus, *Beauveria bassiana* for the biological control of the red palm weevil, *Rhynchophorus ferrugineus* Olivier. Egyptian Journal of Biological Pest Control 19: 157-163.

Sharaby, A., Fallata, S.A. and Ibrahim, S.A., 2011. Artificial diet for continuous maintenance of the maritima earwig *Anisolabis maritima* (Bonelli) (Dermaptera: Carcinophoridae). Archives of Phytopathology and Plant Protection 44: 352-355.

Styer, E. and Hamm, J., 1991. Report of a densovirus in a commercial cricket operation in the southeastern United States. Journal of Invertebrate Pathology 58: 283-285.

Szelei, J., Woodring, J., Goettel, M., Duke, G., Jousset, F.-X., Liu, K., Zadori, Z., Li, Y., Styer, E. and Boucias, D., 2011. Susceptibility of North-American and European crickets to Acheta domesticus densovirus (AdDNV) and associated epizootics. Journal of Invertebrate Pathology 106: 394-399.

Tijssen, P., Pénzes, J.J., Yu, Q., Pham, H.T. and Bergoin, M., 2016. Diversity of small, single-stranded DNA viruses of invertebrates and their chaotic evolutionary past. Journal of Invertebrate Pathology 140: 83-96.

Torta, L., Leone, V., Caldarella, G., Lo Verde, G. and Burruano, S., 2009. Microrganismi fungini associati a *Rhynchophorus ferrugineus* (Olivier) in Sicilia e valutazione dell'efficacia entomopatogena di un isolato di *Beauveria bassiana* (Bals.) Vuill, Osservazioni preliminari. Micologia Italiana 2: 49-56.

Van Huis, A., Van Itterbeeck, J., Klunder, H., Mertens, E., Halloran, A., Muir, G. and Vantomme, P., 2013. Edible insects: future prospects for food and feed security. Food and Agriculture Organization of the United Nations (FAO), Rome, Italy, FAO Forestry Paper no. 171, 187 pp. Available at: http://www.fao.org/docrep/018/i3253e/i3253e.pdf.

Vives, O.D.V., 2011. Bio-ecology and integrated management of the red palm weevil, *Rhynchophorus ferrugineus*, PhD thesis, Universitat Jaume, Castellón, Spain.

Wang, Y. and Jehle, J.A., 2009. Nudiviruses and other large, double-stranded circular DNA viruses of invertebrates: new insights on an old topic. Journal of Invertebrate Pathology 101: 187-193.

Weissman, D.B., Gray, D.A., Pham, H.T. and Tijssen, P., 2012. Billions and billions sold: pet-feeder crickets (Orthoptera: Gryllidae), commercial cricket farms, an epizootic densovirus, and government regulations make a potential disaster. Zootaxa 3504: 67-88.

Williams, T., 2008. Natural invertebrate hosts of iridoviruses (Iridoviridae). Neotropical Entomology 37: 615-632.

Williams, T., Barbosa-Solomieu, V. and Chinchar, V.G., 2005. A decade of advances in iridovirus research. Advances in Virus Research 65: 173-248.

Williams, T., Bergoin, M. and Van Oers, M.M., 2017. Diversity of large DNA viruses of invertebrates. Journal of Invertebrate Pathology 147: 4-22.

Industrial production systems

11.

Calliphora vomitoria

Industrialization of insect farming

J.K. Tomberlin
Department of Entomology, Texas A&M University, TAMU 2475, College Station, TX 77843-2475, USA;
jktomberlin@tamu.edu

Arthropods seem to mass-produce in any given scenario that involves them being a nuisance to humanity, either directly (e.g. mosquitoes biting and transmitting pathogens; Coates, 2000), or indirectly (e.g. grasshoppers feeding on grasslands and reducing vegetation for livestock; Joern, 2004), or blow flies causing damage to livestock by colonizing them while alive (Tomberlin *et al.*, 2017), or beetles colonizing stored products (Bosquet, 1990). Thus, assuming they could be easily mass-produced in culture would be a logical conclusion; however, as demonstrated in this chapter, mass-production of the insects of interest, as food and feed, is more complex than one would expect.

The examples provided in this book are of a few model organisms that are currently mass-produced in many locations worldwide. Additionally, the methods presented represent decades of research, requiring patience, due-diligence, trial and error, as well as fortitude. These efforts by countless individuals over the years have resulted in well-established methodologies. Each example provides a glimpse into the complexities of harnessing Nature and bringing it indoors. However, the success of these individuals within their respective system demonstrates the potential of insect-farming. As examples, the black soldier fly (Sheppard *et al.*, 1994) or house fly (Čičková *et al.*, 2012) can potentially be produced at the tonnage scale per day, within a single factory.

With success comes a note of caution: methods used for mass-producing these insects vary dramatically depending on the biology of the insect in question, location, and resources available for use in establishing a mass-production facility. If attempting to apply these methods, one should allow for some flexibility in facility structure or procedural implementation as population differences could require some tweaking for optimal production. These differences in methods (e.g. shifts in temperature or humidity (Holmes *et al.*, 2012), variation in light:dark patterns (Holmes, 2010), feed quality (Nguyen *et al.*, 2013) can be the difference between optimal production and a cycle of repeated crashing of the production line, or even worse, the insect colony. This is the heartbeat of an insect factory, as it is needed to produce the eggs of the individuals harvested for use as food or feed. Therefore, these methods should be considered as a guide, where you, the reader and potential practitioner, will need to refine them to mass-produce the target insect in your region.

References

Bosquet, Y., 1990. Beetles associated with stored products in Canada: an identification guide. Canadian Government Publishing Centre, Quebec, Canada.

Čičková, H., Pastor, B., Kozánek, M., Martínez-Sánchez, A., Rojo, S. and Takáč, P., 2012. Biodegradation of pig manure by the house fly, *Musca domestica*: a viable ecological strategy for pig manure management. PLoS ONE 7: e32798.

Coates, C.J., 2000. Malaria: a mosquito transformed. Nature 405: 900-901.

Holmes, L., 2010. Role of abiotic factors on the development and life history of the black soldier fly, *Hermetia illucens* (L.) (Diptera: Stratiomyidae). MSc thesis, University of Windsor, Windsor, ON, USA.

Holmes, L.A., Vanlaerhoven, S.L. and Tomberlin, J.K., 2012. Relative humidity effects on the life history of *Hermetia illucens* (Diptera: Stratiomyidae). Environmental Entomology 41: 971-978.

Joern, A., 2004. Variation in grasshopper (Acrididae) densities in response to fire frequency and bison grazing in tallgrass prairie. Environmental Entomology 33: 1617-1625.

Nguyen, T.T.X., Tomberlin, J.K. and VanLaerhoven, S.L., 2013. Influences of resources on *Hermetia illucens* (Diptera: Stratiomyidae) larval development. Journal of Medical Entomology 50: 898-906.

Sheppard, D.C., Newton, G.L., Thompson, S.A. and Savage, S., 1994. A value added manure management system using the black soldier fly. Bioresource Technology 50: 275-279.

Tomberlin, J.K., Crippen, T.L., Tarone, A.M., Chaudhury, M.F., Singh, J.B., Cammack, J.A. and Meisel, R.P., 2017. A review of bacterial interactions with blow flies (Diptera: Calliphoridae) of medical, veterinary, and forensic importance. Annals of the Entomological Society of America 110: 19-36.

12.

Hermetia illucens

Black soldier fly: biology and mass production

J.K. Tomberlin and J.A. Cammack*
Department of Entomology, Texas A&M University, TAMU 2475, College Station, TX 77843-2475, USA;
jcammack_07@tamu.edu

Abstract

For the past 30+ years, the black soldier fly *Hermetia illucens* (L.) (Diptera: Calliphoridae) has been studied intensively as a means for addressing two issues of global concern: organic waste management, and alternate protein production. The larvae are voracious, generalist feeders, able to consume a wide variety of materials ranging from plant matter (fruits, vegetables, grains, etc.) to animal matter (tissue, manure, etc.), and transform these materials into body mass that can be upwards of 45% protein and 35% fat. Additionally, this is the only insect species that has been approved for use as an animal feed in the United States (for salmonid fish) and Canada (for salmonid fish and broiler poultry). Maintaining a robust population to supply an operation with the larvae necessary for waste management and protein production is of utmost importance for entities interested in mass producing black soldier flies, while at the same time ensuring consistent larval output (in both quantity and quality). This chapter focuses on methods used for rearing black soldier flies, as well as challenges associated with mass production for waste management and protein production.

Keywords: *Hermetia illucens*, waste management, sustainable agriculture

12.1 Introduction

Since the 1990s, black soldier flies, *Hermetia illucens* (L.) (Diptera: Stratiomyidae) have been proposed as a method for managing the vast amount of organic wastes generated by human activities and agricultural production (Sheppard *et al.*, 1994). Due to the ability of their larvae to consume a wide variety of otherwise wasted materials, the black soldier fly is perhaps our biggest ally when it comes to tackling the problem of organic waste. The larvae can consume fruit, vegetable, and other plant matter, animal tissue, and even manure, and transform these materials into body mass comprised of 45% protein and 35% fat. Worldwide, efforts have been made to utilize this bioconversion to address the challenges associated with producing feed on which livestock are raised. To date, this is the only species of insect that is approved in the United States for use as feed (for salmonid fish species), and the same is true for Canada (for salmonid fish species and broiler chickens), as long as the larvae were raised on feed grade material. This chapter will focus on the biology and maintenance of a black soldier fly colony that is to be used for the application of black soldier fly biology to waste management, with specific emphasis on challenges associated with such mass production.

12.2 Soldier fly biology

The black soldier fly is easily recognized for its unique morphology and behaviour. Adults can range in size from 1.0-2.5 cm in length, are typically black and have a wasp-like appearance due to two translucent sections on the proximal margins of the abdomen, resulting in a thread-waist appearance (Tomberlin *et al.*, 2002). Larvae are vermiform and emerge from eggs at <1 mm in length and less than 0.015 mg in weight; and due to their voracious feeding, can grow to 3.5 cm in length and weigh as much as 0.5 g.

As many have noted, this species is now distributed globally throughout the tropic and temperate zones, likely due to human-mediated transportation (Benelli *et al.*, 2014; Callan, 1974; Marshall *et al.*, 2015; Stone *et al.*, 1965). Historically, *H. illucens* was considered a pest in layer poultry operations due to its association with manure and tendency of the larvae to displace the faecal waste through their dispersal activities (Axtell and Arends, 1990; Axtell and Edwards, 1970). However, research beginning during the 1980s (e.g. Sheppard, 1983) determined that this ability to digest diverse types of animal wastes (Banks *et al.*, 2014; Diener *et al.*, 2011; Oonincx *et al.*, 2015b) and disperse from a substrate could be harnessed to allow for self-harvesting of the larvae (Newton *et al.*, 2005a,b; Sheppard *et al.*, 1994), which could then be used for livestock (Newton *et al.*, 1977), poultry (Hale, 1973; Schiavone *et al.*, 2017), and aquaculture feed (Bondari and Sheppard, 1981, 1987; Kroeckel *et al.*, 2012; Latney and Clayton, 2014; Sealey *et al.*, 2011; St-Hilaire *et al.*, 2007; Van Huis, 2013) (to be discussed more in depth later).

Studies in Canada (Nguyen *et al.*, 2013, 2015), China (Zhou *et al.*, 2013), Italy (Salomone *et al.*, 2017), the Netherlands (Oonincx *et al.*, 2015b), and the USA (Harnden and Tomberlin, 2016; Tomberlin *et al.*, 2002), to name a few, have examined the lifecycle of the black soldier fly. Eggs tend to hatch in approximately four days at 27 °C (Tomberlin *et al.*, 2002), and larvae pass through six stadia (May, 1961). Their growth from first to sixth instar can be quite extraordinary, as they hatch at <1 mm in length and have the potential to grow to 2.5

to 3.5 cm. Development time can be highly variable depending on numerous factors such as temperature (Harnden and Tomberlin, 2016), moisture of the larval diet (Fatchurochim *et al.*, 1989), relative humidity (Holmes *et al.*, 2012), and resource quality (Nguyen *et al.*, 2013, 2015) to name a few. Because of the effects of these factors, larval development can range from two weeks to several months.

Resulting prepupae, which are dark brown/black in appearance (larvae tend to be more white) (May, 1961), disperse from the larval resource to pupate. In general, duration of the pupal stage can take two weeks or more. However, pupation substrate can impact the duration of this stage. Placing prepupae in sand, topsoil, wood shavings or potting soil resulted in the prepupal stage lasting between 7.6 and 8.4 days; however, no substrate resulted in a prepupal duration of 10.3 days (Holmes *et al.*, 2013).

Limited information is available on adult biology. A common myth is adults do not have functioning mouthparts. However, this assertion is simply not accurate and this misconception is due largely to adults not *needing* to feed to survive, as it appears they rely on fat body reserves accrued during the larval stage as the primary source of pupal and adult nutrition. Without food being provided, adults typically live one or two weeks (Tomberlin *et al.*, 2002). Newly-emerged adults require approximately two days before they will mate (Figure 12.1) (Tomberlin and Sheppard, 2002). At that time, adults aggregate (lek), where males compete for females (Tomberlin and Sheppard, 2001). Once mating has occurred, the female will need an additional two days to develop eggs and oviposit (Tomberlin and Sheppard, 2002). During

Figure 12.1. A mating pair of black soldier flies (photo by Spring Yang).

their lifetime, females will only deposit a single clutch of approximately 400 eggs (Tomberlin and Sheppard, 2002; Tomberlin *et al.*, 2002).

12.3 Industrialization of black soldier fly production

Mass production of the black soldier fly is heavily reliant on published studies conducted in the laboratory. This approach is a major limiting factor towards the development of an industry still in its infancy, as scale is an important factor to consider when extrapolating information from studies where 300 larvae, which are a few grams of product, were used to model the mass-production of tons of larvae daily. Consequently, one of the major limitations present today is the development of industry-scaled studies to determine the true biometrics of this species.

Other limitations as related to industry-scale production include determining the impact of rearing substrates (e.g. food waste, animal manure, etc.) that varies significantly in terms of nutrition, surface area to volume ratio, moisture, salinity, and pH on development of the larvae. When animal manure from a confined animal feeding operation is used as a rearing substrate, these concerns are reduced, as the animals producing the manure are likely receiving similar or constant diets over time. However, if a waste-management facility is to receive wastes from multiple animal species, the above concerns remain. Furthermore, variation in population genetics and resulting phenotypes has not been explored beyond a single on-going study. This factor is of major importance, as flies from Texas, USA are potentially quite different from those in Vancouver, Canada, Cape Town, South Africa, Amsterdam, the Netherlands, and Wuhan, China to name a few. These differences can include development time (for example, Zhou *et al.* (2013) observed a difference in egg development of approximately 2.5 days between Texas, USA and two China populations), temperature tolerance (upper and lower thresholds), dry matter conversion rates, and level of protein and lipid production. Consequently, in order to optimize production, more research is needed to understand these differences within populations at sites where potential factories are being constructed.

12.4 Larval production for sale as a product

Presently, there are two approaches being investigated for mass-production of black soldier fly larvae. The initial approach was through a continual batch system. This approach uses large basins that can vary in length from one or two meters to the length of an American football field (approximately 90 m) and up to one meter broad. This system is inoculated with larvae and associated feed continuously over time. The continuous approach relies on the natural dispersal behaviour of the prepupae to self-harvest. Through this system, larvae are provided food over time and allowed to disperse at the time of pupation from the resource into a collection vessel.

Such management could lead to challenges, such as maintaining consistency in nutritional value of the food waste (Table 12.1 and Figure 12.2 for example) being provided to the larvae, moisture content, larval density, while also preventing nuisance arthropods (e.g. house fly (*Musca domestica* L. (Diptera: Muscidae), blow fly species (Diptera: Calliphoridae),

Table 12.1. Nutritional value of some common produce items and organic waste streams, to illustrate the differences in feedstuff that could be provided to black soldier fly larvae.

Nutrient (/100 g)	Data from USDA (2016)					Data from Nguyen *et al.* (2015)			
	Bananas	Tomatoes	Broccoli	Russet potatoes	Romaine lettuce	Swine manure	Restaurant waste	Fruits and vegetables	Fish rendering
Water (g)	74.91	94.52	89.30	78.58	94.61	–	–	–	–
Energy (kcal)	89.00	18.00	34.00	79.00	17.00	295.23	484.32	375.00	502.76
Protein (g)	1.09	0.88	2.82	2.14	1.23	22.66	20.41	20.07	50.00
Total lipid (g)	0.33	0.20	0.37	0.08	0.30	1.40	19.58	1.55	36.18
Fiber (g)	2.60	1.20	2.60	1.30	2.10	–	–	–	–
Sugar (g)	12.23	2.63	1.70	0.62	1.19	47.61	56.79	68.95	0.55

Figure 12.2. Black soldier fly larvae (A) consuming pumpkin, (B) decomposing pork, (C) mixed food waste containing potatoes and fish, and (D) mixed food waste containing fruits and vegetables (photos by Spring Yang).

phorid flies (Diptera: Phoridae), ants (Formicidae) and vertebrates (e.g. mice and rats)) from colonizing and mass-producing as well. Other issues include determining the appropriate longevity of such receptacles with regards to their use for mass production of black soldier fly larvae to avoid mass mortality due to microbial contamination. Anecdotal information indicates a number of pathogens of the black soldier fly exist and severely handicap larval numbers. However, we do not currently know if these pathogens are bacterial, fungal, viral, or a combination. Such impacts can range from reduced larval feeding and subsequent reduced conversion of waste into protein and fat, to complete die-off of associated larvae.

The second approach is to use a batch system where tubs of various shape/size are inoculated with a set amount of larvae and feed. The larvae are allowed to feed for a set amount of time prior to being processed, after which larvae are removed from the waste. This approach avoids many issues associated with within-batch variation in feed nutrition and moisture, as the larvae are fed once; however, the issue surrounding pest insects colonizing the waste still exists and thus requires due diligence to mitigate such circumstances.

12.5 Adult management: heartbeat of an industry

Maintaining a healthy and productive adult colony is crucial for industrialization of this species. Inconsistent production of high-quality (i.e. viable and fertilized) eggs would result in uncertainty with regards to projecting waste conversion rates within a given factory. Consequently, care must be taken to ensure that a robust colony is maintained throughout the year for maximum waste conversion and protein production.

Based on work of Sheppard *et al.* (2002), adult flies are typically housed in greenhouses in cages (Figure 12.3) that range in shape and size; however, mesh cages approximately of 2-3

Figure 12.3. A 1.2×1.2×2.4 m cage constructed of wood and window screen. The vertical sheet hanging inside the cage provides a resting location for flies, as well as a place for water to accumulate from the misting system (photo by Jonathan A. Cammack).

m (l) × 1-2 m (h) × 1-2 m (w) are typically used. Currently, density of adults and sex ratio for optimal mating and oviposition are not known. Based on previous publications, newly-enclosed adults are released into the cages at a 50:50 sex ratio, where they are provided water, usually through an automated system that mists water onto the cage.

Adults do not mate immediately upon release into the cage. As previously mentioned, an additional two to four days are potentially needed to allow for mating and egg production to occur (Tomberlin and Sheppard, 2002). However, it is also important to note that mating is highly dependent on sunlight (Tomberlin and Sheppard, 2002). Because sunlight is critical, researchers have developed artificial lighting systems to remove this dependency (Oonincx *et al.*, 2016; Zhang *et al.*, 2010). This development cannot be undervalued as it allows for mass production of this fly indoors in regions where sunlight and temperatures are sub-optimal during cooler parts of the year. Furthermore, this approach allows for mass production in facilities where temperature and humidity can be controlled, thus allowing for a reduction of associated expenses and an increase in profit margins.

Methods for collecting eggs are based on strategies previously described in Tomberlin *et al.* (2002). Booth and Sheppard (1984) were the first to describe methods for collecting eggs from wild populations, on which current techniques are based. They reviewed previous reports of black soldier oviposition behaviour, conducted field observations, and determined that black soldier flies prefer to oviposit in dry areas, such as around the edges of decaying organic matter, in crevices of poultry manure, or in the cracks of beehives. They also reported that in commercial fish bait production, black soldier fly eggs were collected in corrugated cardboard (Figure 12.4).

Figure 12.4. Female black soldier flies ovipositing in strips of corrugated cardboard (photo by Spring Yang).

In field experiments conducted by Booth and Sheppard (1984), they suspended strips of corrugated cardboard on posts above fresh chicken manure; females readily oviposited in the cardboard, as long as the cardboard was dry, supporting the hypothesis of preferential oviposition in dry crevices above a moist substrate. Tomberlin and Sheppard (2002) recorded a number of metrics associated with oviposition behaviour. In this study, they placed rolls of corrugated cardboard (2.54 cm diameter × 4 cm length) on the inside of a 5 l bucket containing wet Gainesville diet (Hogsette, 1992); the cardboard was placed approximately 3 cm above the diet. Additionally, they tested the influence of the presence of conspecific larvae on oviposition. From these studies, they found that egg clutches were deposited when temperatures were above 26 °C, flies preferred to oviposit when relative humidity was above 60%, and the presence of conspecific larvae in the oviposition medium had no effect on the number of eggs deposited (Tomberlin and Sheppard, 2002). Tomberlin *et al.* (2002) were the first to use three-layer cardboard blocks (5.0×2.5×1.0 cm), which are commonly used in current production facilities. They also tested the effect of cardboard flute size on oviposition, and found that cardboard with larger flutes (2×5 mm) contained more eggs than smaller flutes (1×3 mm), and that female flies were able to deposit eggs approximately 40% deeper into the larger flutes. In each of these studies, the oviposition medium was attractive to the gravid adults, but the high moisture content prevented oviposition directly in the substrate, and in the cardboard flutes instead. Other methods for egg collection exist in industry, however none are published.

Resulting eggs are moved to a rearing room where they are partitioned either into colony maintenance or used for waste conversion. Once eggs are collected, the flies are allowed to die, resulting dead adults are removed, and the cage broken down and cleaned. A current issue is managing the waste from the adult colonies. As the adults are high in chitin, they might be valuable as a product if the chitin can be efficiently and economically harvested.

12.6 Larval production for colony management

Multiple methods exist for managing larvae in a colony. Environmental conditions (including temperature (25-28 °C), humidity (60-75%), photoperiod (12-16 h of light)) and larval diet (such sources as Gainesville house fly diet or chicken feed have been used) vary slightly across studies. The method presented here evolved from Sheppard *et al.* (2002). Resulting eggs can remain in the cardboard; however, it is important to keep a log of production. Thus, counting flutes containing eggs, or individual egg masses if corrugated cardboard is not used, is critical for informed colony management. Eggs can be partitioned into 1 g allotments and kept in small (500 ml) plastic containers in a room maintained at 27 °C, 60-70% RH, and 12:12 L:D. Cups should be covered with a paper towel secured with an elastic band. Doing so will prevent contamination from pest insects (e.g. house fly and other muscids, phorid flies, etc.) and keep moisture and the developing black soldier fly larvae in the cup.

Eggs typically hatch in four days at 27 °C. Resulting larvae are provided 50-100 g moist Gainesville diet per 1 g of eggs, and allowed to feed for 3-4 days. At this time, the larvae and associated materials (i.e. feed and frass) are moved into larger containers, which are covered with a breathable cloth (such as tulle or silkscreen) secured with an elastic band. This technique, like with the egg cups, reduces the risk of contamination and of larvae escaping from the pans. These larvae are then provided 500-1000 g of feed per day for an additional 7-10 days or when 40% of the larvae have reached the prepupal stage.

Prepupae can remain in these materials until adult emergence begins, or sieved prior to emergence. At this time, the pans can be monitored daily and taken to the colony cages and released. Those applying these methods should keep in mind that males typically emerge first (a phenomenon known as protandry (Buck, 2001)), with females following two days later. This information is important when releasing adults into cages so as to maintain a 50:50 sex ratio.

The diet used for colony maintenance must also be investigated to ensure it is sufficient. Gobbi *et al.* (2013) found that larval and pupal stage duration and mortality, as well as adult size and ovarian development were all affected by larval diet. Cleanliness is crucial for maintaining a colony. Thus, the larval colony room should be removed from the industrialized area where larvae are being fed waste for protein production. Due to potential microbial pathogens that can kill off a colony quickly (within days based on experiences of the authors), all pans containing larvae should be monitored. Any signs of lethargy, bloating or larval death should trigger an immediate evacuation and sterilization of the associated pan. These materials should be removed from the larval colony room and disposed. Associated pans should be cleaned with hot soapy water, and wiped with a 10% bleach solution, followed by a 70% ethanol solution. With this in mind, practitioners should wear disposable gloves and avoid cross contamination risks. All countertops should be frequently wiped down with a 10% bleach solution to reduce contamination issues.

12.7 Hurdles to be addressed with using black soldier fly prepupae as animal feed

Much progress has been made with the mass production of the black soldier fly. The impact of many biotic and abiotic factors on the life-history of this insect have been examined and many companies around the world are utilizing these data to develop industrialized facilities for converting waste into protein, oil, and other products. However, to be cliché, these data are only the tip of the iceberg as a number of hurdles still wait to be addressed that will truly allow the full potential of this species to be reached.

12.7.1 System design

The mass-rearing system first described by Sheppard *et al.* (1994) was a continuous system based on the presence of wild populations in caged layer facilities, and required no special equipment for production or harvesting, as its success was based on the biology and behaviour of prepupal black soldier flies. The floor beneath layer cages was modified with a slope of approximately 40° to the outer wall of the house and at the top of the slope was a PVC pipe with a slit cut along the entire length. When the immature black soldier flies were finished feeding and entered the prepupal stage, they dispersed from the wet larval food source, crawled up the ramp, entered the pipe, and fell into a container at the end of the pipe.

In a continuous/single unit system, one potential downfall is the presence of a mixed-age cohorts. The larvae present in such a system would vary in age, and might not allow for the most efficient digestion of waste material. Many larval Diptera rely on microorganisms for survival, which likely aid in their digestion of resources. Zheng *et al.* (2013) studied the bacterial diversity within black soldier flies and found that the microbiome differs between life stages. The bacteria associated with 7-d-old larvae differ from prepupae, and one could hypothesize that additional differences exist between other age groups/development stages as well. A number of bacteria have been identified from the gut of the black soldier fly (Yu *et al.*, 2011), and the presence of these bacteria in the larval diet has different effects on growth and development. It is possible that certain bacteria species may positively or negatively affect different life stages of the black soldier fly. Therefore, the bacteria associated with a mixed-age cohort might have differential effects on the insects present, resulting in increased or decreased performance of some life stages. Additionally, in a continuous system, harvesting would be limited to prepupae dispersing from the resource. Although this is an effective method in a small-scale agricultural setting (e.g. that described by Sheppard *et al.*, 1994), this method is not as likely to yield consistent numbers of prepupae over time, as many prepupae also remained in the substrate. When using such a system, it is also necessary to completely empty and clean (sterilize) the larval rearing area to reduce the likelihood of larval disease. In the study by Sheppard *et al.* (1994), the manure basin was emptied in April and August, but for commercial production, cleaning should probably occur more than twice annually. Batch systems reduce the likelihood of mixed-age cohorts, as one would expect all larvae from a concurrent egg collection to be close in chronological and physiological age, as long as adequate resources are provided and competition between the larvae is reduced/eliminated.

12.7.2 Producing a consistent larval product

Effects of the diet being provided to the larvae must be addressed to ensure consistent quality and nutritional value of the larval product. Those running a facility accepting waste for bioconversion by black soldier flies should be cognizant of the potential impacts these different waste streams can have on the resulting larval product, and if possible, tailor these 'diets' to produce a product that will meet it's intended use. Nguyen *et al.* (2013, 2015) found that the waste type being fed to larvae had significant impacts on development rate, larval length and weight, and nutritional content of the resulting prepupae. Larvae developed fastest on fish offal and kitchen waste, followed by fruits and vegetables, and swine manure. Additionally, larvae were the heaviest when reared on kitchen waste, followed by fish offal, fruits and vegetables, and swine manure. Rearing on kitchen waste resulted in prepupae that were 21.2% protein, and fish offal and fruits and vegetables resulted in prepupae that were 19.4 and 12.9% protein, respectively; each of these values was lower than expected based on previous studies. Oonincx *et al.* (2015a) fed black soldier flies different mixtures of various food by-products and found that diets high in both protein and fat resulted in faster development than diets low in fat or protein, but feed conversion ratio and survival did not differ across the diet treatments. However, crude protein content of the larvae ranged from 38-46% dry matter, and fat content ranged from 21-35% dry matter, but these values fall within the ranges presented in previous studies. Rearing black soldier fly larvae on different animal manures also resulted in effects on development and nutrient value of the associated prepupae (Oonincx *et al.*, 2015b). Larval survival was highest on pig manure (97%), followed by cow (89%) and chicken (82%) manure, and there was approximately a 70 day difference in development time between larvae reared on pig and chicken (144 days each) manure, and those reared on cow manure (215 days). However, more biomass of larvae was produced on cow manure, and these larvae had a significantly higher N:P ratio in comparison to the other manure types.

St-Hilaire *et al.* (2007) were the first to study the effects of mixing different waste types (fish offal and dairy cow manure) on soldier fly development and bioconversion. As little as 10% inclusion of fish offal with dairy manure resulted in prepupae that were significantly heavier than those reared on dairy manure alone, and contained 43% more lipid, and approximately 10× higher concentrations of omega-3 fatty acids. Rehman *et al.* (2017) investigated the impact of different ratios of dairy manure and soybean curd residue (a byproduct of soy processing) on bioconversion. They found that intermediate ratios (3:2, 2:3, and 1:4) resulted in the fastest larval development, highest larval biomass production, and highest bioconversion of the wastes. Based on the aforementioned effects and differences, operators of waste conversion facilities should ensure they have optimized the rearing parameters associated with the types of waste being received so that subsequent larval production and waste production can also be optimized.

12.7.3 Adult colony maintenance and mating

Establishment of a successful mating population of black soldier fly adults is perhaps the biggest challenge facing those wishing to enter this industry. The narrow temperature range at which the adults are active means they must be reared indoors if colonies are to be kept continuously through the year, especially during winter months. Aside from temperature, light is perhaps the most important factor to consider when maintaining a black soldier fly colony. Sheppard *et al.* (2002) stated that the identification of an artificial light for maintaining a colony could allow for maintenance of a controlled environment with less expense. Tingle *et al.* (1975) established colonies in the spring/summer in Florida, USA, but observed no mating and collected few eggs in a greenhouse setting. When adults were kept in a cage outdoors, mating was frequently observed when sunlight was on the cage. Tomberlin and Sheppard (2002) observed no mating when light intensity was below 63 $\mu mol/m^2/s$ and 75% of the mating pairs observed occurred when light intensity was greater than 200 $\mu mol/m^2/s$. When artificial lights (40 watt Sylvania Gro Lux® (Osram Sylvania, Inc., Danvers, MA, USA) or 430 watt Pro Ultralight Light System® (Hydrofarm Inc., Petaluma, CA, USA)) were used for maintaining a colony indoors, no mating was observed and flies produced infertile eggs (Tomberlin and Sheppard, 2002). Zhang *et al.* (2010) observed mating under sunlight when light intensity was approximately 110 $\mu mol/m^2/s$, observed no mating under a 450 watt rare earth light (intensity of 160 $\mu mol/m^2/s$; Engineering University Infrared Technology Research Institute, Harbin, Heilung-kiang, China), and observed approximately 39% less mating pairs under a 500 watt quartz-iodine lamp (intensity of 135 $\mu mol/m^2/s$; model QVF135, Philips Lighting Ltd., Eindhoven, the Netherlands) than under sunlight. Nakamura *et al.* (2016) achieved successful mating and oviposition when flies were kept under a combination of 40 watt fluorescent (FLR40SEX-W/M/36-HG; NEC, Tokyo, Japan) and 20 watt LED (JLM-LTG20W; Jiaxing Rantoon Co. Ltd., Jiaxing, China) lamps. Recently, Oonincx *et al.* (2016) assessed spectral sensitivity of different portions of the compound eyes, and found that photoreceptors were most sensitive to light in the UV-green range (360-530 nm λ). They used this information identify different light sources for testing effects on reproduction. However, they found no effect of light source/wavelength on oviposition, but did see increased larval survival when adults were exposed to LED lights. These authors suggested that future work on optimizing artificial light for black soldier fly reproduction should investigate different ratios of green:blue/UV light, as well as investigating the effects on light intensity (Oonincx *et al.*, 2016).

12.7.4 Biosafety of a black soldier fly product

A common theme in the literature is the use of whole or ground black soldier flies (larvae, prepupae, or pupae) as a protein source in animal feeds (for review see Barroso *et al.*, 2014; Henry *et al.*, 2015; Jozefiak *et al.*, 2016; Makkar *et al.*, 2014; Sanchez-Muros *et al.*, 2014); however, the biosafety of these products has not been assessed. Soldier flies have been reared on numerous waste streams, ranging from human and livestock faeces to pre- and post-consumer food wastes, but the number of pathogenic bacteria and other microorganisms that can thrive in these materials is quite broad, including *Escherichia coli*, *Staphylococcus aureus*, *Campylobacter jejuni*, *Salmonella*, *Listeria*, and *Shigella* spp. In a survey of approximately 12,00 respondents, the biosafety of black soldier fly larvae that had been used to degrade food waste, as a subsequent animal feed ingredient, was given top priority (Makkar and Ankers, 2014). When reared on poultry manure inoculated with *Salmonella enterica* serovar Enteridis, black soldier fly larvae reduce bacteria populations by approximately 2.5 log, but the larvae themselves harboured populations ranging from 1.9-3.3 log cfu/g depending on the feeding duration (Erickson *et al.*, 2004). In swine manure, populations of *E. coli* O157:H7 were higher in the presence of black soldier fly larvae, but no effect on the bacteria was seen in cow manure (Erickson *et al.*, 2004). Black soldier fly larvae are able to significantly reduce *E.* coli populations in dairy manure, and this activity is temperature dependent: as temperature increases, *E. coli* populations decrease; however, it is not known if the larvae became contaminated with the bacteria (Liu *et al.*, 2008). Lalander *et al.* (2015) studied the effects of black soldier fly feeding on mixed human and swine faeces inoculated with the pathogens *Ascaris suum*, *Enterococcus faecalis*, *S. enterica* serovar Typhimurium, a bacteriophage (ϕX174) and a mixture of three viruses. During their nine week study simulating a continuous rearing system, soldier flies reduced the concentration of *S. enterica*, the bacteriophage, and all three viruses, but had no effect on the other pathogens. Although a number of studies have investigated the effects of black soldier flies on pathogens in waste material, the combined results are inconclusive, and further work is warranted to determine the safety of the black soldier flies themselves. An additional concern with such a system is the accumulation of pesticides and pharmaceuticals in the larvae or the material that remains after digestion. Lalander *et al.* (2016) investigated the fate of two pesticides (azoxystrobin and propiconazole) and three pharmaceuticals (carbamazepine, roxithromycin, and trimethoprim) in this system. They found that the half-life of these compounds was reduced in the presence of black soldier fly larvae, and that larvae did not accumulate any of the compounds tested. The authors suggested that black soldier fly processing of wastes containing these types of contaminants could prevent their spread in the environment (Lalander *et al.*, 2016).

12.8 Conclusions

Worldwide, the black soldier fly shows promise to address a number of issues of global concern, particularly managing organic waste and producing protein to be used as feed or food. Although some progress has been made to address the biosecurity of this system, more studies are needed to ensure we are producing safe products (both the larvae and remaining waste), as one mistake could cripple this industry. Some specific hurdles remain to be addressed, especially the identification of a light source (or further testing of those recently identified) that will ensure consistent mating and egg production in indoor environments.

To date, the microbial pathogen(s) that can result in complete die-off of larval populations remain unidentified and their source and subsequent prevention and control are unknown. Even after these unknowns become known, cleanliness will still be key in maintaining healthy populations of black soldier flies. To date, research suggests that black soldier fly larvae, prepupae, or meal made of the insect can be used as partial or complete replacements for soybean and fish meal used in animal feeds. However, the results of these studies are inconsistent, but promising, so further work is needed that can be used to drive legislation aimed at legally incorporating the black soldier fly into animal feeds.

References

Axtell, R.C. and Arends, J.J., 1990. Ecology and management of arthropod pests of poultry. Annual Review of Entomology 35: 101-126.

Axtell, R.C. and Edwards, T.D., 1970. *Hermetia illucens* control in poultry manure by larviciding. Journal of Economic Entomology 63: 1786-1787.

Banks, I.J., Gibson, W.T. and Cameron, M.M., 2014. Growth rates of black soldier fly larvae fed on fresh human faeces and their implication for improving sanitation. Tropical Medicine and International Health 19: 14-22.

Barroso, F.G., De Haro, C., Sanchez-Muros, M.J., Venegas, E., Martinez-Sanchez, A. and Perez-Banon, C., 2014. The potential of various insect species for use as food for fish. Aquaculture 422: 193-201.

Benelli, G., Canale, A., Raspi, A. and Fornaciari, G., 2014. The death scenario of an Italian Renaissance princess can shed light on a zoological dilemma: did the black soldier fly reach Europe with Columbus? Journal of Archaeological Science 49: 203-205.

Bondari, K. and Sheppard, D.C., 1981. Black soldier fly larvae as a feed in commercial fish production. Aquaculture 24: 103-109.

Bondari, K. and Sheppard, D.C., 1987. Soldier fly *Hermetia illucens* L., as feed for channel catfish, *Ictalurus punctatus* (Rafinesque), and blue tilapia, *Oreochromis aureus* (Steindachner). Aquaculture and Fisheries Management 18: 209-220.

Booth, D.C. and Sheppard, C., 1984. Oviposition of the black soldier fly, *Hermetia illucens* (Diptera, Stratiomyidae) – eggs, masses, timing, and site characteristics. Environmental Entomology 13: 421-423.

Buck, M., 2001. Protogyny, protandry, and bimodal emergence patterns in necrophagous Diptera. Canadian Entomologist 133: 521-531.

Callan, E.M. 1974. *Hermetia illucens* (L.) (Dipt., Stratiomyidae), a cosmopolitan American species long established in Australia and New-Zealand. Entomologist's Monthly Magazine 109: 232-234.

Diener, S., Studt Solano, N., Roa Gutiérrez, F., Zurbrügg, C. and Tockner, K., 2011. Biological treatment of municipal organic waste using black soldier fly larvae. Waste Biomass Valor 2: 357-363.

Erickson, M.C., Islam, M., Sheppard, C., Liao, J. and Doyle, M.P., 2004. Reduction of *Escherichia coli* O157:H7 and *Salmonella enterica* serovar Enteritidis in chicken manure by larvae of the black soldier fly. Journal of Food Protection 67: 685-690.

Fatchurochim, S., Geden, C.J. and Axtell, R.C., 1989. Filth fly (Diptera) oviposition and larval development in poultry manure of various moisture levels. Journal of Entomological Science 24: 224-231.

Gobbi, P., Martinez-Sanchez, A. and Rojo, S., 2013. The effects of larval diet on adult life-history traits of the black soldier fly, *Hermetia illucens* (Diptera: Stratiomyidae). European Journal of Entomology 110: 461-468.

Hale, O.M. 1973. Dried *Hermetia illucens* larvae (Stratiomyidae) as a feed additive for poultry. Journal of the Georgia Entomological Society 8: 16-20.

Harnden, L.M. and Tomberlin, J.K., 2016. Effects of temperature and diet on black soldier fly, *Hermetia illucens* (L.) (Diptera: Stratiomyidae), development. Forensic Science International 266: 109-116.

Henry, M., Gasco, L., Piccolo, G. and Fountoulaki, E., 2015. Review on the use of insects in the diet of farmed fish: past and future. Animal Feed Science and Technology 203: 1-22.

Hogsette, J.A., 1992. New diet for production of house-flies and stable flies (Diptera, Muscidae) in the laboratory. Journal of Economic Entomology 85: 2291-2294.

Holmes, L.A., Vanlaerhoven, S.L. and Tomberlin, J.K., 2012. Relative humidity effects on the life history of *Hermetia illucens* (Diptera: Stratiomyidae). Environmental Entomology 41: 971-978.

Holmes, L.A., Vanlaerhoven, S.L. and Tomberlin, J.K., 2013. Substrate effects on pupation and adult emergence of *Hermetia illucens* (Diptera: Stratiomyidae). Environmental Entomology 42: 370-374.

Jozefiak, D., Jozefiak, A., Kieronczyk, B., Rawski, M., Swiatkiewicz, S., Dlugosz, J. and Engberg, R.M., 2016. Insects – a natural nutrient source for poultry – a review. Annals of Animal Science 16: 297-313.

Kroeckel, S., Harjes, A.G.E., Roth, I., Katz, H., Wuertz, S., Susenbeth, A. and Schulz, C., 2012. When a turbot catches a fly: evaluation of a pre-pupae meal of the black soldier fly (*Hermetia illucens*) as fish meal substitute – growth performance and chitin degradation in juvenile turbot (*Psetta maxima*). Aquaculture 364-365: 345-352.

Lalander, C., Senecal, J., Calvo, M.G., Ahrens, L., Josefsson, S., Wiberg, K. and Vinneras, B., 2016. Fate of pharmaceuticals and pesticides in fly larvae composting. Science of the Total Environment 565: 279-286.

Lalander, C.H., Fidjeland, J., Diener, S., Eriksson, S. and Vinneras, B., 2015. High waste-to-biomass conversion and efficient *Salmonella* spp. reduction using black soldier fly for waste recycling. Agronomy for Sustainable Development 35: 261-271.

Latney, L. and Clayton, L.A., 2014. Updates on amphibian nutrition and nutritive value of common feeder insects. Veterinary Clinics of North America Exotic Animal Practice 17: 347-367.

Liu, Q.L., Tomberlin, J.K., Brady, J.A., Sanford, M.R. and Yu, Z.N., 2008. Black soldier fly (Diptera: Stratiomyidae) larvae reduce *Escherichia coli* in dairy manure. Environmental Entomology 37: 1525-1530.

Makkar, H.P.S. and Ankers, P., 2014. Towards sustainable animal diets: a survey-based study. Animal Feed Science and Technology 198: 309-322.

Makkar, H.P.S., Tran, G., Heuze, V. and Ankers, P., 2014. State-of-the-art on use of insects as animal feed. Animal Feed Science and Technology 197: 1-33.

Marshall, S.A., Woodley, N.E. and Hauser, M., 2015. The historical spread of the black soldier fly, *Hermetia illucens* (L.) (Diptera: Stratiomyidae, Hermetiinae), and its establishment in Canada. Journal of the Entomological Society of Ontario 146: 51-54.

May, B.M., 1961. The occurrence in New Zealand and the life-history of the soldier fly *Hermetia illucens* (L.) (Diptera: Stratiomyidae). New Zealand Journal of Science 4: 55-65.

Nakamura, S., Ichiki, R.T., Shimoda, M. and Morioka, S., 2016. Small-scale rearing of the black soldier fly, *Hermetia illucens* (Diptera: Stratiomyidae), in the laboratory: low-cost and year-round rearing. Applied Entomology and Zoology 51: 161-166.

Newton, G.L., Booram, C.V., Barker, R.W. and Hale, O.M., 1977. Dried *Hermetia illucens* larvae meal as a supplement for swine. Journal of Animal Science 44: 395-400.

Newton, G.L., Sheppard, C., Watson, D.W., Burtle, G. and Dove, R., 2005a. Using the black soldier fly, *Hermetia illucens*, as a value-added tool for the management of swine manure.

Newton, G.L., Sheppard, D.C., Watson, D.W., Burtle, G.J., Dove, C.R., Tomberlin, J.K. and Thelen, E.E., 2005b. The black soldier fly, *Hermetia illucens*, as a manure management/resource recovery tool. In: Proceedings of the Symposium on the State of the Science of Animal Manure and Waste Management, January 5-7, 2005, San Antonio, TX, USA.

Nguyen, T.T.X., Tomberlin, J.K. and VanLaerhoven, S.L., 2013. Influences of resources on *Hermetia illucens* (Diptera: Stratiomyidae) larval development. Journal of Medical Entomology 50: 898-906.

Nguyen, T.T.X., Tomberlin, J.K. and Vanlaerhoven, S., 2015. Ability of black soldier fly (Diptera: Stratiomyidae) larvae to recycle food waste. Environmental Entomology 44: 406-410.

Oonincx, D., Van Broekhoven, S., Van Huis, A. and Van Loon, J.J.A., 2015a. Feed conversion, survival and development, and composition of four insect species on diets composed of food by-products. PLoS ONE 10: 20.

Oonincx, D.G.A.B., Van Huis, A. and Van Loon, J.J.A., 2015b. Nutrient utilisation by black soldier flies with chicken, pig, or cow manure. Journal of Insects as Food and Feed 1: 131-139.

Oonincx, D.G.A.B., Volk, N., Diehl, J.J.E., Van Loon, J.J.A. and Belušič, G., 2016. Photoreceptor spectral sensitivity of the compound eyes of black soldier fly (*Hermetia illucens*) informing the design of LED-based illumination to enhance indoor reproduction. Journal of Insect Physiology 95: 133-139.

Rehman, K.U., Rehman, A., Cai, M., Zheng, L., Xiao, X., Somroo, A.A., Wang, H., Li, W., Yu, Z. and Zhang, J., 2017. Converison of mixtures of dairy manure and soybean curd residue by black soldier fly larvae (*Hermetia illucens* L.). Journal of Cleaner Production 154: 366-373.

Salomone, R., Saija, G., Mondello, G., Giannetto, A., Fasulo, S. and Savastano, D., 2017. Environmental impact of food waste bioconversion by insects: application of life cycle assessment to process using *Hermetia illucens*. Journal of Cleaner Production 140: 890-895.

Sanchez-Muros, M.J., Barroso, F.G. and Manzano-Agugliaro, F., 2014. Insect meal as renewable source of food for animal feeding: a review. Journal of Cleaner Production 65: 16-27.

Schiavone, A., Cullere, M., De Marco, M., Meneguz, M., Biasato, I., Bergagna, S., Dezzotto, D., Gai, F., Dabbou, S., Gasco, L., Zotte, A.D., 2017. Partial of total replacement of soybean oil by black soldier fly larvae (*Hermetia illucens* L.) fat in broiler diets: effect on growth performances, feed-choice, blood traits, carcass characteristics and meat quality. Italian Journal of Animal Science 16: 93-100.

Sealey, W.M., Gaylord, T.G., Barrows, F.T., Tomberlin, J.K., McGuire, M.A., Ross, C. and St-Hilaire, S., 2011. Sensory analysis of rainbow trout, *Oncorhynchus mykiss*, fed enriched black soldier fly prepupae, *Hermetia illucens*. Journal of the World Aquaculture Society 42: 34-45.

Sheppard, C., 1983. House fly and lesser fly control utilizing the black soldier fly in manure management systems for caged layer hens. Environmental Entomology 12: 1439-1442.

Sheppard, D.C., Newton, G.L., Thompson, S.A. and Savage, S., 1994. A value added manure management system using the black soldier fly. Bioresource Technology 50: 275-279.

Sheppard, D.C., Tomberlin, J.K., Joyce, J.A., Kiser, B.C. and Sumner, S.M., 2002. Rearing methods for the black soldier fly (Diptera: Stratiomyidae). Journal of Medical Entomology 39: 695-698.

St-Hilaire, S., Sheppard, C., Tomberlin, J.K., Irving, S., McGuire, M.A., Mosley, E.E., Hardy, R.W. and Sealey, W., 2007. Fly prepupae as a feedstuff for rainbow trout, *Oncorhynchus mykiss*. Journal of the World Aquaculture Society 38: 59-67.

Stone, A., Sabrozky, C.W., Wirth, W.W., Foote, R.H. and Coulson, J.A., 1965. A catalog of the diptera of America North of Mexico. US Department of Agriculture, Agricultural Handbook 276. U.S. Government Printing Office, Washington, DC, USA, 1698 pp.

Tingle, F.C., Mitchell, E.R. and Copeland, W.W., 1975. The soldier fly, *Hermetia illucens*, in poultry houses in north central Florida. Journal of the Georgia Entomological Society 10: 179-183.

Tomberlin, J.K. and Sheppard, D.C., 2001. Lekking behavior of the black soldier fly (Diptera: Stratiomyidae). Florida Entomologist 84: 729-730.

Tomberlin, J.K. and Sheppard, D.C., 2002. Factors influencing mating and oviposition of black soldier flies (Diptera: Stratiomyidae) in a colony. Journal of Entomological Science 37: 345-352.

Tomberlin, J.K., Sheppard, D.C. and Joyce, J.A., 2002. Selected life-history traits of black soldier flies (Diptera: Stratiomyidae) reared on three artificial diets. Annals of the Entomological Society of America 95: 379-386.

Van Huis, A., 2013. Potential of insects as food and feed in assuring food security. Annual Review of Entomology 58: 563-583.

Yu, G.H., Cheng, P., Chen, Y.H., Li, Y.J., Yang, Z.H., Chen, Y.F. and Tomberlin, J.K., 2011. Inoculating poultry manure with companion bacteria influences growth and development of black soldier fly (Diptera: Stratiomyidae) larvae. Environmental Entomology 40: 30-35.

Zhang, J., Huang, L., He, J., Tomberlin, J.K., Li, J., Lei, C., Sun, M., Liu, Z. and Yu, Z., 2010. An artificial light source influences mating and oviposition of black soldier flies, *Hermetia illucens*. Journal of Insect Science 10: 202.

Zheng, L.Y., Crippen, T.L., Singh, B., Tarone, A.M., Dowd, S., Yu, Z.N., Wood, T.K. and Tomberlin, J.K., 2013. A survey of bacterial diversity from successive life stages of black soldier fly (Diptera: Stratiomyidae) by using 16S rDNA pyrosequencing. Journal of Medical Entomology 50: 647-658.

Zhou, F., Tomberlin, J.K., Zheng, L., Yu, Z. and Zhang, J., 2013. Developmental and waste reduction plasticity of three black soldier fly strains (Diptera: Stratiomyidae) raised on different livestock manures. Journal of Medical Entomology 50: 1224-1230.

13.

Musca domestica

Industrialization of house fly production for livestock feed

J.K. Tomberlin[1*], C. Holcomb[1] and S.K. van Leeuwen[2]
[1]Department of Entomology, Texas A&M University, TAMU 2475, College Station, TX 77843-2475, USA; [2]Laboratory of Entomology, Wageningen University, P.O. Box 16, 6700 AA Wageningen, the Netherlands; jktomberlin@tamu.edu

Abstract

The house fly colonizes and consumes plant and animal waste resulting in the production of an alternate protein source for use as livestock, poultry, and aquaculture feed. While historically this species has been viewed as a pest of human society, today, it is viewed a key model for insect farming in many nations throughout the world. This chapter provides a general overview of the biology of the house fly along with an assessment of its nutritional composition, use as feed, and production methodologies.

Keywords: *Musca domestica*, insect farming, animal waste, food waste, sustainable agriculture

13.1 Introduction

The house fly, *Musca domestica* L. (Diptera: Muscidae) (Figure 13.1) is a cosmopolitan species associated with people, livestock, and associated waste. Historically, this species was primary viewed as a pest due to its nature as a nuisance, association with decomposing material, and ability to transmit pathogens. However, due to its ability to recycle decomposing plant material and animal waste, the house fly is now mass-produced as a means of converting these materials to protein and fat. The life cycle of the house fly undergoes a complete metamorphosis with a distinct egg, larva (three instars), pupa, and adult stage. The fly can complete this cycle in one to two weeks depending on environmental conditions. What is truly remarkable is with enhancements in mass-production technology this species can be produced at the tonnage level. What this means is items considered of no value (e.g. food waste, animal waste not used as fertilizer) can now be converted to protein which can serve as a feed for many different livestock species, and the remaining waste can potentially be used as a compost. This chapter provides a general overview of what is known to date.

13.2 Nutritional composition of house flies

13.2.1 Protein and lipid content

The idea of cultivating house flies for feedstuffs was proposed almost 100 years ago by Lindner (1919), who suggested using house flies to recycle human waste into valuable protein and fat by-products. However, it was not until several decades later when research began to delve into the protein and lipid content of house flies fed different diets. Both larvae and pupae may be harvested for protein and lipids, and it has been suggested that both may be suitable

Figure 13.1. House fly, *Musca domestica*, collected from Snook, Texas (photo courtesy of Dr Heo Chong Chin, Texas A&M University, USA).

for inclusion in animal diets. However, research has demonstrated that diet (Van Huis *et al.*, 2013), age of larvae at harvest (Aniebo and Owen, 2010), drying method (Aniebo and Owen, 2010) and methods used to calculate protein content (Janssen *et al.*, 2017) may influence the value of the nutritional content of house flies.

For example, Pieterse and Pretorius (2014) fed larvae pig blood and bran mixture and suggested the protein and lipid contents of 1.5 day-old larvae to be 60.4 and 14.1%, respectively. Additionally, Aniebo *et al.* (2008) fed house fly larvae bovine blood and wheat bran and determined the protein and lipid contents of 3 day-old larvae to be 47.1 and 7.5%, respectively. Likewise, a more recent study fed larvae an entirely different substrate (dairy manure) and reported the protein and lipid contents of 5-6 day-old larvae to be 60 and 20%, respectively (Hussein *et al.*, 2017). Therefore, it appears both diet and age of larvae at harvest may influence protein and lipid content.

Similarly, larval diet may impact protein and lipid content of pupae. For example, Calvert *et al.* (1969) determined house fly larvae fed a diet consisting of brewers dried grains, wheat bran, and alfalfa meal resulted in pupae with 63.1% protein and 15.5% lipid content. However, Teotia and Miller (1974) fed larvae poultry manure and reported a similar protein content of 61.4%, but a lower lipid content of 9.3% for pupae. On the other hand, Pieterse and Pretorius (2014) reported the protein and lipid content of pupae to be 76.2 and 14.4%, respectively.

A likely explanation for differences among the studies discussed may be attributed to differences in nutrient quality of the rearing medium (Van Huis *et al.*, 2013). Equally, Aniebo and Owen (2010) suggested that protein content reduces and lipid content increases in aging larvae. In the same study, oven drying was found to produce higher protein and less fat while sun drying produces less protein and more fat. Furthermore, recent research has indicated that studies using the Kjeldahl method (Kjeldahl, 1883) may overestimate the actual protein content of some species (Janssen *et al.*, 2017). All studies discussed, except for Calvert *et al.* (1969) who did not disclose the method of protein analysis, used the Kjeldahl method. However, due to the findings of Janssen *et al.* (2017) previous reports on protein analysis may be inaccurate.

13.2.2 Amino acid profile

The differences between the studies mentioned above have also led to different conclusions for the amino acid profile of house fly larvae and pupae. The amino acid composition of house fly larva meal is higher in some essential amino acids (methionine and phenylalanine) than soybean meal and fishmeal (Aniebo *et al.*, 2008; Hussein *et al.*, 2017). Though, contradicting results have been reported. For example, Pieterse and Pretorius (2014) reported low methionine and isoleucine in larvae, but the lysine content was comparable to soybean meal and fishmeal. Also, Hussein *et al.* (2017) reported isoleucine in larva meal to be lower than in soybean meal and fishmeal, while lysine was lower than in fishmeal. Conversely, Aniebo *et al.* (2008) found a higher isoleucine and lysine content in larva meal than in soybean meal and fishmeal. When harvested as pupae, house flies are high in limiting amino acids, such as arginine, lysine and methionine. However, Calvert *et al.* (1969), Teotia and Miller (1974), and Pieterse and Pretorius (2014) reported different percentages for the amino acids constituents of pupae meal. Differences in results may be due to variances in diet, harvested stage and methods used. Still, most studies agree that house flies offer a well-balanced amino acid profile and high protein quality.

13.3 Use as feed

Either house fly larvae or pupae can potentially be used as feed for fish or chickens. To assess their suitability as feed, growth performance of fish and chickens on a diet containing house fly protein has been studied. For example, Ogunji *et al.* (2008) investigated supplementing the diets of Nile tilapia (*Oreochromis niloticus*) fingerlings with house fly larva meal. Test diets consisted of fishmeal, sunflower oil, vitamin and mineral mix, potato starch, and chromium oxide at various ratios with and without larva meal. This study determined that fingerlings showed the best growth performance when fed a diet containing 52% fishmeal (without house fly larva meal). However, good results were obtained with diets formulated with 43% fishmeal and 15% house fly larva meal or 33% with fishmeal and 30% larva meal (Ogunji *et al.*, 2008). In another study, growth of *O. niloticus* increased when 50% of the fishmeal was replaced by house fly larva meal (Ezewudo *et al.*, 2015). Suitability of larva meal as supplementary feed has also been reported for African catfish (*Clarias gariepinus*) (Arong and Eyo, 2017).

Interestingly, when house fly pupae homogenate formed 5% of the diet of red sea breams (*Pagrus major*), not only their body weight increased, but also the survival time of sea breams infected with the bacterial pathogen *Edwardsiella tarda*. This suggests that the immune response was enhanced (Ido *et al.*, 2015). Anti-bacterial and anti-tumour activity of house fly larva extract have been found *in vitro* (Hou *et al.*, 2007). Kawasaki and Andoh (2017) found that haemolymph from house fly larvae demonstrated potent antimicrobial activities against both Gram-positive and Gram-negative bacteria, including *Staphylococcus aureus* and *Pseudomonas aeruginosa*. The antimicrobial activity was retained after heat-treatments, suggesting that pasteurization of animal feed prepared from fly larvae would be a useful sterilization method. These observations indicate that house fly larvae are a source of antimicrobial agents, and indicate the utility of preparing animal feed from these larvae. Further research should be conducted to see if there are indeed beneficial health effects of using house fly as animal feed.

In broiler chickens, carcass weight significantly increased in chickens supplemented with 10% house fly larva meal compared to those fed a control diet containing only soybean meal or a diet with 10% fishmeal (Hwangbo *et al.*, 2009; Pieterse *et al.*, 2014). Furthermore, larvae-fed chickens produced better meat quality (sustained juiciness, low level of mealiness and low thawing loss) (Pieterse *et al.*, 2014).

Past research has demonstrated that house fly larvae or pupae are suitable to form at least part of the diets of chickens and fish. This is important as the use of house fly as feed may provide both environmental and social advantages. For example, house flies can be reared on manure or food waste, transforming side-streams into a useful product. Not only can they be utilized as animal feed, but they also transform the rearing substrate into good quality compost (Čičková *et al.*, 2012; Zhu *et al.*, 2012). However, it should be noted that rearing house flies on animal wastes raises concerns about food and feed safety. Heavy metals, veterinary medications, and pesticide residues are major concerns that impede the development of incorporating insects fed manure in animals diets (Charlton *et al.*, 2015). Additionally, there are concerns about pathogens in manure. For example, Schuster *et al.* (2013) fed house flies bovine manure inoculated with *Escherichia coli* at various concentrations (0, 3, 5, 8 $\log_{10}$ colony forming unit (cfu)/ml) and demonstrated the bacterium persisted through metamorphosis. Results from this study determined that the concentration of the bacterium decreased with each successive life stage. However, flies emerged with up to 2.2 $\log_{10}$ cfu/ml (Schuster *et al.*, 2013) suggesting that such concentrations could initiate illness in humans. Similar trends for bacterial carriage across life stages were observed for other bacteria by Zurek and Nayduch (2016). Still, house fly rearing may require less resources than fish or soybean production. Yet in order to advance the idea of utilizing insects fed manure as a protein source, these concerns must be addressed (Van Huis *et al.*, 2015).

To date, there is one life-cycle assessment published on the use of house fly larvae as feed, when grown on food waste. It was reported that the production of one ton larva meal, resulted in higher energy use and global warming potential than the production of 0.5 ton fishmeal and 0.5 ton soybean meal combined. Land use is significantly lower for larva meal. However, the system to produce house fly larvae is still under development. Energy use and global warming potential may be improved upon, whereas reducing land use for fishmeal and soybean meal production seems difficult. Thus, using house fly larvae as feed may have future environmental advantages (Van Zanten *et al.*, 2015).

Social advantages of using house fly as feed are especially important in poor, rural areas. Prices of fishmeal and maize are rising (Van Huis *et al.*, 2013) and commercial products, such as Coppens feed, can be difficult to obtain in distant locations (Arong and Eyo, 2017). If the house fly can be used to supplement diets of fish and chickens, food security may increase.

13.4 House fly production

13.4.1 Type of facility

House flies are capable of biodegradation of manure (Barnard *et al.*, 1998); however, because they are vectors of numerous pathogens and a nuisance to animals (Malik *et al.*, 2007), they would not be ideal candidates for on-farm waste management and protein production, which has been proposed for other dipterans (Sheppard *et al.*, 1994). Therefore, they should be reared in a controlled facility. However, despite this, house flies may be considered useful as they can be employed to degrade food (Niu *et al.*, 2017) and agricultural wastes (Khan *et al.*, 2012; Larrain and Salas, 2008), both major environmental challenges throughout the world. Cortez Ortiz *et al.* (2016) suggested a facility with separate rearing rooms for larvae and adults; however, both can be maintained at 25-32 °C and 65-75% relative humidity. Additional requirements for mass-producing house flies are discussed below.

13.4.2 Colony for egg production

Cortez Ortiz *et al.* (2016) suggested rearing house flies in rearing rooms and denoted that the optimal density was approximately 5,625 adults (1:1 sex ratio) per cubic metre of rearing room space with eight hours of light. However, house flies can also be reared in 50×50×50 cm mesh cages (Bioquip® Products, Rancho Domingo, CA, USA) with up to 750 flies per cage (Hussein *et al.*, 2017). Additionally, adults must feed to prompt reproductive maturation (Morrison and Davies, 1964) and the nutritional quality of the diet influences fecundity (Pastor *et al.*, 2011). For example, females will lay 50% more eggs when provided a mixture of sugar and powdered milk (2:1) compared to those fed sugar and yeast of the same proportion (Pastor *et al.*, 2011). Adults also require water, which can be provided in a reservoir or placed in a cup covered and secured with filter paper and inverted over a plate (Cortez Ortiz *et al.*, 2016) or provided by a moist sponge (Čičková *et al.*, 2013). Three days after emergence, adults should be provided with an oviposition substrate that includes a wet cotton ball placed on top of fermenting wheat bran saturated to 60-70% moisture for 5-10 days (Cortez Ortiz *et al.*, 2016). Lastly, eggs are collected and placed in rearing pans in the larvae rooms (Cortez Ortiz *et al.*, 2016).

13.4.3 Feeding

Larvae rearing rooms can be maintained within the same range of temperature and humidity as those for the adult flies. However, the required room space depends on the placement of rearing pans (Cortez Ortiz *et al.*, 2016). Hogsette (1992b) reared 30,000 larvae in pans that were 50×40×10 cm. However, to use the area efficiently, pans should be stacked so they minimize space between pans, while also allowing for adequate regulation of the environmental conditions (Cortez Ortiz *et al.*, 2016). House fly larvae are generalist decomposers and can be reared on a wide variety of wastes from poultry (Barnard *et al.*, 1998), dairy (Hussein *et al.*, 2017) and swine (Wang *et al.*, 2013) manures to food wastes (Niu *et al.*, 2017). Additionally, they can also develop on premixed diets consisting of alfalfa, wheat bran, and cornmeal moistened with water (1:1 ratio) (Hogsette, 1992b). Past research suggests one g of eggs/kg of premixed diet (Cortez Ortiz *et al.*, 2016) or one ml eggs/kg of manure (Čičková *et al.*, 2012).

13.4.4 Harvesting

Larval age (Aniebo and Owen, 2010) and the rearing medium (Čičková *et al.*, 2012, Cortez Ortiz *et al.*, 2016) will govern when to collect house flies for harvesting. Larvae are known to self-harvest during their post-feeding stage (Hogsette, 1992a). However, house fly larvae prefer to wander in search of the perfect microhabitat, which may depend on light, moisture, or texture of the rearing medium. Thus, methods can be employed to encourage larvae to migrate from the medium including negative phototaxy and lowering the oxygen concentration in the rearing room (Cortez Ortiz *et al.*, 2016). Another way to collect pupae involves using fine sand moistened with water (3.2 ml sand/1 ml water), which can be placed in 240 ml clear plastic cups with the bottom of the cup removed and replaced with 16-mesh window screen (Hogsette, 1992a). The cup can then be placed on top of the rearing medium to collect pupae (Hogsette, 1992a). Pupae not intended for harvest can be placed in cages in the adult rearing room to maintain the colony. Larvae or pupae intended for protein production should be dried, e.g. microwave oven (Cortez Ortiz *et al.*, 2016) or drying oven, to 5% moisture (Awoniyi *et al.*, 2004), ground into a powder and stored in vacuumed-sealed containers at 25 °C for approximately 4-6 months (Cortez Ortiz *et al.*, 2016).

13.5 Future hurdles and conclusions

While many advancements have been made to mass-produce the house fly for converting food and animal waste to protein, some challenges still exist. Most importantly, current methods need to be refined further in order to optimize conversion ratios of waste to protein and to reduce energy use. Furthermore, containment of resulting adults in order to minimize distribution into surrounding areas is still a concern – especially in industrialized nations where this species is considered a pest. Regardless, this facet of farming the house fly is well established in China and most likely will continue to grow and diversify throughout the world to represent a key model for producing protein from waste.

References

Aniebo, A. and Owen, O., 2010. Effects of age and method of drying on the proximate composition of housefly larvae (*Musca domestica* Linnaeus) meal (HFLM). Pakistan Journal of Nutrition 9: 485-487.

Aniebo, A., Erondu, E. and Owen, O., 2008. Proximate composition of housefly larvae (*Musca domestica*) meal generated from mixture of cattle blood and wheat bran. Livestock Research for Rural Development 20: 1-5.

Arong, G.A. and Eyo, V.O., 2017. Evaluation of house fly (*Musca domestica*) maggot meal and termite (*Macrotermes subhyalinus*) meal as supplementary feed for African catfish *Clarias gariepinus* (Burchell, 1822). International Journal of Entomology and Nematology 3: 42-50.

Awoniyi, T., Adetuyi, F. and Akinyosoye, F., 2004. Microbiological investigation of maggot meal, stored for use as livestock feed component. Journal of Food Agriculture and Environment 2: 104-106.

Barnard, D., Harms, R. and Sloan, D., 1998. Biodegradation of poultry manure by house fly (Diptera: Muscidae). Environmental Entomology 27: 600-605.

Calvert, C.C., Martin, R.D. and Morgan, N.O., 1969. House fly pupae as food for poultry. Journal of Economic Entomology 62: 938-939.

Charlton, A., Dickinson, M., Wakefield, M., Fitches, E., Kenis, K., Han, R., Zhu, F., Kone, N., Grant, M. and Devic, E., 2015. Exploring the chemical safety of fly larvae as a source of protein for animal feed. Journal of Insects as Food and Feed 1: 7-16.

Čičková, H., Kozánek, M. and Takáč, P., 2013. Improvement of survival of the house fly (*Musca domestica* L.) larvae under mass-rearing conditions. Bulletin of entomological research 103: 119-125.

Čičková, H., Pastor, B., Kozánek, M., Martínez-Sánchez, A., Rojo, S. and Takáč, P., 2012. Biodegradation of pig manure by the housefly, *Musca domestica*: a viable ecological strategy for pig manure management. PLoS ONE 7: e32798.

Cortez Ortiz, J., Torres Ruiz, A., Morales Ramos, J., Thomas, M., Rojas, M. and Tomberlin, J., 2016. Insect mass production technologies. Insects as Sustainable Food Ingredients 6: 154-201.

Ezewudo, B.I., Monebi, C.O. and Ugwumba, A.A.A., 2015. Production and utilization of *Musca domestica* maggots in the diet of *Oreochromis niloticus* (Linnaeus, 1758) fingerlings. African Journal of Agricultural Research 10: 2363-2371.

Hogsette, J.A., 1992a. New diets for production of house flies and stable flies (Diptera, Muscidae) in the laboratory. Journal of Economic Entomology 85: 2291-2294.

Hogsette, J.A., 1992b. Autoseparation method for harvesting house fly (Diptera: Muscidae) pupae of known age. Journal of economic entomology 85: 2295-2297.

Hou, L., Shi, Y., Zhai, P. and Le, G., 2007. Inhibition of foodborne pathogens by Hf-1, a novel antibacterial peptide from the larvae of the housefly (*Musca domestica*) in medium and orange juice. Food Control 18: 1350-1357.

Hussein, M., Pillai, V.V., Goddard, J.M., Park, H.G., Kothapalli, K.S., Ross, D.A., Ketterings, Q.M., Brenna, J.T., Milstein, M.B. and H. Marquis, H., 2017. Sustainable production of housefly (*Musca domestica*) larvae as a protein-rich feed ingredient by utilizing cattle manure. PloS ONE 12: e0171708.

Hwangbo, J., Hong, E.C., Jang, A., Kang, H.K., Oh, J.S., Kim, B.W. and Park, B.S., 2009. Utilization of house fly-maggots, a feed supplement in the production of broiler chickens. Journal of Environmental Biology 30: 609-614.

Ido, A., Iwai, T., Ito, K.,, Ohta, T., Mizushige, T., Kishida, T., Miura, C. and Miura, T., 2015. Dietary effects of housefly (*Musca domestica*) (Diptera: Muscidae) pupae on the growth performance and the resistance against bacterial pathogen in red sea bream (*Pagrus major*) (Perciformes: Sparidae). Applied Entomology and Zoology 50: 213-221.

Janssen, R.H., Vincken, J.-P., Van den Broek, L.A., Fogliano, V. and Lakemond, C.M., 2017. Nitrogen-to-protein conversion factors for three edible insects: *Tenebrio molitor*, *Alphitobius diaperinus*, and *Hermetia illucens*. Journal of Agricultural and Food Chemistry 65: 2275-2278.

Kawasaki, K. and M. Andoh, M. 2017. Properties of induced antimicrobial activity in *Musca domestica* larvae. Drug Discoveries & Therapeutics 11: 156-160.

Khan, H.A.A., Shad, S.A. and Akram, W., 2012. Effect of livestock manures on the fitness of house fly, *Musca domestic*a L. (Diptera: Muscidae). Parasitology Research 111: 1165-1171.

Kjeldahl, J., 1883. Neue Methode zur Bestimmung des Stickstoffs in organischen Körpern. Zeitschrift für analytische Chemie 22: 366-383.

Larrain, P.S. and Salas, C.F., 2008. House fly (*Musca domestica* L.) (Diptera: Muscidae) development in different types of manure. Chilean Journal of Agricultural Research 68: 192-197.

Lindner, P., 1919. Extraction of fat from small animals. Zootechnica Biologica 7: 213-220.

Malik, A., Singh, N. and Satya, S., 2007. House fly (*Musca domestica*): a review of control strategies for a challenging pest. Journal of Environmental Science and Health Part B 42: 453-469.

Morrison, P.E. and Davies, D.M., 1964. Feeding of dry chemically defined diets, and egg production in the adult house-fly. Nature 201: 104-105.

Niu, Y., heng, D., Yao, B., Cai, Z., Zhao, Z., Wu, S., Cong, P. and Yang, D., 2017. A novel bioconversion for value-added products from food waste using *Musca domestica*. Waste Management 61: 455-460.

Ogunji, J., Toor, R., Schulz, C. and Kloas, W., 2008. Growth performance, nutrient utilization of Nile tilapia *Oreochromis niloticus* fed housefly maggot meal (magmeal) diets. Turkish Journal of Fisheries and Aquatic Sciences 8: 141-147.

Pastor, B., Čičková, H., Kozanek, M., Martinez-Sanchez, A., Takáč, P. and Rojo, S., 2011. Effect of the size of the pupae, adult diet, oviposition substrate and adult population density on egg production in *Musca domestica* (Diptera: Muscidae). European Journal of Entomology 108: 587-596.

Pieterse, E. and Pretorius, Q., 2014. Nutritional evaluation of dried larvae and pupae meal of the housefly (*Musca domestica*) using chemical-and broiler-based biological assays. Animal Production Science 54: 347-355.

Pieterse, E., Pretorius, Q., Hoffman, L. and Drew, D., 2014. The carcass quality, meat quality and sensory characteristics of broilers raised on diets containing either *Musca domestica* larvae meal, fish meal or soya bean meal as the main protein source. Animal Production Science 54: 622-628.

Schuster, G.L., Donaldson, J.R., Buntyn, J.O., Duoss, H.A., Callaway, T.R., Carroll, J.A., Falkenberg, S.M. and Schmidt, T.B., 2013. Use of bioluminescent *Escherichia coli* to determine retention during the life cycle of the housefly, *Musca domestica* (Diptera: Muscidae, L). Foodborne Pathogens and Disease 10: 442-447.

Sheppard, D.C., Newton, G.L., Thompson, S.A. and Savage, S., 1994. A value-added manure management-system using the black soldier fly. Bioresource Technology 50: 275-279.

Teotia, J.S. and Miller, B.F., 1974. Nutritive content of house-fly pupae and manure residue. British Poultry Science 15: 177-182.

Van Huis, A., Dicke, M. and Van Loon, J.J.A., 2015. Insects to feed the world. Journal of Insects as Food and Feed 1: 3-5.

Van Huis, A., Van Itterbeeck, J., Klunder, H., Mertens, E., Halloran, A., Muir, G. and Vantomme, P., 2013. Edible insects: future prospects for food and feed security. Food and Agriculture Organization of the United Nations (FAO), Rome, Italy, FAO Forestry Paper no. 171, 187 pp. Available at: http://www.fao.org/docrep/018/i3253e/i3253e.pdf.

Van Zanten, H.H., Mollenhorst, H., Oonincx, D.G.A.B., Bikker, P., Meerburg, B.G. and De Boer, I.J.M., 2015. From environmental nuisance to environmental opportunity: housefly larvae convert waste to livestock feed. Journal of Cleaner Production 102: 362-369.

Wang, H., Zhang, Z., Czapar, G.F., Winkler, M.K. and J. Zheng, J., 2013. A full-scale house fly (Diptera: Muscidae) larvae bioconversion system for value-added swine manure reduction. Waste Management & Research 31: 223-231.

Zhu, F.X., Wang, W.P., Hong, C.L., Feng, M.G., Xue, Z.Y., Chen, X.Y., Yao, Y.L. and Yu, M., 2012. Rapid production of maggots as feed supplement and organic fertilizer by the two-stage composting of pig manure. Bioresource Technology 116: 485-491.

Zurek, K. and Nayduch, D. 2016. Bacterial associations across house fly life history: evidence for transstadial carriage from managed manure. Journal of Insect Science 16: 2.

14.

Tenebrio molitor

Mealworms, promising beetles for the insect industry

N. Berezina
Ynsect, 1 Rue Pierre Fontaine, 91058 Evry, France; nbe@ynsect.com

Abstract

Mealworms are actually beetles, and they are given the name 'worms' because they spend the most important part of their life as larvae, thus being more similar to the 'worms' than to the actual 'beetles'. Mealworm species are common throughout much of the world and due to their specific diet some of them are considered as pests of food stocks. Mealworms, however, might be species of choice for the further development of the insect industry. Here the pros and cons of using two species of common mealworms, *Alphitobius diaperinus* (Panzer) and *Tenebio molitor* (Linnaeus) mainly for food and feed are discussed and an overview of optimal rearing conditions, potential health-related implications of consuming mealworms, and the potential nutritional value they provide is given.

Keywords: mealworms, proteins, feed and food, poultry, aquaculture

14.1 Introduction

Insects are increasingly seen as a promising source for food and feed (Van Huis *et al.*, 2013). Among the different species typically utilized, mealworms should also be further considered for this purpose. Although they contain less protein than grasshoppers and crickets, they do contain more than flies and other flying insects (Berezina *et al.*, 2016). Some mealworm species are also easily propagated under captive rearing conditions. For example, adult mealworms such as *Tenebrio molitor* (Linneaus) cannot fly, thus making them easier to handle during reproduction efforts.

Mealworms could someday represent a significant market share of the insect food and feed industry; however, several other aspects need to be considered. Several mealworm species, such as *Alphitobius diaperinus* (Panzer), are considered as pests; thus, any captive rearing should be made with caution. Another potential issue is a safety concern due to the possibility of transferring pathogenic microorganisms and other toxins when consuming these insects. Industrialization is also an important challenge for the whole insect food and feed field, and management of these new farms needs to consider issues related to regulation, supply.

Currently, the most studied species of mealworms are *A. diaperinus* and *T. molitor.* In this chapter, we have detailed several of the most stringent questions concerning using these species in the industry, such as rearing, transformation, applications, safety and life cycle assessment.

14.2 *Alphitobius diaperinus* – lesser mealworm

A. diaperinus was first described by Panzer in 1797. This small beetle, with adults measuring less than 6.5 mm, are important pests, ravaging stocks of wheat and capable of transmitting pathogens to humans and poultry (Dunford and Kaufmann, 2006). *A. diaperinus* is a member of the tenebrionid tribe *Alphitobiini*, which comprises four genera worldwide. These genera represent up to eleven known *Alphitobius* spp., some additional species are encountered in stored products (Dunford and Kaufmann, 2006).

Due to this species' high negative impact in storage of food reserves and on farm animals such as poultry, most work on *A. diaperinus* has been conducted in order to minimize their proliferation, or to control the possibility of transfer of pathogens to other species. Some of these studies are detailed in this chapter.

14.2.1 Development control

Biological control

Even though several natural enemies of *A. diaperinus* have been identified, there are no efficient biological control agents of this beetle (Dunford and Kaufmann, 2006). The most promising approach, however, seems to be the utilization of the fungal pathogen *Beauveria bassiana* (Bals.-Criv.) Vuill.

This specific pathogen was extensively studied by Geden *et al.* (1998). These authors have reported that the young larvae of *A. diaperinus* were more sensitive to the fungal pathogen than later instars, with LC_{90} changing in order of magnitude from 9.86×10^3 to 4.74×10^4, respectively. The LC_{90} for adults increased up to 6.41×10^6, i.e. 1000× more than the 5-day-old larvae. Based on these observations, the authors suggested different treatments, and one consisting of soil treatment with corn starch containing up to 2.5×10^{11} conidia/m^2 provided 100% control of beetle larvae; whereas the pupal trap was found to be 100% efficient with only 4×10^5 conidia/cm^2 on the cardboard.

Chemical control

Utilization of pesticides has also not demonstrated significant control of *A. diaperinus* (Dunford and Kaufmann, 2006). Several pesticides such as pyrethroids have been tested; however, *A. diaperinus* has evolved varying levels of resistance against those chemicals, and the effectiveness of some treatments was found to be limited by the accumulation of dust on treated surfaces and manure levels in buildings (Dunford and Kaufmann, 2006). Some other compounds have demonstrated efficacy in the laboratory, e.g. lime hydrate (calcium hydroxide), but were not explored in the field (Dunford and Kaufmann, 2006).

Szczepanik *et al.* (2016) have recently studied a class of δ-halolactones with a damascone skeleton as potential control substances for *A. diaperinus*. Those compounds have shown anti-feeding potential at 1% concentration. The effect of γ-bromo-δ-lactone was visible even at a lower concentration of 0.5%. Under these conditions, only 37.5% of adult emergence was observed compared to 90% in the controls. However, even if the microbiological pathway of biodegradation of these compounds has already been described, those molecules are not yet authorized for field trials (Szczepanik *et al.*, 2016).

Thermal control

Several studies (Renault *et al.*, 2006; Rueda and Axtell, 1996) have reported that *A. diaperinus* is sensitive to temperature. Thus, no egg hatching or larval development were observed at temperature lower than 17 °C (Rueda and Axtell, 1996). Some temporal changes in free amino acids and survival have been observed in cold-exposed beetles (10 °C), it was noticed that the females have a better ability to survive at cold temperature comparing to the males (Renault *et al.*, 2006). It is, however, worthwhile to mention, that here again, the authors were mainly seeking the limitation of the *A. diaperinus* propagation than the optimal rearing conditions.

Other rearing considerations

Rice and Lambkin (2008) published a study on the optimization of rearing conditions and the synchronization of generations for production of *A. diaperinus* colonies for the purpose of testing of pesticides. The conditions described in this work do not focus on the increased production of the insects for an industrial production but on controlled synchronized generations of beetles for maintaining populations in laboratory colonies. The adults were allowed to reproduce for four days and the resulting generations from hatching of these eggs and subsequent rearing (56 days, 32 °C, 55% RH) resulted in 4,500 adults per experiment.

One recent study has evaluated the possibility of rearing *A. diaperinus* on organic by-products (Van Broekhoven *et al.*, 2015). It was found that this species can perfectly adapt to feeding on these substrates which may be suitable for the industrialization of *A. diaperinus* for food and feed purposes. Also a specific design of rearing units dedicated to *A. diaperinus* was recently described (Calis, 2016), this trend shows a change in the paradigm of the consideration of this insect, from pest to a valuable source of nutriments.

14.2.2 Vectors of pathogens

As *A. diaperinus* is a pest in several places close to the human food or animal feed, an important work has been done to understand how this beetle vectors different pathogens. The mycotoxin F-2, produced by the fungus *Fusarium roseum* was found to be ingested and harboured by *A. diaperinus* with no traces of the deactivation despite starvation and death of the host (Eugenio *et al.*, 1970). Similar conclusions were drawn after studies of carriage and conservation of several bacterial strains such as *Salmonella typhimurium* (De las Casas *et al.*, 1968), *Salmonella enteritidis* (Davis and Wray, 1995), *Salmonella enterica* (Crippen *et al.*, 2009) or *Escherichia coli* (De las Casas *et al.*, 1968).

It was also shown that the horizontal transfer of antimicrobial resistance can occur between *Salmonella* and *E. coli* within the gut of *A. diaperinus*, and therefore these insects may be used as *in vivo* model to study resistance gene transfer (Poole and Crippen, 2009).

Finally, to better understand the dynamics of movement of bacteria within the environment and species, specific methods have been designed to successfully disinfect the surface of *A. diaperinus* to distinguish the internal and external sources of contamination (Crippen and Sheffield, 2006).

14.2.3 Utilization for feed and other aspects

We have found one study dealing with the growth of broiler chicks fed with *A. diaperinus* (Despins and Axtell, 1995). The authors have reported that, whatever the age of the considered chicks was, their growth was altered by consumption of *A. diaperinus* larvae compared to the designed mix. Here again, it is worthwhile to mention, that the study was designed to see how the chicks can be affected by accidental consumption of undesirable pest, rather than to optimize the chicks diet with the supplementation based on insect larvae or meal.

Along with the studies mentioned above, one work has noted that *A. diaperinus* can also have some beneficial effect in poultry houses, namely it appears to be a valuable predator of another pest capable of transferring several pathogens, the house fly, *Musca domestica* L. (Despins *et al.*, 1988). This study showed how the interactions within the ecological niches are complex and that trying to manage *A. diaperinus*, can result in an increase of the even more deleterious *M. domestica*.

14.3 *Tenebrio molitor* – yellow mealworm

T. molitor was first described by Linnaeus in 1758. It is a pest of moderate importance and is present throughout the world (Ghaly and Alkoik, 2009). Adults measure about 2-2.5 cm and adults cannot fly, thus they cannot spread as fast as lesser mealworms, which may explain why they are usually considered pests of moderate concern compared to *A. diaperinus*.

These two considerations, size and non-invasive behaviour, has favoured extensive studies for improvement of rearing of *T. molitor* for feed and food applications. Non-food/feed applications of *T. molitor* oil have also been considered. Using *T. molitor* as feed and food has also originated some safety concerns, especially related to microbiological and allergenic effects.

14.3.1 Rearing

Studies on the rearing of *T. molitor* can be divided into different categories, ones focused on feed quality, its nutritional value and the impact of diet on the quality of the produced proteins; and, studies focused on aspects of water supply, heat emission and overall animal behaviour. In this section we summarize these different approaches.

Diets and their impacts

Early studies on diets for *T. molitor* were mainly focused on the design of the most suitable artificial medium and understanding the nutritional needs of this species; whereas more recent studies have focused on the utilization of by-products and the overall lowering of the diets' costs.

One of the most significant studies on the *T. molitor* nutritional needs was performed by Fraenkel in 1950. It was found that these beetles require 80-85% carbohydrates in their diet, the carbohydrate source being indifferently starch or glucose; the protein supply could be made by different sources such as casein, lactalbumin, etc., whereas some others were found deleterious (zein, gliadin, etc.). Some important supplements were also identified, such as cholesterol at 1% and vitamin B. Those observations were then confirmed (Martin *et al.*, 1976) and additional discoveries, especially on the calcium intake (Hunt *et al.*, 2001; Klasing *et al.*, 2000) were added.

More recent studies, focused on the utilization of by-products such as brewer's grain, bread/cookie baking, potato processing, biodiesel production (Van Broekhoven *et al.*, 2015), boccaiuva pulp floor (Alves *et al.*, 2016) or even polystyrene (Yang *et al.*, 2015), have shown these by-products to be feeding sources by highly adaptable *T. molitor.* The only substrate inducing significant change in the growth of *T. molitor* was found to be T-2 toxin, known for its inhibitor capacity for the synthesis of proteins (Davis and Shieffer, 1982).

The impact of different diets on the composition of *T. molitor* was also evaluated and the impact on the protein content and quality has been found to be minimal. *T. molitor* were reported to possess a range of 45.1% on wet mass basis to 65% on dry mass basis (Alves *et al.*, 2016; Ghaly and Alkoik, 2009; Jones *et al.*, 1972; Martin *et al.*, 1976; Van Broekhoven *et al.*, 2015) of proteins; the diets have, however, significant influence on fat content, 18.9% (wet mass basis) to 50% (dry mass basis), and composition (Alves *et al.*, 2016; Ghaly and Alkoik, 2009; Jones *et al.*, 1972; Martin *et al.*, 1976; Van Broekhoven *et al.*, 2015). It remains very difficult to explain the whole metabolism of their transformation and to predict the fat content and composition of *T. molitor* based on diet substrates (Van Broekhoven *et al.*, 2015).

The induction of the immune response was also studied (Lee *et al.*, 2015), and has shown a transgenerational immune priming in *T. molitor* (Moret, 2006), thus increasing its value as animal feed additive.

Other aspects of rearing *T. molitor*

Other aspects have also been considered for the optimization of rearing *T. molitor*, such as water supply, which has been found to have a significant importance on the growth rate (Hansen *et al.*, 2004), heat emission (Hansen *et al.*, 2004), egg harvesting (Ghaly and Alkoik, 2009) and the design of specific rearing units (Comparat *et al.*, 2016a,b).

All of these studies demonstrate that rearing *T. molitor* should be considered as a whole system, and it is very difficult to optimize only one aspect (e.g. diet) without taking into consideration these other factors.

14.3.2 Utilization for feed

T. molitor dried larvae and meal have been tested as feed for different monogastric animals such as pigs, poultry and fishes. Dried mealworms comprised up to 6% in the diet of weaning pigs (Jin *et al.*, 2016). The supplementation of mealworms was found to linearly improve body weight, the average daily gain and the average daily feed intake. Also, as mealworm proportion was increased, nitrogen retention and digestibility of dry matter as well as crude protein were linearly increased, thus showing a clear beneficial effect of feeding weaning pigs on a diet composed up to 6% dried *T. molitor*.

Several trials were performed on feeding poultry *T. molitor* meal. It was shown that *T. molitor* larvae (TML) meal improved the food conversion ratio compared to soybean meal and the albumin to globulin ratio, suggesting a higher immune response in broilers fed on TML meal (Bovera *et al.*, 2015). Consistent results were obtained with defatted TML meal, in this case 2% increase of carcass yield was reported as well as 15% decrease of water consumption by broilers (Van Huis, 2016). Another study (Klasing *et al.*, 2000) showed that *T. molitor* enriched in calcium contributes to the bone mineralization of growing chicks, with the calcium bioavailability being measured at a level of 76% in this study.

Different species of fish have also been fed TML meal. Rainbow trout were shown to perform with a 30% weight gain and growth performance (Hubert, 2016); different results were observed for catfishes, however no precise meal preparation description was given (Roncarati *et al.*, 2015).

14.3.3 Utilization for food

Safety concerns

Numerous studies and programs have promoted the utilization of edible insects for human nutrition; the short list of suitable insects usually includes *T. molitor* along with grasshoppers, crickets and a few others. However, in order to introduce such an uncommon food source in Western countries, several regulations and public concerns should be well documented. Safety is one of the most important concerns, and toxicology, allergenic properties and microbial charge have been extensively studied:

Toxicological aspects. The evaluation of toxicity of freeze-dried TML was studied for different periods of time, 28 (Han *et al.*, 2014) and 90 (Han *et al.*, 2016) days, and evaluated on Sprague-Dawley rats. These studies have shown that intake of freeze-dried TML for 28 or 90 days had no adverse effects up to the maximum tested dose, i.e. 3,000 mg/kg/day intake. Therefore, TML were considered to be non-genotoxic and the no observed adverse effect level was determined to be 3,000 mg/kg/day.

Allergenic aspects. The first study reported on the evaluation of allergies induced by ingestion and inhalation of *T. molitor* was performed on volunteers from the New York Entomological Society (Freye *et al.*, 1996). It showed that allergic reaction was possible for patients already with other known allergies to trees, dust, mites or cats. More recent studies, made with immunoglobulin E have confirmed the increased sensitivity of patients with known allergies to shellfishes or house dust mites (Verhoeckx *et al.*, 2013, 2014). The most important allergen in aqueous phase was identified to be arginine kinase, whereas tropomyosin was found in the solid part (Verhoeckx *et al.*, 2013). Moreover it was shown that even if thermal processing changes the solubility of *T. molitor* allergenic substances (Broekman *et al.*, 2015), it does not affect their allergenic potential (Broekman *et al.*, 2015; Van Broekhoven *et al.*, 2016).

Microbiological aspects. A microbial community assessment of *T. molitor* was performed (Stoops *et al.*, 2016) and results suggested total aerobic counts were close to 8±0.3 log cfu/g, with Enterobacteriaceae and lactic bacteria close to 7.23±0.4 and 7.47±0.15 cfu/g, respectively, and yeasts to 5.37±0.29 cfu/g. These totals were close to those found for grasshoppers, however the bacterial endospores were much less frequent in *T. molitor* compared to grasshopper (Stoops *et al.*, 2016). Several techniques have thus been evaluated to reduce or eliminate this microbial presence in TML, the most efficient were found to be either high pressure, 600 MPa, or thermal, 90 °C, treatments (Rumpold *et al.*, 2014).

Preparations

The nutritional value of whole larvae was evaluated (Siemianowska *et al.*, 2013), as well as the extraction techniques to separate the most nutritional part, proteins, from fats and chitin. The main difference between the discussed methods was the utilization (Zhao *et al.*, 2016) or not (Armenjon *et al.*, 2016) of organic solvents as defatting agents. Drying methods to preserve the most initial nutritional value were also evaluated; it was found that for the preservation of food quality and the inhibition of polyphenol oxidase, a well-known alteration enzyme, heat treatment was also recommended (Azzollini *et al.*, 2016).

Several recipes and utilization were also described, including some pastry preparations (Hwang and Choi, 2015), and some more peculiar applications such as space diet for long-term space exploration trips (Jones, 2015).

14.3.4 Utilization of oil

For some food applications, it might be worthwhile to separate *T. molitor* proteins from oil, as oil becomes another product which requires proper valorisation. Different valorisation studies were recently performed.

Studies focusing on the inner composition and possible utilization of oil for food and feed purposes have found that it was mainly composed by unsaturated fatty acids, with a very interesting ω3 to ω6 ratio, moreover it exhibits a superior oxidative stability (Jeon *et al.*, 2016).

All of these aspects makes this oil a valuable source of fatty acids for human and animal nutrition.

Also the potential of *T. molitor* as biodiesel source has been evaluated. It was found that this oil was of great value for the biodiesel industry, exhibiting several characteristics such as a cetane number of 58 which is reportedly much better quality according to the EN 14214 norm than, for example, the rapeseed biodiesel with the cetane number of 45, and satisfying the requirement of being at least 51 (Zheng *et al.*, 2013).

14.3.5 Life cycle assessment analysis

The evaluation of the environmental impact of *T. molitor* and processing is essential to understand the actual benefice from the use of this insect as a valuable source for food and feed. When evaluating the life cycle assessment of *T. molitor* production and transformation it was found that production of one kg of edible protein from this insect produces less greenhouse emissions and require less land (Oonincx and De Boer, 2012) and water (Miglietta *et al.*, 2015).

However, it is worthwhile to mention that this evaluation was made considering TML as food, and some figures might change if *T. molitor* is not the final product but an intermediate, for example when used as feed.

Another point to be considered is that the insect food and feed industry remains in its infancy and several improvements in rearing and mass production are to be expected in the coming years, which will also modify the overall life cycle analysis consideration.

14.4 Outlook

Mealworms are very promising insects for the insect food and feed industry. They present interesting characteristics in terms of nutritional composition, are easier to rear than many other insect species, and their safety aspects are well studied. These might be reasons why several companies across Europe are focusing their attention on mealworms, such as Proti-Farm (Ermelo, the Netherlands) working mainly with *A. diaperinus* for food applications or Ynsect, working with *T. molitor* for feed applications.

However, several challenges remain, such as rearing at industrial scale, regulatory issues, the Novel Food directive, and social acceptance for food applications. Feed market production appears to be more accessible at present, as insects have been authorized for pet food use and for aquaculture in Europe since December 2016. To enter this market, companies need to overcome the price issues and discover ways of sustainable and economically viable models of insect production and transformation.

In conclusion, it is apparent that due to the high quality of products obtained with mealworms detailed by several food and feed studies noted above, the industrialization of mealworm production is an increasing reality. Even with several challenges remaining and important research, development and optimization work yet to be completed, we do think that mealworm-based feed and food will be widespread in markets within the coming years.

References

Alves, A.V., Sanjinez-Argandona, E.J., Linzmeier, A.M., Cardoso, C.A.L. and Macedo, M.L.R., 2016. Food value of mealworm grown on *Acrocmania aculeate* pulp floor. PLoS ONE 11: e0151275.

Armenjon, B., Berezina, N., Laurent, S., Socolsky, C., Sanchez, L. and Hubert, A., 2016. Beetle powder. Patent WO2016108037. Description available at: http://www.google.dm/patents/WO2016108037A1?cl=en.

Azzollini, D., Derossi, A. and Severini, C., 2016. Understanding the drying kinetic and hygroscopic behavior of larvae of yellow mealworm (*Tenebrio molitor*) and the effects on their quality. Journal of Insects as Food and Feed 2: 233-243.

Berezina, N., Hubert, A., Berro, F., Levon, J-G., Le Roux, K., Socolsky, C., Sanchez, L. and Laurent, S., 2016. Chitin, hydrolysate and method for the production of one or more desired products from insects by means of enzymatic hydrolysis. Patent WO2016108033 (A1). Description available at: http://www.google.com.au/patents/WO2016108033A1?cl=en.

Bovera, F., Piccolo, G., Gasco, L., Marono, S., Loponte, R., Vassaloti, G., Mastellone, V., Lombardi, P., Attia, Y.A. and Nizza, A., 2015. Yellow mealworm larvae as possible alternative to soybean meal in broiler diets. British Poultry Science 56: 569-575.

Broekman, H., Knulst, A., Jager, Sd.H., Monteleone, F., Gaspari, M., De Jong, G., Houben, G. and Verhoeckx, K., 2015. Effect of thermal processing on mealworm allergenicity. Molecular Nutrition and Food Research 59: 1855-1864.

Calis, H., 2016. Method and facility for breeding insects. Patent WO2016153338. Description available at: https://www.google.com/patents/WO2016153338A1?cl=es.

Comparat, S., Hubert, A., Berro, F. and Levon, J.G., 2016a. Method for insect breeding. Patent WO2016166465. Description available at: http://www.google.com.np/patents/WO2016166465A1?cl=en.

Comparat, S., Hubert, A., Berro, F., Levon, J.G., Launay, F. and Sarton du Jonchay, T., 2016b. Farm for rearing insects. Patent WO2016166471. Description available at: http://www.google.com/patents/WO2016166471A1?cl=en.

Crippen, T.L. and Sheffield, C., 2006. External surface disinfection of the lesser mealworm. Journal of Medical Entomology 43: 916-923.

Crippen, T.L., Sheffield, C.L., Esquivel, S.V., Droleskey, R.E. and Esquivel, J.F., 2009. The acquisition and internalization of *Salmonella* by the lesser mealworm, *Alphitobius diaperinus*. Vector-Borne and Zoonotic Diseases 9: 65-71.

Davies, R.H. and Wray, C., 1995. Contribution of the lesser mealworm beetle (*Alphitobius diaperinus*) to carriage of *Salmonella enteridis* in poultry. Veterinary Record 137: 407-408.

Davis, G.R.F. and Schieffer, H.B., 1982. Effects of dietary T-2 toxin concentrations fed to larvae of yellow mealworm at three dietary protein levels. Comprehensive Biochemistry Physiology 73: 13-16.

De las Casas, E., Pomeroy, B.S. and Harein, P.K., 1968. Infection and quantitative recovery of *Salmonella typhymurium* and *Escherichia coli* from within the lesser mealworm, *Alphitobius diaperinus*. Poultry Science 47: 1871-1875.

Despins, J.L. and Axtell, R.C., 1995. Feeding behavior and growth of broiler chicks fed larvae of the darkling beetle, *Alphitobius diaperinus*. Poultry Science 74: 331-336.

Despins, J.L., Vaughan, J.A. and Turner, E.C., 1988. Role of the lesser mealworm, *Alphitobius diaperinus*, as a predator of the house fly, *Musca domestica*, in poultry houses. Coleopterists Bulletin 42: 211-216.

Dunford, J.C. and Kaufman, P.E., 2006. Lesser mealworm, litter beetle, *Alphitobius diaperinus* (panzer) (Insecta: Coleoptera: Tenebrionidae). Department of Entomology and Nematology, University of Florida. Available at: http://tinyurl.com/yctoraj6.

Eugenio, C., De las Casas, E., Harein, P.K. and Mirocha, G.J., 1970. Detection of the mycotoxin F-2 in the confused flour beetle and the lesser mealworm. Journal of Economical Entomology 63: 412-415.

Fraenkel, G., 1950. The nutrition of the mealworm, *Tenebrio molitor*. Physiology Zoology 23: 92-108.

Freye, H.B., Esch, R.E., Litwin, C.M. and Sorkin, L., 1996. Anaphylaxis to the ingestion and inhalation of *Tenebrio molitor* (mealworm) and *Zophobas morio* (superworm). Allergy Asthma Proceedings 17: 215-219.

Geden, C.J., Arends, J.J., Rutz, D.A. and Steinkraus, D.C., 1998. Laboratory evaluation of *Beauveria bassiana* against the lesser mealworm, *Alphitobius diaperinus*, in poultry litter, soil and pupal trap. Biological Control 13: 71-77.

Ghaly, A.E. and Alkoik, F.N., 2009. The yellow mealworm as a novel source of proteins. American Journal of Agriculture Biological Science 4: 319-331.

Han, S.-R., Lee, B.-S., Jung, K.-J., Yu, H.-J., Yun, E.-Y., Hwang, J.S. and Moon, K.-S., 2016. Safety assessment of freeze-dried powdered *Tenebrio molitor* larvae (yellow mealworm) as novel food source: evaluation of 90-day toxicity in Sprague-Dawley rats. Regulatory Toxicology and Pharmacology 77: 206-212.

Han, S.-R., Yun, E.-Y., Kim, J.-Y., Hwang, J.S., Jeong, E.J. and Moon, K.-S., 2014. Evaluation of genotoxicity and 28-day oral dose toxicity on freeze-dried powder of *Tenebrio molitor* larvae (yellow mealworm). Toxicological Research 30: 121-130.

Hansen, L.L., Ramlov, H. and Westh, P., 2004. Metabolic activity and water vapour absorption in the mealworm *Tenebrio molitor*: real-time measurement by two-channel microcalorimetry. Journal of Experimental Biology 207: 545-552.

Hubert, A., 2016. Mealworm meal shows promise for fish. All About Feed, 17 March 2016. Reed Business Information, Doetinchem, the Netherlands. Available at: http://tinyurl.com/y8cfukwa.

Hunt, A.S., Ward, A.M. and Ferguson, G., 2001. Effects of a high calcium diet on gut loading in varying ages of crickets and mealworms. In: Proceedings of the Nutrition Advisory group, 4th Conference on Zoo and Wildlife Nutrition, Lake Buena Vista, FL, USA, pp. 94-98. Available at: http://tinyurl.com/ycpzgg7y.

Hwang, S.-Y. and Choi, S.-K., 2015. Quality characteristics of muffins containing mealworm (*Tenebrio molitor*). Korean Journal of Culinary Research 21: 104-115.

Jeon, Y.H., Son, Y.J., Kim, S.H., Yun, E.Y., Kang, H.J. and Hwang, I.K., 2016. Physicochemical properties and oxidative stabilities of mealworm (*Tenebrio molitor*) oils under different roasting conditions. Food Science and Biotechnology 25: 105-110.

Jin, X.H., Heo, P.S., Hong, J.S., Kim, N.J. and Kim, Y.Y., 2016. Supplementation of dried mealworm on growth performance, nutrient digestibility and blood profiles in weaning pigs. Asian Australasian Journal of Animal Science 29: 979-986.

Jones, R.S., 2015. Space diet: daily mealworm (*Tenebrio molitor*) harvest on a multigenerational spaceship. Journal of Interdisciplinary Science Topics 4: 28-29.

Jones, L.D., Cooper, R.W. and Harding, R.S., 1972. Composition of mealworm *Tenebrio molitor* larvae. Journal of Zoo Animal Medicine 3: 34-41.

Klasing, K.C., Thacker, P., Lopez, M.A. and Calvert, C.C., 2000. Increasing the calcium content of mealworms to improve their nutritional value for bone mineralization of growing chicks. Journal of Zoo Wildlife Medicine 31: 512-517.

Lee, S.J., Yu, J.S., Parker, B.L., Skinner, M., Je, Y.H. and Kim, J.S., 2015. Production of antibacterial *Bombyx mori* cecropin A in mealworm-pathogenic *Beauveria bassiana* ERL 1170. Journal of Industrial Microbiology Biotechnology 42: 151-156.

Martin, R.D., Rivers, J.P.W. and Cowgill, U.M., 1976. Culturing mealworms as food for animals in captivity. Principles of Zoo Animals Feeding 16: 63-70.

Miglietta, P.P., Leo, F.D., Ruberti, M. and Massari, S., 2015. Mealworms for food: a water footprint perspective. Water 7: 6190-6203.

Moret, Y., 2006. Trans-generational immune priming: specific enhancement of the antimicrobial immune response in the mealworm beetle, *Tenebrio molitor*. Proceedings of Royal Society of Biology 273: 1399-1405.

Oonincx, D.G.A.B. and De Boer, I.J.M., 2012. Environmental impact of the production of mealworms as a protein source for humans – a life cycle assessment. PLoS ONE 7: e51145.

Poole, T. and Crippen, T., 2009. Conjugative plasmid transfer between *Salmonella enterica* Newport and *Escherichia coli* within the gastrointestinal tract of the lesser mealworm beetle, *Alphitobius diaperinus*. Poultry Science 88: 1553-1558.

Renault, D., Bouchereau, A., Delettre, Y.R., Hervant, F. and Vernon, P., 2006. Changes in free amino acids in *Alphitobius diaperinus* (Coleoptera: Tenebrionidae) during thermal and food stress. Comparative Biochemistry Physiology Part A: Molecular and Integrative Physiology 143: 279-285.

Rice, S.J. and Lambkin, T.A., 2008. A new culture method for lesser mealworm, *Alphitobius diaperinus*. Journal of Applied Entomology 133: 67-72.

Roncarati, A., Gasco, L., Parisi, G. and Terova, G., 2015. Growth performance of common catfish fingerlings fed mealworm diet. Journal of Insect as Food and Feed 1: 233-240.

Rueda, L.M. and Axtell, R.C., 1996. Temperature-dependent development and survival of the lesser mealworm, *Alphitobius diaperinus*. Medicinal and Veterinary Entomology 10: 80-86.

Rumpold, B.A., Fröhling, A., Reineke, K., Knorr, D., Boguslawski, S., Ehlbeck, J. and Schlüter, O., 2014. Comparison of volumetric and surface decontamination techniques for innovative processing of mealworm larvae (*Tenebrio molitor*). Innovative Food Science and Emerging Technologies 26: 232-241.

Siemianowska, E., Kosewska, A., Aljewicz, M., Skibniewska, K.A., Polak-Juszczak, L., Jarocki, A. and Jędras, M., 2013. Larvae of mealworm (*Tenebrio molitor* L.) as European novel food. Agricultural Sciences 4: 287-291.

Stoops, J., Crauwels, S., Waud, M., Claes, J., Lievens, B. and Van Campenhout, L., 2016. Microbial community assessment of mealworm larvae (*Tenebrio molitor*) and grasshoppers (*Locusta migratoria* migratorioides) sold for human consumption. Food Microbiology 53B: 122-127.

Szczepanik, M., Gliszczynska, A., Hnatejko, M. and Zawitowska, B., 2016. Effects of halolactones with strong feeding deterrent activity on the growth and development of larvae of the lesser mealworm, *Alphitobius diaperinus*. Applied Entomology Zoology 51: 393-401.

Van Broekhoven, S., Bastiaan-Net, S., De Jong, N.W. and Wichers, H.J., 2016. Influence of processing and *in vitro* digestion on the allergic cross-reactivity of three mealworm species. Food Chemistry 196: 1075-1083.

Van Broekhoven, S., Oonincx, D.G.A.B., Van Huis, A. and Van Loon, J.J.A., 2015. Growth performance and feed conversion efficiency of three edible mealworm species on diets composed of organic by-products. Journal of Insect Physiology 73: 1-10.

Van Huis, A., Van Itterbeeck, J., Klunder, H., Mertens, E., Halloran, A., Muir, G. and Vantomme, P., 2013. Edible insects: future prospects for food and feed security. Food and Agriculture Organization of the United Nations (FAO), Rome, Italy, FAO Forestry Paper no. 171, 187 pp. Available at: http://www.fao.org/docrep/018/i3253e/i3253e.pdf.

Verhoeckx, K.C.M., Van Broekhoven, S., Gaspari, M., Den Hartog-Jager, C.F., De Jong, G.A.H., Wichers, H.J., Van Hoffen, E., Houben, G.F. and Knulst, A.C., 2013. House dust mite (Der p 10) and shellfish allergic patients may be at risk when consuming food containing mealworm proteins. Clinical and Translational Allergy 3(Suppl 3): 48.

Verhoeckx, K.C.M., Van Broekhoven, S., Den Hartog-Jager, C.F., Gaspari, M., De Jong, G.A.H., Wichers, H.J., Van Hoffen, E., Houben, G.F. and Knulst, A.C., 2014. House dust mite (Der p 10) and crustacean allergic patients may react to food containing yellow mealworm proteins. Food and Chemical Toxicology 65: 364-373.

Yang, Y., Yang, J., Wu, W., Zhao, J., Song, Y., Gao, L., Yang, R. and Jiang, L., 2015. Biodegradation and mineralization of polystyrene by plastic-eating mealworms. Chemical and physical characterization and isotopic tests. Environmental Science and Technology 49: 12080-12086.

Zhao, X., Vázquez-Gutiérrez, J.L., Johansson, D.P., Landberg, R. and Langton, M., 2016. Yellow mealworm protein for food purposes – extraction and functional properties. PLoS ONE 11: e0147791.

Zheng, L., Hou, Y., Li, W., Yang, S., Li, Q. and Yu, Z., 2013. Exploring the potential of grease from yellow mealworm beetle (*Tenebrio molitor*) as a novel biodiesel feedstock. Applied Energy 101: 618-621.

15.

Acheta domesticus

Cricket rearing

G. Mott
Aspire Food Group LTD, 24 Desbarats, Kirkland, QC, H9J 2P2, Canada; gm@aspirefg.com

Abstract

Orthoptera have been a valuable food source for humanity for millennia. In the past century, farming of Order Gryllidae (crickets) first as bait and later as pet food became a common practice in Europe and North America. More recently there has been an explosion of small and medium cricket farms in Southeast Asia to produce crickets for human food. Some estimates put the number in excess of 20,000 independent facilities in Thailand alone. Given the increasing recognition of the value of insects as a source of food it is vital to establish and disseminate cricket farming best practices. Accordingly, this chapter begins by summarizing the lifecycle of crickets and describing optimal habitat design – this information is presented as the foundation for the development of a cricket farm. Further, the chapter offers guidance on how to navigate logistical considerations and operational trade-offs that prospective farmers will encounter when planning their approach to rearing crickets.

Keywords: *Acheta domesticus*, *Gryllodes sigillatus*, insect farming, entomophagy, insect protein, environment, food security, food sustainability, cricket farming

15.1 Farming crickets

Wild-caught Gryllidae (Orthoptera; Gryllidae) have long been a popular food item in many geographies. Some countries, including Thailand, Cambodia, and Vietnam, have a history of captive-rearing cricket species for food to avoid the challenges and risks of wild harvest and to reap the benefits of stability and reduced cost. Interest in stabilizing supply and reducing cost through farming has been growing, and there are now thousands of successful cricket rearing facilities around the world.

There are several approaches to rearing crickets that vary depending on the objective. Thus, it is important to first determine the purpose of the rearing operation and the species to be bred. Once this has been determined, an appropriate space can be prepared, a starter population obtained, and rearing can commence.

15.2 Species selection

There is a vast and colourful array of cricket species around the world. However, there are a limited number that have a history of captive rearing and are well represented in the scientific literature. While other species may present interesting captive-rearing opportunities, this chapter focuses on the most common species currently found in captive-rearing operations:

- *Acheta domesticus* (house cricket; Figure 15.1);
- *Gryllodes sigillatus* (banded or decorated cricket).

There is a range of characteristics to consider when selecting a species of cricket to rear. First though, a prospective farmer must ensure that they can actually obtain a population with which to start. While in some countries these species may not be present, in many

Figure 15.1. Adult female *Acheta domesticus* (photo by Gabe Mott).

countries one or both are readily available. Once a farmer is able to identify the species available to rear, then they should consider biological factors that will determine rearing and farming efficiency (Table 15.1), as well as nutritional characteristics that will be of interest to prospective consumers (Table 15.2 and 15.3).

There is significant variation in the results shown in Table 15.1. Some of this can be accounted for as the result of natural variation as seen in any population, and farmers should expect to see individual crickets vary in these characteristics in a roughly normal distribution. However that bell curve can be right- or left-shifted through the manipulation of the environment, feed, water or other conditions.

Table 15.1. Biological characteristics of *Acheta domesticus* and *Gryllodes sigillatus* that should be of interest to prospective farmers.[1]

Characteristic	***A. domesticus***	***G. sigillatus***
Fecundity (eggs per female)	340 to 1,060 [a]	470 to 950 [b]
Incubation period (days from lay to hatch)	11 to 15 [c]	~12 [b]
Time to maturity (days from hatch to max body weight)	32-49 [a,d]	33-40 [e]
Adult weight (mg per adult)	190-393 [f]	125-312 [e]
Vulnerability to densovirus	high [g]	very low [h]

[1] [a] = Ghouri and McFarlane (1958); Murtaugh and Denlinger (1985); [b] = Ivy and Sakaluk (2005); [c] = Nowosielski and Patton (1965); [d] = Booth and Kiddell (2007), note that shorter times to maturation have been reported by commercial operations; [e] = McFarlane (1964); [f] = McFarlane *et al.* (1959); [g] = Szelei *et al.* (2011); [h] = Weissman *et al.* (2012).

Table 15.2. Nutritional characteristics (by dry weight) of *Acheta domesticus* and *Gryllodes sigillatus* that may be of interest to prospective consumers.

Characteristic	***A. domesticus* (Finke, 2002)**	***G. sigillatus* (Zielińska *et al.*, 2015)**
Protein	67%	70%
Fat	22%	18%
Ash	4%	5%
Calcium (mg/100 g)	132	130
Phosphorus (mg/100 g)	957	–
Magnesium (mg/100 g)	109	101
Sodium (mg/100 g)	435	330
Potassium (mg/100 g)	1,127	1,190
Iron (mg/100 g)	6.27	4.23
Zinc (mg/100 g)	21.8	13.9
Copper (mg/100 g)	2.01	4.79

Table 15.3. Amino acid composition (g/kg of dry weight) of *Acheta domesticus* and *Gryllodes sigillatus*.

Characteristic	*A. domesticus* (Finke, 2002)	*G. sigillatus* (Zielińska *et al.*, 2015)
Alanine	58.4	40.6
Aspartic acid	55.8	51.0
Arginine	40.6	32.6
Cystine	5.5	7.8
Glycine	33.8	28.5
Glutamic acid	69.8	74.6
Histidine	15.6	12.0
Isoleucine	30.5	18.6
Leucine	66.6	40.5
Lysine	35.7	26.9
Methionine	9.7	11.1
Methionine+cystine	15.3	18.9
Phenylalanine	21.1	15.4
Proline	37.3	37.9
Serine	33.1	28.3
Threonine	24.0	25.8
Tryptophan	4.2	–
Tyrosine	32.5	22.3
Valine	34.7	32.9

There remains a range of other characteristics to consider before finalizing species selection, though some of these are more subjective. For example, the different cricket species have different behaviours and calls (Weissman *et al.*, 2012); crucially, informal reports suggest different flavour profiles for these species. It is critical to make an informed decision, so personal exposure to multiple species is recommended.

15.3 Life cycle

When 'nymph' or juvenile crickets emerge from the egg, they already appear physiologically similar to adult Gryllidae. These youngest of crickets are often referred to as 'pinheads' as they are approximately the size of the head of a pin for the first few days of their lives (Figure 15.2). Crickets experience 'hemimetabolism' or incomplete metamorphosis, as opposed to many invertebrates that have two distinct and morphologically different forms separated by a pupation phase. Crickets gain in size and slowly accumulate more adult-like features as they proceed through 7-10 moults, or instars, without ever pupating. Individuals of the same species may moult more or fewer times, dependant on the time it takes them to reach maturity (Clifford *et al.*, 1977).

At the fourth or fifth instar, females develop a clearly visible ovipositor (though it can occasionally be observed in earlier stages). The ovipositor is a dark, needle-like protrusion from the base of the female's abdomen that grows to two thirds the length of the body by

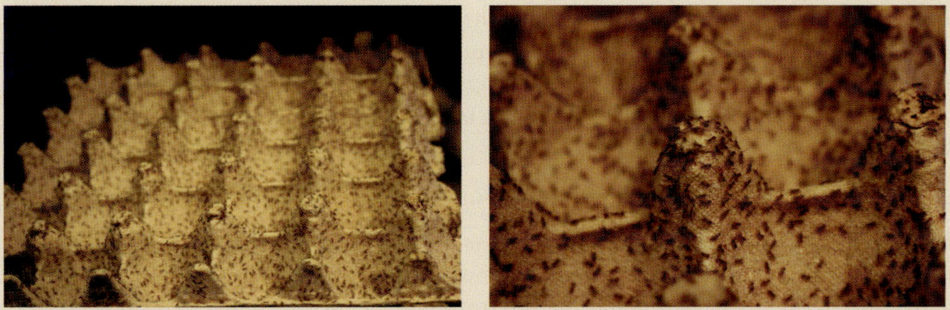

Figure 15.2. Day-old *Acheta domesticus*; 'pinheads' (photo by Gabe Mott).

adulthood. After mating the ovipositor is used to deposit eggs into the laying substrate, much like a syringe.

In the last two nymphal instars, wings begin to develop. At adulthood, these are fully developed and the males use them to produce the characteristic mating chirps. At adulthood females are 15-25% larger than the males. Much of this extra weight is accounted for by eggs the females carry.

A colony of mostly penultimate instar crickets will not all transition to adulthood at once, but tend to mature over a window of 3-10 days (with some outliers in either direction). Because adults can easily be identified by their fully developed wings, the proportion of adults in a colony is often referred to as the 'percentage winged'. For example, a glance at a 40-day old population of crickets might indicate that they are about 60% winged.

Mating activity between females and males begins within a day after maturation and is followed by laying, though maximum laying activity is usually 4-8 days after maturation (Murtaugh and Denlinger, 1985). Within 24-48 hours after mating, female crickets will lay their eggs in an appropriate surface. Given the right conditions, these eggs will hatch in 11-15 days and begin the cycle again.

15.4 Preparations for rearing crickets

15.4.1 Determining the purpose of the colony

Currently it is possible to find cricket farmers maintaining colonies of all sizes, from a single small vivarium used to support a pet lizard or amphibian to 10,000 square meters of rearing space across multiple buildings. The efficiency of the farm and the yield per square meter increases at greater scale.

Any prospective farmer must carefully consider the purpose of the colony. This will inform a host of forthcoming decisions to be made throughout the course of farming. Accordingly, Table 15.4 provides a general guide for determining the size and output of a farm necessary for achieving various farming objectives. While it is outside the scope of this chapter, prospective farmers must also ensure that they have the resources available to establish and maintain their colony.

15.4.2 Maintaining the appropriate environment

The growth rates of crickets (as with all insects) are heavily influenced by temperature. Environmental conditions in most locations will not conform to the optimal range. If the area experiences extreme temperatures, environmental control will be necessary. Even if the environmental conditions fall in the optimal range, exercise caution when selecting a location open to air or lacking climate control. While temperature may not be a concern, there are other reasons to enclose and secure an insect colony, including pest control and preventing livestock escape.

15.4.3 Selecting suitable housing

When rearing insects, the colony will rarely be in one single contiguous space. Instead, it is common to use multiple self-contained spaces, each with its own population, water supply, food sources, and space maximizers.

There are a range of container options that each present their own advantages and challenges. A few of the more common options are as follows:

- Off-the-shelf plastic storage tub; 100-200 litres (Figure 15.3A).
- Larger cardboard, fiberglass or custom container; 500-700 litres (Figure 15.3B).
- Trough (usually either concrete construction or a wood or metal frame draped in plastic sheeting); 3,000-8,000 litres (Figure 15.3C).
- Room (rather than building containers for the crickets, all materials are simply placed on the floor of a large room, with spaces provided so that farmers can walk through and deliver food and water); 10,000 litres or more (Figure 15.3D).

Table 15.4. Common objectives for cricket farms sorted by size.

Size of farm	Output of live crickets	Common objectives
Small (~1 m^2)	1 kg/month	• School or children's project for education • A low-yield colony to feed a few pet reptiles or amphibians • Enough to provide a meal a week for human food
Medium (4-40 m^2)	10-250 kg/month	• A colony to provide the protein requirements for an individual or a family • A small colony for economic benefit • A trial colony
Large (50-100 m^2 or more)	400-1000 kg/month or more	• A scale colony for commercial production

Figure 15.3. Suitable housing for crickets: (A) plastic storage tub (photo by Austin Miller); (B) plastic-lined cardboard boxes (photo by Gabe Mott); (C) trough at cricket farm in Chiang Mai, Thailand (photo by Arnold van Huis); and (D) a room-sized farming area (photo courtesy of Entomo Farms).

Trials have also been performed with a variety of less common containers, including custom stackable plastic boxes, collapsible nets, modified two-litre soft drink bottles and many others.

The advantage of moving from smaller to larger units is a direct reduction of capital and operating expenses per unit of product. The capital cost per cubic meter of material is higher with smaller units. Also, the labour per cubic meter of container space is reduced as the size of the individual containers increase. However, there is a concurrent loss of control and loss of visibility into the wellbeing of the colony. There is also a reduction in protection against pathogen spread. Deciding on the appropriate balance requires a careful cost-benefit analysis that will be unique to the farming context.

One final note on containers. Most adult crickets are able to fly to some degree. However strains bred in captivity for long periods of time often see a significant reduction in those traits – crickets that fly, often remove themselves from their containers and thus the gene pool. Even in the populations that do not exhibit flying behaviour, crickets are enthusiastic climbers and jumpers. If the container is not properly secured (by either a lid or an un-climbable surface such as slick packing tape), an aspiring farmer will soon find that their colony has engineered a mass exodus. This is an especially unfortunate occurrence in small, home-based colonies.

15.4.4 Maximizing surface area and use of vertical space

When farming anything, from cattle to corn to crickets, density is an important consideration. A higher density will reduce both fixed costs (i.e. rent and utilities) as well as variable costs (i.e. labour) per unit of output. There will be a reasonable maximum number of individuals that can be supported in a given amount of surface area, and this number will also vary by size. A container of a given size will be able to support far more pinheads than adults. Since crickets do like to climb, increasing available surface area in their containers is best accomplished by maximizing the vertical space. Depending on the material chosen and husbandry practices, this will increase the maximum population by an order of magnitude or more.

Options for increasing surface area include off-the-shelf products such as egg flats and custom-made products such as cardboard 'stackers' (Figure 15.4). There are also more esoteric products in use, including foam weather-stripping or plastic re-useable egg flats.

Figure 15.4. Examples of surface area maximizers: (A) egg flats; and (B) cardboard stackers (photos by Gabe Mott).

When considering what surface area maximizer to choose, there are essentially two broad options; limited use or repeated-use. Limited use-options, like cardboard products, are convenient and easy to obtain. Further, cardboard is very fibrous, which makes it very easy for crickets to climb. However, they rapidly accumulate waste and degrade. They are also difficult to disinfect between cycles. This increases the risk of build-up and transfer of microbes between generations. Thus, it would be irresponsible for a food facility to reuse cardboard products and highly inadvisable for a feed facility to do so.

Repeated-use surfaces are a better choice for a farmer aspiring to a zero-waste facility, but are more difficult to manage. They require infrastructure to properly sanitize the material, otherwise the risk of contamination remains. Further, finding a surface that crickets adapt to as well as cardboard is a difficult endeavour.

Finally, crickets are voracious, but not particularly discerning feeders. Expect that they will attempt to eat anything in their environment. Any material surface that is consumable must be food safe; otherwise, there is a real risk to the end consumer, be they people, pets, or macro-livestock.

15.5 Raising crickets

Once the decisions that drive the physical structure have been determined, there remains the task of actually raising the crickets. There is a great deal of trial-and-error in this process. While this chapter provides a summary of many industry best practices, a new cricket farmer must expect a learning curve that will last as long as they work with crickets

15.5.1 Considerations on output

Any commercial operation must be concerned with 'output' or 'yield'. With regard to any livestock operation, output is a measure of weight of product generated from a colony in a unit of time (i.e. kilogrammes a month, or pounds a year). Maximizing this is a question of increasing weight per cycle and decreasing cycle time. Thus, increasing weight and decreasing cycle time are central factors to consider throughout the process of raising crickets. In the pursuit of maximizing output, it is important to remember that the success of a commercial operation will often be driven by product quality. Decisions that reduce quality in exchange for increased output risk eroding customer demand.

15.5.2 Sourcing initial stock

While many cricket species are cosmopolitan (found around the world), that does not mean that it is easy or desirable to source from the wild. Wild-caught crickets may carry disease, or, if already adult, may be past peak rearing age. There are also several additional challenges associated with wild-caught stock. First, it is difficult to acquire an adequate initial population without considerable expenditure of effort in the field. A further challenge is that wild cricket populations may only be available for capture during a brief window every year. If the population collapses (as may happen during the first few attempts at rearing), it may prove impossible to restart the colony for close to a year while waiting for crickets to come back into season.

An alternative is to source a domestic or international source of live crickets, as eggs, nymphs, or adults. Many countries currently have farmers that produce live crickets for feed or for food. Accessing these as a source for a starter population is, in many ways, superior to wild sources. One significant disadvantage is that sourcing from farmers will often be more capital-intensive (unless the capital cost of the labour to wild-craft the founder populations is taking into account). Note that sourcing eggs, if possible, is often superior as they are less sensitive to transportation trauma and are likely the least expensive of all commercial options.

If domestic wild and captive-bred crickets are not viable options, a final choice is to identify an international source. Due to the cosmopolitan nature of many species of cricket, it may be possible to obtain an import permit from the government. However, as most countries are justifiably concerned about invasions from alien species, and as foreign species may carry diseases not currently in the host country, this option is highly inadvisable.

15.5.3 Maintaining populations

Crickets are hardy insects, and will often survive conditions well outside what is optimal. However, if a strong and vibrant colony is an objective, controlling conditions will be important.

15.5.4 Environmental conditions

Crickets tend to seek warm, dark, humid environments. In general, research indicates a target temperature of 28 °C, although there are suggestions that even higher temperatures may generate faster growth (Booth and Kiddell, 2007; Ghouri and McFarlane, 1958; Nowosielski and Patton, 1965). However, with increased temperature comes increased stress and the risk of mortality or reduced fecundity.

Crickets thrive with a relative humidity of approximately 50-60% (Booth and Kiddell, 2007; Ghouri and McFarlane, 1958; Nowosielski and Patton, 1965). However, note that at a high temperature and humidity, surface mould and mite invasion in a container or a facility becomes a risk both to humans and crickets. The ability to vent the air will help manage these issues.

Finally, while crickets are nocturnal, they accept a light/dark cycle characterized by 12 hours of light followed by 12 hours of dark. (Booth and Kiddell, 2007; Ghouri and McFarlane, 1958; Nowosielski and Patton, 1965). There is significant opportunity for farmers to manipulate these variables and optimize the systems, but maintaining conditions within these parameters will be sufficient for most cricket species.

15.5.5 Food and water

Food

One of the great promises of insects as a source of feed or food is that they are more efficient than conventional livestock. This is already the case with crickets, and can be expected to further improve as more resources are directed towards research (Halloran *et al.*, 2017). A goal of this drive towards efficiency is to feed crickets entirely on waste streams. However, populations fed exclusively on those sources tend to develop slowly, if they are able to survive at all (Lundy and Parrella, 2015). For the moment, redirecting grain from poultry feed is the most common diet for crickets, and this seems to maximize output of commercial facilities

(Figure 15.5). Fully achieving the potential of insects as a superior source of sustainable protein will depend on identifying the appropriate mechanism to deliver feed from pre- and post-consumer waste while achieving maximized output.

Water

While crickets can generally be considered a hardy family, this does not extend to the pinheads. Nymphal crickets have a host of natural enemies, including open water. They are so small that they can become trapped under the surface tension of a large drop of water and drown. Thus, any water delivery system must involve a wicking agent, be it paper towel, foam, sand, soil, or some other alternative that permits access to moisture without exposing the pinheads to open water (Figure 15.6). Condensation in the container must also be avoided for the same reasons.

Viable water delivery systems include both automated and manual chicken waterers, open trays, modified 2-litre bottles and a host of other options. Note that larger reservoirs of water will reduce the need for maintenance, but that crickets will foul their water, so changes will be required.

15.5.6 Reproducing vs harvesting

Reproduction (the process by which further generations of a colony are produced) and harvesting (the process by which individuals of a colony are culled for consumption) becomes a trade-off. Crickets tend to mature over a 3-10 day window (with some outliers in either direction). At the same time, there is natural mortality in the colony that depletes the population. If the objective is to harvest the maximum biomass in the minimum time, this usually occurs when approximately 85% of the population is adult (winged). Waiting longer risks increasing mortality in the population reducing biomass faster than it increases.

Figure 15.5. *Acheta domesticus* feeding on chicken feed (photo by Gabe Mott).

Figure 15.6. Three different ways of providing water to crickets: (A) paper towel; (B) plastic ring; (C) coco-fiber (photos by Gabe Mott).

This 85-95% window often occurs between 28-45 days after hatch, depending on conditions. However, this is not the optimal time to mate the crickets. Maximum laying activity is usually 4-8 days after mating (Murtaugh and Denlinger, 1985), so a mating container should usually be maintained at least one week longer than a harvest container. The smallest proportion of the population necessary to either sustain or grow the colony should be set aside for mating, and the rest harvested at the 85-95% winged state.

15.6 Mating crickets

The actual mating component of farming crickets takes virtually no work to induce. A mixed population of males and females will begin to mate within 24-72 hours after achieving maturity. However, translating this activity into the next generation in an efficient, high-yield manner is somewhat more complex.

15.6.1 Establishing mating populations

When farming crickets it is valuable to partition the population into two segments; those containers destined for mating and those destined for harvest, as mating containers will need to be retained and supported for at least a week longer. This ratio should be selected ahead of time, and has a significant impact on the scale, stability, and profitability of any farming operations.

As mentioned in Table 15.1, crickets can lay hundreds and hundreds of eggs per female. However, this does not mean that the average farmer can support 1000 (or even 100) containers destined for harvest with a single mating container. In fact, most farmers find they need a ratio of 10 harvest containers to each mating container, and sometimes the ratio needs to be even smaller.

The need to set aside such a significant proportion for mating is driven by three primary factors. First, while a female can lay many eggs over her lifetime, this period spans well over a month, with a significant drop-off in fecundity over time (one-to-two weeks) (Behrens *et al.*, 1983; Murtaugh and Denlinger, 1985). As there is a daily cost in labour, rent, electricity, depreciation, and materials, there comes a point of diminishing returns, where the average monthly production of eggs is maximized by culling older mating containers and replacing them with a fresh population well before the females reach the end of their lives or have laid the majority of eggs they are capable of producing.

The second factor that forces farmers to allocate many of their newly-hatched crickets to mating is simply general mortality. A farmer who sets aside 1000 individuals for mating (approximately half female) can find that, by the time the population has fully matured, the number has dropped by 20, 40 or even 80%. When mating commences, mortality will continue in the population, further reducing the number of females able to lay eggs.

The final issue that drives farmers to maintain higher mating populations than seems necessary is the simple natural variance in the population of any container. Specifically, the container-to-container differences in population size can be quite significant, up to a 50% fluctuation around the mean. This is caused by a host of factors, including variation in initial population size, mortality, fecundity and environmental conditions. This in turn leads to fluctuations in the mating container-to-container production of eggs. At the same time, farmers desire stability in their colony as a whole. It is better to err on the side of over-production, even if that forces over-allocation of individual containers to mating, than to risk experiencing a weak mating cycle and seeing the total output of the farm drop. In summary, the greater the size of the mating population, the greater the stability of the colony as a whole. The smaller the size of the mating population, the more bins can be allocated for harvesting and the greater the total output generally, but stability of egg output becomes impaired.

While it depends on the farm, the conditions, the experience of the farmer, and the species of cricket, many farms find they observe a generational multiple of 3-10 containers of crickets generated from one mating container. Thus they require between 10-33% of their population dedicated to mating stock.

15.6.2 Providing suitable laying material

Gravid females will seek suitable material into which to lay their eggs (Figure 15.7). Typically, farmers will only introduce a container of laying material into the container once the crickets are fully mature. When the females identify suitable material, they will insert their ovipositors and deposit fertilized eggs. Egg survival is determined by the suitability of the substrate that surrounds them. There are many different materials that can be used to incubate eggs, so long as they meet a few simple requirements. Eggs need to be kept moist (though standing water must be avoided) throughout their development period, so the material must support this (Clifford *et al.*, 1977). Also, the material must not be too hard to be penetrated by the ovipositor. If the female cannot penetrate the surface, she is unlikely to lay eggs. Materials

Figure 15.7. Laying crickets (photo by Gabe Mott).

that have been used successfully include sand, potting soil, perlite, and peat moss (with peat moss being the most popular). Note that different materials will retain moisture for different periods of time. Maintaining the required level of water in the substrate will necessitate more maintenance if the container of laying substrate cannot be covered post-lay.

15.6.3 Incubating eggs

Once the crickets have laid their eggs in the substrate the container should be removed from the container and placed in a separate location for the incubation period (Table 15.1). The temperature requirement for eggs is the same as for live crickets. Additionally, if the substrate begins to dry out, adding water is critical to prevent the death of the eggs. Typically, there is little interaction or care required until the eggs hatch.

15.6.4 Transferring 1st instar crickets

As the eggs hatch, thousands of tiny pinheads will begin to crawl and hop around on the surface of the substrate. For the first 24 hours after emergence they need little to sustain their lives besides the temperature and humidity in the incubation space. However, the lay container is designed with lay in mind, not long-term survival for crickets, so the pinheads need to be transferred to a container to grow and thrive for the rest of their lives. As was mentioned earlier, first instar crickets are quite fragile. Mechanical damage from manipulation or transportation can significantly increase mortality. To minimize mechanical damage there are two conventional approaches to transferring newly hatched crickets.

The first and most common method for transferring newly hatched crickets is volumetric measurements. The pinheads are shaken or blown out of their incubation container into a large (50 cm × 50 cm) plastic storage container. Avoid using tools or brushes as they can cause mass mortality among the nymphs. They are then dumped through a funnel into a graduated cylinder, measured to a certain volume, and then decanted into a container. This has the advantage of ensuring that every container has approximately the same number of nymphal crickets from the outset, generating greater predictability at harvest. However, this approach risks mechanical damage and greater mortality. Further, determining the correct volume of pinheads to transfer is a challenge. Each container will have a carrying capacity based on volume and total climbable surface area. Every farmer must experiment and calibrate the volume at hatch that provides the maximum population at harvest.

The alternative to volumetric measurement Is direct hatching. The container with the laying substrate (or some fraction of the substrate) is put directly into the container, and the pinheads are permitted to hatch directly into their new home. This is a simpler approach from the perspective of labour, calibration, and mechanical injury but leads to a much greater variation in population from container to container.

15.7 Harvest

Harvesting is the end-point of any container, and there is a range of possible approaches, but one major factor must be carefully monitored. No corpses or cricket waste should be permitted past the harvest step. Decaying animal mater cannot be put into the food supply, so precautions must be taken to only harvest live, vibrant individuals.

While hand-selecting crickets for harvest is one way of ensuring that no corpses are inadvertently harvested, some larger operations harvest millions of adults a day. At smaller scales hand-sorting is a tedious process, at larger it would require a huge team and is impractical.

Instead of hand-sorting, most farms will transfer the individuals from the container into a larger container, leaving all surface area maximizers behind. They then introduce 'sorters' (vertical material like stackers, cardboard tubes, egg trays) that the individual adults will begin to climb. Any insects that climb are clearly alive, thus the sorters can be picked up and the crickets shaken off into a final container for freezing.

Freezing is a humane, food-safe approach to killing crickets that preserves them for storage and maintains their structural integrity thereby allowing for a variety of later uses (e.g. cooking whole versus grinding). As crickets are ectotherms, introduction to sub-zero temperatures rapidly results in *torpor*, a naturally occurring low-activity state that they maintain until the temperature becomes warmer. If the crickets are maintained in sub-zero conditions for 24 hours or greater they will not revive.

15.8 Processing

As insects become more accepted into the western diet, recipes, cookbooks, and videos are becoming more prevalent. There is a staggering array of approaches to processing frozen crickets, but that is beyond the scope of this chapter. One final caveat: as a whole animal, the crickets should be rinsed and then cooked thoroughly before consumption to ensure that all possible biological contaminants are heat-killed. Food safety is critical.

15.9 Acheta domesticus densovirus

In 1977 researchers in France reported on the first occurrence of a densovirus infecting Orthoptera (Meynadier *et al.*, 1977). The infection was found in commercially raised *Acheta domesticus* and was both highly contagious and, in many cases, lethal.

Fast-forward to 1988 and a similar outbreak occurs in the southern USA, all but entirely wiping out a commercial rearing facility. (Styer and Hamm, 1991). The facility saw mortality of 20%-80% per container. After selectively mating only those crickets that survived in the 80% mortality containers and generally increasing cleanliness protocols, the population recovered, although it took close to a year. While samples from other commercial facilities tested positive for the virus, no major outbreaks were reported.

Starting in September 2009 outbreaks occurred across Canada and the USA (Szelei *et al.*, 2011). They were often devastating, approaching 100% mortality. Many facilities failed to recover. Some tried to sanitize and restart their operations, but saw the disease return and eradicate their populations a second time. In fact, one group of researchers proposed 'nuclear destruction' as the only solution to eliminate infection (Weissman *et al.*, 2012).

The Acheta domesticus densovirus (AdDNV) causes the highest rates of mortality in the last three nymphal instars and the adult phase (Szelei *et al.*, 2011). The crickets suffer reduced mass, lessened (or completely absent) feeding behaviour and partial or complete paralysis. While a natural reservoir for the virus has not been identified, it appears to be airborne and can also be transmitted through feeding or contact. Various species of cricket as well as other captive-raised invertebrates have tested positive for AdDNV, providing a number of possible vectors for the disease.

Subsequent research and sequencing of the virus has revealed several key insights. The second, more severe North American outbreak did not originate from the same source as the earlier European or American outbreaks (Liu *et al.*, 2011). Further investigation determined that the two variants diverged in 2006 (Szelei *et al.*, 2011). Finally, a number of investigators have attempted to infect a diverse selection of crickets other than *A. domesticus*, but without any pathology developing (Szelei *et al.*, 2011; Weissman *et al.*, 2012). So far, *A. domesticus* remains the only species that demonstrates any susceptibility to AdDNV.

Recently populations in Japan have seen AdDNV-related collapses, suggesting that there are few places safe from the virus (Pham *et al.*, 2013). However, several North American commercial facilities have raised and continue to raise *A. domesticus* without any infections. While there is clearly a risk involved, for now it appears that careful management of materials and strict biological quarantine are sufficient to prevent infection of a population.

15.10 Conclusions

Research regarding the scale farming of crickets remains in its infancy. Over time, efficiencies will increase and crickets will become significantly more sustainable and efficient to produce relative to poultry. However, this will only happen as the result of a concerted effort. Using this chapter as a starting point, new farmers will add to the volume of data available and will enable the optimization that other livestock industries have required centuries to achieve.

References

Behrens, W., Hoffmann, K.-H., Kempa, S., Gäβler, S. and Merkel-Wallner, G., 1983. Effects of diurnal thermoperiods and quickly oscillating temperatures on the development and reproduction of crickets, *Gryllus bimaculatus*. Oecologia 59: 279-287.

Booth, D.T. and Kiddell, K., 2007. Temperature and the energetics of development in the house cricket (*Acheta domesticus*). Journal of Insect Physiology 53: 950-953.

Clifford, C.W., Roe, R.M. and Woodring, J., 1977. Rearing methods for obtaining house crickets, *Acheta domesticus*, of known age, sex, and instar. Annals of the Entomological Society of America 70: 69-74.

Finke, M.D., 2002. Complete nutrient composition of commercially raised invertebrates used as food for insectivores. Zoo Biology 21: 269-285.

Ghouri, A. and McFarlane, J., 1958. Observations on the development of crickets. Canadian Entomologist 90: 158-165.

Halloran, A., Hanboonsong, Y., Roos, N. and Bruun, S., 2017. Life cycle assessment of cricket farming in north-eastern Thailand. Journal of Cleaner Production 156: 83-94.

Ivy, T.M. and Sakaluk, S.K., 2005. Polyandry promotes enhanced offspring survival in decorated crickets. Evolution 59: 152-159.

Liu, K., Li, Y., Jousset, F.-X., Zadori, Z., Szelei, J., Yu, Q., Pham, H.T., Lépinem F., Bergoin, M. and Tijssen, P., 2011. The Acheta domesticus densovirus, isolated from the European house cricket, has evolved an expression strategy unique among parvoviruses. Journal of Virology 85: 10069-10078.

Lundy, M.E. and Parrella, M.P., 2015. Crickets are not a free lunch: protein capture from scalable organic side-streams via high-density populations of *Acheta domesticus*. PLoS ONE 10: e0118785.

McFarlane, J., 1964. Factors affecting growth and wing polymorphism in *Gryllodes sigillatus* (Walk.): dietary protein level and a possible effect of photoperiod. Canadian Journal of Zoology 42: 767-771.

McFarlane, J., Neilson, B. and Ghouri, A., 1959. Artificial diets for the house cricket, *Acheta domesticus* (L.). Canadian Journal of Zoology 37: 913-916.

Meynadier, G., Matz, G., Veyrunes, J.-C. and Bres, N., 1977. Virose de type densonucleose chez les orthopteres. Annales de la Société Entomologique de France 13: 487-493.

Murtaugh, M.P. and Denlinger, D.L., 1985. Physiological regulation of long-term oviposition in the house cricket, *Acheta domesticus*. Journal of Insect Physiology 31: 611-617.

Nowosielski, J.W. and Patton, R.L., 1965. Life-tables for the house cricket, *Acheta domesticus* L., and the effect of intra-specific factors on longevity. Journal of Insect Physiology 11: 201-209.

Pham, H.T., Iwao, H., Szelei, J., Li, Y., Liu, K., Bergoin, M. and Tijssen, P., 2013. Comparative genomic analysis of Acheta domesticus densovirus isolates from different outbreaks in Europe, North America, and Japan. Genome Announcements 1: e00629-13.

Styer, E. and Hamm, J., 1991. Report of a densovirus in a commercial cricket operation in the southeastern United States. Journal of Invertebrate Pathology 58: 283-285.

Szelei, J., Woodring, J., Goettel, M., Duke, G., Jousset, F.-X., Liu, K., Zadori, Z., Li, Y., Styer, E. and Boucias, D., 2011. Susceptibility of North-American and European crickets to *Acheta domesticus* densovirus (AdDNV) and associated epizootics. Journal of Invertebrate Pathology 106: 394-399.

Weissman, D.B., Gray, D.A., Pham, H.T. and Tijssen, P., 2012. Billions and billions sold: pet-feeder crickets (Orthoptera: Gryllidae), commercial cricket farms, an epizootic densovirus, and government regulations make for a potential disaster. Zootaxa 3504: 67-88.

Zielińska, E., Baraniak, B., Karaś, M., Rybczyńska, K. and Jakubczyk, A., 2015. Selected species of edible insects as a source of nutrient composition. Food Research International 77: 460-466.

Nutritional quality and processing

16.

Blaptica dubia

Nutrient content of insects

M.D. Finke[1*] and D.G.A.B. Oonincx[2]

[1]Mark Finke LLC, 17028 E Wildcat Dr., Rio Verde, AZ 85263, USA; [2]Laboratory of Entomology, Wageningen University and Research, P.O. Box 16, 6700 AA Wageningen, the Netherlands; markfinke@desertinet.com

Abstract

This chapter reports on the nutrients present in insects. Data on the protein content and amino acid profiles of a variety of insect species are discussed and their amino acid profiles compared to nutrient requirements of growing broiler chicks, catfish, trout, swine, and human adults and young children. Both *in vitro* and *in vivo* protein digestibility data for a variety of insect species fed to several animal species is presented and factors affecting these data are discussed. Furthermore, the fat content and fatty acid profiles of a variety of insect species is reviewed, with special attention on both omega-6 and omega-3 fatty acids. Information on carbohydrates, fibre and chitin in insects is shown along with potential effects on nutrient availability. This is followed by a discussion of essential minerals in insects with an emphasis on calcium and phosphorus content. Data on insect vitamin content is shown along with a discussion of the presence of thiaminase in certain insects which can have adverse effects in humans consuming these species. Dietary effects on insect nutrient composition are reviewed with an emphasis on essential minerals, heavy metals, vitamin E, and carotenoids. Lastly, the effects of processing, including protein extraction and various cooking methods on insect composition are discussed. In summary, this chapter provides an overview of the nutrient content of insects and how select nutrients can be altered.

Keywords: amino acids, fatty acids, vitamins, minerals, digestibility

16.1 Introduction

Insects are important sources of nutrients for humans and a wide variety of other animal species. Hence studies on insect nutrient composition can be found in disciplines ranging from anthropology to zoology. Comprehensive literature reviews of insect nutrient content have been published (Bukkens, 1997; Finke, 2004; Payne *et al.*, 2016; Raubenheimer and Rothman, 2013; Rumpold and Schlüter, 2013a). While the first mention of using insects to feed production animals was in 1919, it was not until the 1960s and 1970s that research started in earnest (Calvert *et al.*, 1969; Teotia and Miller, 1973, 1974; Ueckert *et al.*, 1972). The renewed interest in this area of study in the last ten years has resulted in numerous papers and some recent comprehensive reviews regarding the safety of insects as food and feed as well as several overviews of feeding trials for production animals (Henry *et al.*, 2015; Makkar *et al.*, 2014; Riddick, 2014; Rumpold and Schlüter, 2013b; Sánchez-Muros *et al.*, 2014, 2016).

Given the comprehensive nature of these recent reviews this chapter will briefly summarize the data for a variety of nutrients in insects and insect products and focus on some specific aspects of nutrients in insects that have received little attention.

16.2 Nutrient content of insects

16.2.1 Proteins and amino acids

The protein content of insects varies between 25 and 75% on a dry matter basis (Barker *et al.*, 1998; Bukkens, 1997; Cerda *et al.*, 2001; Finke, 2002, 2013, 2015a; Oonincx and Dierenfeld, 2012; Oonincx and Van der Poel, 2011). Proteins are composed of amino acids and are generally grouped as either nutritionally indispensable (essential) amino acids or nutritionally dispensable (non-essential) amino acids. While all amino acids are required, nutritionally indispensable amino acids cannot be synthesized by most animal species and must be provided in the diet. The amino acid pattern of an insect at various life stages partially depends on whether the species undergoes incomplete metamorphosis (hemimetabolous) or complete metamorphosis (holometabolous). Comparison of the amino acid pattern (mg amino acid/g crude protein) for a hemimetabolous insect, the house cricket (*Acheta domesticus*), from a number of papers suggests that amino acid composition is fairly constant between crickets fed different diets as well as between nymphs and adults (Table 16.1). Prepupae from black soldier fly larvae (*Hermetia illucens*) fed four different diets had similar amino acid patterns (Spranghers *et al.*, 2017), as did tobacco hornworm larvae (*Manduca sexta*) fed two different diets (Landry *et al.*, 1986). These data also suggest that for holometabolous species the relative abundance of amino acids is fixed within a specific life stage. Body parts, such as wings, legs or mandibles, have specific physical requirements in order to function properly; therefore, differences in amino acid composition between insects fed different diets are unlikely. Amino acid patterns between different life stages of holometabolous insects would however be expected to differ because larvae and adults are morphologically so dissimilar. This difference was seen for yellow mealworms (*Tenebrio molitor*) where the amino acid pattern for larvae was different than for the harder bodied adults (Finke, 2002). The adults contain more glycine (52%) and tryptophan (37%) than mealworm larvae while larvae contained more leucine (29%), phenylalanine (35%) and tyrosine (120%) than adults (Finke, 2002). The amino acid

Table 16.1. Amino acid patterns (mg/g crude protein) of house crickets (*Acheta domesticus*) as reported by various authors.[1]

Amino acid	Finke (2002) nymphs	Finke (2007) nymphs	Finke (2015) nymphs	Finke (2002) adults	Finke (2007) adults	Yi *et al.* (2013) adults	Bosch *et al.* (2014) adults	Nakagaki *et al.* (1987)
Alanine	89.0	101.1	90.9	87.8	76.9	81.0	83.6	95.0
Arginine	61.0	70.9	78.8	61.0	57.3	65.0	56.8	60.0
Aspartic acid	70.8	79.4	82.4	83.9	84.9	76.0	76.4	88.0
Cystine	8.4	9.1	9.8	8.3	9.8	NR	11.2	NR
Glutamic acid	103.9	117.1	114.5	104.9	104.4	110.0	105.9	117.0
Glycine	52.6	60.6	53.5	50.7	45.3	51.0	51.3	59.0
Histidine	22.1	25.7	22.1	23.4	22.7	21.0	33.9	26.0
Isoleucine	42.9	40.6	40.3	45.9	36.4	36.0	40.0	42.0
Leucine	95.5	72.6	70.9	100.0	66.7	66.0	66.4	73.0
Lysine	53.9	62.3	57.9	53.7	51.1	53.0	58.0	56.0
Methionine	13.0	15.4	16.6	14.6	19.6	NR	15.8	15.0
Phenylalanine	27.9	32.0	35.6	31.7	30.2	NR	31.8	22.0
Proline	55.2	61.1	59.8	56.1	54.2	54.0	54.4	62.0
Serine	41.6	42.9	40.4	49.8	52.0	38.0	37.1	49.0
Threonine	35.7	38.9	37.6	36.1	31.1	35.0	36.2	35.0
Tryptophan	5.2	6.3	8.7	6.3	7.6	9.0	NR	6.0
Tyrosine	55.2	62.9	64.8	48.8	44.0	NR	62.1	41.0
Valine	49.4	60.0	59.6	52.2	48.4	55.0	57.2	60.0

[1] NR = not reported.

patterns of house fly (*Musca domestica*) larvae meal and house fly pupae meal from larvae raised on the same diet also differed (Pieterse and Pretorius, 2014). In this case differences were seen for the amino acids alanine, arginine, aspartic acid, methionine, serine, and tyrosine.

The first limiting amino acid is the one at the lowest concentration relative to the requirement of the animal. When used in diets the first limiting amino acid in insect protein depends on both the species and life stage used and the animal being fed. Table 16.2 shows the amino acid pattern of four commonly raised insect species and compares them to the amino acid requirements of growing broiler chicks (poultry), catfish, trout, swine, and adult humans and young children (NRC, 1994, 2011, 2012; WHO/FAO/UNU, 2007). These data suggest that methionine and cystine are generally the first limiting amino acids for most insect species when fed to production animals or humans. The few exceptions appear to be for house crickets where threonine and/or tryptophan is first limiting for catfish and swine. These calculations are supported by data from animal feeding trials. Methionine and arginine are the first limiting amino acids for broiler chicks when Mormon cricket meal (*Anabrus simplex*) is the sole source of dietary protein in purified diets (Finke *et al.*, 1985) or when house cricket meal is the source of protein in a corn based diet (Nakagaki *et al.*, 1987). Similarly, methionine is

Table 16.2. Amino acid patterns (mg/g crude protein) of four commonly raised insects and amino acids scores and first limiting amino acid for various species.[1]

Amino acid	*Acheta domesticus*	*Tenebrio molitor* larvae	*Zophbas mori* larvae	*Hermetia illucens* larvae/prepupae
Alanine	87.8	80.2	72.7	62.7
Arginine	65.7	60.0	57.4	52.8
Aspartic acid	79.1	81.0	83.1	88.3
Glutamic acid	109.2	112.1	127.0	103.7
Glycine	52.3	53.1	48.7	55.0
Histidine	22.8	30.2	31.1	32.4
Isoleucine	40.3	46.1	46.9	43.3
Leucine	78.6	84.9	80.4	69.9
Lysine	55.3	55.4	54.6	59.1
Methionine	15.8	13.3	12.2	18.8
Methionine+cystine	24.9	23.3	21.8	24.4
Phenylalanine	31.5	35.2	37.2	41.5
Phenylalanine+tyrosine	87.5	102.6	108.4	112.3
Proline	56.7	68.1	55.9	55.4
Serine	44.1	47.2	44.1	38.3
Threonine	35.7	40.3	39.9	39.0
Tryptophan	7.2	10.5	11.4	15.1
Valine	54.1	62.9	60.5	63.8
Amino acid score/first limiting amino acid for				
Poultry	64/met+cys	60/met+cys	56/met+cys	62/met+cys
Catfish	55/thr+try	54/met+cys	50/met+cys	56/met+cys
Trout	52/met+cys	49/met+cys	46/met+cys	51/met+cys
Swine	59/try	56/met+cys	52/met+cys	58/met+cys
Children	96/met+cys	90/met+cys	84/met+cys	94/met+cys
Adults	113/met+cys	106/met+cys	99/met+cys	111/met+cys

[1] The amino acid data shown for insects are an average of data from published sources (Finke, 2002, 2007, 2015b; Spranghers *et al.*, 2017; Yi *et al.*, 2013).

the first limiting amino acid in a corn-soy-meat meal diet for growing chickens when dried maggot meal substitutes for meat meal (Bamgbose, 1999). Methionine is also the first limiting amino acid when adult Mormon cricket meal (Finke *et al.*, 1987), yellow mealworm larvae meal (Goulet *et al.*, 1978), and house fly larvae meal are fed to growing rats (Onifade *et al.*, 2001). The high amino acid scores of all four insect species for both children and adults suggest insects are an excellent source of high quality protein for humans.

Besides concentrations, amino acid availability is critical in determining insect protein quality. A summary of studies on insect meals fed to poultry is shown in Table 16.3. These data show that the amino acids from insect meals are readily available with values equal

Table 16.3. Amino acid digestibilities of various insect meals when fed to poultry.

Insect species and life stage	Amino acid digestibility			Study
	Low	High	Mean	
Teleogryllus mitratus adults	85 (cys)	96 (ala)	93	Wang *et al.* (2005)
Acrida cinerea adults	85 (cys)	99 (thr)	94	Wang *et al.* (2007)
Musca domestica larvae	92 (ile)	98 (lys)	95	Hwangbo *et al.* (2009)
M. domestica larvae	83 (gly)	96 (tyr)	91	Pieterse and Pretorius (2014)
M. domestica pupae	86 (ala)	100 (ser+asp)	95	Pieterse and Pretorius (2014)
Tenebrio molitor larvae	80 (met)	93 (ala)	86	DeMarco *et al.* (2015)
Hermetia illucens pupae	42 (met)	89 (pro)	68	DeMarco *et al.* (2015)

to or better than those from more conventional protein sources such as soybean meal or fish meal. The only exception is black solder fly larvae meal which has lower amino acid digestibilities. The apparent protein digestibility of black soldier prepupae meal when fed to turbot is also low (63.1%) (Kroeckel *et al.*, 2012) and similar to the average amino acid digestibility value in poultry (DeMarco *et al.*, 2015). This might be because some protein is bound to the mineralized exoskeleton in black soldier fly larvae and prepupae. In black soldier fly prepupae the mineral content is influenced by diet (Spranghers *et al.*, 2017). Therefore the amino acid and protein digestibilities observed for poultry and fish may be higher for black soldier fly larvae and prepupae raised on diets with a lower mineral content. DeMarco *et al.* (2015) reported an apparent protein digestibility of 60% for yellow mealworm and 51% for black soldier fly larvae meal fed to broilers while Bovera *et al.* (2016) reported an ileal protein digestibility of 80% for yellow mealworms when fed to broilers.

Slightly higher (65.5 to 68.7) *in vitro* protein digestibilities were reported for six commercial samples of yellow mealworm larvae meal and black soldier fly larvae meal (drying conditions were unknown) based on a monogastric digestibility model (Marono *et al.*, 2015). Using a porcine model, low *in vitro* protein digestibilities were obtained for raw house crickets (65.5%) and yellow mealworms (72.5%) although values were similar to those obtained for beef muscle and chicken breast (Poelaert *et al.*, 2017). These low initial values declined after oven cooking or autoclaving. Higher *in vitro* protein digestibilities (76.4 to 91.5%) are reported for 10 species of freeze-dried insects using a method based on the canine digestive system including 91.7% for house crickets and 91.3% for yellow mealworms (Bosch *et al.*, 2014). This study obtained a higher *in vitro* protein digestibility for black soldier fly larvae compared to black soldier fly pupae (89.7 vs 77.7%). In a later study using the same method Bosch *et al.* (2016) reported high *in vitro* protein (87.7 to 93.3%) and amino acid (86.6 to 100.0%) digestibilities for three species of freeze-dried insect larvae. Kinyuru *et al.* (2010a) also obtained high *in vitro* protein digestibilites (82.3 to 90.5%) for freshly caught termites (*Macrotermes subhylanus*) and grasshoppers (*Ruspolia differens*). These values declined slightly after solar drying at approximately 30 °C. These data suggest that until *in vitro* digestibility values can be correlated to those obtained *in vivo*, *in vitro* protein digestibility data should be viewed with caution. The digestibility model and the methods used to process and dry samples likely affect *in vitro* protein digestibility values.

There is little amino acid availability data for insects fed to species other than poultry. However, Table 16.4 summarizes the results of a number of protein digestibility studies in rats. The lower protein digestibility of freeze-dried yellow mealworms (Goulet *et al.*, 1978), dried termites (*Macrotermes falciger*) (Phelps *et al.*, 1975) and silkworm (*Bombyx mori*) pupae (Rao, 1994) is in part because the values shown in Table 16.4 are for apparent not true digestibility. True digestibility measurements are higher than apparent digestibility values because the true digestibility method corrects for endogenous protein/nitrogen loss. The true protein digestibility of dried honey bees (*Apis melifera*) (71.5%) was higher than their apparent protein digestibility (62.0%) suggesting the true protein digestibility of the three previously mentioned species is higher (Ozimek *et al.*, 1985). Additionally feed consumption for rats fed either dried termites (Phelps *et al.*, 1975) or silkworm pupae (Rao, 1994) was much lower than for rats fed casein which would also adversely affect protein digestibility values. Other studies using termite and silkworm meal obtained much higher protein digestibility values. This suggests these low values are not representative. The milk protein casein is typically used as a standard in protein digestibility and protein quality studies. In all of these studies casein digestibility was determined, and the digestibility of insect protein was in most cases equal to or slightly lower than that for casein. The wide range of protein digestibilities of insects when fed to rats is likely a function of both the insect species and life stage and how the material was processed prior to use in feeding trials.

Table 16.4. True protein digestibility of various insect proteins when fed to rats.

Insect species and life stage	Order	Insect protein digestibility (%)	Method of processing	Protein digestibility of casein (%)	Study
Tenebrio molitor larvae meal[1]	Coleoptera	75.1	freeze-dried	88.4	Goulet *et al.* (1978)
T. molitor larvae meal + methionine[1]	Coleoptera	78.9	freeze-dried	88.4	Goulet *et al.* (1978)
Rhynchophorus phoenicis larvae	Coleoptera	91.8	raw	92.6	Ekpo (2011)
R. phoenicis larvae	Coleoptera	92.6	boiled	92.6	Ekpo (2011)
R. phoenicis larvae	Coleoptera	92.1	fried	92.6	Ekpo (2011)
Oryctes rhinoceros larvae	Coleoptera	89.3	raw	92.6	Ekpo (2011)
O. rhinoceros larvae	Coleoptera	90.1	boiled	92.6	Ekpo (2011)
O. rhinoceros larvae	Coleoptera	89.7	fried	92.6	Ekpo (2011)
Musca domestica larvae meal[1]	Diptera	90.6	dried at 65 °C	93.3	Iñiguez-Covarrubias *et al.* (1994)
M. domestica larvae meal	Diptera	89.0-91.0	unknown	89.0-91.0	Kouamé *et al.* (2011)

>>>

Table 16.4. Continued.

Insect species and life stage	Order	Insect protein digestibility (%)	Method of processing	Protein digestibility of casein (%)	Study
Dried *Apis mellifera* adults	Hymenoptera	71.5	dried at 70 °C	96.8	Ozimek *et al.* (1985)
A. mellifera adult concentrate	Hymenoptera	94.3	alkaline extraction of bees dried at 70 °C	96.8	Ozimek *et al.* (1985)
Macrotermes falciger meal[1]	Isoptera	45.3-50.7	dewinged, dried and lightly fried	84.2	Phelps *et al.* (1975)
Macrotermes bellicosus adults	Isoptera	90.2	dewinged, raw	92.6	Ekpo (2011)
M. bellicosus adults	Isoptera	91.0	dewinged, boiled	92.6	Ekpo (2011)
M. bellicosus adults	Isoptera	90.9	dewinged, fried	92.6	Ekpo (2011)
Bombyx mori chrysalid meal	Lepidoptera	88.9	dried, conditions unknown	99.8	Lin *et al.* (1983)
B. mori chrysalid meal	Lepidoptera	89.4-89.8	dried, conditions unknown, water extracted	99.8	Lin *et al.* (1983)
B. mori pupae meal[1]	Lepidoptera	67.0	unknown	84.0	Rao (1994)
Samia ricinii pupae meal	Lepidoptera	87.0	dried at 60-70 °C and defatted	92.0	Longvah *et al.* (2011)
S. ricinii pupae meal	Lepidoptera	87.0	dried at 60-70 °C and defatted	92.0	Longvah *et al.* (2011)
Imbrasia belina larvae meal	Lepidoptera	85.5	gastrointestinal contents removed then dried; drying conditions unknown	98.0	Drewer and Wehmeyer (1982)
I. belina larvae meal	Lepidoptera	83.9	gastrointestinal contents removed then dried; drying conditions unknown then soaked in water and canned	98.0	Drewer and Wehmeyer (1982)
I. belina larvae	Lepidoptera	86.0	raw	92.6	Ekpo (2011)
I. belina larvae	Lepidoptera	88.4	boiled	92.6	Ekpo (2011)
I. belina larvae	Lepidoptera	87.8	fried	92.6	Ekpo (2011)
Clanis bilineata larvae meal	Lepidoptera	95.8	heads removed and dried at 60 °C	96.1	Xia *et al.* (2012)

[1] Apparent protein digestibility.

These data suggest the protein digestibility of insects is generally high and that the variation between studies is likely a result of several factors: (1) insects that have a larger proportion of their amino acids in cuticular proteins complexed with chitin or which are highly sclerotized would likely have a lower protein digestibility than those that do not (Finke, 2007; Ozimek *et al.*, 1985); (2) protein digestibility of some of the insects shown in Table 16.4 had certain body parts (typically wings, or heads) removed before being processed and fed to rats. This likely increases the variability between studies; (3) protein digestibility of insect meals may be affected by both the processing methods and the time and temperature used for drying the material (Dreyer and Wehmeyer, 1982; Pieterse and Pretorius, 2014).

16.2.2 Fats and fatty acids

The fat content of insects varies between 10 and 70% on a dry matter basis (Barker *et al.*, 1998; Bukkens, 1997; Finke, 2002, 2013; Lease and Wolf, 2011; Oonincx and Dierenfeld, 2012; Yang and Siriamornpun, 2006). Insect fat is typically estimated by crude fat determination, which is based on the total weight of all fat-soluble molecules. These not only include glycerides but also waxes, sterols, fat soluble vitamins and some other compounds.

Fat is composed of fatty acids. Two or three fatty acids are coupled to glycerol and form diglycerides and triglycerides, respectively. Fatty acids are divided into saturated fatty acids, mono-unsaturated fatty acids and poly-unsaturated fatty acids based on their degree of saturation. Poly-unsaturated fatty acids are subdivided into omega 3, 6 or 9 poly-unsaturated fatty acids based on the relative position of the first double bond.

The fatty acid composition of insects depends on the species and life stage as well as environmental factors such as diet, temperature and light. In most species males have smaller fat reserves than females (Lease and Wolf, 2011; Nestel *et al.*, 2005; Zhou *et al.*, 1995). Commercially produced insects seem to have a higher fat content than those collected from the wild (Finke, 2002, 2013; Lease and Wolf, 2011; Oonincx and Dierenfeld, 2012; Yang and Siriamornpun, 2006). This might be due to decreased energy expenditure in captivity, ready access to diets with a high energy content or a combination of both.

In general wild caught insects contain relatively high amounts of linoleic acid (18:2 n-6) and linolenic acid (18:3 n-3) (Fast, 1970; Thompson, 1973). Commercially raised insects also contain high levels of linoleic acid but much lower levels of linolenic acid than wild caught insects because their diet often contains large amounts of grains and grain by-products which have low levels of linolenic acid (Dreassi *et al.*, 2017; Finke, 2002, 2013; Jones *et al.*, 1972; Oonincx *et al.*, 2015a,b; Paul *et al.*, 2017). An unusual fatty acid profile can be found in larvae of the black soldier fly which contain high levels of lauric acid (C12:0) irrespective of the diet (Finke, 2013; Oonincx *et al.*, 2015a; Spranghers *et al.*, 2017; St-Hilaire *et al.*, 2007; Surendra *et al.*, 2016). Similar to vertebrates, insects can synthesize saturated and monounsaturated fatty acids (Beenakkers *et al.*, 1985, Tietz and Stern, 1969). Both linoleic acid and linolenic acid are essential nutrients because most species are unable to synthesize them. For most insects this also appears to be true, although the American cockroach (*Periplaneta americana*) and the house cricket appear to be able to synthesize these fatty acids (Borgeson and Blomquist, 1993; Borgeson *et al.*, 1991).

In three species of insects with an aquatic larval stage, the concentration of linolenic acid increased, while both arachidonic acid and eicosapentaenoic acid decreased in the adults after the insect left the water (Hanson *et al.*, 1985). This is likely a result of differences in the fatty acid content of their diets. Several studies that report the fatty acid content of collected insect species find higher levels of eicosapentaenoic acid (20:5 n-3) in aquatic species than in terrestrial species, which rarely contain these fatty acids (Fontaneto *et al.*, 2011; Ghioni *et al.*, 1996; Sushchik *et al.*, 2003, 2013). In aquatic ecosystems certain microalgae produce eicosapentaenoic acid, which is transferred to higher trophic levels including insects (Gladyshev *et al.*, 2011). Eicosapentaenoic acid is not produced by higher plants in terrestrial ecosystems, therefore herbivorous terrestrial insects would be less likely to contain this fatty acid.

16.2.3 Carbohydrates

In general, carbohydrates as measured by nitrogen free extract, are present in small amounts in insects (Barker *et al.*, 1998; Finke, 2002, 2013, Oonincx and Dierenfeld, 2011; Pennino *et al.*, 1991). The carbohydrate content of yellow mealworm larvae can vary between 1 and 7% (Ramos-Elorduy *et al.*, 2002). However these differences are likely the result of the food remaining in the gastro-intestinal tract.

16.2.4 Fibre and chitin

Insects contain significant amounts of fibre as measured by crude fibre, acid detergent fibre or neutral detergent fibre (Barker *et al.*, 1998; Finke, 2002, 2007, 2013; Lease and Wolf, 2010; Madibela *et al.*, 2007; Marono *et al.*, 2015; Oonincx and Dierenfeld, 2011; Pennino *et al.*, 1991). Oven cooking or autoclaving of raw house crickets or yellow mealworms increases neutral detergent fibre but not acid detergent fibre (Poelaert *et al.*, 2017). The components in these fibre fractions are not well known, although in whole insects they include sclerotized proteins and proteins, minerals and other compounds bound to chitin (Finke, 2007; Kramer *et al.*, 1995; Madibela *et al.*, 2007; Marono *et al.*, 2015). Chitin is a polymer of N-acetyl-β-D-glucosamine units, which provides rigidity to the insect's exoskeleton.

The cuticle, the outmost part of the insect, is a matrix of proteins, lipids, minerals and other compounds (Kramer *et al.*, 1995). Chitin is only present in the insect's procuticle, the two innermost layers of the cuticle (Moussian, 2010), and therefore likely makes up a small amount (by weight) of an insect. Quantitative data on the chitin content of whole insects is limited and comparisons between studies are difficult due to the use of a variety of analytical methods (Cauchie, 2002; Finke, 2007; Kaspari, 1991; Lease and Wolf, 2010). Traditional fibre assays (crude, neutral detergent or acid detergent) are sometimes used to approximate chitin even though these methods overestimate true chitin levels (Finke, 2007, 2013; Madibela *et al.*, 2007; Marono *et al.*, 2015). Several animal species including humans have chitinases, which can digest chitin (Fujimoto *et al.*, 2002; Paoletti *et al.*, 2007; Whitaker *et al.*, 2004). However to what extent these enzymes are effective *in vivo* is currently unknown.

The predominant compound in the cuticle of most insects is protein not chitin (Kramer *et al.*, 1995). 'Harder bodied' insects such as adult yellow mealworm beetles contain higher fibre levels than softer bodied insects such as the larvae of yellow mealworms, silkworms, and house cricket nymphs. This is due to higher levels of amino acids in the acid detergent fibre fraction (Finke, 2007). The amino acid pattern of whole insects differs from the pattern in the acid detergent fibre fraction, and it also differs between species (Finke, 2007). In particular concentrations of valine, histidine, and glycine are higher in the acid detergent fibre fraction than in the whole insect. These amino acids likely contribute to the relative strength, stiffness, elasticity and other physical properties of sclerotized and cuticular proteins. Insects with 'harder' cuticles do not necessarily contain more chitin but contain more sclerotized protein and proteins cross-linked to chitin than softer bodied insects.

16.2.5 Minerals

Essential minerals can be classified as macro-minerals (calcium, phosphorus, magnesium, sodium, potassium and chloride) and micro or trace minerals (iron, zinc, copper, manganese, iodine and selenium). This classification is based on the amount needed to meet dietary requirements. The requirements for macro-minerals for most species are typically measured in g per kg while micro-mineral requirements are typically measured in mg per kg.

Insects generally contain low levels of calcium because they lack a mineralized skeleton. Calcium levels are typically less than 0.3% dry matter although higher levels have been reported for stoneflies (Plecoptera) and some other species (Barker *et al.*, 1998; Finke, 2002, 2013; Oonincx and Dierenfeld, 2012; Oonincx and Van der Poel, 2011; Punzo, 2003; Studier and Sevick, 1992). The relatively high levels of calcium occasionally reported for feeder crickets is likely due to dietary calcium remaining in the gut (Barker *et al.*, 1998; Hatt *et al.*, 2003; Punzo, 2003). While the exoskeleton of most insects is primarily composed of protein and chitin, black solder fly larvae (Finke, 2013; Spranghers *et al.*, 2017) and face fly larvae (*Musca autumnalis*) (Dashefsky *et al.*, 1976; Koo *et al.*, 1980; Roseland *et al.*, 1985) have a mineralized exoskeleton in which calcium and other minerals are incorporated into the cuticle. Therefore they can contain high levels of calcium.

Most insects contain more phosphorus than calcium with the exception of species with a mineralized exoskeleton (i.e. face fly and black soldier fly larvae). In plants significant amounts of phosphorus are present as phytate and are hence unavailable for digestion. In contrast the phosphorus in insects is likely to be a good source of bioavailable phosphorus as was reported when face fly pupa were fed to poultry (Dashefsky *et al.*, 1976).

Magnesium levels in insects seem sufficient to meet the dietary requirements of most species (Barker *et al.*, 1998; Finke, 2002, 2013; Oonincx and Dierenfeld, 2012; Studier and Sevick, 1992). Black soldier fly larvae contain 3 to 10 times more magnesium than most other insects (Finke, 2013; Spranghers *et al.*, 2017). This is likely the result of their mineralized exoskeleton in which both calcium and magnesium probably form a complex with chitin (Diener *et al.*, 2015).

Only a few studies report the sodium and potassium content of insects (Finke, 2002, 2013; Oonincx and Dierenfeld, 2012; Oonincx and Van der Poel, 2011; Reichle *et al.*, 1969; Studier and Sevick, 1992). Insects typically contain more potassium than sodium whereas chloride levels are intermediate. Most insect species likely contain adequate amounts of these three minerals to meet the dietary requirements of most animal species.

Most insects contain relatively high levels of the trace minerals iron, zinc, copper, manganese and selenium, which would meet the dietary requirements for most animals (Barker *et al.*, 1998; Finke, 2002, 2013; Oonincx and Dierenfeld, 2012; Oonincx and Van der Poel, 2011; Punzo, 2003; Studier and Sevick, 1992). However these studies suggest possible species-specific differences for certain trace minerals. For example adult fruit flies (*Drosophila melanogaster*) (Barker *et al.*, 1998; Oonincx and Dierenfeld, 2012) and house flies (Finke, 2013) contain relatively high levels of iron (125 to 454 mg/kg dry matter) compared to most other insect species. The contents of the gastrointestinal tract can be a significant percentage of the total weight of an insect (Finke, 2003); and, therefore the diet can also significantly influence the trace mineral content of fully fed insects.

Mineral availability can be inhibited by so-called anti-nutritional factors such as phytate, oxalate and tannins. For example, caterpillars of the Pallid emperor moth (*Cirina forda*) contain both phytic acid (10 mg/kg) and oxalate (40 mg/kg), but no tannins (Omotoso, 2006). Tannins (10.4 mg/kg), phytic acid (13.5 mg/kg) and oxalates (0.8 mg/kg) were found in larvae of the African palm weevil (*Rhynchophorus phoenicis*) (Ekpo, 2010). The source of these anti-nutritional factors is likely from the food present in the gastrointestinal tract. Evidence of accumulation of these substances in insects is lacking. The concentrations found seem to be relatively low; in seeds and cereal grains concentrations up to 7% dry matter have been found (Zhou and Erdman, 1995).

16.2.6 Vitamins and carotenoids

Vitamin A

Vitamin A is actually a group of compounds composed of both retinoids and carotenoids. Similar to most vertebrates, insects obtain retinoids via the cleavage of various carotenoids (Von Lintig, 2012). However in vertebrates carotenoid cleavage primarily takes place in the intestine, and the resulting retinoids are stored in the liver (Olson, 1989), whereas insects convert carotenoids to retinoids only in the compound eye (Von Lintig, 2012). In fruit flies (*D. melanogaster*) and honey bees (*Apis mellifera*) retinoids are only found in the compound eye while other parts of the adult insect do not contain retinoids nor do fruit fly larvae (Giovannucci and Stephenson, 1999; Goldsmith and Warner, 1964). This explains why adult insects contain very low levels of vitamin A/retinoids (Barker *et al.*, 1998; Finke, 2002, 2013; Oonincx and Dierenfeld, 2012; Oonincx and Van der Poel, 2011; Pennino *et al.*, 1991) and that holometabolous insect larvae, which lack compound eyes, do not contain preformed retinoids (retinal or 3-hydroxyretinal).

In many animal species certain carotenoids can be converted to vitamin A/retinal (Olson, 1989). High levels of carotenoids, including those that can be converted to retinal, are found in various wild insect species whereas commercially produced insects contain far lower quantities (Eeva *et al.*, 2010; Finke, 2002, 2013, 2015a; Isaksson and Andersson, 2007; Oonincx and Van der Poel, 2011; Seki *et al.*, 1998; Ssepuuya *et al.*, 2017).

Vitamin D

Data on vitamin D in insects are limited. Most commercial insects contain low levels of vitamin D (typically <400 IU/kg dry matter; Finke, 2002, 2013; Oonincx *et al.*, 2010). Data from three wild caught species were highly variable with values ranging from undetectable

(<100 IU vitamin D3/kg dry matter) to 1,288 IU vitamin D3/kg dry matter (Finke, 2015b). More research is needed to better understand the vitamin D content of insects and how it might be affected by diet and environmental conditions.

Vitamin E

Vitamin E content of insects, including house crickets and yellow mealworm larvae, varies. Values for house crickets ranged from 7 to 195 IU vitamin E/kg dry matter, while yellow mealworm larvae ranged from <22 to 116 IU vitamin E/kg dry matter (Barker *et al.*, 1998; Finke, 2002, 2015a; Hatt *et al.*, 2003; Pennino *et al.*, 1991). This variation is likely due to dietary differences resulting in different amounts being incorporated in the insect's tissues and dietary vitamin E in the insect's gastrointestinal tract. While there is significant variation between species the vitamin E content of most commercially raised insects is low (<37 IU/kg dry matter) (Barker *et al.*, 1998; Finke, 2002, 2013; Oonincx and Dierenfeld, 2012; Punzo, 2003). There is little data on the vitamin E content of wild insects, although they appear to contain higher levels than those found in commercially raised insects (Barker *et al.*, 1998; Cerda *et al.*, 2001; Finke, 2015b; M.D. Finke, unpublished data; Fogang *et al.*, 2017; Pennino *et al.*, 1991).

B-vitamins

There is a significant amount of information regarding the B-vitamin content of commercially raised insects (Finke, 2002, 2013, 2015a; Jones *et al.*, 1972). However, the B-vitamin content of unprocessed wild-caught insects is limited (Finke, 2015b; Igwe *et al.*, 2011; Kinyuru *et al.*, 2010a,b) whereas some information on the B-vitamin content of wild caught insects that are consumed by humans is available (Banjo *et al.*, 2006; Dreyer and Wehmeyer, 1982; Igwe *et al.*, 2011; Kinyuru *et al.*, 2010a; Kodondi *et al.*, 1987; Santos Oliveira *et al.*, 1976; Teffo *et al.*, 2007). Insects consumed by humans are usually dried, boiled, fried, or roasted and specific body parts are often removed. This is important as some B-vitamins are unstable and can be destroyed by exposure to heat, light or oxygen. Differences in sample preparation and analytical methods (microbiological versus chemical techniques) further complicate data comparisons.

In spite of these issues some general observation regarding B-vitamins in insects can be made. Most studies show unprocessed insects are a very good source of riboflavin (vitamin B2), niacin, pantothenic acid, pyridoxine (vitamin B6), biotin, folic acid, and cyanocobalamin (vitamin B12). Low levels for several of these vitamins occasionally reported in the literature are likely due to losses during processing and storage. So for insects processed for human consumption, the B-vitamin levels in these insects likely decreased due to processing (Kinyuru *et al.*, 2010a).

The one B-vitamin that appears to be low in many species of insects is thiamine (vitamin B_1). Many species of commercially raised or wild caught insects including house crickets, adult yellow mealworms, superworms (*Zophobas morio*), butterworms, (*Chilecomadia moorei*) turkistan roaches (*Blatta lateralis*), pallid wing grasshoppers (*Trimerotropis pallidipennis*) and rhinoceros beetles (*Oxygrylius ruginasus*) contain relatively low levels of thiamine (<3.0 mg/kg dry matter) although others (black soldier fly larvae, adult house flies, silkworms, yellow mealworm larvae, waxworms (*Galleria mellonella*) and white-lined sphinx moths

(*Hyles lineata*) contain much higher levels (5.0 to 45.0 mg thiamine/kg dry matter; Finke, 2002, 2013, 2015b). This variation is similar to that observed for 14 insect species collected in Nigeria (0.3 to 32.4 mg thiamine/kg dry mater; Banjo *et al.*, 2006) and three species collected in Angola (1.3 to 36.7 mg thiamine/kg dry matter; Santos Oliveira *et al.*, 1976). Dried and smoked Attacidae caterpillars (*Nudaurelia melanops*, *Imbrasia truncata*, *Imbrasia epimethea*) from Zaire contain low thiamine level (1.5 to 2.7 mg/kg dry product; Kodondi *et al.*, 1987) and unprocessed longhorned grasshoppers (*R. differens*) from East Africa contained no detectable thiamine (Kinyuru *et al.*, 2010b). Because thiamine is relatively unstable and most of the values reported are for insects processed using various traditional methods (i.e. body parts removed and dried, smoked or fried) it is unclear how representative these values are for raw whole insects. That said the range of values for thiamine in processed insects is similar to that observed for raw insects.

Two species of edible insects (domesticated silkworm larvae and African silkworm pupae (*Anaphe* spp.)) contain the enzyme thiaminase (Nishimune *et al.*, 2000). Unless inactivated during processing, thiaminase destroys any thiamine present. Thiamine deficiency due to the consumption of *Anaphe* pupae is suggested as a cause of a seasonal ataxia in local populations in Nigeria (Adamolekun, 1993, 1997). Most enzymes are inactivated by heat, but some thiaminase activity remains after brief exposure of the extract of *Anaphe* pupae to 100 °C suggesting this enzyme is very heat tolerant. Only after exposure to 100 °C for 15 minutes is its activity almost completely destroyed (Nishimune *et al.*, 2000). Whether other insect species contain thiaminases is currently unknown.

16.2.7 Choline

Although data are limited most research shows insects contain high levels of choline (1,570 to 7,258 mg/kg dry matter; Finke, 2002, 2013, 2015a,b; Fogang *et al.*, 2017; Noland and Baumann, 1949). Choline is a component of both lecithin and the neurotransmitter acetylcholine. It also plays an important role in one-carbon metabolism and as such can spare the need for dietary methionine (Pesti *et al.*, 1979), which usually is the first-limiting amino acid in insects. Because choline is abundant in most insects this is an important issue to consider when using insects as food or feed.

16.3 Life stage effects on nutrient composition

Fat stores are usually highest in the final larval or nymphal stage prior to metamorphosis (Fast, 1970). This pattern however depends to some extent on whether the species is holometabolous or hemimetabolous.

Larvae of holometabolous species have a higher fat content than adults (Church and Robertson, 1966; Cookman *et al.*, 1984; Finke, 2002; Lease and Wolf, 2011). The fat content of the larvae of fruit flies (Church and Robertson, 1966), house flies (Pearincott, 1960), and yellow mealworms (Finkel, 1948) increases as they develop but drops significantly in the adult because fat is used as an energy source during pupation. A similar decrease is observed in insects undergoing diapause or hibernation (Ali and Ewiess, 1977; Downer and Matthews, 1976).

Similarly to holometabolous insects, nymphs of hemimetabolous insects slowly increase their fat content as they develop as was shown for house crickets (Lipsitz and McFarlane, 1971). However unlike holometabolous insects, the fat content of newly emerged hemimetabolous adults is similar to that of late stage nymphs and then slowly declines as they age (Lipsitz and McFarlane, 1971).

The taurine content of fruit flies (Massie *et al.*, 1989), bertha armyworms (*Mamestra configurata*) (Bodnaryk, 1981) and yellow mealworms (Finke, 2002), all holometabolous species, increases after metamorphosis. In the hemimetabolous house cricket, the taurine content of adults (4,578 mg/kg dry matter) was higher than in nymphs (3,537 mg/kg dry matter; Finke, 2002).

16.4 Dietary effects on insect nutrient composition

Diet can affect the composition of fats/fatty acids, minerals, vitamins and carotenoids in insects. Insect fat content (both the amount and the fatty acid composition) is highly variable and affected by both life stage and diet. This subsequently affects the levels of other nutrients most notably protein and moisture. If insect fat content increases, and the amount of both moisture and protein stay the same this results in a decrease in their concentrations as was shown for house fly larvae (Pearincott, 1960).

As stated previously, both the crude fat content and the fatty acid composition of insects can be altered by the diet. The extent to which this changes is highly variable depending on the species. Yellow mealworms fed four different diets had fat contents ranging from 23 to 29% dry matter, while Argentinean cockroaches (*Blaptica dubia*) fed the same diets had fat content ranging from 16 to 40% dry matter (Oonincx *et al.*, 2015a). The fat content of velvetbean caterpillars (*Anticarsia gennatalis*) is 40 to 65% higher when fed an artificial diet than when fed leaves from three plant species (Cookman *et al.*, 1984). This increase in fat content carried through to the adult stage; moths from larvae fed the artificial diet contained 71 to 105% more fat than those fed plant leaves (Cookman *et al.*, 1984). Similarly, higher fat contents were found in dried insect meals from larvae of the tobacco hornworm (20.7 vs 17.3% dry matter) and the fall armyworm (*Spodoptera frugiperda*) (20.6 vs 11.7% dry matter) fed artificial diets compared to those fed leaves or grasses (Landry *et al.*, 1986).

Besides the variability in total fat content, large differences are seen in the fatty acid composition of insects when different diets are fed, which is similar to what is seen in most vertebrates. Numerous studies indicate that in both larval and adult insects the fatty acid composition partially reflects the fatty acid composition of the diet (Cookman *et al.*, 1984; Dreassi *et al.*, 2017; Finke, 2015a; Madariaga *et al.*, 1971; Schaefer, 1968; Spranghers *et al.*, 2017; St-Hilaire *et al.*, 2007; Van Broekhoven *et al.*, 2015). However the fatty acid composition of the insect does not directly match that of the diet due to selective accumulation, catabolism and/or fatty acid synthesis (Cookman *et al.*, 1984; Madariaga *et al.*, 1971; Oonincx *et al.*, 2015a; Schaefer, 1968).

The omega-3 fatty acid content of house fly larvae (Hussein *et al.*, 2017), black soldier fly larvae (St-Hilaire *et al.*, 2007), Jamaican field crickets (*Gryllus assimilis*) (Starčević *et al.*,

2017), house crickets, yellow mealworms, superworms, and waxworms (Finke, 2015b) can be increased by providing a diet enriched with these fatty acids. House fly larvae, black soldier fly larvae, and Jamaican field crickets fed diets containing fish by-products or fish oil accumulate both eicosapentaenoic acid and docosahexaenoic acid. In contrast house crickets, superworms and waxworms fed diets containing fish oil accumulate eicosapentaenoic acid but not docosahexaenoic acid even though both were present in the diets.

In addition to diet having an effect on insect fat and fatty acid composition, insect mineral composition can to some extent be altered by diet. The calcium content of black soldier fly prepupae fed four different diets ranged from 1.23 to 66.15 g/kg dry matter (Spranghers *et al.*, 2017). Iron, zinc and manganese also varied while phosphorus and potassium were similar between larvae fed different diets. The mineral content of black soldier fly larvae fed three different diets also varied although in this case since larvae were analysed some of the effects are likely a result of the diet remaining in the gastrointestinal tract (Tschirner and Simon, 2015). The mineralized exoskeleton of black soldier fly larvae might be in part the reason for these extreme variations. It is unclear if other insect species would show these variations in mineral content when fed different diets. Zinc, which is both a required nutrient and a heavy metal accumulates in black soldier fly larvae, prepupae and adults when zinc is added to the larval diet (Diener *et al.*, 2015). Zinc also accumulated in both the larval and prepupal exuvia.

Heavy metal accumulation has been studied in a variety of insect species. Black solder fly larvae, prepupae, and adults accumulate both lead and cadmium when these heavy metals are present in the larval diet (Diener *et al.*, 2015; Purschke *et al.*, in press; Van der Fels-Klerx *et al.*, 2016). In larvae this is due in part, to the residual food in the gut but these metals also accumulate in the body. Arsenic and chromium do not accumulate in black soldier flies but cadmium seems to accumulate in several dipteran species (Gao *et al.*, 2017; Kazimírová and Ortel, 2000; Maroni and Watson, 1985; Purschke *et al.*, in press; Van der Fels-Klerx *et al.*, 2016; Wu *et al.*, 2006). In contrast yellow mealworm larvae accumulate arsenic but not lead or cadmium (Van der Fels-Klerx *et al.*, 2016). Wild caught grasshoppers accumulate both lead and cadmium from their environment (Devkota, 2000; Hunter *et al.*, 1987; Zhang *et al.*, 2012). Grasshopper lead and cadmium levels are highly correlated with levels in both plants and soil in the area they are collected. Also the orthopteran *G. assimilis* accumulates cadmium from the diet, whereas their zinc levels remain constant at different dietary levels (Bednarska *et al.*, 2015). It seems that accumulation and excretion patterns show similarities at the order level.

There are little data regarding the effect of diet on the vitamin content in insects. Supplementing the diets of house crickets, yellow mealworms, superworms, and waxworms with high levels of vitamin E during growth resulted in high levels in the insects (116-440 IU/kg dry matter) (Finke, 2015a). Because the insects were fasted prior to analysis these elevated levels are mainly due to incorporation of vitamin E into the tissue of the insect. This is similar to swine where elevated levels of dietary vitamin E increase the vitamin E concentration of various tissues (Asghar *et al.*, 1991).

Carotenoid content can also be increased via the insect's diet. Feeding a diet containing β-carotene to species that use retinal as their chromophore such as house crickets, yellow mealworm larvae, and superworm larvae results in β-carotene accumulation in their tissues

(Finke, 2015a). Providing a diet containing β-carotene to species that use 3-hydroxyretinal as their chromophore such as Diptera and Lepidoptera does not result in detectable levels of β-carotene but elevated levels of zeaxanthin as was shown for blow flies (*Calliphora*) (Vogt and Kirschfeld, 1984), fruit flies (Giovannucci and Stephenson, 1999), and waxworm larvae (Finke, 2015b). Fruit fly larvae accumulate carotenoids and convert these to retinoids during pupation when they form compound eyes (Seki *et al.*, 1998; Von Lintig, 2012).

16.5 Environmental effects on insect composition

Environmental factors, such as temperature, light, and humidity, can affect growth, development and chemical composition. These effects are all not well studied, but mostly affect growth rate and body size (Akman Gündüz and Gülel, 2002; Ali and Ewiess, 1977; Ali *et al.*, 2011; Han *et al.*, 2008; Roe *et al.*, 1985). However, bean beetles (*Acanthoscelides obtectus*) raised at 30 °C had a similar fat but higher protein content than beetles raised at 20 °C (Sönmez and Gülel, 2008). In contrast the fat content of field crickets (*Gryllus bimaculatus*) and the relative abundance of saturated fatty acids is higher when raised at 20 vs 27 °C. The effects of photoperiod on the nutrient content of insects is likely to be indirect, acting through other processes, such as preparation for diapause.

16.6 Processing effects on insect nutrient composition

Although specific information of the effects of various processing methods on the nutrient content and availability of insect products is limited it is expected that these are similar to other human foods or animal feeds (Ssepuuya *et al.*, 2017). Heat and processing methods effects on protein/amino acid availability, the effect of traditional cooking methods on trace mineral, fatty acid and fat content and the effects of processing on B-vitamins on insect and insect products have been studied. These three subjects are discussed below.

Processing using high pressures and temperatures negatively affects both amino acid availability and protein quality (Batterham *et al.*, 1986). As such the protein quality of insect meals is also likely affected by the processing conditions especially drying temperatures. The lower average amino acid availability of house fly larvae meal (91%) compared to house fly pupae meal (95%) might be in part a result of their longer drying times (Pieterse and Pretorius, 2014). Likewise dried mopane caterpillars (*Imbrasia belina*), exposed to high temperatures during canning have a slightly reduced apparent protein digestibility compared to those that are not canned (83.9 and 85.8% respectively) (Dreyer and Wehmeyer, 1982). In contrast no effect of processing (either boiling or frying) on the protein digestibility or the biological value was seen for four species of insects when fed to rats (Ekpo, 2011).

In addition to processing insects into meals for animal feed, protein extraction from raw insects for use in human food has also been evaluated (Lin *et al.*, 1983; Ozimek *et al.*, 1985; Yi *et al.*, 2013). Dried and ground silkworm chrysalid protein was compared to an aqueous extract of ground silkworm chrysalid protein with and without pretreatment with hydrogen peroxide (Lin *et al.*, 1983). Amino acid patterns were similar for the three treatments; however, sodium hydroxide treatment prior to extraction decreased cysteine levels. Whereas protein digestibility was similar between treatments, net protein utilization was lower for the protein extract pretreated with sodium hydroxide. The lower net protein value of the material pretreated with sodium hydroxide is likely a result of the destruction of cysteine in this sample.

While many processing methods negatively affect protein/amino acid availability some can improve protein quality. Ozimek *et al.* (1985) compared the amino acid pattern, amino acid availability, and the net protein value of dried adult honey bees to an alkaline extract of dried adult honey bees when fed to rats. Alkaline extraction of honey bees increased all measures of protein quality. Amino acid patterns (mg of amino acid/g protein) of the concentrate showed much higher levels of the essential amino acids leucine (54%) and methionine (59%) compared to dried honey bees. Amino acid availability also improved by the extraction process (90%) compared to dried bees (65.4%) and was similar to casein (92.2%). Because the concentrate did not contain any measurable chitin, versus 11.1% chitin in whole dried honey bees, the improved amino acid availability of this material may partially be due to the removal of cuticular proteins complexed with chitin by the alkaline extraction process. The improved amino acid profile and increased amino acid availability of the bee protein concentrate improved net protein utilization by 46% (60.6 vs 41.5) compared to dried adult honey bees.

The nutrient content of insects prepared for human consumption may be altered by traditional cooking methods in two ways. Depending on the type of cookware used, iron, zinc, copper, and nickel can leach from the cookware and into the insect being prepared for consumption. The high values for copper, iron and zinc reported for some insects prepared using traditional methods may be due in part to the cookware used in their preparation (Payne *et al.*, 2015; Santos Oliveira *et al.*, 1976). Additionally many insects consumed by humans are either broiled, roasted, or fried prior to consumption. Broiled or roasted insects contain less fat than the raw insect as fat is lost during the cooking process. For fried insects, the fat content increases as the fat from the cooking oil is absorbed by the insect (Santos Oliveira *et al.*, 1976). This would also change the fatty acid profiles because this material contains both the fat naturally present in the insect and the fat in the cooking oil (Santos Oliveira *et al.*, 1976).

As previously mentioned, some B-vitamins are relatively unstable when exposed to heat, light, or oxygen. The vitamin content of insects processed for human consumption is lower than raw unprocessed insects, possibly due to losses during cooking and drying. A study by Kinyuru *et al.* (2010a) directly evaluated the effect of toasting (frying at approximately 150 °C for 5 minutes) and drying (solar drying at 30 °C) on the niacin, riboflavin, pyridoxine and folic acid content of termites and grasshoppers. Toasting decreased both the folic acid content of these species by 37 and 43% respectively and riboflavin by 23 and 34%, respectively. In contrast there was little loss of both pyridoxine and niacin. Drying of termites and grasshoppers also decreased both folic acid by 47 to 66% and riboflavin by 29 to 46%. There was little loss of pyridoxine during drying and a significant loss of niacin for termites (26%) but not for grasshoppers. The reason for these differences is unclear. Of note is that thiamine (vitamin B1) which is known to be unstable during heating was not measured in this study.

Dreyer and Wehmeyer (1982) showed that thiamine, riboflavin and niacin levels in mopane worm meal were affected by the canning process most likely as a result of the heat used to sterilize the cans. Thiamine, riboflavin and niacin losses due to canning were 84, 20 and

42%, respectively. In contrast to the loss of B-vitamins, β-carotene was unaffected by the canning process.

16.7 Conclusions

Insects have a high protein content, and their amino acid profiles are suitable for production animals and humans alike. This protein is highly digestible; hence, insects are generally a good source of dietary protein. In most cases methionine is the first limiting amino acid when insects are used as food or feed. Fat is the next most prominent nutrient in insects. Large variations in fat content and fatty acid composition occur due to species, life stage, diet and gender.

Carbohydrates are virtually absent in the insect body but fibre is present in significant amounts. This fibre consists primarily of chitin and sclerotized proteins, which are components of the insect exoskeleton. Whereas calcium levels in insects are lower than in animals with an endoskeleton, most other minerals are present in adequate concentrations to meet nutrient requirements of animals and humans. Mineral concentrations can be especially high in certain fly larvae. Vitamin A levels in insects are low, however carotenoids can be present in relevant concentrations depending on the provided diet. Also vitamin E concentrations are highly variable and depend on dietary levels. B vitamins are present in adequate concentrations in most insect species.

Several factors such as species and life stage influences the composition of insect species. Diet can strongly affect the concentration of most nutrients, although some, especially amino acids and minerals, tend to be less flexible. Heavy metals can accumulate in insects, however accumulation patterns vary between species. Insect processing can affect their nutritional value in many ways but the most common are destruction of vitamins and denaturation of proteins due to heat. Although large variations in insect composition are known, in general they are to be considered a prime source of many important nutrients.

References

Adamolekun, B., 1993. *Anaphe venata* entomophagy and seasonal ataxic syndrome in southwest Nigeria. Lancet 341: 629.

Adamolekun, B., McCandless, D.W. and Butterworth, R.F., 1997. Epidemic of seasonal ataxia in Nigeria following ingestion of the African silkworm *Anaphe venata*: role of thiamine deficiency? Metabolic Brain Disease 12: 251-258.

Akman Gündüz, N.E. and Gülel, A., 2002. Effect of temperature on development, sexual maturation time, food consumption and body weight of *Schistocerca gregaria* Forsk. (Orthoptera: Acrididae). Türk Zooloji Dergisi 26: 223-227.

Ali, M. and Ewiess, M.A., 1977. Photoperiodic and temperature effects on rate of development and diapause in the green stink bug, *Nezara viridula* L. (Heteroptera: Pentatomidae). Zeitschrift Fur Angewandte Entomologie 84: 256-264.

Ali, M.F., Mashaly, A.M.A., Mohammed, A.A. and El-Magd Mahmoud Mohammed, A., 2011. Effect of temperature and humidity on the biology of *Attagenus fasciatus* (Thunberg) (Coleoptera: Dermestidae). Journal of Stored Product Research 47: 25-31.

Asghar, A., Gray, J.I., Miller, E.R., Ku, P.K., Booren, A.M. and Buckley, D.J., 1991. Influence of supranutritional vitamin E supplementation in the feed on swine growth performance and deposition in different tissues. Journal of the Science of Food and Agriculture 57: 19-29.

Bamgbose, A.M., 1999. Utilization of maggot-meal in cockerel diets. Indian Journal of Animal Sciences 69: 1056-1058.

Banjo, A.D., Lawal, O.A. and Songonuga, E.A., 2006. The nutritional value of fourteen species of edible insects in southwestern Nigeria. African Journal of Biotechnology 5: 298-301.

Barker, D., Fitzpatrick, M.P. and Dierenfeld, E.S., 1998. Nutrient composition of selected whole invertebrates. Zoo Biology 17: 123-134.

Batterham, E.S., Darnell, R.E., Herbert, L.S. and Major, E.J., 1986. Effect of pressure and temperature on the availability of lysine in meat and bone meal as determined by slope-ratio assays with growing pigs, rats and chicks and by chemical techniques. British Journal of Nutrition 55: 441-453.

Bednarska, A.J., Opyd, M.O., Zurawicz, E. and Laskowski, R., 2015. Regulation of body metal concentrations: toxicokinetics of cadmium and zinc in crickets. Ecotoxicology and Environmental Safety 119: 9-14.

Beenakkers, M.T.A., Van der Horst, D.J. and Van Marrewijk, W.J.A., 1985. Insect lipids and lipoproteins, and their role in physiological processes. Progress in Lipid Research 24: 19-67.

Bodnaryk, R.P., 1981. The biosynthesis function, and fate of taurine during the metamorphosis of the noctuid moth *Mamestra configurata* Wlk. Insect Biochemistry 11: 199-205.

Borgeson, C.E. and Blomquist, G.J., 1993. Subcellular location of the Δ^{12} desaturase rules out bacteriocyte contribution to linoleate biosynthesis in the house cricket and the American cockroach. Insect Biochemistry and Molecular Biology 23: 297-302.

Borgeson, C.E., Kurtti, T.J., Munderloh, U.G. and Blomquist, G.J., 1991. Insect tissues, not microorganisms, produce linoleic-acid in the house cricket and the American cockroach. Experientia 47: 238-241.

Bosch, G., Vervoort, J.J.M. and Hendricks, W.H., 2016. *In vitro* digestibility and fermentability of selected insects for dog foods. Animal Feed Science and Technology 221: 174-184.

Bosch, G., Zhang, S., Oonincx, D.G.A.B. and Hendriks, W.H., 2014. Protein quality of insects as potential ingredients for dog and cat foods. Journal of Nutritional Science 3: e29.

Bovera, F., Loponte, R., Marono, S., Piccolo, G., Parisi, G., Iaconisi, V., Gasco, L. and Nizza, A., 2016. Use of *Tenebrio molitor* larvae meal as protein source in broiler diet: effect on growth performance, nutrient digestibility, and carcass and meat traits. Journal of Animal Science 94: 639-647.

Bukkens, S.G.F., 1997. The nutritional value of edible insects. Ecology of Food and Nutrition 36: 287-319.

Calvert, C.C., Martin, R.D. and Morgan, N.O., 1969. House fly pupae as food for poultry. Journal of Economic Entomology 62: 938-939.

Cauchie, H.M., 2002. Chitin production by arthropods in the hydrosphere. Hydrobiologia 470: 63-96.

Cerda, H., Martinez, R., Briceno, N., Pizzoferrato, L., Manzi, P., Tommaseo Ponzetta, M., Marin, O. and Paoletti, M.G., 2001. Palm worm: (*Rhynchophorus palmarum*) traditional food in Amazonas, Venezuela – nutritional composition, small scale production and tourist palatability. Ecology of Food and Nutrition 40: 13-32.

Church, R.B. and Robertson, F.W., 1966. A biochemical study of the growth of *Drosophila melanogaster*. Journal of Experimental Zoology 162: 337-351.

Cookman, J.E., Angelo, M.J., Slansky Jr., F., Nation, J.L., 1984. Lipid content and fatty acid composition of larvae and adults of the velvetbean caterpillar, *Anticarsia gemmatalis*, as affected by larval diet. Journal of Insect Physiology 30: 523-527.

Dashefsky, H.S., Anderson, D.L., Tobin, E.N., Peters, T.M., 1976. Face fly pupae: a potential feed supplement for poultry. Environmental Entomology 5: 680-682.

DeMarco, M., Martinez, S., Hernandez, F., Madrid, J., Gai, F., Rotolo, L., Belforti, M., Bergero, D., Katz, H., Dabbou, S., Kovitvadhi, A., Zoccarato, I., Gasco, L. and Schiavone, A., 2015. Nutritional value of two insect larval meals (*Tenebrio molitor* and *Hermetia illucens*) for broiler chickens: apparent nutrient digestibility, apparent ileal amino acid digestibility and apparent metabolizable energy. Animal Feed Science and Technology 209: 211-218.

Devkota, B. and Schmidt, G.H., 2000. Accumulation of heavy metals in food plants and grasshoppers from the Taigetos Mountains, Greece. Agriculture, Ecosystems and Environment 78: 85-91.

Diener, S., Zurbrugg, C., Tockner, K., 2015. Bioaccumulation of heavy metals in the black soldier fly, *Hermetia illucens* and effects on its life cycle. Journal of insects and Food and Feed 1: 261-270.

Downer, R.G.H. and Matthews, J.R., 1976. Patterns of lipid distribution and utilisation in insects. American Zoologist 16: 733-745.

Dreassi, E., Cito, A., Zanfini, A., Materozzi, L., Botta, M. and Francardi, V., 2017. Dietary fatty acids influence the growth and fatty acid composition of the yellow mealworm *Tenebrio molitor* (Coleoptera: Tenebrionidae). Lipids 52: 285-294.

Dreyer, J.J. and Wehmeyer, A.S., 1982. On the nutritive value of mopane worms. South African Journal of Science 78: 33-35.

Eeva, T., Helle, S. and Salminen, J.P., 2010. Carotenoid composition of invertebrates consumed by two insectivorous bird species. Journal Chemical Ecology 36: 608-613.

Ekpo, K.E., 2011. Effect of processing on the protein quality of four popular insects consumed in Southern Nigeria. Archives of Applied Science Research 3: 307-326.

Fast, P.G., 1970. Insect lipids. Progress in the Chemistry of Fats and other Lipids 11: 181-242.

Finke, M.D., 2002. Complete nutrient composition of commercially raised invertebrates used as food for insectivores. Zoo Biology 21: 269-285.

Finke, M.D., 2003. Gut loading to enhance the nutrient content of insects as food for reptiles: a mathematical approach. Zoo Biology 22: 147-162.

Finke, M.D., 2004. The nutrient content of insects. In: Capinara, J.L. (ed.) Encyclopedia of entomology (Vol. 2). Kluwer Academic Publishers, Dordrecht, the Netherlands, pp. 1562-1575.

Finke, M.D., 2007. Estimate of chitin in raw whole insects. Zoo Biology 26: 105-115.

Finke, M.D., 2013. Complete nutrient content of four species of feeder insects. Zoo Biology 32: 27-36.

Finke, M.D., 2015a. Complete nutrient content of three species of wild caught insects, pallid-winged grasshopper *Trimerotropis pallidipennis*, rhinoceros beetles *Oxygrylius ruginasus* and white-lined sphinx moth *Hyles lineata*. Journal of Insects as Food and Feed 1: 281-292.

Finke, M.D., 2015b. Complete nutrient content of four species of commercially available feeder insects fed enhanced diets during growth. Zoo Biology 34: 554-564.

Finke, M.D., DeFoliart, G.R. and Benevenga, N.J., 1987. Use of a four-parameter logistic model to evaluate the protein quality of mixtures of Mormon cricket meal and corn gluten meal in rats. Journal of Nutrition 117: 1740-1750.

Finke, M.D., Sunde, M.L. and DeFoliart, G.R., 1985. An evaluation of the protein quality of Mormon crickets (*Anabrus simplex* Haldeman) when used as a high protein feedstuff for poultry. Poultry Science 64: 708-712.

Finkel, A.J., 1948. The lipid composition of *Tenebrio molitor* larvae. Physiological Zoology 21: 111-133.

Fogang, A.R., Kansci, G., Viau, M., Hafnaoui, N., Meynier, A., Demmano, G. and Genot, C., 2017. Lipid and amino acids profiles support the potential of *Rhynchophorus phoenicis* larvae for human nutrition. Journal of Food Composition and Analysis 60: 64-73.

Fontaneto, D., Tommaseo-Ponzetta, M., Galli, C., Risé, P., Glew, R.H. and Paoletti, M.G., 2011. Differences in fatty acid composition between aquatic and terrestrial insects used as food in human nutrition. Ecology of Food and Nutrition 50: 351-367.

Fujimoto, W., Suzuki, M., Kimura, K. and Iwanaga, T., 2002. Cellular expression of the gut chitinase in the stomach of frogs *Xenopus laevis* and *Rana catesbeiana*. Biomedical Research-Tokyo 23: 91-99.

Gao, Q., Wang, X., Wang, W., Lei, C. and Zhu, F., 2017. Influences of chromium and cadmium on the development of black soldier fly larvae. Environmental Science and Pollution Research 24: 8637-8644.

Ghioni, C., Bell, J.G. and Sargent, J.R., 1996. Polyunsaturated fatty acids in neutral lipids and phospholipids of some freshwater insects. Comparative Biochemistry and Physiology B 114: 161-170.

Giovannucci, D.R. and Stephenson, R.S., 1999. Identification and distribution of dietary precursors of the *Drosophila* visual pigment chromophore: analysis of carotenoids in wild type and ninaD mutants by HPLC. Vision Research 39: 219-229.

Gladyshev, M.I., Kharitonov, A.Y., Popova, O.N., Sushchik, N.N., Makhutova, O.N. and Kalacheva, G.S., 2011. Quantitative estimation of dragonfly role in transfer of essential polyunsaturated fatty acids from aquatic to terrestrial ecosystems. Doklady Biochemistry and Biophysics 438: 141-143.

Goldsmith, T.H. and Warner, L.T., 1964. Vitamin A in the vision of insects. Journal of General Physiology 47: 433-441.

Goulet, G., Mullier, P., Sinave, P. and Brisson, G.J., 1978. Nutritional evaluation of dried *Tenebrio molitor* larvae in the rat. Nutrition Reports International 18: 11-15.

Han, R.D., Parajulee, M., Zhong, H. and Feng, G., 2008. Effects of environmental humidity on the survival and development of pine caterpillars, *Dendrolimus tabulaeformis* (Lepidoptera: Lasiocampidae). Insect Science 15: 147-152.

Hanson, B.J., Cummins, K.W., Cargill, A.S. and Lowry, R.R., 1985. Lipid content, fatty acid composition, and the effect of diet on fats of aquatic insects. Comparative Biochemistry and Physiology B 80: 257-276.

Hatt, J.M., Hung, E. and Wanner, M., 2003. The influence of diet on the body composition of the house cricket (*Acheta domesticus*) and consequences for their use in zoo animal nutrition. Der Zoologische Garten 73: 238-244.

Henry, M., Gascob, L., Piccoloc, G. and Fountoulakia, E., 2015. Review on the use of insects in the diet of farmed fish: past and future. Animal Feed Science and Technology 203: 1-22.

Hunter, B.A., Hunter, L.M., Johnson, M.S. and Thompson, D.J., 1987. Dynamics of metal accumulation in the grasshopper *Chorthippus brunneus* in contaminated grasslands. Archives of Environmental Contamination and Toxicology 16: 711-716.

Hussein, M., Pillai, V.V., Goddard, J.M., Park, H.G., Kothapalli, K.S., Ross, D.A., Ketterings, Q.M., Brenna, J.T., Milstein, M.B., Marquis, H., Johnson, P.A., Nyrop, J.P. and Selvaraj, V., 2017. Sustainable production of housefly (*Musca domestica*) larvae as a protein-rich feed ingredient by utilizing cattle manure. PLoS ONE 12: e0171708.

Hwangbo, J., Hong, E.C., Jang, A., Kang, H.K., Oh, J.S., Kim, B.W. and Park, B.S., 2009. Utilization of house fly-maggots, a feed supplement in the production of broiler chickens. Journal of Environmental Biology 30: 609-614.

Igwe, C.U., Ujowundu, C.O., Nwaogu, L.A. and Okwu, G.N., 2011. Chemical analysis of an edible African termite, *Macrotermes nigeriensis*, a potential antidote to food security problem. Biochemistry and Analytical Biochemistry 1: 1-4.

Iñiguez-Covarrubias, G., De Franco-Gömez, M.J., Del R Andrade-Maldonado, G., 1994. Biodegradation of swine waste by house-fly larvae and evaluation of their protein quality in rats. Journal of Applied Animal Research 5: 65-74.

Isaksson, C. and Andersson, S., 2007. Carotenoid diet and nestling provisioning in urban and rural great tits *Parus major*. Journal of Avian Biology 38: 564-572.

Jones, L.D., Cooper, R.W. and Harding, R.S., 1972. Composition of mealworm *Tenebrio molitor* larvae. Journal of Zoo Animal Medicine 3: 34-41.

Kaspari, M., 1991. Prey preparation as a way that grasshopper sparrows (*Ammodramus savannarum*) increase the nutrient composition of their prey. Behavioral Ecology 2: 234-241.

Kazimírová, M. and Ortel, J., 2000. Metal accumulation by *Ceratitis capitata* (Diptera) and transfer to the parasitic wasp *Coptera occidentalis* (Hymenoptera). Environmental Toxicology and Chemistry 19: 1822-1829.

Kinyuru, J.N., Kenji, G.M., Njoroge, S.M. and Ayieko, M., 2010a. Effect of processing methods on the *in vitro* protein digestibility and vitamin content of edible winged termite (*Macrotermes subhylanus*) and grasshopper (*Ruspolia differens*). Food Bioprocess Technology 3: 778-782.

Kinyuru, J.N., Kenji, G.M., Muhoho, S.N. and Ayieko, M., 2010b. Nutritional potential of longhorn grasshopper (*Ruspolia differens*) consumed in Siaya district Kenya. Journal of Agriculture, Science and Technology 12: 32-46.

Kodondi, K.K., Leclercq, M. and Gaudin-Harding, F., 1987. Vitamin estimations of three edible species of Attacidae caterpillars from Zaire. International Journal of Vitamin and Nutrition Research 57: 333-334.

Koo, S.I., Currin, T.A., Johnson, M.G., King, E.W. and Turk, D.E., 1980. The nutritional value and microbial content of dried face fly pupae (*Musca autumnalis* (De Geer)) when fed to chicks. Poultry Science 59: 2514-2518.

Kouamé, B., Marcel, G., Brou André, K., Alassane, M., Koffi Gabouet, K. and Coulibally Séraphi, K., 2011. Détermination du taux optimal de farine d'asticots séchés dans le régime du rat en croissance. Journal of Animal and Plant Sciences 12: 1553-1559.

Kramer, K.J., Hopkins, T.L., Schaefer, J., 1995. Applications of solids NMR to the analysis of insect sclerotized structures. Insect Biochemistry and Molecular Biology 25: 1067-1080.

Kroeckel, S., Harjes, A.-G.E., Roth, I., Katz, H., Wuertz, S., Susenbeth, A., Schulz, C., 2012. When a turbot catches a fly: evaluation of a pre-pupae meal of the black soldier fly (*Hermetia illucens*) as fish meal substitute – growth performance and chitin degradation in juvenile turbot (*Psetta maxima*). Aquaculture 364-365: 345-352.

Landry, S., DeFoliart, G.R. and Sunde, M.L., 1986. Larval protein quality of six species of Lepidoptera (Saturniidae, Sphingidae, Noctuidae). Journal of Economic Entomology 79: 600-604.

Lease, H.M. and Wolf, B.O., 2010. Exoskeletal chitin scales isometrically with body size in terrestrial insects. Journal of Morphology 271: 759-768.

Lease, H.M., Wolf, B.O., 2011. Lipid content of terrestrial arthropods in relation to body size, phylogeny, ontogeny and sex. Physiological Entomology 36: 29-38.

Lin, S., Njaa, L.R., Eggum, B.O. and Shen, H., 1983. Chemical and biological evaluation of silk worm chrysalid protein. Journal of the Science of Food and Agriculture 34: 896-900.

Lipsitz, E.Y. and McFarlane, J.E., 1971. Analysis of lipid during the life cycle of the house cricket, *Acheta domesticus*. Insect Biochemistry 1: 446-460.

Longvah, T., Mangthya, K. and Ramulu, P., 2011. Nutrient composition and protein quality evaluation of eri silkworm (*Samia racinii*) prepupae and pupae. Food Chemistry 128: 400-403.

Madariaga, M.A., Mata, F., Municio, A.M. and Ribera A., 1971. Effect of the lipid composition of the larval diet on the fatty acid composition during development of *Ceratitis capitata*. Insect Biochemistry 2: 249-256.

Madibela, O.R., Seitiso, T.K., Thema, T.F. and Letso, M., 2007. Effect of traditional processing methods on chemical composition and *in vitro* true dry matter digestibility of the mopane worm (*Imbrasia belina*). Journal of Arid Environments 68: 492-500.

Makkar, H.P.S., Tran, G., Heuze, V. and Ankers, P., 2014. State-of-the-art on use of insects as animal feed. Animal Feed Science and Technology 197: 1-33.

Maroni, G. and Watson, D., 1985. Uptake and binding of cadmium, copper and zinc by *Drosophila melanogaster* larvae. Insect Biochemistry 15: 55-63.

Marono, S., Piccolo, G., Loponte, R., Di Meo, C., Attia, Y.A., Nizza, A. and Bovera, F., 2015. *in vitro* crude protein digestibility of *Tenebrio molitor* and *Hermetia illucens* insect meals and its correlation with chemical composition traits. Italian Journal of Animal Sciences 14: 338-343.

Massie, H.R., Williams, T.R. and DeWolfe, L.K., 1989. Changes in taurine in aging fruit flies and mice. Experimental Gerontology 24: 57-65.

Moussian, B., 2010. Recent advances in understanding mechanisms of insect cuticle differentiation. Insect Biochemistry and Molecular Biology 40: 363-375.

Nakagaki, B.J., Sunde, M.L. and Defoliart, G.R., 1987. Protein quality of the house cricket, *Acheta domesticus*, when fed to broiler chicks. Poultry Science 66: 1367-1371.

National Research Council (NRC), 1994. Nutrient requirements of poultry (9th Ed.). National Academy Press, Washington, DC, USA, 157 pp.

National Research Council (NRC), 2011. Nutrient requirements of fish and shrimp. National Academy Press, Washington, DC, USA, 376 pp.

National Research Council, (NRC), 2012. Nutrient requirements of swine (11th Ed.). National Academy Press, Washington, DC, USA, 420 pp.

Nestel, D., Papadopoulos, N.T., Liedo, P., Gonzales-Ceron, L. and Carey, J.R., 2005. Trends in lipid and protein contents during medfly aging: an harmonic path to death. Archives of Insect Biochemistry and Physiology 60: 130-139.

Nishimune, T., Watanabe, Y., Okazaki, H. and Akai, H., 2000. Thiamin is decomposed due to *Anaphe* spp. entomophagy in seasonal ataxia patients in Nigeria. Journal of Nutrition 130: 1625-1628.

Noland, J.L. and Baumann, C.A., 1949. Requirement of the German cockroach for choline and related compounds. Experimental Biology and Medicine 70: 198-201.

Olson, J.A., 1989. Provitamin A function of carotenoids: the conversion of beta-carotene into vitamin A. Journal of Nutrition 119: 105-108.

Omotoso, O.T., 2006. Nutritional quality, functional properties and anti-nutrient compositions of the larva of *Cirina forda* (Westwood) (Lepidoptera: Saturniidae). Journal of Zhejiang University Science B 7: 51-55.

Onifade, A.A., Oduguwa, O.O., Fanimo, A.O., Abu, A.O., Olutunde, T.O., Arije, A. and Babatunde, G.M., 2001. Effects of supplemental methionine and lysine on the nutritional value of housefly larvae meal (*Musca domestica*) fed to rats. Bioresource Technology 78: 191-194.

Oonincx, D., Van Leeuwen, J.P., Hendriks, W.H. and Van der Poel, A.F.B., 2015b. The diet of free-roaming Australian central bearded dragons (*Pogona vitticeps*). Zoo Biology 34: 271-277.

Oonincx, D.G.A.B. and Van der Poel, A.F., 2011. Effects of diet on the chemical composition of migratory locusts (*Locusta migratoria*). Zoo Biology 30: 9-16.

Oonincx, D.G.A.B. and Dierenfeld, E.S., 2012. An investigation into the chemical composition of alternative invertebrate prey. Zoo Biology 31: 40-54.

Oonincx, D.G.A.B., Stevens, Y., Van den Borne, J.J.G.C., Van Leeuwen, J.P.T.M. and Hendriks, W.H., 2010. Effects of vitamin D3 supplementation and UVb exposure on the growth and plasma concentration of vitamin D3 metabolites in juvenile bearded dragons (*Pogona vitticeps*). Comparative Biochemistry and Physiology B 156: 122-128.

Oonincx, D.G.A.B., Van Broekhoven, S., Van Huis, A. and Van Loon, J.J.A., 2015a. Feed conversion, survival and development, and composition of four insect species on diets composed of food by-products. PLoS ONE 10: e0144601.

Ozimek, L., Sauer, W.C., Kozikowski, V., Ryan, J.K., Jorgensen, H. and Jelen, P., 1985. Nutritive value of protein extracted from honey bees. Journal of Food Science 50: 1327-1329.

Paoletti, M.G., Norberto, L., Damini, R. and Musumeci, S., 2007. Human gastric juice contains chitinase that can degrade chitin. Annals of Nutrition and Metabolism 51: 244-251.

Paul, A., Frederich, M., Megido, R.C., Alabi, T., Malik, P., Uyttenbroeck, R., Francis, F., Blecker, C., Haubruge, E., Lognay, G. and Danthine, S., 2017. Insect fatty acids: a comparison from three Orthopteran and *Tenebrio molitor* L. larvae. Journal of Asia-Pacific Entomology 20: 337-340.

Payne, C.L.R., Scarborough, P., Rayner, M. and Nonaka, K., 2016. A systematic review of nutrient composition data available for twelve commercially available edible insects and comparison with reference values. Trends in Food Science and Technology 47: 69-77.

Payne, C.L.R., Umemura, M., Dube, S., Azuma, A., Takenaka, C. and Nonaka, K., 2015. The mineral composition of five insects as sold for human consumption in southern Africa. African Journal of Biotechnology 14: 2443-2448.

Pearincott, J.V., 1960. Changes in lipid content during growth and metamorphosis of the house fly *Musca domestica* Linnaeus. Journal of Cellular and Comparative Physiology 55: 167-174.

Pennino, M., Dierenfeld, E.S. and Behler, J.L., 1991. Retinol, alpha-tocopherol, and proximate nutrient composition of invertebrates used as feed. International Zoo Yearbook 30: 143-149.

Pesti, G.M., Harper, A.E. and Sunde, M.L., 1979. Sulfur amino acid and methyl donor status of corn-soy diets fed to starting broiler chicks and Turkey poults. Poultry Science 58: 1541-1547.

Phelps, R.J., Struthers, J.K. and Moyo, S.J., 1975. Investigations into the nutritive value of *Macrotermes falciger* (Isoptera: Termitidae). Zoologica Africana 10: 123-132.

Pieterse, E. and Pretorius, Q., 2014. Nutritional evaluation of dried larvae and pupae meal of the housefly (*Musca domestica*) using chemical and broiler-based biological assays. Animal Production Science 54: 347-355.

Poelaert, C., Despret, X., Sindic, M., Beckers, Y., Francis, F., Portetelle, D., Soyeurt, H., Théwis, A. and Bindelle, J., 2017. Cooking has variable effects on the fermentability in the large intestine of the fraction of meats, grain legumes, and insects that is resistant to digestion in the small intestine in an *in vitro* model of the pig's gastrointestinal tract. Journal of Agricultural and Food Chemistry 65: 435-444.

Punzo, F., 2003. Nutrient composition of some insects and arachnids. Florida Scientist 66: 84-98.

Purschke, B., Scheibelberger, R., Axmann, S., Adler, A. and Jäger, H., in press. Impact of substrate contamination with mycotoxins, heavy metals and pesticides on growth performance and composition of black soldier fly larvae (*Hermetia illucens*) for use in the feed and food value chain. Food Additives and Contaminants, Part A, DOI: https://doi.org/10.1080/19440049.2017.1299946.

Ramos-Elorduy, J., Gonzalez, E.A., Hernandez, A.R. and Pino, J.M., 2002. Use of *Tenebrio molitor* (Coleoptera: Tenebrionidae) to recycle organic wastes and as feed for broiler chickens. Journal of Economic Entomology 95: 214-220.

Rao, P.U., 1994. Chemical composition and nutritional evaluation of spent silk worm pupae. Journal Agriculture and Food Chemistry 42: 2201-2203.

Raubenheimer, D. and Rothman, J.M., 2013. Nutritional ecology of entomophagy in humans and other primates. Annual Review of Entomology 58: 141-160.

Reichle, D.E., Shanks, M.H. and Crossley, D.A., 1969. Calcium, potassium, and sodium content of forest floor arthropods. Annals of the Entomological Society of America 62: 57-62.

Riddick, E.W., 2014. Insect protein as a partial replacement for fishmeal in the diets of juvenile fish and crustaceans. In: Morales-Ramos, J.A., Rojas, M.G. and Shapiro, D.A. (eds.) Mass production of beneficial organisms invertebrates and entomopathogens. Elsevier, Waltham, MA, USA, pp. 565-582.

Roe, R.M., Clifford, C.W. and Woodring, J.P., 1985. The effect of temperature on energy distribution during the last-larval stadium of the female house cricket, *Acheta domesticus*. Journal of Insect Physiology 31: 371-378.

Roseland, C.R., Grodowitz, M.J., Kramer, K.J., Hopkins, T.L. and Broce, A.B., 1985. Stabilization of mineralized and sclerotized puparial cuticle of muscid flies. Insect Biochemistry 15: 521-528.

Rumpold, B.A. and Schlüter, O.K., 2013a. Nutritional composition and safety aspects of edible insects. Molecular Nutrition Food Research 57: 802-823.

Rumpold, B.A. and Schlüter, O.K., 2013b. Potential and challenges of insects as an innovative source for food and feed production. Innovative Food Science and Emerging Technologies 17: 1-11.

Sánchez-Muros, M.J., Barroso, F.G. and De Haro, C., 2016. Brief summary of insect usage as an industrial animal feed/feed ingredient. In: Dorsey, A.T., Morales-Ramos, J.A. and Rojas, M.G. (eds.) Insects as sustainable food ingredients production, processing and food applications. Elsevier, Waltham, MA, USA, pp. 273-309.

Sánchez-Muros, M.J., Barroso, F.G. and Manzano-Agugliaro, M., 2014. Insect meal as renewable source of food for animal feeding: a review. Journal of Cleaner Production 65: 16-27.

Santos Oliveira, J.F., Passos De Carvalho, J., Bruno De Sousa, R.F.X. and Simao, M.M., 1976. The nutritional value of four species of insects consumed in Angola. Ecology of Food and Nutrition 5: 91-97.

Schaefer, C.H., 1968. The relationship of the fatty acid composition of *Heliothis zea* larvae to that of its diet. Journal of Insect Physiology 14: 171-178.

Seki, T., Isono, K., Ozaki, K., Tsukahara, Y., Shibata-Katsuta, Y., Ito, M., Irie, T. and Katagiri, M., 1998. The metabolic pathway of visual pignment chrnomophore formation in *Drosophila melanogaster* All-trans (3S)-3-hydroxyretinal is formed from all-trans retinal via (3R)-3-hydroxyretinal in the dark. European Journal of Biochemistry 257: 522-527.

Sönmez, E. and Gülel, A., 2008. Effects of different temperatures on the total carbohydrate, lipid and protein amounts of the bean beetle, *Acanthoscelides obtectus* Say (Coleoptera: Bruchidae). Pakistan Journal of Biological Sciences 11: 1803-1808.

Spranghers, T., Ottoboni, M., Klootwijk, C., Ovyn, A., Deboosere, S., De Meulenaer, B., Michiels, J., Eeckhout, M., De Clercq, P. and De Smet, S., 2017. Nutritional composition of black soldier fly (*Hermetia illucens*) prepupae reared on different organic waste substrates. Journal of the Science of Food and Agriculture 97: 2594-2600.

Ssepuuya, G.V., Mukisa, I.M. and Nakimbugwe, D., 2017. Nutritional composition, quality, and shelf stability of processed *Ruspolia nitidula* (edible grasshoppers). Food Science and Nutrition 5: 103-112.

Starčević, K., Gavrilović, A., Gottstein, Z. and Mašek, T., 2017. Influence of substitution of sunflower oil by different oils on the growth, survival rate and fatty acid composition of Jamaican field cricket (*Gryllus assimilis*). Animal Feed Science and Technology 228: 66-71.

St-Hilaire, S., Cranfill, K., McGuire, M.A., Mosley, E.E., Tomberlin, J.K., Newton, L., Sealey, W., Sheppard, C. and Irving, S., 2007. Fish offal recycling by the black soldier fly produces a foodstuff high in omega-3 fatty acids. Journal of the World Aquaculture Society 38: 309-313.

Studier, E.H. and Sevick, S.H., 1992. Live mass, water content, nitrogen and mineral levels in some insects from south-central lower Michigan. Comparative Biochemistry and Physiology A 103: 579-595.

Surendra, K.C., Olivier, R., Tomberlin, J.K., Jha, R. and Khanal, S.K. 2016. Bioconversion of organic wastes into biodiesel and animal feed via insect farming. Renewable Energy 98: 197-202.

Sushchik, N.N., Gladyshev, M.I., Moskvichova, A.V., Makhutova, O.N. and Kalachova, G.S., 2003. Comparison of fatty acid composition in major lipid classes of the dominant benthic invertebrates of the Yenisei river. Comparative Biochemistry and Physiology B 134: 111-122.

Sushchik, N.N., Yurchenko, Y.A., Gladyshev, M.I., Belevich, O.E., Kalachova, G.S. and Kolmakova, A.A., 2013. Comparison of fatty acid contents and composition in major lipid classes of larvae and adults of mosquitoes (Diptera: Culicidae) from a steppe region. Insect Science 20: 585-600.

Teffo, L.S., Toms, R.B. and Eloff, J.N., 2007. Preliminary data on the nutritional composition of the edible stink-bug, *Encosternum delegorguei* Spinola, consumed in Limpopo province, South Africa. South African Journal of Science 103: 434-436.

Teotia, J.S. and Miller, B.F., 1973. Fly pupae as a dietary ingredient for starting chicks. Poultry Science 52: 1830-1835.

Teotia, J.S. and Miller, B.F., 1974. Nutritive content of house fly pupae and manure residue. British Poultry Science 15: 177-182.

Thompson, S.N., 1973. Review and comparative characterization of fatty-acid compositions of seven insect orders. Comparative Biochemistry and Physiology 45: 467-482.

Tietz, A. and Stern, N., 1969. Stearate desaturation by microsomes on the locust fat-body. FEBS Letters 2: 286-288.

Tschirner, M. and Simon, A., 2015. Influence of different growing substrates and processing on the nutrient composition of black soldier fly larvae destined for animal feed. Journal of Insects as Food and Feed 1: 249-259.

Ueckert, D.N., Yang, S.P. and Albin, R.C., 1972. Biological value of rangeland grasshoppers as a protein concentrate. Journal of Economic Entomology 65: 1286-1288.

Van Broekhoven, S., Oonincx, D.G.A.B., Van Huis, A. and Van Loon, J.J.A., 2015. Growth performance and feed efficiency of three edible mealworm species (Coleoptera: Tenebrionidae) on diets composed of organic by-products. Journal of Insect Physiology 73: 1-10.

Van der Fels-Klerx, H.J., Camenzuli, L., Van der Lee, M.K. and Oonincx, D.G.A.B., 2016. Uptake of cadmium, lead and arsenic by *Tenebrio molitor* and *Hermetia illucens* from contaminated substrates. PLoS ONE 11: e0166186.

Vogt, K. and Kirschfeld, K., 1984. Chemical identity of the chromophore of fly visual pigment. Naturwissenschaften 71: 211-213.

Von Lintig, J., 2012. Metabolism of carotenoids and retinoids related to vision. Journal of Biological Chemistry 287: 1627-1634.

Wang, D., Zhai, S.W., Zhang, C.X., Bai, Y.Y., An, S.H. and Xu, Y.N., 2005. Evaluation on nutritional value of field crickets as a poultry feedstuff. Asian-Australasian Journal of Animal Sciences 18: 667-670.

Wang, D., Zhai, S.W., Zhang, C.X., Zhang, Q. and Chen, H., 2007. Nutrition value of the Chinese grasshopper *Acrida cinerea* (Thunberg) for broilers. Animal Feed Science and Technology 135: 66-74.

Whitaker Jr., J.O., Dannelly, H.K. and Prentice, D.A., 2004. Chitinase in insectivorous bats. Journal of Mammalogy 85: 15-18.

World Health Organization/Food and Agriculture Organization of the United Nations/United Nations University (WHO/FAO/UNU), 2007. Protein and amino acid requirements in human nutrition. Expert consultation. WHO Technical Reports Series 935. WHO, Geneva, Switzerland, 265 pp.

Wu, G.X., Ye, G.E., Hu, C. and Cheng, J.A., 2006. Accumulation of cadmium and its effects on growth, development and hemolymph biochemical compositions in *Boettcherisca peregrina* larvae (Diptera: Sarcophagidae). Insect Science 13: 31-39.

Xia, Z., Wu, S., Pan, S. and Kim, J.M., 2012. Nutritional evaluation of protein from *Clanis bilineata* (Lepidoptera), an edible insect. Journal of the Science of Food and Agriculture 92: 1479-1482.

Yang, L.F., Siriamornpun, S. and Li, D., 2006. Polyunsaturated fatty acid content of edible insects in Thailand. Journal of Food Lipids 13: 277-285.

Yi, L., Lakemond, C.M.M., Sagis, L.M.C., Eisner-Schadler, V., Van Huis, A. and Van Boekel, M.A.J.S., 2013. Extraction and characterisation of protein fractions from five insect species. Food Chemistry 141: 3341-3348.

Zhang, Z., Song, X., Wang, Q. and Lu, X., 2012. Cd and Pb contents in soil, plants, and grasshoppers along a pollution gradient in Huludao City, Northeast China. Biological Trace Element Research 145: 403-410.

Zhou, J.R. and Erdman, J.W., 1995. Phytic acid in health and disease. Critical Reviews in Food Science and Nutrition 35: 495-508.

Zhou, X., Honek, A., Powell, W. and Carter, N., 1995. Variations in body length, weight, fat content and survival in *Coccinella septempunctata* at different hibernation sites. Entomologia Experimentalis et Applicata 75: 99-107.

17.

Galleria mellonella

Insect processing

B.A. Rumpold[1*], S. Bußler[2], H. Jäger[3] and O. Schlüter[2]

[1]Technische Universität Berlin, Institute of Vocational Education and Work Studies, Department of Education for Sustainable Nutrition and Food Science, Marchstraße 23, 10587 Berlin, Germany; [2]Leibniz Institute for Agricultural Engineering and Bioeconomy (ATB), Research Program 'Quality and Safety of Food and Feed', Max-Eyth-Allee 100, 14469 Potsdam, Germany; [3]Universität für Bodenkultur Wien, Institute of Food Technology (LMT), Muthgasse 18, 1190 Vienna, Austria; rumpold@tu-berlin.de

Abstract

Utilization of insects for food and feed requires a systemic approach in the area of bioeconomic production. In order to deliver raw materials and ingredients for food and feed applications that are safe and of high quality the processing of insects is a major pre-requisite. In addition, it can be applied to improve the properties of insect derived ingredients by process induced optimization of ingredient functionality and the production of tailored functional fractions. Processing pathways may include several unit operations currently already used in food and feed processing. In this section an overview is given of decontamination, drying and comminution methods as well as subsequent fractionation methods for protein, fat or chitin recovery. The impact of the different processing steps on product properties is discussed and examples for whole insect processing pathways are presented.

Keywords: postharvest processing, extraction technologies, product properties, quality aspects, safety aspects

17.1 Introduction

The aim of food and feed processing is the production of products out of plant- and animal-based raw materials with desired and/or tailor-made properties (chemical and nutrient composition, digestibility and bioavailability of components, texture, size, colour, taste) (Tscheuschner, 1996), while increasing the shelf-life and ensuring product safety at optimum product quality. Accordingly, the development of insect processing concepts is necessary for the production of high-quality and safe insect-based products. Furthermore, the development of sustainable and cost-effective production processes is required (Rumpold and Schlüter, 2013b).

As shown in Figure 17.1, possible unit operations for insect processing potentially include decontamination, drying and comminution. Subsequent separation or extraction includes protein, lipid, and chitin extraction. For a sustainable holistic approach it should be aimed to reduce processing waste, while using all obtained fractions.

17.2 Decontamination

Before marketing insects and fractions thereof as feed and food, technological processing is required for decontamination, preservation, and to guarantee microbial safety (Rumpold and Schlüter, 2013b) including the inactivation of pathogens and spoilage microorganisms with particular consideration of the inactivation of the gut microbiota since whole insects are generally not degutted. Possible processes for decontamination may include thermal processes such as blanching, pasteurization and sterilization, or nonthermal processes including

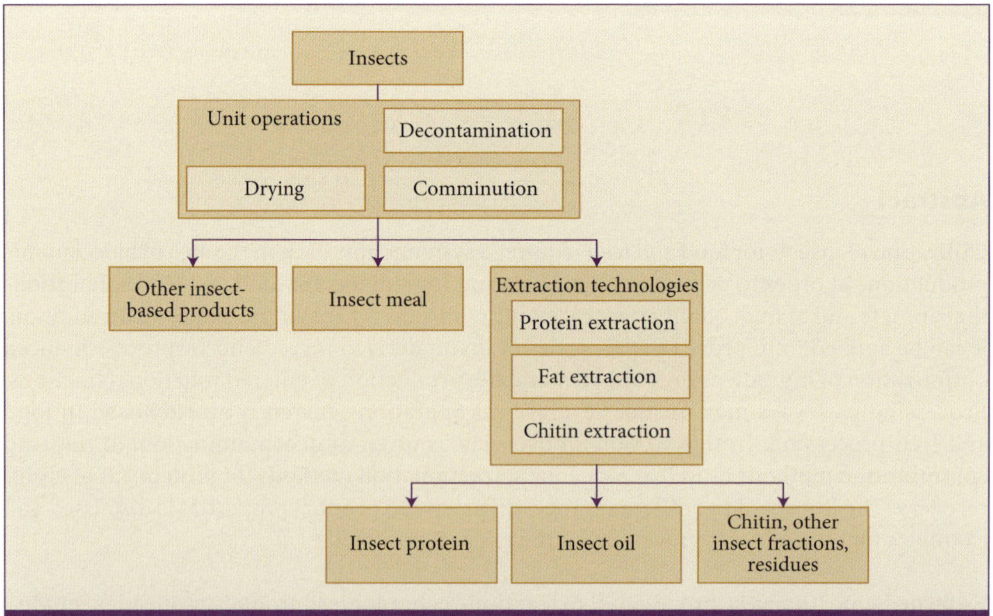

Figure 17.1. Processing pathways for insects including unit operations and resulting products.

radiation processes such as UV irradiation, electron beam processing, and γ-radiation (Tscheuschner, 1996). Furthermore, novel innovative decontamination technologies such as pulsed electric fields, cold plasma and high hydrostatic pressure are of relevance (Knorr *et al.*, 2011). In addition to the aforementioned decontamination technologies, further preservation can be achieved using cool and frozen storage, processes resulting in moisture removal such as drying, freeze-drying, concentration, salting and pickling; application of smoking, lowering the pH or reducing the oxygen content by controlled atmosphere storage.

Insects are well-known to carry and spread pathogenic and non-pathogenic micro-organisms and have the potential to cause food spoilage or function as vectors transmitting zoonoses. Total viable aerobic counts of 7.2-8.3 log cfu/g (Klunder *et al.*, 2012), 7.3-8.4 log cfu/g (Rumpold *et al.*, 2014), 7.6-8.2 log cfu/g (Vandeweyer *et al.*, 2017) and 7.7-8.6 log cfu/g (Stoops *et al.*, 2016) of fresh insects were reported in literature. An investigation of live mealworm larvae and grasshoppers sold for human consumption revealed they were highly contaminated with microorganisms including spoilage microorganisms and potential food pathogens (Stoops *et al.*, 2016). Klunder *et al.* (2012) investigated the microbiological aspects of processing and storage on selected edible insects and came to the conclusion they need to be processed and stored properly. Furthermore, spore-forming bacteria were identified as a potential risk of edible insects that – unlike Enterobacteriaceae – could not be eliminated by boiling. This result was confirmed by Vandeweyer *et al.* (2017) who investigated the effect of blanching and subsequent microwave drying or refrigerated storage on the microbial load of mealworm (*Tenebrio molitor* L.) larvae. They also identified spore-forming bacteria as a potential risk in edible insects since bacterial endospores were found to be the most resistant and their number only reduced slightly at the conditions chosen. For the inactivation or handling of potential spore-forming bacteria in insect-based food and feed, the same hygienic measures are to be taken than for similar food groups e.g. meat products.

Furthermore, indirect cold plasma treatment was determined to be a very effective method for the surface decontamination of whole *T. molitor* larvae (i.e. reduction of 7 log cycles was observed). However, for the inactivation of the overall microbial count including the gut microbiota of whole larvae, plasma had no noticeable effect and a complemental decontamination treatment had to be applied. The best results for the reduction of the overall microbial count had been achieved by thermal treatments at 90 °C (Rumpold *et al.*, 2014). Investigating the impact of cold atmospheric pressure plasma (CAPP) on the microbial load of *T. molitor* flour in comparison to dry heat treatment resulted in a reduction of the total microbial load from 7.7 to 4.3 log cfu/g for CAPP treatment for 15 min and from 6.8 log cfu/g to below detection limit upon thermal treatment for 15 min in an oven at temperatures of 120 °C and above (Bußler *et al.*, 2016a). Generally there is still scarce data on microbial safety of raw and processed insects, especially on species other than *T. molitor*, and the development of effective but economic decontamination processes are needed.

In addition to the microbial inactivation, decontamination processes could also affect other insect-based product properties. For example, the impact of thermal processing on the allergenicity of mealworm has been investigated comparing raw, freeze-dried, blanched, boiled, baked, and fried mealworm extracts. This study determined the allergenicity of mealworm proteins was not decreased by heat treatments, but the solubility of the proteins was modified (Broekman *et al.*, 2015). The immunoglobulin (Ig)E cross-reactivity of raw, freeze-dried, fried and boiled mealworms (*Zophobas atratus* Fab, *Alphitobius diaperinus* Panzer) and subsequent *in vitro* digestion was also investigated, and it was observed that heat processing and *in vitro* digestion reduced, but did not eliminate house dust mite or tropomyosin IgE cross-reactivity (Broekhoven *et al.*, 2016). Consequently, allergens present in mealworms and possibly other insects appear to be resistant to heat treatments.

It was furthermore shown that plasma treatment and thermal treatment influence techno-functional properties such as water binding and oil binding capacity, protein solubility and protein composition of *T. molitor* flour (Bußler *et al.*, 2016a). However, decontamination of insect fractions might be required if the microbial load cannot be sufficiently lowered by previous processing steps or recontamination occurs. The microbial load of proteins after different extraction steps was shown to be 3 log cycles lower than the starting material, but still too high for a safe food product. Hence, a decontamination step is of outmost importance (Bußler *et al.*, 2015a,b).

17.3 Drying

The thermal drying process consists of three fundamental sub-processes: the heat transfer from the environment to the drying material, the phase transition of the solvent, and the evacuation of the solvent vapour. By the mode of heat supply, drying is classified into contact drying by hot surfaces, convective drying by a hot gas flow and radiation drying, where the energy is supplied via electromagnetic waves. In many drying processes, more than one kind of heat supply takes place. Table 17.1 gives an overview of dryers and their process and product properties.

The properties of the starting material to be dried determine the drying method. Is the material free-flowing and pourable or is it pasty and could stick to contact surfaces of the dryer and form films? Is the product easily removable from the dryer and facilitates easy cleaning of the drying equipment? What are initial and targeted final water contents? In addition, the energy and time requirement of the drying process need to be considered as well as the desired throughput of material and/ capacity of the dryer.

For example, the moisture content of fly larvae amounts to 61.2% for black soldier flies (Finke, 2012), 67.1 for fruit flies (Barker *et al.*, 1998) and the moisture content of mealworm larvae ranges from 61.9 to 62.9% (Barker *et al.*, 1998; Finke, 2002). After drying, the water content must not exceed 10% to ensure safe storage and microbial safety. Consequently, for the exemplary fly and mealworm larvae, 1.05 to 1.33 kg of water must be removed to obtain 1 kg of dried insects with a moisture content of 10%. Regarding such high amounts of water to be removed, an energy efficient drying process is essential for an economic production of dried insects and insect products.

To date, economic aspects of drying insects have not been a focus in research and industry. On a laboratory scale for insect-based diet preparation and feeding trials, insect drying was performed by sun drying (Awoniyi *et al.*, 2003; Ijaiya and Eko, 2009; Odesanya *et al.*, 2011; Oyegoke *et al.*, 2006) or by drying in an oven such as a laboratory scale convection oven at

Table 17.1. Overview of dryers (adapted from Mujumdar, 2011; www.drying.de).

Mode of heat supply Dryer	Process and product properties
Convective drying	
spray dryer	continuous, for pumping, heat-sensitive pastes, slurries and solutions, drying time seconds to minutes
fluidized bed dryer	continuous or batch, for fluidizable particulate materials, drying time minutes to hours
band dryer with forced air circulation (belt dryer/conveyor dryer)	continuous, for coarse, fragmented materials, drying time minutes to hours
carrousel dryer	continuous, for fluid particulate materials, drying time a few minutes, space saving alternative for the band dryer
cabinet dryer with forced air circulation (drying oven)	batch, for pastes or particulate fluid materials, drying time hours, for small amounts (e.g. laboratory)
flow dryer (pneumatic conveyor flash dryer)	continuous, for particulate pourable material, drying time seconds, most simple and cost-effective convective short-time dryer
Contact drying	
disc dryer	continuous, for particulate, fluid material, drying time minutes
drum dryer	continuous, for liquid or pasty materials that will afterwards be ground to a powder, drying time seconds to minutes
contact-belt dryer (conveyor dryer)	continuous, for fluid, pasty or liquid materials, drying time minutes to hours
freeze-dryer	batch, for all materials, long drying time of hours plus freezing time, no thermal effects
Radiation drying	
microwave-vacuum dryer	batch/ quasi-continuous, for heat-sensitive solid, liquid or pasty materials, drying time minutes
infrared-rotary dryer	continuous, for fluid materials, drying time minutes

50 °C for three days (Ramos-Elorduy *et al.*, 2002), at 55 °C for one day (Hwangbo *et al.*, 2009), at 60 °C until constant weight is reached (Xia *et al.*, 2012), at 65 °C for 12-24 hours (Pretorius, 2011), at up to 85 °C (Newton *et al.*, 1977) or at 100-105 °C for 48 hours (Koo *et al.*, 1980). In these investigations it was mainly focused on the feeding trials and not on the insect processing. The applied drying methods are generally not designed in a resource-efficient way and can be optimized regarding energy consumption, drying time as well as resulting product quality.

Agriprotein Technologies in South Africa is drying fly larvae on a fluidized bed dryer after which the dried larvae are milled into flakes (Veldkamp *et al.*, 2012). The 'Insect rearing centre' of the Mississippi State University (Starkville, MS, USA) and Neptune Industries Inc.

(Boca Raton, FL, USA) have developed a patented production protocol for the production of dry protein meal out of insects where the drying methods depend on the insect species and include belt and tumble dryers (Papadoyanis and Cherch, 2008). In a patented method air- and spray-drying are applied (Arsiwalla and Aarts, 2015) to convert insects into nutrient streams and compositions obtained thereby. In the food market, mealworm larvae and other insects such as grasshoppers and crickets are sold freeze-dried (e.g. www.eatgrub.co.uk; www.bugsoriginals.nl). The Dutch company Bugs Originals (Oostzaan, the Netherlands) states on its web site that the insects are blanched before they are freeze-dried.

In a student engineering competition of the European Union project PROteINSECT with the topic 'engineering solutions for maggot drying' it was aimed to develop a concept for drying one tonne of fly larvae per day so that they can be stored at room temperature for one year. The winner Simon Schantl came up with the concept 'ENTODRYA' (http://tinyurl.com/ydxktx3u), which consisted of a steam-heated drum dryer adapted from the milk industry, that succeeded in drying 1 t of larvae to a moisture content of 5% within 3 h.

An investigation of drying kinetics and hygroscopic behaviour of freeze-dried and air-dried *T. molitor* larvae subsequent to a blanching step showed the Page model had the best fit to predict the drying kinetics of mealworm larvae at 50, 60 and 70 °C. The Arrhenius equation described the temperature dependence of water diffusion rate (correlation coefficient r^2=0.95) and the Guggenheim-Anderson-de Boer model was suitable for interpolating the sorption isotherms (determination coefficient >0.993). Air-drying was performed in a forced-air oven on a laboratory scale (Azzolini *et al.*, 2016).

Since insects are animal products with an intestinal flora, not only processing properties but also food safety measures differ immensely from plant products. While low drying temperatures can lead to a product with native proteins and a high nutritional value, safety problems may arise. Investigating the microbial load of mealworm larvae upon blanching for 40 s and microwave-drying of up to 20 min, the water activity of the larvae was reduced from 0.95 to 0.23 and the moisture content from 63-67% to around 1%. A drying time of 13 min was sufficient to achieve a water content of 11% but the water activity was slightly above 0.6 (0.67) and therefore too high for a microbiologically stable food product (Vandeweyer *et al.*, 2017).

17.4 Comminution

Grinding or milling is done using a mill in order to cut, crush and/or grind a solid into smaller particles using mechanical forces such as impact, compression, shearing, cutting, friction or combinations thereof. Exemplary, mills include ball mill, roller mill, rotor mills, cutting mill, knife mill, disc mill, jaw crushers, and mortar grinders. The choice of grinding equipment mostly depends on the raw material properties to be ground (hard, soft, brittle, pasty, elastic, fibrous). A company from Nebraska (Pleasant Hill Grain, Hampton, NE, USA) recommends a disc mill for grinding dehydrated crickets into cricket flour (http://tinyurl.com/yd3x285w). As alternate options, they suggest meat grinders and coffee grinders.

In the blog https://www.cricketflours.com/how-to-make-cricket-flour the grinding process of crickets is described as follows: 'Generally, the dried crickets are then ground using two different grinding or milling machines. The first machine is set to a coarse grind. Once the crickets have been placed in the first machine, the coarse cricket flour is then sifted to remove the lighter content, which consists of legs, wings, etc. and is removed from the final cricket flour product. Next, the remaining coarse cricket flour is placed in the second milling machine, which is set to a fine grain size to produce a smooth and fine cricket flour.'. Although there are several web shops selling insect flours (bizarrefood.com, thailandunique.com, and others) there are no detailed indications on the grinding process. When grinding insects, the fat content of the insect has to be taken into consideration. Grinding insects with high fat contents, smearing upon grinding can make the handling of the insect product and of the grinding equipment (cleanability) more difficult and/or time and labour consuming. In order to facilitate the grinding process, defattening prior to the milling step or cooling/freezing (freeze-grinding) is possible. The latter increases the production costs, requires special milling equipment and complicates handling. On a laboratory scale, a knife mill seems to be a promising alternative for grinding whole insects according to the appropriate material properties (for dry, fibrous, soft or medium-hard food products containing oil and water). A coffee mill was also mentioned for milling freeze-dried insect puree (Bußler *et al.*, 2016a).

17.5 Protein extraction

Plant proteins can be divided in different protein fractions according to their different solvent solubility. For example, rice proteins are divided in water-soluble, salt-soluble, alkali-soluble and alcohol-soluble protein fractions (Ju *et al.*, 2001). Industrial plant protein extraction is usually done via isoelectric precipitation. This process was patented for the extraction of soy proteins by Anson and Pader in 1957 (U.S. Pat. No. 2,785,155). Proteins are solubilized under alkaline conditions, the insoluble material is removed by centrifugation and the alkali-solubilized proteins in the supernatant are precipitated isoelectrically by lowering the pH using hydrochloric acid resulting in a protein isolate. A dry protein extraction process for example for starchy legume seeds such as peas and faba beans is also possible. After a pin milling step, air classification techniques can be applied to separate the light or fine fraction (the protein concentrate) from the heavy or coarse fraction (the starch fraction) (Gueguen, 1983). Other physical protein separation methods include centrifugation and filtration.

The average protein content based on dry matter of insects per order range from 35.34% for Isoptera (termites) being rich in fat to 61.32% for Orthoptera (grasshoppers, crickets, locusts). This assessment shows that insects generally are rich in protein (Rumpold and Schlüter, 2013a). Previous research determined protein extraction increased the protein digestibility in comparison to whole honey bees used in feeding trials on rats. This variation had been attributed to the removal of chitin. The protein was extracted by homogenising whole frozen honey bees with 1.5 parts of water and addition of NaOH solution (10%) in order to adjust the pH to 10.5. Twenty minutes later, the resulting homogenate was pressed through cheese cloth for separation of the chitin. The filtrate was adjusted to pH 5.5 using HCl solution (1 N) and centrifuged at 3,350×*g* at 10 °C for 10 min. The resulting precipitate (protein concentrate) contained 64% protein and showed higher lysine content, higher availability of amino acids and higher protein digestibility than dried honey bees. The honey bee protein concentrate had a corrected protein efficiency ratio (PER) of 2.47 (98.8% in comparison to casein) and a corrected net protein utilization (NPU) of 62 (88.6% in comparison to casein) (Ozimek *et al.*,

1985). Accordingly, a protein extraction might be beneficial from a nutritional point of view. In addition, it facilitates the tailor-made production of food and feed with targeted properties and/or nutritional profiles.

There is increasing but still little scientific data published on protein extraction from insects. Valle *et al.* (1982) performed an alkaline protein extraction from the Mexican fruit fly larvae *Anastrepha ludens*. Subsequent to the determination of protein solubility depending on the pH, the insect sample was washed with water at the least solubility of the proteins (pH 5, 60 °C) in order to remove water-soluble substances and fat and obtain a protein concentrate. For a protein isolate, the washing step was performed at maximum protein solubility (room temperature, pH 10) with subsequent protein precipitation at pH 5. A protein concentrate with a protein content of 65.4% and a protein isolate with a protein content of 86.6% were obtained (based on dry matter). The larvae protein quality was comparable to oil seed and legume proteins with a corrected PER of 1.63 (65% in comparison to casein) and a NPU of 38.0 (75% in comparison to casein). In comparison to conventional animal proteins, the investigated fly protein had a generally lower quality. However, the high contents of lysine and methionine are remarkable. An investigation of the functional properties of the proteins from the Mexican fruit fly resulted in considerably lower foaming and emulsion capacities compared to egg white protein. The solubility of the extracted fly proteins was with 95% the highest at a pH of 10 and with 8% the lowest at the isoelectric point at a pH of 5 (Valle *et al.*, 1982).

Accordingly, a protein extraction protocol for an alkaline protein extraction from freeze-dried, ground and defatted *T. molitor* larvae was optimized varying the four factors NaOH concentration (0.1 vs 0.5 M), ratio of NaOH solution to defatted larvae (6:1 ml/g vs 20:1 ml/g), temperature (20 °C and 80 °C) and time (30 min and 120 min) (Zhao *et al.*, 2016). Optimum parameters found were 0.25 M NaOH, NaOH solution-to-defatted larvae ratio of 15:1, temperature of 40 °C and time of 60 min and a v-shaped protein solubility curve was obtained in accordance with other authors (Azagoh *et al.*, 2016; Bußler *et al.*, 2015a,b; Valle *et al.*, 1982). An alkaline extraction of *T. molitor* protein was also carried out by Azagoh *et al.* (2016). Beforehand, sample preparation of frozen larvae included blanching to prevent browning, pressing in a screw press for fat and water removal, drying at 75 °C for 6 h and grinding. The resulting meal as well as whole larvae were defatted using hexane and the protein was extracted at pH 10 at 45 °C for 60 min, centrifuged, and the supernatant freeze-dried. Interestingly, the protein solubility in dependence of pH of the larvae was higher than of the larvae meal resulting in protein yields of 59.9% for whole larvae and 26.4% for meal. As mentioned above, characteristic v-shaped solubility curves were obtained for both (Azagoh *et al.*, 2016).

After a systematic investigation of the dependence of pH, temperature and solvent molarity of protein solubility of *T. molitor* and black soldier fly (*Hermetia illucens*), a two-step extraction process first at alkaline conditions and then at acidic conditions in order to extract the maximum amount of proteins was suggested. V-shaped protein solubility curves similar to plant proteins were obtained and the required processing parameters seemed to be similar to those of plant proteins enabling the application of traditional methods of protein processing. The solubility properties were shown to be specific for each insect species. For example, in case of *T. molitor*, a fat removal by hexane increased protein solubility at low and medium pH, whereas it did not have an impact on *H. illucens* protein extraction. The sample preparations before the extraction process were performed at low temperatures in order to minimize thermal effects and obtain native proteins. Therefore, fresh larvae were frozen, pureed with water at 4 °C, frozen again, freeze-dried and ground before the protein extraction process (Bußler *et al.*, 2016b).

For an investigation of functional properties of the proteins of the five insect species *T. molitor* (larvae), *Zophobas morio* (larvae), *A. diaperinus* (larvae), *Acheta domesticus* (adult) and *Blaptica dubia* (adult), an aqueous protein extraction was performed by mixing 400 g of insects frozen in liquid nitrogen with 1,200 g distilled water and 2 g ascorbic acid in a mixer, followed by a filtration through a steel sieve with a pore size of 500 μm, and centrifugation of the filtrate at 15,000×*g* for 30 min at 4 °C resulting in a separation in filter residue, supernatant, precipitate/sediment (pellet) and fat fraction of which the filter residue, the supernatant and the pellet contained protein. The protein purity based on dry matter ranges from 50-61% for the supernatant, 58-69% for the filter residue to 65-75% for the pellet (Yi *et al.*, 2013). The protein fractions obtained were characterized including their protein purity, and the protein yield of the fractions and their respective foaming ability and ability to form a gel at pH 3, 5, 7, and 10 were investigated. The insect proteins investigated had the ability to form gels depending on their concentration and on the pH and had the potential to be used as gelling agents or texturizers in food (Yi *et al.*, 2013).

Mariod *et al.* (2011) extracted the protein gelatin from the two defatted, dried and ground Sudanese beetles *Aspongubus viduatus* and *Agonoscelis pubescens* either after an acidic pretreatment with 50 mM acetic acid in distilled water (45 °C), after an alkaline pretreatment with distilled water (50 °C), or after an alkaline pretreatment with hot to boiling water. The solutions obtained were filtrated, respectively. Except for the hot water extraction, the filtrates obtained were centrifuged at 10,000×*g* and 15 °C for 30 min. Subsequently, the extracted gelatins were freeze-dried. The microstructure of the proteins from different insect species differed and the extraction method also had an impact on protein properties (Mariod *et al.*, 2011). These results were confirmed by Cilia *et al.* (2009) who extracted proteins from aphids by comparing three protein extraction methods for proteomic studies. A comparison of the three protein extracts obtained by mass spectrometry and gel electrophoresis resulted in three unique characteristics and protein bands, different protein compositions, and different isoforms of single proteins. Their obtained results suggested that the protein extraction methods influence the properties of the extracted protein. The applicability of the gelatin extracted from insects in comparison to commercially available gelatin as a stabilizer in the ice cream production was also investigated. The properties of the ice cream containing insect gelatin as a stabilizer were rated as acceptable by a panel. In addition, there were no significant differences by the general preferences between ice cream produced with insect gelatin and produced with commercial gelatin (Mariod, 2013).

17.6 Lipid extraction

In industrial applications for the fat extraction from plant materials, a hexane extraction optionally preceding a pressing step predominates with oil yields of 95% and solvent recovery of over 95%. The solvent-based oil extraction process usually consists of extraction by continuous countercurrent washing with hexane of the previously cracked, flaked, ground, or pressed oleaginous material (e.g. oil seeds). The solvent hexane is recovered from the extracted meal in enclosed vessels by using jacket or sparged steam. Hexane is removed from the oil in rising film evaporators and by vacuum distillation (Rosenthal *et al.*, 1996).

Another possible method is an aqueous extraction. Advantages of an aqueous extraction process compared with solvent-based processes include: (1) simultaneous production of edible oil and protein isolate or concentration in the same process; (2) lower protein damage during extraction; and (3) improved process safety due to the lower risk of fire and explosion. Aqueous extraction processes may be more cost effective since the solvent recovery step is eliminated. The main limitations of this process appear to be: (1) lower efficiency of oil extraction as evident in earlier studies; (2) demulsification requirements to recover oil when emulsions are formed; and (3) treatment of the resulting aqueous effluent. Enzymes have been used to facilitate oil release resulting in oil yields of up to 90% (Rosenthal *et al.*, 1996). In addition, supercritical carbon dioxide (CO_2) is a possible solvent for a solvent-based lipid extraction.

Mechanical pressing is another fat extraction method resulting in comparable low yields. In addition, thermal impact on product properties and quality can occur. Another physical fat extraction is a mechanical separation of a lipid and an aqueous phase in a separator, decanter or centrifuge. Also, fat recovery by heating causing the fat to melt and to be released from the matrix can be applied.

The average fat content based on dry matter of insects per order range from 13.41% for Orthoptera (grasshoppers, crickets, locusts) being rich in protein to 33.40% for Coleoptera (beetles, grubs). In addition, bugs (Hemiptera), termites (Isoptera), Blattodea (cockroaches) and in some cases caterpillars (Lepidoptera) are also rich in fat (Rumpold and Schlüter, 2013a). There is evidence that lipid content as well as lipid composition of insects is dependent on the insect diet (Alves *et al.*, 2016; Beenakkers *et al.*, 1985; Oonincx *et al.*, 2015; Ramos-Elorduy *et al.*, 2002; St-Hilaire *et al.*, 2007; Tschirner and Simon, 2015; Van Broekhoven *et al.*, 2015) and developmental stage (Pearincott, 1960).

There is still comparably little scientific data available on the oil extraction from insects, although an oil extraction for the application of insect oils in food is interesting due to their fatty acid spectra and their partly high contents in unsaturated fatty acids (Rumpold and Schlüter, 2013a). For example, the potential of the oil from spent and desilked silkworm pupae *Bombyx mori* L., a by-product of the silk production, for the production of α-linolenic acid was shown by Wang *et al.* (2010). An oil extraction by supercritical CO_2 of dried larvae of *Clanis bilineata* (Lepidoptera), of which six thousand tons are consumed in China each year, resulted in oil yields of up to 97% at the chosen extraction parameters (25 °C, 60 min, 35 MPa, 20 l/min flow rate CO_2) (Wu, 2012). Pan *et al.* (2012) investigating the optimization of the oil extraction process by supercritical CO_2 of the Chinese oak tussah moth *Antheraea pernyi* (Lepidoptera) observed highest oil yields at extraction conditions of 28 MPa, 1.83 h, 35.3 °C und 20.26 l/h flow rate CO_2. Extracting oil from the beetle *A. pubescens* by supercritical CO_2 at 15 g/min flow rate for 30 min and at different pressures (20-40 MPa) and temperatures (50-70 °C) resulted in higher α-tocopherol contents, unaltered fatty acid compositions and lower yields in comparison to the conventional hexane extraction. The highest yield of 45%

by supercritical CO_2 extraction was gained at 30 MPa and 60 °C, whereas a hexane extraction yielded up to 60% (Mariod *et al.*, 2010). Purschke *et al.* (2016) furthermore stated that oil yields and oil composition were not substantially different when comparing hexane and supercritical CO_2 as solvents for *T. molitor* lipids and additionally concluded that selective extraction conditions enabled simple time-dependent fractionation and deacidification yielding fractions with tailored fatty acid profiles or facilitating subsequent refining processes when using supercritical CO_2. Optimum extraction parameters for a maximum defatting of 95% were pressures of 40 or 25 MPa at 45 °C for 105 min.

In contrast, a comparison of oil extraction methods with aqueous and organic solvents via Soxhlet and Folch extraction from the four insect species *T. molitor*, *A. diaperinus* (buffalo worm), *A. domesticus* (house cricket) and *B. dubia* (cockroach) resulted in differences in oil yields and lipid properties in dependence of the extraction method and showed the highest oil yields by Folch extraction and the lowest yields by aqueous extraction. However, the aqueous extraction resulted in lipids with low ω-6/ω-3 ratios and less polar lipids than the Soxhlet and Folch extractions, which is desired in a food oil (Tzompa-Sosa *et al.*, 2014).

This variance indicates that the oil extraction method might influence the quality and composition of the insect oil. Furthermore, the sample pretreatment influenced lipid quality and composition. A roasting step for up to 15 min of freeze-dried mealworm larvae (*T. molitor*) and subsequent lipid extraction via hexane resulted in oils with increasing browning, a slight shift of unsaturated fatty acids to saturated fatty acids and decreasing tocopherol contents but increasing oxidative stability as indicated by induction time with increasing roasting time (Jeon *et al.*, 2016).

In addition, the further treatment steps of the insect oil or fat affect its quality. For example, the purification of the oil from the beetle *A. viduatus* after oil extraction affected its antibacterial properties. The raw oil and the phenol free oil showed high antibacterial properties whereas after a purification on a silicic acid column, the oil showed no antibacterial activity anymore (Mustafa *et al.*, 2008). An investigation of the influence of several thermal treatments and storage scenarios on lipid properties of the oil of the palm weevil larvae *Rhynchophorus phoenicis* showed that by boiling, cool storage for three days, freezing, sun drying and oven drying the acidity of the oil was increased and by boiling followed by smoking the acidity of the oil was decreased (Tiencheu *et al.*, 2013).

Investigating the frying properties and oxidative stability of the oil from the beetle *A. pubescens* resulted in a general applicability of the insect oil as frying oil for potatoes. However after utilization for 6-12 hours at 175 °C, sensory evaluations of the fried potatoes revealed an insufficient sensory quality. After utilization of the oil for 18 hours at 175 °C, the content of free fatty acids, oligomeric triacylglycerides and polar portions exceeded the maximum values according to the guidelines of the Deutsche Gesellschaft für Fettwissenschaft e.V. (Münster, Germany) (Mariod *et al.*, 2006).

Lipid removal also influences the properties of the non-fat fraction and is species-specific. Prior to a protein extraction, a fat extraction of *H. illucens* and *T. molitor* flour was performed using hexane. Although insect flour from *Hermetia* and *Tenebrio* contained nearly the same quantity of fat, defatting of the *H. illucens* flour was less effective. In both cases processability (e.g. grinding/milling) and analyzability of the flours was improved substantially by removal of the fat. In the case of *H. illucens*, hexane extraction of the fat did not lead to increased yields in soluble proteins; on the contrary, protein solubility was even decreased at low pH values. Besides the composition of the insect flour, defatting via hexane extraction process also influences the techno-functional properties of the raw materials (Bußler *et al.*, 2015a,b, 2016b).

It can be concluded that while for the oil extraction from insects, conventional extraction methods can be adapted, the lipid quality needs to be carefully monitored and the extraction yield needs to be optimized.

17.7 Chitin extraction

The polysaccharide chitin (N-acetyl-glucosamine) is utilized for the production of chitosan via deacetylation. Chitin extraction is either performed chemically or biologically via fermentation with microorganisms or enzymes. Quality parameters include purity, degree of acetylation (DA), molecular weight (MW) and polydispersity index (Kaur and Dhillon, 2015), which is a heterogeneity index that indicates the distribution of molecular masses in a non-uniform, polydisperse polymer. Also, the MW is of interest since it influences physicochemical and functional properties of the resulting chitosan (Kaya *et al.*, 2014a). Chitin sources include marine food processing waste such as shrimp waste or crab shells (crustacean shells), mollusks, fungi, and insects (Kaur and Dhillon, 2015).

On an industrial scale, the chemical extraction method including an acidic demineralization and an alkaline deproteination step is applied. The biological extraction of chitin from waste is performed by lactic acid fermentation using *Lactobacillus plantarum* (Rao and Stevens, 2005) or using a mixture of lactic acid and proteolytic enzyme producing bacteria such as *L. plantarum*, *Lactobacillus salivarus*, *Streptococcus faecium* and *Pediococcus acidilactici* (Healy *et al.*, 2003).

While the chemical process isolates chitin effectively, the protein is rendered useless (Healy *et al.*, 2003). The biological extraction also yields a solubilized protein. Advantages and disadvantages of chemical and biological extraction methods are summarized in Table 17.2. It can be concluded that for a holistic approach on the utilization of insects as feedstuff, a chitin extraction from insects should be done after the protein and fat extraction.

In insects, chitin is located in the exoskeleton where it is bound to skleroprotein. Information on chitin contents of insects in scientific literature is not conclusive. Chitin contents were estimated as low as 1.16% for bee brood from *Apis mellifera*, 3.81% for *Galleria mellonella* larvae, 5.57% for giant *T. molitor* larvae, 6.66% for *B. mori* larvae, 6.76% for *A. domesticus* adults, 8.15% for *A. domesticus* nymphs to 13.72% for *T. molitor* adults (on a dry matter basis) (Finke, 2007). In addition, values as high as 16.5 to 20.5% for two Orthopteran species (based on dry matter) (Kaya *et al.*, 2015a) and 54% for flies and 64% for butterflies (based on the dry weight of the organic cuticle) (Kaur and Dhillon, 2015) are mentioned.

Table 17.2. Pros and cons of traditional chemical extraction and biological extraction methods of chitin (adapted from Kaur and Dhillon, 2015).

	Chemical extraction	Biological extraction
Parameter		
	• Demineralization treatment with concentrated acids (HCl, HNO_3, H_2SO_4, CH_3COOH and HCOOH), temperature >100 °C and long processing time	• Demineralization carried out in situ by lactic acid produced by bacteria through the conversion of added carbon
	• Deproteination with alkali (NaOH, KOH) 30-50% w/v, up to 100 °C, up to 72 h	• Deproteination carried out by proteases secreted into the fermentation medium or by adding exo-proteases and/or proteolytic bacteria
Advantages/disadvantages		
	• Well established	• Processes limited to date to laboratory scale studies
	• Not environmentally friendly, generation of huge amounts of waste	• Environmentally safe, no generation of waste
	• High costs, laborious process	
	• Difficult to extract by-products due to concentrated acids and alkali	
Quality of extracted chitin/other remarks		
	• Inconsistent levels of degree of acetylation and high molecular weight which negatively affects the intrinsic properties of the purified chitin	• Final product homogeneous and of high quality
		• Extraction cost can be brought down using abundant waste biomass as carbon source
	• This method allows complete removal of organic salts, but at the same tie deacetylation and depolmerisation may occur	• Solubilized protein and minerals may be used as human or animal nutrients
	• Effluent treatment of acid and alkaline extraction of chitin may cause increase in the cost of chitin	• Co-culture or consortium of bacteria producing lactic acid and proteases can be used for single step extraction of chitin.

In all investigations on insect chitin, the chitin was extracted chemically on a laboratory scale (Kaya *et al.*, 2015a,c; Liu *et al.*, 2012; Paulino *et al.*, 2006; Zhang *et al.*, 2000, 2011). For example, Zhang *et al.* (2011) prepared and characterized food-grade chitosan from house fly larvae applying an alkaline extraction method. The extraction procedure resulted in chitosan with a degree of deacetylation of 83.1%.

Liu *et al.* (2012) isolated and characterized chitin from the adult beetles *Holotrichia parallela* using standard chemical methods (1 M HCl, then 1 M NaOH, followed by 1% potassium permanganate solution for decolorization). The obtained α-chitin from the beetles had a DA of 93.1% and a crystallinity of 89.05% with chemical structures and physico-chemical properties comparable to shrimp chitin. The yield of chitin was around 15%.

Chitosan and chitin from dried silkworm chrysalides, a by-product from the silk industry, was also produced and characterized (Paulino *et al.*, 2006). The chitin was extracted using a chemical extraction method, and the chitosan was produced by deacetylation of the chitin using NaOH containing $NaBH_4$. Chitin and chitosan with a high purity were obtained but yields did not exceed 4.23%. The degree of deacetylation of the chitosan averaged 83%.

Chitin and chitosan obtained from larvae and adults of the Colorado potato beetle *Leptinotarsa decemlineata* were also investigated by Kaya *et al.* (2014a). The obtained chitins had an α-structure and low MWs (2.72 kDa for adults and 2.68 kDa for larvae). In addition, chitin yields of 72% (adults) and 67% (larvae) were obtained, where the chitin content of the adults was with 20% (based on dry weight) almost three times higher than the chitin content of the larvae with 7% (based on dry weight). Crystalline index values of the chitins were determined by x-ray diffraction analyses to be 76% (adults) and 72% (larvae). Elemental analyses resulted in degrees of acetylation of the chitins of 108% (adults) and 232% (larvae) and degrees of deacetylation of chitosans of 82% (adults) and 76% (larvae) Whereas the DA above 100% were attributed to mineral residues, the degree of deacetylation close to 100% were considered to represent chitosans with a high purity.

An extraction of chitin from seven *Orthoptera* (grasshopper) species also resulted in α-chitin with low molecular masses ranging between 5.2 to 6.8 kDa (Kaya *et al.*, 2015c). Chitin and chitosan of two additional orthopteran species with chitin contents of 16.5% (*Oedalus decorus*) and 20.5% (*Calliptamus barbarus*) (based on dry weight), respectively, were also characterized. α-Chitins were obtained and their DA amounted to 90.64% (*O. decorus*) and 87.80% (*C. barbarus*), and their degree of crystallinity was 76.8% (*O. decorus*) and 70.9% (*C. barbarus*). Chitosan yields of 74-76% were obtained with a degree of deacetylation of 70-75% (Kaya *et al.*, 2015a).

Investigations on the structure and morphology of chitin derived from insects revealed a new uncommon morphology found for chitin from wings of the cockroach *Periplaneta americana* (Kaya and Baran, 2015) and differing nanofibers and pore structures observed for chitin from the beetle *Melolontha melolontha* in comparison to a crustacean species (Kaya *et al.*, 2014b). In addition, different surface morphologies were determined for chitins of different body parts of the same insect species, that depended on different functions of the chitin relevant in the different parts of the body (Kaya *et al.*, 2015b). Different surface morphologies were also found for chitins of female and male insects of the same species (Kaya *et al.*, 2015d) and surface characteristics of chitin altered along the developmental stages of the insect (Kaya *et al.*, 2016).

17.8 *Tenebrio molitor* and browning

A highly discussed phenomenon is the browning of *T. molitor* upon processing and/or (cold) storage. It was observed that the browning is dependent on the pretreatment of the larvae (e.g. blanched or not, frozen, freeze-dried, microwave-dried) (L. Van Campenhout, personal communication). For example, Vandeweyer *et al.* (2017) observed mealworm larvae to have a dark colour upon blanching for 40 s followed by industrial microwave-drying for 16 and 20 min and thus considered them as over-processed due to the long drying times. This was contradicted by Azzolini *et al.* (2016) who observed profound differences in product colour of mealworm larvae after drying at up to 70 °C with and without blanching applied before drying to prevent the darkening in colour. It was suggested this darkening in colour was due to melanosis analogous to crustaceans, which is an enzymatic oxidative reaction leading to darkening of crustaceans such as lobster, crabs and shrimp (Azzolini *et al.*, 2016). However, it is assumed the browning of *T. molitor* is not solely due to enzymatic reactions since strong browning was observed after treatments where all enzymes were expected to be inactivated. For instance, unblanched larvae stored for one week at -80 °C still showed browning after they were thawed (L. Van Campenhout, personal communication).

The oxidative stability of raw and pan-fried *T. molitor* larvae during storage at 4 °C was also investigated. The pan frying itself led to profound darkening of colour (browning index 1.042 at day 0; see Box 17.1) that did not change considerably upon storage at 4 °C for 18 days, whereas the browning index of raw larvae increased from 0.276 at day 0 to 1.001 after 18 days of storage (Kim *et al.*, 2015). These results suggest that a heat treatment at 140 °C in the presence of oxygen does not prevent browning of mealworm larvae and that melanoidins formed during a Maillard reaction between amino acids and reducing sugars could be responsible for the darkening in colour.

For an investigation of the physicochemical properties and oxidative stabilities of *T. molitor* oils, freeze-dried mealworm larvae were roasted for 0, 5, 10, and 15 min at 200 °C before a lipid extraction using hexane. It was observed that the browning index of the oil increased with increasing roasting time from 0.205 for oil from unroasted larvae to 0.839 for oil from larvae roasted for 15 min. Furthermore, while the tocopherol content decreased, the oxidative stability, as indicated by induction time, increased with increasing roasting time (Jeon *et al.*, 2016). This also indicates a formation of lipophilic melanoidins due to Maillard reaction since melanoidins have antioxidant properties.

Investigating the impact of drying temperature on quality and safety attributes of mealworm larvae and flour, increasing the drying temperature up to 60 °C was found to increase darkening of the flour (Figure 17.2 and 17.3) that was produced by (1) boiling fresh larvae for decontamination and inactivation purposes; (2) oven drying; (3) mechanical fat removal via pressing; and (4) milling during own investigations performed by the authors (B.A. Rumpold *et al.*, unpublished data). This is mainly attributed to accelerated enzymatic browning reactions. It was shown that the activity of polyphenoloxidase (PPO), an enzyme in crustaceans responsible for browning, is temperature-dependent and decreases with

Box 17.1. Browning index.

After spectrophotometric measurement of L*-values (brightness), a*-values (green-red axis) and b*-values (blue-yellow axis), the browning index is calculated as follows:

$$\text{browning index} = \frac{100\,(x - 0.31)}{0.17}$$

$$\text{where } x = \frac{a^* + 1.75L^*}{5.645L^* + a^* - 0.3012b^*}$$

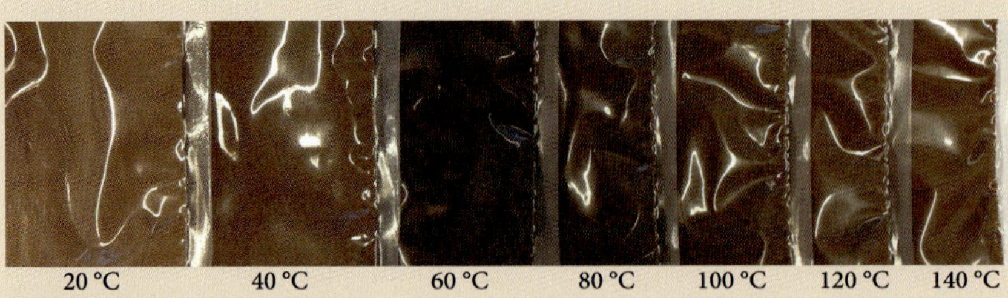

Figure 17.2. Impact of drying temperature on the colour of *Tenebrio molitor* flour after drying whole larvae for 25 h and subsequent milling (B.A. Rumpold *et al.*, unpublished data).

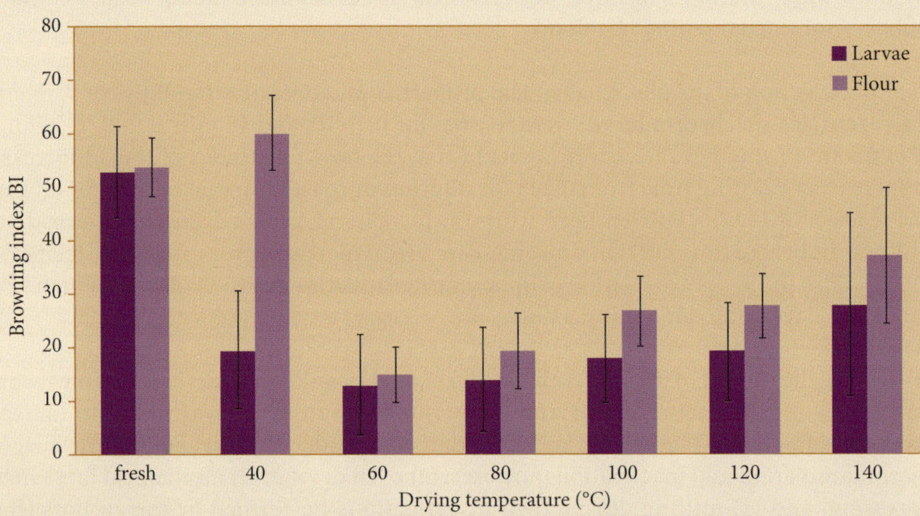

Figure 17.3. Impact of drying temperature on browning index of larvae and meal of *Tenebrio molitor* after drying for 25 h along the dry processing route (B.A. Rumpold *et al.*, unpublished data).

increasing temperature. After 2 min at 60 °C or higher temperatures, the activity of PPO converges towards zero (Verhaeghe *et al.*, 2016). Further increase of the drying temperature to up to 140 °C induced less pronounced darkening of the flours. It is assumed that enzymatic browning is prevented whereas Maillard reactions occur at higher temperatures above 60 °C (B.A. Rumpold *et al.*, unpublished data).

Upon optimizing protein extraction conditions, it was observed that the colour of the protein extract became darker with increasing temperature and treatment time. In addition, a noticeable darkening of the protein extracts at high pH was observed during protein extraction at 60 °C; however, the protein concentrations in the extracts differ (B.A. Rumpold *et al.*, unpublished data; Figure 17.4). By contrast, Zhao *et al.* (2016) determined that 2% (w/v) dispersions of protein extracts from freeze-dried *T. molitor* larvae became clearer, brighter and more yellow with increasing pH between 7 and 13, which was confirmed by colour measurements. It can nevertheless be concluded that the colour of *Tenebrio* proteins is pH dependent.

17.9 Potential insect processing pathways

Processing insects into a microbiologically safe food or feed products usually involves several unit operations and process steps depending on the intended product properties as shown in the previous sections. Especially the high fat content of certain insect species such as *H. illucens* may require a lipid removal step. Three possible processing pathways for the production of an insect lipid fraction and a protein-rich insect meal are introduced in this section. Ethical aspects on humane inactivation of insects are not considered or discussed in this section.

17.9.1 Insect processing pathway according to Arsiwalla and Aarts (2015)

A method to convert insects into nutrient streams consisting of a fat, aqueous protein and solids fraction was patented (patent number US 20150374005 A1; Arsiwalla and Aarts, 2015). The method consists of the following steps: (1) size reduction by squashing the insects into a pulp and possibly enzymatic hydrolysis; (2) heating the pulp to 70-100 °C; (3) physical separation e.g. by centrifugation and/or decanting; and (4) drying e.g. by spray drying or air-drying (Figure 17.5).

Critical steps along this pathway could be the heat treatment, where burning and caking of the pulp has to be prevented and microbial safety has to be ensured. Also, the lipid extraction needs to be taken into account as a critical step, since a physical separation could be impeded by emulsifying agents occurring in the insect.

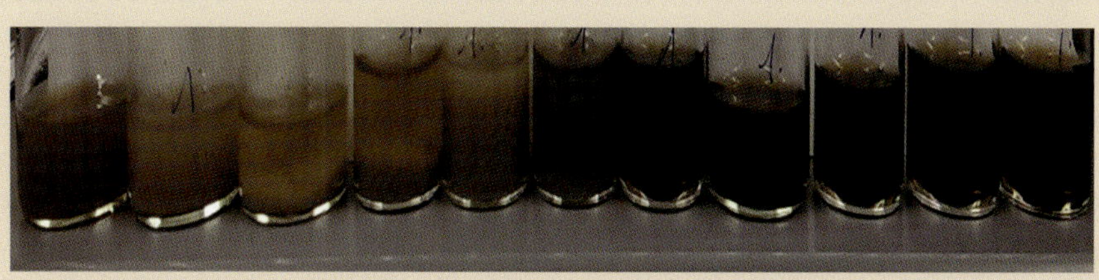

Figure 17.4. Impact of pH from 2 to 12 (in single steps from left to right) during protein extraction of *Tenebrio molitor* on the colour of the extracts (B.A. Rumpold *et al.*, unpublished data).

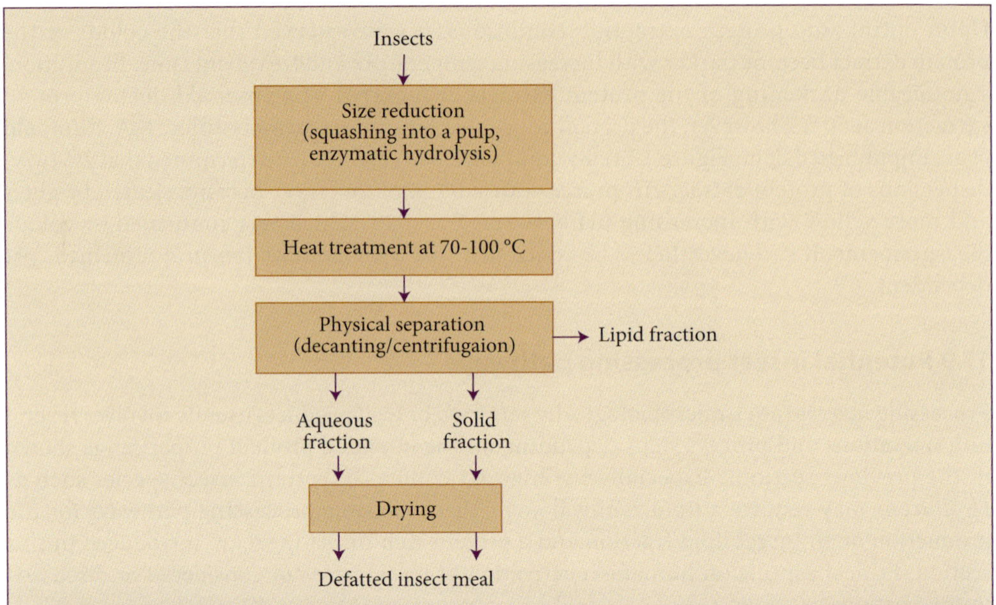

Figure 17.5. Potential insect processing pathway according to Arsiwalla and Aarts (2015).

17.9.2 Aqueous insect processing pathway analogous to the fish meal production process

Another possible processing pathway that is analogous to the fish meal production process consists of the following steps: (1) boiling of the insects with excess water (blanching); (2) mechanical pressing of the insect-water-mixture resulting in the press cake; which is (3) dried to an insect meal, and the press liquid; from which (4) the lipid fraction is removed mechanically via e.g. centrifugation or decanting; and subsequently (5) the soluble solids including soluble proteins are recovered by concentration and (3) drying (Figure 17.6).

This aqueous pathway consumes a considerable amount of water, which is environmentally unfriendly and economically inefficient, also since the drying step is more energy consuming due to an increased amount of water to be removed. Except for the addition of water and the squashing into a pulp before the heat treatment, this process pathway and the one introduced before do not differ essentially. As mentioned before, the lipid removal could be impeded by emulsifying agents present in the insect when lipid is removed without the use of solvents. This processing pathway was applied by Azagoh *et al.* (2016) on *T. molitor* on a laboratory scale and additionally included a grinding step after the drying, but not a recovery of soluble proteins. Here, the blanching step was applied in order to prevent the aforementioned browning.

17.9.3 Dry insect processing pathway

In this processing pathway, a precautionary decontamination step is preferably performed by microwave- or steam blanching in order to avoid the addition of access water commonly applied during hot water blanching. Omission of hot water blanching would also prevent the occurring loss of soluble proteins, the consumption of considerable amounts of water and additional energy needed for the drying. After a possible decontamination step, the insects

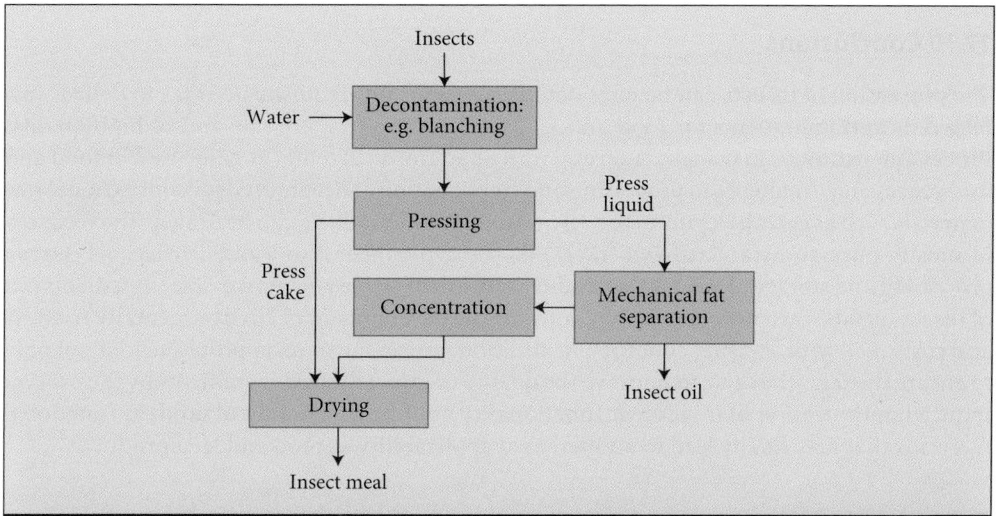

Figure 17.6. Potential insect processing pathway analogous to fish meal production.

are (1) dried e.g. air-dried; (2) lipids are removed mechanically e.g. via pressing; and (3) the resulting press cake is ground into a defatted insect meal (Figure 17.7).

The major difference between the dry pathway and the other two pathways is the drying step before the mechanical lipid separation. The lipid removal was interposed to facilitate the grinding step that was impeded by smearing due to fat. Comparing all three pathways, it can be observed that they all include a heat treatment before the mechanical lipid removal.

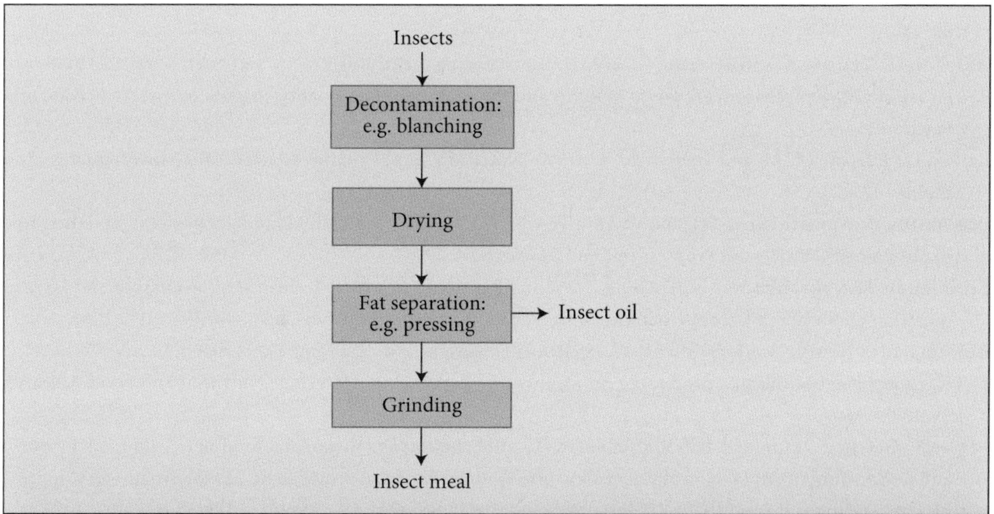

Figure 17.7. Potential dry insect processing pathway.

17.10 Conclusions

The processing of insects can be considered as a major pre-requisite in order to deliver raw materials and ingredients for food and feed applications that are safe and of high quality. Processing pathways may include several unit operations currently already used in food and feed processing. Analogies to processing and fractionation concepts used for conventional raw materials such as cereals, legumes but also animal based products can be drawn. The recovery of mostly pure and standardized fractions, the avoidance of unwanted reactions during processing and storage as well as the production of shelf stable and marketable ingredients out of insects can be seen as major motivations for the development of environmentally friendly and cost efficient processing concepts. In addition, processing is an appropriate tool not only to ensure the safety but also to improve the quality of insect derived ingredients by the process induced optimization of ingredient functionality and the production of tailored functional fractions that may finally lead to an increased applicability as food and feed products.

Acknowledgements

We thank Bühler AG (Uzwil, Switzerland) for supporting part of this work.

References

Alves, A.V., Sanjinez-Argandona, E.J., Linzmeier, A.M., Cardoso, C.A.L. and Macedo, M.L.R., 2016. Food value of mealworm grown on *Acrocomia aculeata* pulp flour. PLoS ONE 11: e0151275.

Arsiwalla, T. and Aarts, K., 2015. Method to convert insects or worms into nutrient streams and compositions obtained thereby. Patent: WO2014123420 A1. Available at: http://www.google.com/patents/WO2014123420A1?cl=en.

Awoniyi, T.A.M., Aletor, V.A. and Aina, J.M., 2003. Performance of broiler – chickens fed on maggot meal in place of fishmeal. International Journal of Poultry Science 2: 271-274.

Azagoh, C., Ducept, F., Garcia, R., Rakotozafy, L., Cuvelier, M.E., Keller, S., Lewandowski, R. and Mezdour, S., 2016. Extraction and physicochemical characterization of *Tenebrio molitor* proteins. Food Research International 88: 24-31.

Azzolini, D., Derossi, A. and Severini, C., 2016. Understanding the drying kinetic and hygroscopic behaviour of larvae of yellow mealworm (*Tenebrio molitor*) and the effects on their quality. Journal of Insects as Food and Feed 2: 233-243.

Barker, D., Fitzpatrick, M. and Dierenfeld, E., 1998. Nutrient composition of selected whole invertebrates. Zoo Biology 17: 123-134.

Beenakkers, A., Van der Horst, D.J. and Van Marrewijk, W.J.A., 1985. Insect lipids and lipoproteins, and their role in physiological processes. Progress in Lipid Research 24: 19-67.

Broekhoven, S.V., Bastiaan-Net, S., De Jong, N.W. and Wichers, H.J., 2016. Influence of processing and *in vitro* digestion on the allergic cross-reactivity of three mealworm species. Food Chemistry 196: 1075-1083.

Broekman, H., Knulst, A., Jager, S.D.H., Monteleone, F., Gaspari, M., De Jong, G., Houben, G. and Verhoeckx, K., 2015. Effect of thermal processing on mealworm allergenicity. Molecular Nutrition and Food Research 59: 1855-1864.

Bußler, S., Guérin, L., Rumpold, B.A. and Schlüter, O., 2015a. Insects as a sustainable food source for food: recovery and techno-functionality of meal worm (*Tenebrio molitor*) flour and proteins. In: 12th International Congress on Engineering and Food (ICEF12), June 14-18, 2015, Quebec-City, Canada.

Bußler, S., Guérin, L., Rumpold, B.A. and Schlüter, O., 2015b. Protein extraction from edible insects: techno-functional properties of flours and proteins from meal worm and black soldier fly larvae. In: IFT15: Where Science Feeds Innovation, July 11-14, 2015, Chicago, IL, USA.

Bußler, S., Rumpold, B.A., Fröhling, A., Jander, E., Rawel, H.M. and Schlüter, O.K., 2016a. Cold atmospheric pressure plasma processing of insect flour from *Tenebrio molitor*: impact on microbial load and quality attributes in comparison to dry heat treatment. Innovative Food Science and Emerging Technologies 36: 277-286.

Bußler, S., Rumpold, B.A., Jander, E., Rawel, H.M. and Schlüter, O.K., 2016b. Recovery and techno-functionality of flours and proteins from two edible insect species: meal worm (*Tenebrio molitor*) and black soldier fly (*Hermetia illucens*) larvae. Heliyon 2: e00218.

Cilia, M., Fish, T., Yang, X., McLaughlin, M., Thannhauser, T. and Gray, S., 2009. A comparison of protein extraction methods suitable for gel-based proteomic studies of aphid proteins. Journal of Biomolecular Techniques 20: 201-215.

Finke, M.D., 2002. Complete nutrient composition of commercially raised invertebrates used as food for insectivores. Zoo Biology 21: 269-285.

Finke, M.D., 2007. Estimate of chitin in raw whole insects. Zoo Biology 26: 105-115.

Finke, M.D., 2012. Complete nutrient content of four species of feeder insects. Zoo Biology 32: 27-36.

Gueguen, J., 1983. Legume seed protein extraction, processing, and end product characteristics. Qualitas Plantarum-Plant Foods for Human Nutrition 32: 267-303.

Healy, M., Green, A. and Healy, A., 2003. Bioprocessing of marine crustacean shell waste. Acta Biotechnologica 23: 151-160.

Hwangbo, J., Hong, E.C., Jang, A., Kang, H.K., Oh, J.S., Kim, B.W. and Park, B.S., 2009. Utilization of house fly-maggots, a feed supplement in the production of broiler chickens. Journal of Environmental Biology 30: 609-614.

Ijaiya, A.T. and Eko, E.O., 2009. Effect of replacing dietary fish meal with silkworm (Anaphe infracta) caterpillar meal on growth, digestibility and economics of production of starter broiler chickens. Pakistan Journal of Nutrition 8: 845-849.

Jeon, Y.-H., Son, Y.-J., Kim, S.-H., Yun, E.-Y., Kang, H.-J. and Hwang, I.-K., 2016. Physicochemical properties and oxidative stabilities of mealworm (*Tenebrio molitor*) oils under different roasting conditions. Food Science and Biotechnology 25: 105-110.

Ju, Z.Y., Hettiarachchy, N.S. and Rath, N., 2001. Extraction, denaturation and hydrophobic properties of rice flour proteins. Journal of Food Science 66: 229-232.

Kaur, S. and Dhillon, G.S., 2015. Recent trends in biological extraction of chitin from marine shell wastes: a review. Critical Reviews in Biotechnology 35: 44-61.

Kaya, M. and Baran, T., 2015. Description of a new surface morphology for chitin extracted from wings of cockroach (*Periplaneta americana*). International Journal of Biological Macromolecules 75: 7-12.

Kaya, M., Baran, T., Asan-Ozusaglam, M., Cakmak, Y.S., Tozak, K.O., Mol, A., Mentes, A. and Sezen, G., 2015a. Extraction and characterization of chitin and chitosan with antimicrobial and antioxidant activities from cosmopolitan Orthoptera species (Insecta). Biotechnology and Bioprocess Engineering 20: 168-179.

Kaya, M., Baran, T., Erdogan, S., Mentes, A., Ozusaglam, M.A. and Cakmak, Y.S., 2014a. Physicochemical comparison of chitin and chitosan obtained from larvae and adult Colorado potato beetle (*Leptinotarsa decemlineata*). Materials Science and Engineering C-Materials for Biological Applications 45: 72-81.

Kaya, M., Baublys, V., Can, E., Satkauskiene, I., Bitim, B., Tubelyte, V. and Baran, T., 2014b. Comparison of physicochemical properties of chitins isolated from an insect (*Melolontha melolontha*) and a crustacean species (*Oniscus asellus*). Zoomorphology 133: 285-293.

Kaya, M., Bitim, B., Mujtaba, M. and Koyuncu, T., 2015b. Surface morphology of chitin highly related with the isolated body part of butterfly (*Argynnis pandora*). International Journal of Biological Macromolecules 81: 443-449.

Kaya, M., Erdogan, S., Mol, A. and Baran, T., 2015c. Comparison of chitin structures isolated from seven *Orthoptera* species. International Journal of Biological Macromolecules 72: 797-805.

Kaya, M., Lelesius, E., Nagrockaite, R., Sargin, I., Arslan, G., Mol, A., Baran, T., Can, E. and Bitim, B., 2015d. Differentiations of Chitin content and surface morphologies of Chitins extracted from male and female grasshopper species. PLoS ONE 10: e0115531.

Kaya, M., Sofi, K., Sargin, I. and Mujtaba, M., 2016. Changes in physicochemical properties of chitin at developmental stages (larvae, pupa and adult) of *Vespa crabro* (wasp). Carbohydrate Polymers 145: 64-70.

Kim, S.-Y., Son, Y.-J., Kim, S.-H., Kim, A.-N., Lee, G.-Y. and Hwang, I.-K., 2015. Studies on oxidative stability of *Tenebrio molitor* larvae during cold storage. Korean Journal of Food and Cookery Science 31: 62-71.

Klunder, H.C., Wolkers-Rooijackers, J., Korpela, J.M. and Nout, M.J.R., 2012. Microbiological aspects of processing and storage of edible insects. Food Control 26: 628-631.

Knorr, D., Froehling, A., Jaeger, H., Reineke, K., Schlueter, O. and Schoessler, K., 2011. Emerging technologies in food processing. Annual Review of Food Science and Technology 2: 203-235.

Koo, S.I., Currin, T.A., Johnson, M.G., King, E.W. and Turk, D.E., 1980. The nutritional value and microbial content of dried face fly pupae (*Musca autumnalis* (Degeer)) when fed to chicks. Poultry Science 59: 2514-2518.

Liu, S., Sun, J., Yu, L., Zhang, C., Bi, J., Zhu, F., Qu, M., Jiang, C. and Yang, Q., 2012. Extraction and characterization of Chitin from the beetle *Holotrichia parallela motschulsky*. Molecules 17: 4604-4611.

Mariod, A., Matthaeus, B., Eichner, K. and Hussein, I.H., 2006. Frying quality and oxidative stability of two unconventional oils. Journal of the American Oil Chemists Society 83: 529-538.

Mariod, A.A., 2013. Insect oil and protein: biochemistry, food and other uses: review. Agricultural Sciences 4: 76-80.

Mariod, A.A., Abdelwahab, S.I., Gedi, M.A. and Solati, Z., 2010. Supercritical carbon dioxide extraction of sorghum bug (*Agonoscelis pubescens*) oil using response surface methodology. Journal of the American Oil Chemists Society 87: 849-856.

Mariod, A.A., Abdelwahab, S.I., Ibrahim, M.Y., Mohan, S., Abd Elgadir, M. and Ain, N.M., 2011. Preparation and characterization of gelatins from two Sudanese edible insects. Journal of Food Science and Engineering 1: 45-55.

Mujumdar, A., 2011. Drying. In: Kirk-Othmer Encyclopedia of Chemical Technology. Wiley, Hoboken, NJ, USA, pp. 1-53.

Mustafa, N.E.M., Mariod, A.A. and Matthaeus, B., 2008. Antibacterial activity of *Aspongopus viduatus* (melon bug) oil. Journal of Food Safety 28: 577-586.

Newton, G.L., Booram, C.V., Barker, R.W. and Hale, O.M., 1977. Dried *Hermetia illucens* larvae meal as a supplement for swine. Journal of Animal Science 44: 395-400.

Odesanya, B.O., Ajayi, S.O., Agbaogun, B.K.O. and Okuneye, B., 2011. Comparative evaluation of nutritive value of maggots. International Journal of Scientific & Engineering Research 2: 1-5.

Oonincx, D.G.A.B., Van Broekhoven, S., Van Huis, A. and Van Loon, J.J.A., 2015. Feed conversion, survival and development, and composition of four insect species on diets composed of food by-products. PLoS ONE 10: e0144601.

Oyegoke, O.O., Akintola, A.J. and Fasoranti, J.O., 2006. Dietary potentials of the edible larvae of *Cirina forda* (Westwood) as a poultry feed. African Journal of Biotechnology 5: 1799-1802.

Ozimek, L., Sauer, W.C., Kozikowski, V., Ryan, J.K., Jorgensen, H. and Jelen, P., 1985. Nutritive value of protein extracted from honey bees. Journal of Food Science 50: 1327-1332.

Pan, W.-J., Liao, A.-M., Zhang, J.-G., Dong, Z. and Wei, Z.-J., 2012. supercritical carbon dioxide extraction of the oak silkworm (*Antheraea pernyi*) pupal oil: process optimization and composition determination. International Journal of Molecular Sciences 13: 2354-2367.

Papadoyanis, E.D. and Cherch, X.T., 2008. Production and processing of insects for transformation into protein meal for fish and animal diets. Patent: US20080075818 A1. Available at: http://www.google.ch/patents/US20080075818.

Paulino, A.T., Simionato, J.I., Garcia, J.C. and Nozaki, J., 2006. Characterization of chitosan and chitin produced from silkworm crysalides. Carbohydrate Polymers 64: 98-103.

Pearincott, J.V., 1960. Changes in the lipid content during growth and metamorphosis of the house fly, *Musca domestica* Linnaeus. Journal of Cellular and Comparative Physiology 55: 167-174.

Pretorius, Q., 2011. The evaluation of larvae of *Musca domestica* (common house fly) as protein source for broiler production. MSc thesis, University of Stellenbosch, Stellenbosch, South Africa.

Purschke, B., Stegmann, T., Schreiner, M. and Jäger, H., 2016. Pilot-scale supercritical CO_2 extraction of edible insect oil from *Tenebrio molitor* L. larvae – influence of extraction conditions on kinetics, defatting performance and compositional properties. European Journal of Lipid Science and Technology 119: 1600134.

Ramos-Elorduy, J., Gonzalez, E.A., Hernandez, A.R. and Pino, J.M., 2002. Use of *Tenebrio molitor* (Coleoptera: Tenebrionidae) to recycle organic wastes and as feed for broiler chickens. Journal of Economic Entomology 95: 214-220.

Rao, M.S. and Stevens, W.F., 2005. Chitin production by *Lactobacillus fermentation* of shrimp biowaste in a drum reactor and its chemical conversion to chitosan. Journal of Chemical Technology and Biotechnology 80: 1080-1087.

Rosenthal, A., Pyle, D.L. and Niranjan, K., 1996. Aqueous and enzymatic processes for edible oil extraction. Enzyme and Microbial Technology 19: 402-420.

Rumpold, B.A., Fröhling, A., Reineke, K., Knorr, D., Boguslawski, S., Ehlbeck, J. and Schlüter, O., 2014. Comparison of volumetric and surface decontamination techniques for innovative processing of mealworm larvae (*Tenebrio molitor*). Innovative Food Science and Emerging Technologies 26: 232-241.

Rumpold, B.A. and Schlüter, O.K., 2013a. Nutritional composition and safety aspects of edible insects. Molecular Nutrition and Food Research 57: 802-823.

Rumpold, B.A. and Schlüter, O.K., 2013b. Potential and challenges of insects as an innovative source for food and feed production. Innovative Food Science and Emerging Technologies 17: 1-11.

St-Hilaire, S., Cranfill, K., McGuire, M.A., Mosley, E.E., Tomberlin, J.K., Newton, L., Sealey, W., Sheppard, C. and Irving, S., 2007. Fish offal recycling by the black soldier fly produces a foodstuff high in omega-3 fatty acids. Journal of the World Aquaculture Society 38: 309-313.

Stoops, J., Crauwels, S., Waud, M., Claes, J., Lievens, B. and Van Campenhout, L., 2016. Microbial community assessment of mealworm larvae (*Tenebrio molitor*) and grasshoppers (*Locusta migratoria migratorioides*) sold for human consumption. Food Microbiology 53, Part B: 122-127.

Tiencheu, B., Womeni, H.M., Linder, M., Mbiapo, F.T., Villeneuve, P., Fanni, J. and Parmentier, M., 2013. Changes of lipids in insect (*Rhynchophorus phoenicis*) during cooking and storage. European Journal of Lipid Science and Technology 115: 186-195.

Tscheuschner, H.-D., 1996. Grundzüge der Lebensmitteltechnik. Behr's Verlag, Hamburg, Germany.

Tschirner, M. and Simon, A., 2015. Influence of different growing substrates and processing on the nutrient composition of black soldier fly larvae destined for animal feed. Journal of Insects as Food and Feed 1: 249-259.

Tzompa-Sosa, D.A., Yi, L., Van Valenberg, H.J.F., Van Boekel, M.A.J.S. and Lakemond, C.M.M., 2014. Insect lipid profile: aqueous versus organic solvent-based extraction methods. Food Research International 62: 1087-1094.

Valle, F.R., Mena, M.H. and Bourges, H., 1982. An investigation into insect protein. Journal of Food Processing and Preservation 6: 99-110.

Van Broekhoven, S., Oonincx, D.G.A.B., Van Huis, A. and Van Loon, J.J.A., 2015. Growth performance and feed conversion efficiency of three edible mealworm species (Coleoptera: Tenebrionidae) on diets composed of organic by-products. Journal of Insect Physiology 73: 1-10.

Vandeweyer, D., Lenaerts, S., Callens, A. and Van Campenhout, L., 2017. Effect of blanching followed by refrigerated storage or industrial microwave drying on the microbial load of yellow mealworm larvae (*Tenebrio molitor*). Food Control 71: 311-314.

Veldkamp, T., Van Duinkerken, G., Van Huis, A., Lakemond, C.M.M., Ottevanger, E., Bosch, G. and Van Boekel, M., 2012. Insects as a sustainable feed ingredient in pig and poultry diets – a feasibility study. Wageningen UR Livestock Research, Lelystad, the Netherlands.

Verhaeghe, T., Vlaemynck, G., De Block, J., Van Weyenberg, S. and Hendrickx, M., 2016. Thermal inactivation kinetics of proteases and polyphenoloxidase in brown shrimp (*Crangon crangon*). Food Chemistry 197: 641-647.

Wang, J., Wu, F.-A., Liang, Y. and Wang, M., 2010. Process optimization for the enrichment of alpha-linolenic acid from silkworm pupal oil using response surface methodology. African Journal of Biotechnology 9: 2956-2964.

Wu, S., 2012. Supercritical carbon dioxide extraction of oil from *Clanis bilineata* (Lepidoptera), an edible insect. African Journal of Biotechnology 11: 4607-4610.

Xia, Z.Q., Wu, S.J., Pan, S.K. and Kim, J.M., 2012. Nutritional evaluation of protein from *Clanis bilineata* (Lepidoptera), an edible insect. Journal of the Science of Food and Agriculture 92: 1479-1482.

Yi, L., Lakemond, C.M.M., Sagis, L.M.C., Eisner-Schadler, V., Van Huis, A. and Van Boekel, M.A.J.S., 2013. Extraction and characterisation of protein fractions from five insect species. Food Chemistry 141: 3341-3348.

Zhang, A.-J., Qin, Q.-L., Zhang, H., Wang, H.-T., Li, X., Miao, L. and Wu, Y.-J., 2011. Preparation and characterisation of food-grade chitosan from housefly larvae. Czech Journal of Food Sciences 29: 616-623.

Zhang, M., Haga, A., Sekiguchi, H. and Hirano, S., 2000. Structure of insect chitin isolated from beetle larva cuticle and silkworm (*Bombyx mori*) pupa exuvia. International Journal of Biological Macromolecules 27: 99-105.

Zhao, X., Vazquez-Gutierrez, J.L., Johansson, D.P., Landberg, R. and Langton, M., 2016. Yellow mealworm protein for food purposes – extraction and functional properties. PLoS ONE 11: e0147791.

Regulation, ethics and promotion

18.

Omphisa fuscidentalis (dried larvae)

Health risks and EU regulatory framework

N. Meijer and H.J. van der Fels-Klerx*
RIKILT Wageningen University and Research Centre, P.O. Box 230, 6700 AE Wageningen, the Netherlands; ine.vanderfels@wur.nl

Abstract

The most important aspects influencing the food and feed safety of insects are: the type of substrate, the insect species, the life stage of the insects upon harvest, and the nature of subsequent processing. Insects may carry and transmit biological hazards to humans, but active replication of those pathogens does not appear to occur in insects. Most microbes that are pathogenic to insects are not pathogenic to humans. Hazardous chemicals such as heavy metals may accumulate in insects from the substrate. Residues of veterinary drugs and persistent organic pollutants may also be chemicals of concern. Certain pan-allergens can be found in house dust mites and crustaceans, as well as in some insect species. As such, cross-reactivity may occur. A component of insect exoskeletons (chitin) may also affect the immune system, but this is not fully clear yet. Much additional research on the safety of insects is needed before a full risk assessment can be performed. All EU food and feed businesses must comply with the general principles and requirements of food law, and certain hygiene provisions. For all food and feed businesses, except those in primary production, it is mandatory to implement a hazard analysis and critical control points plan. Insects intended to be marketed as food have to undergo premarket authorisation. Compared to other countries, the types of substrates that may be used to rear insects on in the EU are limited. Provided that hygiene requirements are met and maximum limits for hazardous substances are not exceeded; insects may be fed to pets, zoo and circus animals, and aquaculture animals. Heat treatment is seen as necessary to control microbiological hazards in insects. Subsequent freeze drying may also be required. It is expected that with more scientific evidence, some of these legal limitations will be loosened in the future.

Keywords: food safety, risks; chemical contaminants, pathogen, food safety management system

18.1 Introduction

Just as with other animals more traditionally eaten as food, or fed as feed, in the European Union (EU), insects may pose health risks, and they are therefore subject to a large body of legislation. Some of those health risks and accompanying rules are similar or identical to the rules for other food and feed businesses dealing with products of animal origin, while some risks and rules are unique for insects. The aim of this chapter is to provide a summary of the possible health risks associated with consumption of insects (Section 18.2), to describe the general regulatory framework applicable to all food and feed businesses in the EU and the specific regulations for insects (Section 18.3), and to describe the systems that have to be in place to manage the safety of insects for food and feed in the EU (Section 18.4). In the last section of this chapter (Section 18.5) a general conclusion is given.

Attempts have been made to make this chapter as specific to insects as possible, although it is inevitable that some basic underlying concepts need be explained first. This chapter deals only with the possible health risks for human and animal health, while insect welfare during production is outside of the scope. Due to the recent push for insects to be (legally as well as culturally) accepted, the EU regulatory framework is in a state of flux; as such, the framework here described may not be up to date in the near future.

18.2 Health risks

The term risk is defined as, 'a function of the probability of an adverse health effect and the severity of that effect, consequential to a hazard' (Article 3(9) Regulation (EC) No 178/2002 (GFL); EC, 2002a). The term hazard is defined as 'a biological, chemical or physical agent in, or condition of, food or feed with the potential to cause an adverse health effect' (Article 3(14) GFL; EC, 2002a). Hazardous biological agents relevant for all foods are, for instance, pathogenic bacteria, viruses and parasites. Examples of chemical hazards are certain pesticides, dioxins, persistent organic pollutants (POPs), and metals. Physical hazards can be represented by small particles or objects like glass, metal shavings, plastic, or wood that may be present in food/feed products.

Risk assessment of food and feed is done in four steps: (1) hazard identification; (2) hazard characterization; (3) exposure assessment; and (4) risk characterization (Article 3(11) GFL; EC, 2002a). The first step of risk assessment consists of identifying a biological, chemical or physical agent capable of causing an adverse effect in a food or feedstuff, with a qualitative description of those effects. This measure is followed by quantitatively establishing the dose-response relationship for each critical effect of the hazard (hazard characterization) and assessing the likely levels and duration of exposure to the risk source for humans via food (exposure assessment). Finally, the dose-response relationship is compared to the likely level of exposure to characterize the risk in a given population (CAC, 2016).

In general, the occurrences in insects of hazards discussed in this section are primarily influenced by four factors (that may be different per business and venue), namely: the type of substrate, the insect species, the life stage of the insects upon harvest, and the nature of subsequent processing (EFSA, 2015). In this chapter, food and feed safety hazards related to insects will be categorized into three groups: biological hazards (Section 18.2.1), chemical hazards (Section 18.2.2), and allergens (Section 18.2.3.).

18.2.1 Biological hazards

Some typologies of biological hazards may be intrinsically associated with insects, while others can be introduced during farming or processing. High levels of (not necessarily pathogenic) bacteria can be found in many species of insects: total aerobic bacterial count, but also non-aerobic bacterial count and *Enterobacteriaceae*, were found to exceed process hygiene criteria for 'comparable' types of products (since no criteria exist for insects yet) (FASFC, 2014; NVWA, 2014). In general, most microbes that are pathogenic to insects are not pathogenic to humans (FAO, 2013), although they may cause economic losses to insect farms. In this section, the biological hazards represented by bacteria, viruses, parasites, and prions are discussed.

The most important bacteria that are pathogenic to humans, and have been associated with insects, are *Salmonella*, *Campylobacter*, and pathogenic strains of *Escherichia coli* (e.g. O157:H7). While active replication of these pathogens does not appear to occur in insects, and their prevalence is often lower than in other animal derived products, insects may carry and transmit (act as 'vectors') these pathogens to humans (EFSA, 2015; NVWA, 2014). However, Erickson *et al.* (2004) showed that *Salmonella enterica* serovar Enteritidis accumulated in black soldier fly larvae, after feed on manure that was contaminated with this organism during two days. After six days, the concentrations of *Salmonella* Enteritidis decreased again.

The same is true for viruses: those pathogenic to humans do not replicate in insects, but they may be transferred to humans. While parasitic infections have been associated with eating insects, those cases are limited to insects gathered from the wild rather than reared in closed farm environments and in controlled conditions (EFSA, 2015).

Prions are misfolded proteins responsible for causing transmissible spongiform encephalopathies (TSEs), such as bovine spongiform encephalopathy in cattle (BSE, also called 'mad cow disease'), scrapie in sheep, and variant Creutzfeldt-Jakob Disease (vCJD) in humans. The misfolded proteins replicate and aggregate in the cells of the central nervous system, thereby disrupting the normal cell structure. The disease is microscopically characterized by a 'sponge-like' appearance of the brain-tissue and all prion diseases are untreatable and fatal. The primary route of infection is via ingestion of food. While there is no evidence that scrapie is transmissible to humans, consuming BSE-infected meat may cause CJD in humans (ECDC, 2015). While it is expected that no specific prion-diseases can develop in insects, they may act as mechanical vectors of infectious prions by transmitting those prions from infected substrate to humans and/or other animals (EFSA, 2015; Post *et al.*, 1999).

18.2.2 Chemical hazards

Hazardous chemical substances may be introduced into the insect farming chain via three main routes: (1) direct accumulation from substrate; (2) from the use of chemicals during insect rearing (e.g. biocides for cleaning farms and veterinary drugs used to treat insect diseases); and (3) self-production of toxins by some insect species. Currently, there is no evidence that the insect species currently being reared for food or feed in the EU self-produce toxins in the developmental stage in which they are intended to be harvested and consumed (FASFC, 2014). Hazardous substances may accumulate in insects and they may respond differently to those chemicals than other animals.

The following typologies of chemical hazards may theoretically be associated with farmed insects (Charlton *et al.*, 2015; EFSA, 2015; Poma *et al.*, 2017):

- heavy metals (cadmium, lead, mercury, arsenic);
- veterinary drugs and hormones;
- pesticide residues;
- dioxins and dioxin-like PCBs;
- polycyclic aromatic hydrocarbons;
- packaging migration contaminants (e.g. bisphenol A, phthalates) in former foodstuffs;
- mycotoxins.

Charlton *et al.* (2015) explored the chemical safety of fly larvae as a source of protein for animal feed. While the concentrations of chemical contaminants they analysed for (all of the above except packaging migration contaminants and hormones) was generally below the legal limit, cadmium levels in three samples of one type of fly larvae (*Musca domestica*) were above the legal limit for complete feed (0.5 mg/kg) as specified in Directive 2002/32/EC (EC, 2002b). They concluded further work is required to determine the source of the cadmium and to what degree it is transferred to livestock if the insects are used as feed. Van der Fels-Klerx *et al.* (2016) investigated the potential accumulation of the heavy metals cadmium, lead and arsenic in larvae of yellow mealworm (*Tenebrio molitor* L.) and black soldier fly (*Hermetia illucens*). They found arsenic accumulated in three yellow mealworm treatments and lead and cadmium accumulated in all black soldier fly experimental setups. Cadmium concentrations in black soldier fly larvae were found to exceed current EU maximum limits, which is in line with Charlton *et al.* (2015).

Residues of veterinary drugs and hormones might be transferred to the insects from contaminated substrates (e.g. animal manure) but veterinary drugs can also be used directly on insects to prevent or treat insect disease. Currently, there are no maximum residue limits set for veterinary drugs in insects, but it is recommended that 'testing for veterinary drugs and hormones in insects intended for food and feed can be managed in accordance with current practices for other animal products in the food chain' (EFSA, 2015). It can be interpreted that carcinogenic drugs banned for use in other food-producing animals, such as chloramphenicol, are also not allowed to be used in insect farming (Table 2 of the annex of Regulation (EC) No 37/2010; EC, 2009c). In addition, as with other food-producing animals, withdrawal periods for the remaining drugs (the period between administration of the drug and the use of the animal product) have to be taken into account.

Another possible type of chemical hazard associated with insects used for food or feed is represented by mycotoxins. Mycotoxins, such as aflatoxins and ochratoxin A, are toxic compounds produced by fungi. They may cause many different adverse health effects in humans and food producing animals, including cancer (Bennett and Klich, 2003). Mycotoxins may be transferred to insects via contaminated substrate or be produced by fungi growing on processed insects. Van Broekhoven *et al.* (2017) investigated the effects of mycotoxin-contaminated diet (predominantly with deoxynivalenol; DON) on edible yellow mealworm (*T. molitor*) and the possible insect contamination from that diet. They found that the presence of DON did not affect the mealworms' growth or survival rate and no DON was found in the mealworms after harvest. However, more research on other species of insects and mycotoxins is required for a complete risk assessment.

18.2.3 Allergens

The term 'food allergy' is defined as 'an adverse health effect arising from a specific immune-mediated response that occurs reproducibly on oral exposure to a given food' (Boyce *et al.*, 2011; EFSA, 2014). Common allergens are venom from insect bites, medications, and proteins contained in certain food items, such as nuts, crustaceans, and shellfish. Allergic reactions may occur as a result of an established allergy or may be developed as a new allergy. Allergic reactions may be caused not only by the original allergen but also by other allergens (called pan-allergens) if they share structural or sequence similarities – this is called cross-reactivity (Boyce *et al.*, 2011). In the case of insects, Verhoeckx *et al.* (2014) demonstrated the pan-allergens tropomyosin and arginine kinase can be found both in house dust mites, crustaceans as well as in yellow mealworm.

Chitin is a polysaccharide commonly found in some fungi and bacteria, and in the exoskeletons of crustaceans and insects. Although chitin is not believed to be an allergen itself, it has certain immunomodulatory properties. This can affect the way in which susceptible individuals react to certain allergens, mostly depending on the chitin particle size and administration method (Muzzarelli, 2010). While 'intermediate' sized particles appear to cause a pro-inflammatory (allergic) reaction, 'small' sized particles in turn induce the production of anti-inflammatory molecules that will lead to resolution of the immune response (Lee *et al.*, 2008). Much of the recent research on chitin's role in the immune system has focused on lung diseases (e.g. asthma) related to inhalation of chitin particles from, primarily, house dust mite and fungi (Brinchmann *et al.*, 2011; Muzzarelli *et al.*, 2012). Which of those findings are also applicable to insect consumption however is not clear (see also Schlüter *et al.*, 2016).

18.3 EU regulatory framework

As with any business in the EU, companies rearing insects for food or feed must comply with specific rules. Discussing the whole food legislation applicable to all food businesses is beyond the scope of this chapter, but a short summary of the basic provisions is given; focusing on the specifics for the insect production chain.

In Section 18.3.1, the EU framework regulation for food law (general food law) is discussed, followed by the legal requirements for insects as food (Section 18.3.2) and feed (Section 18.3.3). In the second section, attention is paid to the general food hygiene requirements, followed by the legislation concerning premarket authorisation of edible insects (applicable from 2018.)

18.3.1 General food law

The foundation of food and feed legislation in the EU is Regulation (EC) No 178/2002 (EC, 2002a), laying down the general principles and requirements of food law, establishing the European Food Safety Authority (EFSA) and laying down procedures in matters of food safety – also called the general food law (GFL). The definitions given are generally applicable to all other food-related legislation. In this regulation, 'food' is defined as 'any substance or product, whether processed, partially processed or unprocessed, intended to be, or reasonably expected to be ingested by humans' (Article 2). The term 'feed' is defined as 'any substance or product, including additives, whether processed, partially processed or unprocessed, intended to be used for oral feeding to animals' (Article 3(4) GFL; EC, 2002a).

The principles of food law laid down in this regulation are:

- Food law shall be based on a risk analysis approach (Article 6).
- If scientific uncertainty persists, provisional measures may be taken (the precautionary principle, Article 7).
- Food and feed shall not be placed on the market if unsafe (i.e. injurious to health or unfit for human consumption – Article 14, see Article 15 for the safety of feed).
- Food business operators (FBOs) are primarily responsible for the safety of food and feed (Article 17; see also Articles 19 and 20, respectively).

Traceability of all food and feed shall be established (Article 18). Traceability follows the 'one step forward – one step back' approach: FBOs must be able to identify the individuals from which they received their raw materials, and which businesses they have supplied their products to (Article 18(2) and (3) GFL, respectively).

18.3.2 Legal requirements: food

Food hygiene requirements

EU rules on food hygiene apply to all stages of food production, processing, distribution, and placing on the market. The key legislative acts on food hygiene are Regulations (EC) No 852, 853 (EC, 2004c), and 854/2004 (EC, 2004d), and Directive 2004/41/EC (EC, 2004a). Of these, most important for the insect industry is Regulation (EC) No 852/2004 (EC, 2004b) on the hygiene of foodstuffs.

Pursuant to Article 4(1) of Regulation (EC) No 852/2004 (EC, 2004b), all FBOs must comply with general and specific hygiene provisions, so called prerequisite programs (further explained in Section 18.4.1). FBOs carrying out processes in primary production, including insect rearing, are to comply with the general hygiene provisions laid down in part A of Annex I of that regulation. FBOs carrying out processes in any post-harvest stage (i.e. processing) are to comply with the general hygiene requirements laid down in Annex II of that Regulation. For FBOs handling food of animal origin, these general hygiene rules are supplemented by the requirements laid down in Regulation (EC) No 853/2004 (EC, 2004c). However, no provisions specifically applicable to insects have thus far been included in that regulation.

Pursuant to Article 5 of Regulation (EC) No 852/2004 (EC, 2004b), FBOs are to, 'put in place, implement and maintain a permanent procedure or procedures based on the hazard analysis and critical control points (HACCP) principles'. HACCP is discussed in more detail in Section 18.4. For primary production, implementation of HACCP is recommended but not mandatory (Article 5(3) of Regulation (EC) No 852/2004; EC, 2004b). For some types of FBOs (e.g. beekeepers and butcheries), so-called codes to good practice are available (Articles 7-9 of Regulation (EC) No 852/2004; EC, 2004b). These codes, which may substitute a custom food safety management system (FSMS), act as guidelines to compliance with the general obligations (Article 3 of Regulation (EC) No 852/2004; EC, 2004b), and contain provisions on the prerequisite program and HACCP. Codes to good practice may be developed as national or community-wide guides. No code to good practice specifically applicable to the insect-industry has thus far been officially adopted.

Premarket authorisation as novel food

Regulation (EU) 2015/2283 (EC, 2015) will replace Regulation (EC) No 258/97 (EC, 1997) as the new novel food regulation on 1 January 2018. The definition of 'novel food' consists of two cumulative conditions: (1) 'any food that was not used for human consumption to a significant degree within the Union before 15 May 1997'; and (2) food falling under one of the specified categories (Article 3 of Regulation (EU) 2015/2283; EC, 2015). While the first condition in the new Regulation remains unchanged compared to the definition as given in the old one, there are now ten categories compared to the four categories in the last consolidated version of Regulation (EC) No 258/97 (EC, 1997). Examples of food categories now explicitly falling under the definition of novel foods are those consisting of engineered nanomaterials, and food consisting of, isolated from or produced from cell culture or tissue culture derived from animals, plants, micro-organisms, fungi or algae (Article 3.2(a) viii and vi of Regulation (EU) 2015/2283, respectively; EC, 2015). Food classified as novel food may only be placed on the market following premarket authorisation by the European Commission, a process which generally requires a risk assessment to be performed by EFSA (Article 10-12 of Regulation (EU) 2015/2283; EC, 2015).

In the old novel food regulation, one of the specified categories was, 'food ingredients isolated from animals, except for foods and food ingredients obtained by traditional propagating or breeding practices and having a history of safe food use' (Article 1(2.e) of Regulation (EC) No 258/97; EC, 1997). In some EU member states, this was interpreted so as to mean only parts of animals would qualify as novel foods, but whole insects would not fall under this category and would therefore not be classified as novel foods (Belluco, 2013; Lähteenmäki-Uutela *et al.*, 2017). In the new Regulation on the other hand, it is explicitly stated, 'the categories of food which constitute novel foods...should cover whole insects and their parts' (preamble 8 of Regulation (EU) 2015/2283; EC, 2015). The definition of the specified category has been amended accordingly: 'food consisting of, isolated from or produced from animals or their parts...' (Article 3(2.a.V.) of Regulation (EU) 2015/2283; EC, 2015). Thus, only those insects authorized as novel foods may be marketed in the EU after 1 January 2018.

While the EU risk managers ultimately make the decision whether or not to authorize marketing of a novel food, a large role herein is played by EFSA as risk assessor. An application for premarket authorisation may be filed for a completely novel food, or for traditional foods with history of safe use in a third country (Article 3(2)(c) of Regulation (EU) No 2283/2015; EC, 2015). The procedure for the latter type is much less strict. It is expected that most applications for insect species will go via the second route due to the consumption history of insects in South-East Asia and certain parts of Africa. EFSA has published two guidance documents – one document for novel foods and one for traditional foods from third countries (EFSA, 2016a,b) – detailing the information applicants should provide for risk assessment prior to authorisation as a novel food.

18.3.3 Legal requirements: feed

Animal feed may only be marketed and fed to animals if it is safe and it does not have a direct adverse effect on the environment or animal welfare (Article 2 of Regulation (EC) No 767/2009 (EC, 2009a); see also Article 15 GFL; EC, 2002a). With respect to insects, EU feed law applies to the substrates on which they are reared but also to those insects if in turn intended to be used as feed for other food producing animals. The feed hygiene provisions as laid down in Regulation (EC) No 183/2005 (EC, 2005) and the maximum levels for certain contaminants established in Directive 2002/32/EC (EC, 2002b) have to be complied with in both cases. For many of the substances listed in the annex of this Directive, the same maximum levels apply to all feed materials but in some cases a different limit is given for a specific type of feed. No such exceptions are made specifically for insects yet.

EFSA (2015) divided the materials that could potentially be used as insect substrates into seven categories with varying levels of risk. While a wide range of bio-waste products are used as substrates for insect rearing in countries outside of the EU (EFSA, 2015), manure, catering waste or former foodstuff containing meat and fish are not allowed for that purpose in the EU. At this time, feed materials authorized as feed for all food-producing animals (as listed in Regulation (EC) No 68/2013; EC, 2013b), former foodstuffs not containing meat/fish, and 'co-products from primary production of food of non-animal origin' are being used as substrates on which to rear insects (EFSA, 2015).

After the BSE crisis in the 90s (Section 18.2.1), restrictions were placed on the feeding of animals with the remains of other animals. The objective of these restrictions was to prevent, control and eradicate TSEs. Three classes of animal by-products are distinguished according to their potential hazardous effects. Aquatic and terrestrial invertebrates (including insects) are classified as Category 3 material (Article 10(l) of Regulation (EC) No 1069/2009; EC, 2009b). Category 3 materials are in principle can be used as feed as long as they do not pose an unacceptable risk to public or animal health. Provided that the aforementioned requirements are met, the current legislative framework allows insects to be:

- used to produce pet food (Article 14(d)(iii) and (e) of Regulation (EC) No 1069/2009; EC, 2009b);
- used for the production of feed for zoo and circus animals, reptiles and birds of prey other than zoo and circus animals; used as bait for fishing, etc. – although these practices are to be authorized by the competent authority of each Member State (Article 18 of Regulation (EC) No 1069/2009; EC, 2009b);
- fed as (non-hydrolysed) processed animal proteins (PAPs) to aquaculture animals (Regulation (EU) No 56/2013; EC, 2013a, and Annex IV, chapter IV, section F of Regulation (EC) No 999/2001; EC, 2001).

Pending additional research, the rules allowing PAPs to be fed to aquaculture animals may soon be extended to other non-ruminant food-producing animal species such as pig and poultry, and insect substrates may in the future be based on more feed material categories such as former foodstuffs containing meat or fish (IPIFF, 2016).

18.4 Food and feed safety management systems

The two most important components of a FSMS are the prerequisite program and the HACCP plan. The former consists of the more general safeguards that must be universally met in any production facility to be able to produce safe food or feed. The HACCP approach should be used in most facilities in addition to the prerequisite program to ensure that certain hazards, which may exceed limits critical for food safety, are controlled throughout the process. The Codex Alimentarius standard on food hygiene (CAC, 1969) is the authoritative document on this topic.

In this chapter, the two main components of a European FSMS, the prerequisite program and HACCP plan (in Section 18.4.1 and 18.4.2, respectively) are discussed, followed by certification and private standards (Section 18.4.3). These concepts will be then applied to the insect production chain (Section 18.4.4).

18.4.1 Prerequisite program

The prerequisite program consists of general hygiene requirements that must be met for the production of safe food. The program may differ depending on the type of business and location. The fundamental question should always be, 'what is necessary and appropriate on the grounds of the safety and suitability of food for consumption?' (CAC, 1969). The Codex standard on food hygiene (CAC, 1969) has laid down the groundwork for the most general requirements, which may have to be supplemented by:

- different Codex standards applicable to specific industries;
- legal requirements (e.g. those laid down in Regulation (EC) No 852/2004; EC, 2004b; Section 18.3.2);
- and/or those requirements of chosen private standards (Section 18.4.3).

Most prerequisite programs will cover topics like the design and layout of the premises, cleaning procedures, and pest control. Requirements will be more or less strict depending on the complexity of operations.

18.4.2 Hazard analysis of critical control points

The HACCP approach consists of seven principles that have been translated into a twelve-step program. These twelve steps are shown in Figure 18.1. Core to the HACCP approach is the concept of critical control points (CCPs). These are the points, or steps, in the processing of a specific product where control of certain parameters is necessary for safety (Figure 18.2). It is particularly important to consider per step whether any subsequent step will eliminate identified hazards or reduce likely occurrence of them to an acceptable level. If that is not the case, then the step should be a CCP and control measures should be implemented. The term 'control measure' is defined as, 'any action and activity that can be used to prevent or eliminate a food safety hazard or reduce it to an acceptable level'. A CCP common for many products is the cooking step: specific time/temperature criteria may have to be met to reduce bacterial contamination to a safe level.

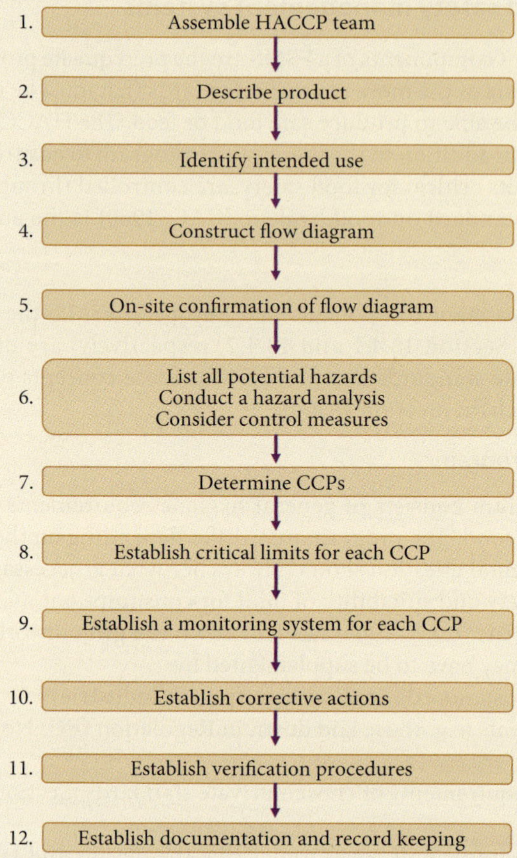

Figure 18.1. Logic sequence for application of hazard analysis of critical control points (HACCPs) (CAC, 1969; p. 21).

18.4.3 Certification and private standards

Private standards in the food industry arose as a way to standardize contracts between suppliers and (primarily) retailers. Prior to the advent of private standards, numerous retailers would check their suppliers' premises and processes to ensure that contractual obligations related to food safety were met. Because this auditing process was very resource-intensive, it has been streamlined to a limited number of standards covering all the important safeguards for safe food. Instead of each retailer having to visit and check all its suppliers, placing a heavy burden on both parties, this auditing process was gradually outsourced to third-party Certification bodies that would inspect a supplier's premises and processes and certify its compliance with a standard. While certification is not a legal requirement for market access

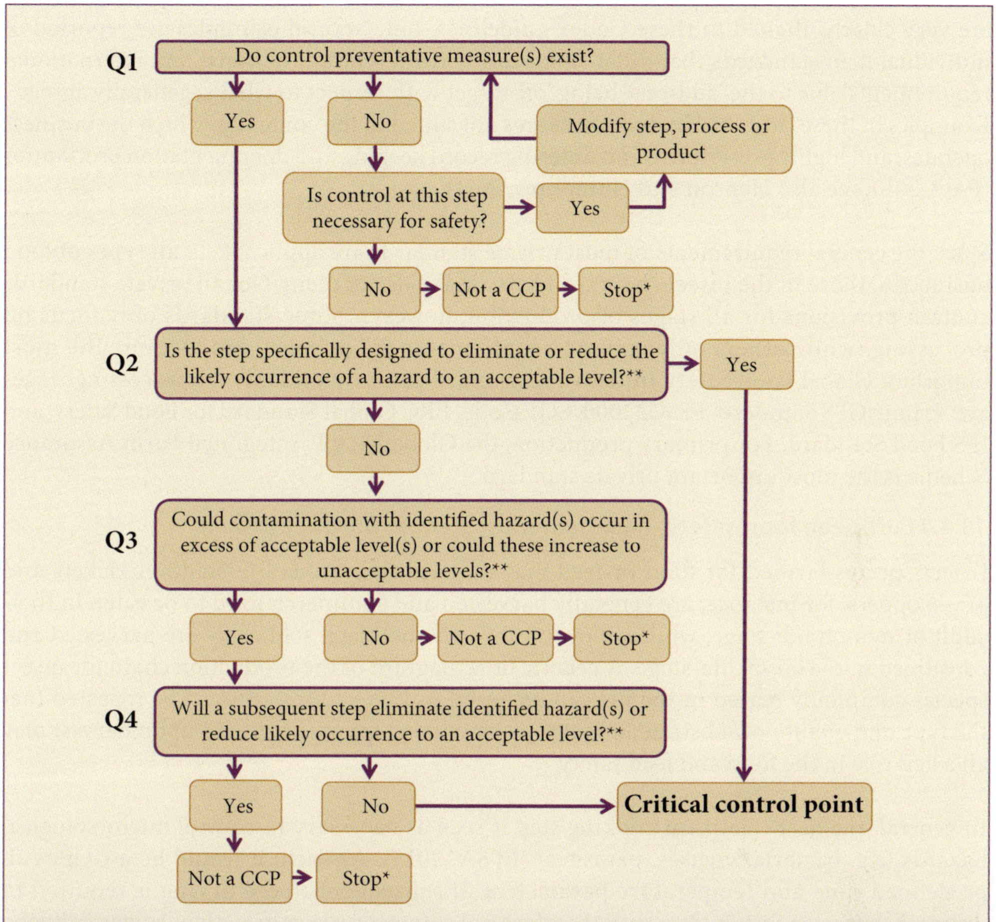

Figure 18.2. Example of decision tree to determine critical control points (CCPs) (CAC, 1969; p. 22). * Proceed to the next identified hazard in the described process. ** Acceptable and unacceptable levels need to be defined within the overall objectives in identifying the CCPs of HACCP plan.

for most foods, most retailers demand certification before a business will be able to supply to them (Van der Meulen, 2011).

Efforts have been made to achieve more consistency for all food safety standards via the global food safety initiative (GFSI), a benchmarking organization that developed 'internationally recognised minimum food safety requirements'. Its objective is to deliver, 'equivalence and convergence among effective FSMSs' (GFSI, 2015). Despite progress made in that area, some retailers still do not accept all GFSI recognized standards or may have their own 'add-ons' or even their own separate standard (FAO, 2010). Compared to the Codex standard on food hygiene (CAC, 1969; Section 18.4.1), most of the major collective private food safety standards are very closely aligned to these Codex guidelines, but, 'several examples are reported of individual firm standards that differ significantly from relevant national and international requirements' due to the 'add-ons' being 'off-target with respect to what is generally agreed'. Examples of these 'add-ons' include measures not suited to the context in which the business operates, and highly prescriptive and differing record keeping and documentation procedures (FAO, 2010; see also Henson and Humphrey, 2011).

Since the general requirements of most private standards are applicable to all types of food businesses, those in the insect chain could also make use of them. Not all private standards contain provisions for all stages of production, however. Some standards only focus on processing, while others only contain requirements for primary production. The most important Global Food Safety Initiative recognized private standards for processing stages are: PrimusGFS Standard, FSSC22000, SQF Code, BRC Global Standard for Food Safety, and IFS Food Standard. For primary production, the Global G.A.P. Integrated Farm Assurance Scheme is the most important private standard.

18.4.4 European food safety management system to produce insects

Insect species farmed for food or feed may be used in various life-stages. Crickets and grasshoppers, for instance, are generally harvested and commercialized to be eaten in their adult or nymph life-stage, while yellow mealworm and black soldier fly are harvested and consumed in the larval life-stage. A generic flow diagram of the production chain for insect species commonly reared in the EU can be found in EFSA (2015). It must be repeated that the type and quality of substrate, insect type, and life-stage of the insects upon harvest play all a key-role in the food and feed safety.

In general, the heat treatment/cooking step is seen as necessary to control microbiological hazards (e.g. bacteria, viruses, parasites) (EFSA, 2015). As such, it would be a CCP with predefined time and temperature parameters. If subsequent freeze drying is required to decrease the water activity to a level that does not support further growth of bacteria during storage, then it may also be a CCP. It is therefore recommended that insects are cooked and/or frozen before consumption (Belluco, 2013; EFSA, 2015). The presence of many contaminants depends on the type and quality of substrate used. As such, verifying that specified contaminants are below the required levels upon receiving of substrates is an important step. Since there are no legislative standards specific to insect substrates, it is up to each business to set requirements in its purchasing specifications to ensure that the level of contaminants accumulated in the insects does not affect the final product safety – and to verify that those criteria are met. It is finally recommended that the labelling accompanying the final product states that insects may cause allergenic reactions in individuals susceptible to crustaceans and mites.

18.5 Conclusions

The most important aspects influencing the food and feed safety of insects are the type of substrate, the insect species, the life stage of the insects upon harvest, and the nature of subsequent processing. While insects generally do not support the growth of pathogenic biological hazards, they may transmit pathogens from affected substrates to the consumer.

Certain hazardous chemicals accumulate in insects from the substrate, such as heavy metals. Veterinary drugs, antibiotics, and POPs may also be chemicals of concern. Finally, insects may share certain allergens with crustaceans and mites. Much additional research on the safety of insects is needed before a full risk assessment can be performed.

Insect farms producing for food or feed must comply with current legislation applying to all producers of products of animal origin. This means that insects as food and feed must be safe for human and animal consumption. To do so, all components of the European FSMS have to be implemented. The two most important components of an FSMS are the prerequisite programme and the HACCP plan.

The marketing of insects for food or feed in the EU is currently restricted by legislation. This is in part due to the fact that using them for food or feed was not foreseen at the time when current legislation was being implemented. Compared to other foods of animal origin, the most important additional requirement for insects as food is that insects must undergo prior authorisation before marketing. For feed, insects may not yet be fed to a number of more conventional livestock species such as pigs and poultry. Finally, at this time insects may only be reared on a limited variety of substrates. It is expected that with more scientific evidence, some of these limitations will be loosened in the future.

References

Belgian Federal Agency for the Safety of the Food Chain (FASFC), 2014. Food safety aspects of insects intended for human consumption. FASFC, Brussels, Belgium. Available at: http://tinyurl.com/ybjn7upc.

Belluco, S., Losasso, C., Maggioletti, M., Alonzi, C.C., Paoletti, M.G. and Ricci, A., 2013. Edible insects in a food safety and nutritional perspectivse: a critical review. Comprehensive Reviews in Food Science and Food Safety 12: 296-313.

Bennett, J.W. and Klich, M., 2003. Mycotoxins. Clinical Microbiology Reviews 16: 497-516.

Boyce, J.A., Assa'a, A., Burks, A.W., Jones, S.M., Sampson, H.A., Wood, R.A., Plaut, M., Cooper, S.F., Fenton, M.J., Arshad, S.H., Bahna, S.L., Beck, L.A., Byrd-Bredbenner, C., Camargo Jr., C.A., Eichenfield, L., Furuta, G.T., Hanifin, J.M., Jones, C., Kraft, M., Levy, B.D., Lieberman, P., Luccioli, S., McCall, K.M., Schneider, L.C., Simon, R.A., Simons, F.E., Teach, S.J., Yawn, B.P., Schwaninger, J.M. and Panel, N.I.-E., 2011. Guidelines for the diagnosis and management of food allergy in the United States: summary of the NIAID-Sponsored Expert Panel Report. Nutrition 27: 253-267.

Brinchmann, B.C., Bayat, M., Brøgger, T., Muttuvelu, D.V., Tjønneland, A. and Sigsgaard, T., 2011. A possible role of chitin in the pathogenesis of asthma and allergy. Annals of Agricultural and Environmental Medicine 18: 7-12.

Charlton, A.J., Dickinson, M., Wakefield, M.E., Fitches, E., Kenis, M., Han, R., Zhu, F., Kone, N., Grant, M., Devic, E., Bruggeman, G., Prior, R. and Smith, R., 2015. Exploring the chemical safety of fly larvae as a source of protein for animal feed. Journal of Insects as Food and Feed 1: 7-16.

Codex Alimentarius Commission (CAC), 1969. General principles of food hygiene, CAC/RCP 1-1969. CAC, Rome, Italy. Available at: http://tinyurl.com/y7gbckm7.

Codex Alimentarius Commission (CAC), 2016. Procedural manual (25th Ed.). CAC, Rome, Italy. Available at: http://tinyurl.com/ybcpvg8o.

Erickson, M.C., Islam, M., Sheppard, C., Liao, J. and Doyle, M.P., 2004. Reduction of *Escherichia coli* O157:H7 and *Salmonella enterica* serovar Enteritidis in chicken manure by larvae of the black soldier fly. Journal of Food Protection 67: 685-690.

European Centre for Disease Prevention and Control (ECDC), 2015. Variant Creutzfeldt-Jakob disease – factsheet. April, 2015. ECDC, Solna, Sweden. Available at: http://tinyurl.com/y8lqdjpf.

European Commission (EC), 1997. Regulation (EC) No 258/97 of the European Parliament and of the Council of 27 January 1997 concerning novel foods and novel food ingredients. Official Journal of the European Communities L43: 1-6.

European Commission (EC), 2001. Regulation (EC) No 999/2001 of the European Parliament and of the Council of 22 May 2001 laying down rules for the prevention, control and eradication of certain transmissible spongiform encephalopathies. Official Journal of the European Communities L147: 1-40.

European Commission (EC), 2002a. Regulation (EC) No 178/2002 of the European Parliament and of the Council of 28 January 2002 laying down the general principles and requirements of food law, establishing the European Food Safety Authority and laying down procedures in matters of food safety. Official Journal of the European Communities L31: 1-24.

European Commission (EC), 2002b. Directive 2002/32/EC of the European Parliament and of the Council of 7 May 2002 on undesirable substances in animal feed. Official Journal of the European Communities L140: 10-22.

European Commission (EC), 2004a. Directive 2004/41/EC of the European Parliament and of the Council of 21 April 2004 repealing certain directives concerning food hygiene and health conditions for the production and placing on the market of certain products of animal origin intended for human consumption and amending council directives 89/662/EEC and 92/118/EEC and council decision 95/408/EC. Official Journal of the European Communities L157: 33-44.

European Commission (EC), 2004b. Regulation (EC) No 852/2004 of the European Parliament and of the Council of 29 April 2004 on the hygiene of foodstuffs. Official Journal of the European Communities L139: 1-54.

European Commission (EC), 2004c. Regulation (EC) No 853/2004 of the European Parliament and of the Council of 29 April 2004 laying down specific hygiene rules for food of animal origin. Official Journal of the European Communities L139: 55-205.

European Commission (EC), 2004d. Regulation (EC) No 854/2004 of the European Parliament and of the Council of 29 April 2004 laying down specific rules for the organisation of official controls on products of animal origin intended for human consumption. Official Journal of the European Communities L139: 206-320.

European Commission (EC), 2005. Regulation (EC) No 183/2005 of the European Parliament and of the Council of 12 January 2005 laying down requirements for feed hygiene. Official Journal of the European Communities L35: 1-22.

European Commission (EC), 2009a. Regulation (EC) No 767/2009 of the European Parliament and of the Council of 13 July 2009 on the placing on the market and use of feed, amending European Parliament and Council Regulation (EC) No 1831/2003 and repealing Council Directive 79/373/EEC, Commission Directive 80/511/EEC, Council Directives 82/471/EEC, 83/228/EEC, 93/74/EEC, 93/113/EC and 96/25/EC and Commission Decision 2004/217/EC. Official Journal of the European Communities L229: 1-28.

European Commission (EC), 2009b. Regulation (EC) No 1069/2009 of the European Parliament and of the Council of 21 October 2009 laying down health rules as regards animal by-products and derived products not intended for human consumption and repealing Regulation (EC) No 1774/2002 (Animal by-products Regulation). Official Journal of the European Communities L300: 1-33.

European Commission (EC), 2009c. Commission Regulation (EU) No 37/2010 of 22 December 2009 on pharmacologically active substances and their classification regarding maximum residue limits in foodstuffs of animal origin. Official Journal of the European Communities L15: 1-72.

European Commission (EC), 2013a. Commission Regulation (EU) No 56/2013 of 16 January 2013 amending Annexes I and IV to Regulation (EC) No 999/2001 of the European Parliament and of the Council laying down rules for the prevention, control and eradication of certain transmissible spongiform encephalopathies. Official Journal of the European Communities L21: 3-16.

European Commission (EC), 2013b. Commission Regulation (EU) No 68/2013 of 16 January 2013 on the Catalogue of feed materials. Official Journal of the European Communities L29: 1-64.

European Commission (EC), 2015. Regulation (EU) 2015/2283 of the European Parliament and of the Council of 25 November 2015 on novel foods, amending Regulation (EU) No 1169/2011 of the European Parliament and of the Council and repealing Regulation (EC) No 258/97 of the European Parliament and of the Council and Commission Regulation (EC) No 1852/2001. Official Journal of the European Communities L327: 1-22.

European Food and Safety Authority Panel on Dietetic Products, Nutrition and Allergies (EFSA NDA Panel), 2014. Scientific Opinion on the evaluation of allergenic foods and food ingredients for labelling purposes. EFSA Journal 12: 3894.

European Food Safety Authority (EFSA) Scientific Committee, 2015. Scientific Opinion on a risk profile related to production and consumption of insects as food and feed. EFSA Journal 13: 4257.

European Food Safety Authority (EFSA) Scientific Committee, 2016a. Guidance on the preparation and presentation of an application for authorisation of a novel food in the context of Regulation (EU) 2015/2283. EFSA Journal 14: 4594.

European Food Safety Authority (EFSA) Scientific Committee, 2016b. Guidance on the preparation and presentation of the notification and application for authorisation of traditional foods from third countries in the context of Regulation (EU) 2015/2283. EFSA Journal 14: 4590.

Food and Agriculture Organization of the United Nations (FAO), 2010. Private food safety standards: their role in food safety regulation and their impact. FAO, Rome, Italy. Available at: http://www.fao.org/docrep/016/ap236e/ap236e.pdf.

Food and Agriculture Organization of the United Nations (FAO), 2013. Edible insects – future prospects for food and feed security. FAO, Rome, Italy. Available at: http://www.fao.org/docrep/018/i3253e/i3253e00.htm.

Global Food Safety Initiative (GFSI), 2015. GFSI governance model and rules of procedure. GFSI, Issy-les-Moulineaux, France. Available at: http://www.mygfsi.com/files/GFSI_Governance_Model.pdf.

Henson, S. and Humphrey, J., 2011. Codex Alimentarius and private standards. In: Van der Meulen, B. (ed.) Private food law. Wageningen Academic Publishers, Wageningen, the Netherlands, pp. 149-174.

International Platform of Insects for Food and Feed (IPIFF), 2016. Position paper on feed. IPIFF, Brussels, Belgium. Available at: http://www.ipiff.org.

Lähteenmäki-Uutela, A., Grmelová, N., Hénault-Ethier, L., Deschamps, M.-H., Vandenberg, G.W., Zhao, A., Zhang, Y., Yang, B. and Nemane, V., 2017. Insects as food and feed: laws of the European Union, United States, Canada, Mexico, Australia, and China. European Food and Feed Law Review 12: 22-36.

Lee, C.G., Da Silva, C.A., Lee, J.-Y., Hartl, D. and Elias, J.A., 2008. Chitin regulation of immune responses: an old molecule with new roles. Current Opinion in Immunology 20: 684-689.

Muzzarelli, R.A.A., 2010. Chitins and chitosans as immunoadjuvants and non-allergenic drug carriers. Marine Drugs 8: 292-312.

Muzzarelli, R.A.A., Boudrant, J., Meyer, D., Manno, N., DeMarchis, M. and Paoletti, M.G., 2012. Current views on fungal chitin/chitosan, human chitinases, food preservation, glucans, pectins and inulin: a tribute to Henri Braconnot, precursor of the carbohydrate polymers science, on the chitin bicentennial. Carbohydrate Polymers 87: 995-1012.

Netherlands Food and Consumer Product Safety Authority (NVWA), 2014. Advies over de risico's van consumptie van gekweekte insecten – advies over de chemische, microbiologische, en parasitologische risico's van de consumptie van verhitte en onverhitte insecten. NVWA, Utrecht, the Netherlands. Available at: http://tinyurl.com/yarbfeg2.

Poma, G., Cuykx, M., Amato, E., Calaprice, C., Focant, J.F. and Covaci., A., 2017. Evaluation of hazardous chemicals in edible insects and insect-based food intended for human consumption. Food and Chemical Toxicology 100: 70-79.

Post, K., Riesner, D., Walldorf, V. and Mehlhorn, H., 1999. Fly larvae and pupae as vectors for scrapie. Lancet 354: 1969-1970.

Schlüter, O., Rumpold, B., Holzhauser, T., Roth, A., Vogel, R.F., Quasigroch, W., Vogel, S., Heinz, V., Jäger, H., Bandick, N., Kulling, S., Knorr, D., Steinberg, P. and Engel, K.-H., 2016. Safety aspects of the production of foods and food ingredients from insects. Molecular Nutrition and Food Research 61: 1-14.

Van Broekhoven, S., Mota Gutierrez, J., De Rijk, T.C., De Nijs, W.C.M. and Van Loon, J.J.A., 2017. Degradation and excretion of the *Fusarium* toxin deoxynivalenol by an edible insect, the yellow mealworm (*Tenebrio molitor* L.). World Mycotoxin Journal 10: 163-169.

Van der Fels-Klerx, H.J., Camenzuli, L., Van der Lee, M.K. and Oonincx, D.G.A.B., 2016. Uptake of cadmium, lead and arsenic by *Tenebrio molitor* and *Hermetia illucens* from contaminated substrates. PLoS ONE 11: e0166186.

Van der Meulen, B., 2011. The anatomy of private food law. In: Van der Meulen, B. (ed.) Private food law. Wageningen Academic Publishers, Wageningen, the Netherlands, pp. 75-112.

Verhoeckx, K.C.M., Van Broekhoven, S., De Hartog-Jager, C.F., Gaspari, M., De Jong, G.A.H., Wichers, H.J., Van Hoffen, E., Houben, G.F. and Knulst, A.C., 2014. House dust mite (Der p 10) and crustacean allergic patients may react to food containing Yellow mealworm proteins. Food and Chemical Toxicology 65: 364-373.

19.

Gryllus assimilis

Ethical issues in insect production

H. Röcklinsberg[1*], C. Gamborg[2] and M. Gjerris[2]
[1]Department of Animal Environment and Health, Swedish University of Agricultural Sciences, P.O. Box 7068, 750 07 Uppsala, Sweden; [2]Department of Food and Resource Economics, University of Copenhagen, Rolighedsvej 25, 1958 Frederiksberg C2, Denmark; helena.rocklinsberg@slu.se

Abstract

Insect rearing is often presented as a promising novel source of protein in many industrialised countries in the West. In this chapter, we will first give an overview of the different ethical concerns insect production for food and feed give rise to. This is followed by an elaboration of two of the ethical issues that have, so far, been least discussed. (1) Animal welfare: What are the typical reasons given for including welfare considerations in animal production system, and to what extent do they apply to insects? In order to answer these questions, we will discuss how one may conceptualise insect welfare and present an account of what is known, or can be inferred, about the capability of insects to experience welfare and where future research needs lie. (2) Animal integrity: Do insects possess integrity and can it be violated through large-scale production systems? To clarify this, we will discuss whether it can be argued there is more to the ethical discussion than how insects are bred, kept, and killed (i.e. large-scale production including domestication of new species and the killing of billions of insects to further human ends may entail ethically relevant dimensions). Further, we discuss these issues in relation to biotechnological changes of insects for different purposes.

Keywords: animal welfare, animal ethics, animal integrity, biotechnology, intensive insect production

19.1 Introduction

Developing large-scale production systems for farmed insects to supplement or replace food and food ingredients from (vertebrate) livestock, is seen by many as a more sustainable way to achieve animal protein products than currently used livestock production methods. However, as with any food production system several ethical issues can be identified – especially when the production system in question includes animals. Regarding large-scale insect production especially environmental impact, human and animal health, human preferences, and social acceptability have so far been at the centre of attention. But as with other livestock production systems (e.g. cattle, pigs, poultry, fish, etc.), animal welfare and broader animal ethics issues connected to integrity are beginning to be discussed as relevant concerns (Gjerris *et al.*, 2016). One of many challenges in this area – given it makes sense at all to talk about insect welfare – is the limited research on the cognitive and emotional capacities of the more than 2,000 edible species of insects (Jongema, 2017).

Until the 1970's, the limits to modern livestock production were mainly seen as scientific/technical: 'How can the production of animal proteins be increased?' However, as technology and animal science succeeded in intensifying the production, the question increasingly became one of, 'What is acceptable to do?' Today, the traditional livestock sector faces a growing number of ethical challenges related in particular to climate change, loss of biodiversity, animal welfare issues, and questions relating to the social context of production systems (EC, 2016; Röcklinsberg, 2009; Röös *et al.*, 2015) Thus, public apprehension about conventional livestock production is triggered by a range of concerns.

When examining the potential concerns related to insects for food and feed production systems, and assessing the underlying ethical issues, we need to consider two sets of issues. First, what is considered an ethically relevant concern? This may differ between stakeholders with different interests and diverging ethical views. Secondly, how should the differing attitudes and values be evaluated and addressed? In this chapter, we focus on the former, in the hope that an understanding of the different issues that can be seen as relevant will impact the development of large-scale production systems for insects for feed and food, and reduce the controversies else connected with animal-based production. Hence, we see a unique situation in terms of considering ethical concerns at the very beginning of a production system establishment, contrary to the more common practice of handling ethical concerns once a certain system has proven to be problematic due to impaired animal welfare or environmental damage.

A note on nomenclature: 'ethics', as the term is understood here, has as its main function to clarify value-based discussions. In general, ethical reflection may help to formulate and discuss the relative importance of potentially conflicting concerns and values (Gjerris *et al.*, 2013). In the specific context of using insects for food and feed, reflection may for example concern the complex trade-offs between production of food and feed, conservation of genetic resources, environmental/climate impact, and animal welfare. The outcome of such ethical reflection may be a better understanding of the value base of various ways of evaluating such insect production systems and related trade-offs, thereby making room for dialogue about the methods and goals of food and feed production systems. Further, ethical reflection may function as an eye-opener by formulating implicit values or showing incongruences in how concepts and values are used. In this chapter, such scrutiny will be relevant in relation to insect welfare, the concept of integrity, and the application of biotechnological breeding methods.

19.2 Social and ethical feasibility

A broadly agreed upon premise for seeing insects as part of a more sustainable, stable and efficient food and feed production is that it will be necessary to move from the currently prevailing hunter-gatherer methods and scattered, unsystematic attempts of small scale semi-cultivation into systematically planned small or large-scale production systems. This is increasingly being attempted in several places globally (FAO, 2013).

Even in a situation where full scale intensive production systems would be technically feasible and legally permitted, they may still not be implemented in many countries as the success of the strategy also depends on convincing decision makers, farmers, and consumers that using insects for food and feed provides an attractive alternative to current methods. From an ethical point of view, it further needs to be shown that the production offers a more sustainable path for animal-based agriculture with (e.g. less environmental and climate impact, better use of resources, and a reduction in animal welfare related problems). A central task in this relation is to compare insect production systems not only with conventional animal production, but also with alternative kinds of protein production based on plants – both traditional, high-tech, and lab-grown meat – from a number of perspectives such as climate impact, biodiversity, and animal welfare.

Evidently, the public's choice of food is governed by many factors, which include economic, social (i.e. public acceptance), and moral concerns. Typical factors used to explain the attitudes of citizens include perceived risk, perceived benefit, trust in food chain actors, consumer autonomy and labelling, level of knowledge, the roles of science and technology in agriculture as well as other distinct moral concerns (Barnett, 2011; Pirscher, 2016; Rundgren, 2016).

Generally speaking, there are two overall dimensions in discussions on the social and ethical feasibility of any practice. One dimension is covered by so-called classical ethical theories (i.e. philosophical theories about what is morally right). The other dimension concerns who or what has moral standing (i.e. who or what should for its own sake be considered when contemplating an action). Hence, it is possible that two proponents of one and the same ethical theory have different views on who or what has moral standing, and vice versa: one can agree on what objects are worthy of moral concern, but differ in view on what is the right ethical theory.

Three prominent examples of classical ethical theories often alluded to are: (1) utilitarianism, a consequentialistic theory, which crudely speaking can be described as advocating the principle that the morally right act (which in this context could be a food policy) is the one maximizing positive actual or foreseen outcome viz a viz a set of available alternatives; (2) deontology, part of which is Kantianism (roughly characterized as ethical theories stemming from the German philosopher Immanuel Kant (1724-1804)) asserting that a person should act in accordance with principles the person can rationally want to be universal laws. To

Kant this means formulating principles of action in a way so that other humans are always treated as ends in themselves, and not merely as means. Based on this kind of thinking other philosophers have expanded Kant's thinking to include animals. Most notable here is Tom Regan's idea (Regan, 1983) the idea of rights should be expanded to cover all subjects-of-a-life; and finally (3) virtue ethics, a set of theories emanating from the ethics of the ancient Greek philosopher Aristotle (384-322 BC). Fundamental to both Kant and Aristotle's views is an anthropology asserting every human wants to be a good person, but virtue ethics elaborates on the question of what kind of person one strives to be, rather than on what principles is needed to manifest it. According to Aristotle each species has its inherent *telos*, aim, which should be fulfilled for a flourishing life. When it comes to humans, wisdom is crucial for flourishing. By developing one's moral character guided by a set of virtues such as prudence, justice, temperance and courage, one continuously improves in moral decision-making trough sincere discernment of all ethically relevant aspects in a given situation. This ability to express the relevant virtues at the right time is known as *phronesis* (practical wisdom).

These three basic theories have been developed into many different ethical positions and providing a full analysis of a subject like insect production from each of them is beyond the scope of this chapter. Rather, we will point to the different ethical issues that we regard most relevant within each of the three over-arching theories. In brief, if one adheres to a utilitarian point of view, as a form of a broader strand of consequentialist ethical theories, emphasis would be on trying to substantiate the purported benefits, risks and other consequences viz a viz relevant alternatives – such as conventional livestock production. One important result of this is that if the insects species reared are considered to have a welfare of their own, the very high number of individuals used in the production comes into view even though the individual welfare 'level' might seem marginal. The welfare of hundreds of thousands of insects would need to be weighed against the welfare of the corresponding amount of edible biomass produced by rearing poultry, pig or ruminants. Such an evaluation is obviously far from unproblematic: how many crickets' suffering matches that of one not-well off cow? Taking a deontological perspective would require us to consider the rights of various individuals with ascribed moral standing, which could, depending on the theory's position on the second dimension, also include insects themselves (see next paragraph). Depending on the type of rights ascribed to insects (e.g. such as the most basic right to life), different duties with regard to insects would persist resulting in prohibitions or restrictions being placed on production systems. Although it might seem clearer than the utilitarian calculus, challenges also exist here: determining which rights an individual should be ascribed and why, and what duties to prioritize in a situation where different rights and duties clash are not straight-forward. Moreover, the fairness of distribution of cost and benefits of a specific insect production system for moral relevant entities should be considered. Within a theory of just distribution, justice could be defined taking the stance of insect's moral standing into account, as it matters for the actual distribution whether insects are seen as individuals (ends in themselves) or merely as resources (means). Similar to a deontological approach a virtue ethics perspective would consider individuals for their own sake, but rather stress the *telos* of each species, often interpreted as concern for species specific needs, and hence argue that *if* insects are to be kept in systems designed for food and feed production, ensuring species-specific rearing conditions would be essential. Further alternatives would need to be evaluated based on their potential to create a flourishing life for all involved morally relevant entities.

In addition to considering how to answer what is morally right, all ethical theories need to address the question of who or what has moral standing (the second dimension). Here a number of different answers can be found among different theories and within the same over-arching theories. Again, it goes beyond the scope of this chapter to give a full account of all the possible answers and their justification. So here we just identify the two main 'camps' within the ethical landscape: one centred on humans as the only organisms having moral standing (anthropocentrism), and another focused on that other beings in addition to humans can have moral standing (non-anthropocentrism). As mentioned both sets of positions can be divided into several sub-positions or theories. Especially relevant in relation to using insects for food and feed, non-anthropocentrism can be divided into: (1) sentientism (also called zoocentrism) ascribing moral standing to animals with the capacity to feel pleasure or pain, thus having some kind of cerebral capacity. This position typically includes animals which are used for conventional livestock production such as cows, pigs, sheep, poultry, etc.; (2) biocentrism including all individual living organisms independent of their capacities but simply due to the fact of being alive; and (3) ecocentrism that suggests a holistic approach not focusing on individuals but wholes, where also non-living entities such as rivers, mountains and landscapes and phenomena such as species, ecosystems and biodiversity matter for the sake of their own.

When looking at the one of the earliest FAO factsheet on edible insects, ethical considerations are mentioned as issues that should be considered in the protection of insect populations in natural environments. However, the sole focus is on the effects of the production directly for humans or indirectly through concerns for the aspects of the natural environment that are of importance for humans: respecting local people, protecting food chains, conserving 'beneficial' (to humans) insect populations and the avoidance of releasing non-endemic (domesticated) insect species into natural environments (FAO, 2013). These kinds of ethical concerns are also sometimes presented under the heading of sustainability, an ideal which has evolved over hundreds of years and which has come to encompass a growing number of concerns as a response to changing societal agendas. Thus, in relation to insect farming, sustainability refers to the tenet of environmental concerns, genetic diversity, short-term and long-term economic values, and other social issues (Gamborg *et al.*, in press). It should be noted, however, determining whether insect production, as an example of mini-livestock (Hardouin, 1995), is a more sustainable protein source than ordinary (vertebrate) livestock such as chickens, pig or cows is a difficult question to answer unanimously (Hardouin, 1995). So far, focus has been on the environmental impact based on various life cycle analyses (Halloran *et al.*, 2016). Here, the outcome depends considerably on how the production system is defined, and what the alternatives are. If plant-based alternatives to animal proteins are considered realistic then traditional protein sources such as seitan, tofu and quorn for example as well as so-called plant-based 'meat' or artificial meat should also be assessed in a sustainability perspective. At the moment, few such studies have been published (e.g. Smetana *et al.*, 2015). Determining how sustainable an insect production could be, also depends on which concerns are seen as relevant and how they are balanced against each other. Different emphasis would give rise to different themes in a debate and bear reference to different concepts, as shown in Table 19.1, which originally was applied to the discussion of the introduction of modern biotechnology in agriculture.

Table 19.1. Concerns, themes and concepts relevant to insect production for food and feed (based on Lassen and Jamison, 2006).

Concerns for discussion	Central themes	Key concepts
Socio-economic	profitability, production	cost-benefit, power
Risk-centred	environment, health	risk, uncertainty
Cultural	moral and religious aspects	rights, integrity

It is striking how – compared to other livestock production systems – animal welfare and animal ethics issues of mini-livestock production have been left out of the considerations so far as, from our point of view, a number of important ethical discussions related to the animal welfare and ethical issues exist. These will be the focus of the rest of the chapter.

19.3 Animal welfare research

Concern for animal well-being and health has a long history in both philosophical and religious thinking, but also among animal owners themselves (Preece, 2001; Walter and Portmess, 1999). That is, people have reasonably believed that animals are sentient, but, similar to today, not necessarily offered due care. The first steps to formalize such concerns in legal terms was taken in Britain, launching the first European legislation against cruelty towards animals through the so-called Martins Act from 1822, followed in 1876 by the Cruelty to Animals Act aiming at preventing harm to animals. As a result of intensification of animal husbandry, in turn a mean to combat food deprivation during and after the Second World War, reduction of animal welfare was taken to be a necessity to increase production levels. The book 'Animal machines' by Ruth Harrison (1964) was an efficient eye-opener to many citizens and politicians, by disclosing practices in livestock farming and arguing for humane practices. Her work led to the establishment of the Brambell committee by the UK government (today UK Farm Animal Welfare council), which launched the now well-known five freedoms, and evoked a citizen awareness leading to the European Convention for the Protection of Animals Kept for Farming Purposes (European Council, 78/923/EEC; EC, 1976). Further, this led to an increased interest in what animals actually experience and need, and hence animal welfare research was required (Fraser, 2008; Webster, 2005).

Within this research field, it has been widely recognised that animal welfare can be understood along three parallel, overlapping and not necessarily mutually excluding definitions: health, affective states/subjective experiences and natural behaviour (Fraser, 1995). An animal might score high on health, but not on the affective state due to recently having lost a companion, or score high on natural behaviour if kept in a suitable environment, but low on health due to a clinical disease. Hence all three aspects need to be considered in any welfare assessment. Which welfare definition is taken to be the most important is a value based choice (Borkfelt *et al.*, 2015; Yeates *et al.*, 2011), which in turn will be mirrored in the focus of, for example, welfare research or choice of welfare assessment protocol. Further, given the genetic advances in livestock farming, animal welfare science has shown that production levels (e.g. high yields or high number of offspring) might well concur with high welfare, but is not as such a valid sign of good welfare; rather the opposite might be true: animals with low welfare might (but need not) have high production levels.

These aspects are relevant when considering insect welfare, not least since mini-livestock is a relatively new production strain where it is still possible to avoid mistakes made in current animal production systems. It might be relevant to compare with fish farming here, an industry expanding enormously over the last 30 years. Whereas environmental concerns were evoked regarding pollution and death of the sea ground below cages, use of antibiotics, escapees potential negative impact on wild fish and energy loss in fish feed production (FAO, 2012; Ziegler, 2006) and have been extensively discussed, animal welfare concerns were not on the agenda until more recently (Röcklinsberg, 2014). This is somewhat surprising given that e.g. salmon (the most farmed species) is a migrating animal and hence there is an inbuilt welfare challenge already by housing them in cages. By excluding natural behaviour aspects of welfare, it is however possible to argue that their welfare might be acceptable from a health perspective. Here problems with lice, water quality and stress levels are in focus. When it comes to understanding fish' subjective experiences it is difficult to design experiments (Braithwaite *et al.*, 2013). In general, fish welfare research is a somewhat split field. A few researchers deny fish the capacity to experience pain or fear (Rose, 2002, 2007), whereas many others argue that studies show a number of capacities and behavioural responses to different circumstances that indicate those capacities (e.g. Ashley *et al.*, 2009; Braithwaite *et al.*, 2013; Huntingford *et al.*, 2012). Further a report from the European Food Safety Authority compiling a large number of research studies concluded that investigated fish species have the capacity to experience pain and fear (Algers *et al.*, 2009). This is, however, far from generally accepted in society due both to traditions of not regarding fish as sentient (Kupsala *et al.*, 2013), and to our difficulties in spontaneous understanding fish, leading to less interest in showing them empathy and acknowledging their vulnerability as a shared experience (Röcklinsberg, 2015).

In parallel with this situation, it might be even more difficult to engage people in insect welfare, as insects are often seen rather as a nuisance than an animal, even less as a being with a welfare of its own. Honey bees have, however, long been known to communicate with each other about sources of nectar and pollen, opening for search of parallels in other 'flock' insects. Also, the threats to crop and vegetable production in some regions due to a radical decrease in number of pollinators, in turn due to unsustainable farming methods (Van der Sluijs and Vaage, 2016), may evoke increased interest in insect needs and behaviour. However, in this chapter we focus on insect farming for food and feed. Given that are there a more than one million insect species and more than 2,000 of these are already part of human diets in various parts of the world, an overview of capacities and welfare assessment for each species is an enormous undertaking. Here we just point to the discussion within animal welfare science on whether insect welfare is a meaningful term with regard to certain species and certain developmental stages.

A report by Erens *et al.* (2012) shows not only a variety of behaviours between insect species, but also between different developmental stages, and concludes that at least some insects have the capability of experiencing welfare. Elwood (2011) points to the capacities of honey bees and spiders as an argument for assigning welfare capacities to at least some species of insects, just as Sherwin (2001) points to a number of studies on different species that makes such a claim valid. Sneddon *et al.* (2014) on the other hand claim that little neurobiological evidence for the existence of pain-like states in insects seemingly exists, supporting the view of Eisemann *et al.* (1984; p. 166), 'the neural organization of insects and observations of their behaviour does not appear to support the occurrence in insects of a pain state, such as occurs in humans'. It does, however, not seem to be fair to dismiss the ethical relevance of insect welfare based on them differing from humans. Rather the question is whether an insect of a certain species has the capacity to experience what is painful, or pleasurable, within its own sphere of existence. Hence, a statement like the one by Broom (2014; p. 66): '...[t]he more different from humans an animal appears to be, the less likely it is to be evaluated as sentient' calls for caution, not only due to our limited imagination, but also on pure methodological scientific grounds: absence of evidence is something else than evidence of absence (of certain capacities). Further, as mentioned above, the sheer number of insect species poses a huge challenge to which the most careful solution seems to be to evaluate the welfare capacities and needs for each species that is considered relevant to a given production system while at the same time remembering that the question of whether an insect can experience welfare is not easily answered, not only because there can be a lack of methodologies to answer the question as Smith (1991) points to, but also as insects, on a more immediate level are harder to relate to than conventional livestock.

It is worth noting that insects have not been included in any welfare legislation of EU so far and that animal welfare guidance specifically related to the production of insects is lacking (Smith and Pryor, 2013). Considering the growing interest in insect production, it therefore makes sense to include insects in animal welfare legislation, if animal welfare science shows that certain insects have capacities of sentience and the ability to experience individual welfare. Pending such research, we suggest a precautionary approach. This could be implemented by including animal welfare scientists and entomologists in the development of production systems to ensure that the facilities meet the demands of the insects. This also includes keeping an open mind to the possibility that such facilities cannot be developed at least for some species of insects and the production therefore might not be initiated. The principle of fairness, from our point of view, should make us avoid what has been termed 'paralytic perfectionism' in relation to sentience (i.e. the demand by scientists for 'absolute certainty before they can accept animal sentience') (Proctor *et al.*, 2013; p. 883).

19.4 Animal death and integrity

The use of insects for food and feed raises more animal ethical issues than just those related to welfare. As with other animal production systems it can also be relevant to discuss issues related to, for example, animal death and integrity. Furthermore, the possible future uses of biotechnologies in insect breeding are also relevant to point to as these may involve both animal welfare issues as well as issues about integrity and naturalness.

Ethical issues related to animal integrity, death, and naturalness have been analysed and discussed with regard to conventional livestock for many years (e.g. Francione, 2008; Regan, 1983; Singer, 1975) and more recently in relation to fishery and aquaculture (e.g. Kupsala *et al.*, 2013; Röcklinsberg, 2015). As these considerations cannot necessarily simply be transferred into the discussion on insect production for feed and food, we here outline the possible issues with relation to insects, acknowledging fully that this this is only an outline and more substantial analysis is needed in the future.

The death of an individual insect is typically regarded as even more trivial than the slaughter of an individual farm animal. However, if insects are regarded as morally relevant in their own right, the killing of insects could raise ethical concerns, not the least when considering the huge number of individuals killed in comparison with conventional livestock production. From an animal rights view causing death is in itself worthy of ethical consideration and does not depend on welfare issues such as whether the animal was caused suffering at the moment of killing (Regan, 1983). According to this theory it hence seems reasonable to ask for an ethical justification regarding not only the way insects are killed, but also the bare fact that they are killed. The outcome of such reflection is dependent on the moral status ascribed to insects, a process which in turn depends on up to date biological knowledge about their capacities. However, from a consequentialist point of view, instant killing of a being without knowledge about its future – whether cow or cicada – is regarded as ethically irrelevant as long as the animal is replaced by an equally happy animal (Singer, 1975). The discussion of the ethical legitimacy of killing insects for food and feed has not been on the legislator's agenda so far, and is just about to be evoked within animal ethics (Gjerris *et al.*, 2016).

Another issue that has escaped much notice in the literature so far is the question of the integrity of insects. However, integrity is a concept, which is increasingly brought up in relation to livestock production and research animals (Hauskeller, 2007; Röcklinsberg *et al.*, 2014). Here we focus on integrity as a way to describe a pre-scientific understanding of living beings in their wholeness or completeness, independently of their usefulness for us (Baranzke, 2002; Röcklinsberg *et al.*, 2014). Beings that have integrity can, from this perspective, be said to possess an inherent value that makes it ethically problematic to use them as means to ends that are not in their own interest. Although respect for animal integrity aim at claiming respect beyond protection against impaired welfare, the reason to ascribe is often related to some mental or physical capacity of the animal like having a good on its own (Taylor, 1986). The question then becomes whether insects can be said to possess the necessary capabilities to be ascribed integrity.

Rutgers and Heeger (1999; p. 44) argue that the term 'inherent worth' includes the claim that: 'the animal's intactness and its species-specific capacities and functions constitute a value towards which an attitude of moral respect is appropriate.' Building on this definition, one may argue that, if and only if insects have 'species-specific capacities and functions', they would qualify for having inherent worth and thereby integrity. Obviously, Rutgers and Heeger's definition can be interpreted to cover other beings as well as, for example, plants. Here, however, we simply use it to show how insects can be seen as beings with integrity. Whether plants are such beings as well is a discussion that goes beyond the scope of this chapter.

If insects are considered to have integrity, it obviously puts into the question the whole idea of utilizing them for human needs and preferences. Further, it raises the question of how to move towards a more respectful attitude towards insects than typically found in Western culture today (Kellert, 1993). Several authors have suggested that this would entail developing a more empathic relationship towards insects (Last, 2014; Looy *et al.*, 2014; Pitt and Shockley, 2014). As Diamond (2001) points out, capability for empathy is an important basis to develop respect for another being – even though it is not in itself enough – starting with the insight of our own vulnerability. Building empathy towards insects can, however, be hard to envision as invertebrates are typically seen as 'largely alien and unfathomable' – viewed with aversion, anxiety, and fear (Kellert, 1993; p. 852). To most it is therefore probably easier to feel empathy with a furry mammal whose face expression is more similar to humans than that of an ant or bee. Another possibility suggested is education understood in a broad sense. Mather (2011) suggests that as invertebrates are understood better, people will be more appreciative and take better care. Moreover, direct experience is deemed important for the formation of nature-related values and attitudes (Sammet *et al.*, 2015). In order to create positive attitudes towards insects among people who are indifferent, afraid or even disgusted by their mere appearance the design of such direct experience occasions is crucial and a research field in itself.

A further challenge with regard to the concept of integrity is if modern biotechnologies such as genetic modification is utilized to produce insects that better fit the production schemes, to produce a strain with special capacities or components, for example, in attempt to create a better protein composition of the insect. Modern 'yellow biotechnology' also aims at creating insects that function as sophisticated vectors in crop pest management, as well as genetically modifying insects to function as vectors in fighting human and animal diseases (Vilcinskas, 2011). Further, the sterile insect technique can be employed as a method to eliminate an insect population and can in combination with transgenic insects be developed to efficiently serve pest management (Schetelig and Wimmer, 2011). This obviously raises animal welfare issues as to how the changes will affect the insects and dependent species, much in parallel to the challenges of assessing welfare in genetically modified laboratory animals (Workman *et al.*, 2010). However, it also raises questions as to whether one can genetically modify a species while at the same time respecting it's integrity understood as 'the basic zone of untouchableness' that follows with having integrity (Röcklinsberg *et al.*, 2014).

The very process of modifying insects this way changes them from what they are to something new and raises the question whether this is compatible with respecting the individual insect. If one argues that insects are not beings worthy of respect as they do not possess integrity whereas for example humans are, other questions follow, for example, whether integrity is dependent on a special biological quality? Taking the basic definition mentioned above by Rutgers and Heeger (1999), the possession of 'species-specific capacities and functions' seems to include insects as beings with integrity. If so, one could argue that by making insects infertile their species-specific capacities and functions are changed and their integrity thus violated. Further, the practice of modifying insects in this way is based on the basic premise that they are merely instruments or tools for our use. Thus, the fundamental question remains: are we ethically justified in changing the species-specific capacities and functions of insects?

From the consequentialist perspective, the question is whether carrying out the modification is that action which will result in the best consequences overall for the entities deemed morally relevant. If the envisaged consequences, as in the above-mentioned examples, are to defeat a pest or limit transmission of a disease, a probable answer is that the end justifies the means.

From a consequentialist perspective, such as the utilitarian view briefly described above, the task is to balance the consequences of the technology as it will affect sentient beings and evaluate whether the modification overall will have more positive than negative consequences. However, from an animal rights' perspective, an individual classified as a moral subject may never be reduced only to an instrument, irrespectively of the ends. Hence the question is whether the insect in question is a moral subject/patient (as opposed to a so-called moral agent which can be held responsible for his or her action and can act with reference to 'right' and 'wrong' and make reasoned choices). According to the 'standard' version by the founder of this theory, Tom Regan, a so-called subject of a life has interests, memory, learn from experiences and have relations (Regan, 1983). Although he targeted inclusion of mammals, into the sphere of moral considerations, also some insects might fulfil these criteria, and some the more general criteria of 'having a good of its own'. If so, from an animal rights' view any use of insects is ethically questionable independently of whether they are genetically modified or not, but the modification process does seem to add an extra layer of violation to the whole process.

From a virtue perspective, it could be argued that the central question in our relationship with insects is what kind of virtues we express in our actions and attitudes towards them. Paring this with the scientific uncertainty and our general lack of knowledge about their ability to perceive pain and more complicated kinds of suffering, it seems likely that from a virtue perspective a general attitude of compassion, carefulness and respect would be recommendable. As Lockwood (2017) suggest this would not only mean that we would use insects as a 'training ground' for treating morally relevant beings adequately by cultivating central virtues for a flourishing life, but also mean that we would give the insects the benefit of the doubt. And, also according to Lockwood (2017), it can also explain why people usually condemn individuals who pull the wings of flies and conversely why people often find it praiseworthy to be the kind of person who would not 'harm a fly'. Looking specifically at genetically modification of insects, the virtuous approach could thus be to consider whether manipulating the genome of an animal species to obtain a novel source of protein when adequate plant based alternatives could be claimed to exist, would be in accordance with the virtues that we usually find recommendable that people incarnate in their relationship with insects.

19.5 Conclusions

Will it be possible to produce sufficient amounts of 'mini-livestock' to meet rising demands of animal protein and to contribute to food security, livelihoods, climate change mitigation and environmental protection in a more sustainable way than current livestock production? Or rather, should such amounts be produced? Evidently, the first question is difficult to answer unanimously as there are many scientific and technical challenges to address before this will be possible. But as the second question points to, there are also a number of social and ethical challenges to consider. The social challenges mainly pertain to changing the cultural perception of insects in the Western world, where insects are typically not seen as food for humans. The ethical challenges pertain to value discussions on how to understand 'sustainability', what to compare insect production with in relation to environmental impact, and issues about animal welfare and animal ethics.

Different ethical theories and perspectives will lead to different answers to the question whether large-scale production of insects for food and feed is acceptable. In this chapter, we have shown some of the most important values at stake in relation to the issues of animal welfare and animal ethics. First and foremost, it is worth noticing the scientific uncertainty surrounding the question of whether insects can experience welfare at all. Next was the question of deciding which definition of animal welfare to apply to possible production systems and then issues about death and integrity where different ethical theories come up with different conclusions. Finally, we have discussed the possible use of biotechnology in insect breeding, for example, genetic modification and the ethical issues pertained to this.

There is little doubt the success of implementing insects into the food and feed system to some degree hinges on how the 'insect community', which includes entomologists and those with scientifically informed views of insect welfare, enters these discussions and evaluate the use of insects on the drawing board now. The opportunity to be pro-active and craft practices and policies that at the outset attend to ethical issues is rare and should be embraced to avoid a reactive scenario as seen in the livestock sector with regard, for example, to environmental sustainability and animal welfare. The better the understanding of the underlying value discussions that is built, the better a chance one stands to contribute to a fruitful discussion on the way forward.

Acknowledgements

We greatly appreciate the comments on an earlier version from two reviewers. The study was partly supported by Mistra, the Swedish Foundation for Strategic Environmental Research (grant 'Mistra Biotech'). Their support is gratefully acknowledged.

References

Algers, B., Blokhuis, H., Boetner, A., Broom, D.M., Costa, P., Domingo, M., Greiner, M., Hartung, J., Koenen, F., Müller-Graf, C., Morton, D.B., Osterhaus, A., Pfeiffer, D.U., Raj, M., Roberts, R., Sanaa, M., Salman, M., Sharp, J.M., Vannier, P. and Wierup, M., 2009. General approach to fish welfare and to the concept of sentience in fish. EFSA Journal 954. Animals 48: 61-71.

Ashley, P.J., Ringrose, S., Edwards, K.L., Wallington, E., McCrohan, C. and Sneddon, L.U., 2009. Effect of noxious stimulation upon antipredator responses and dominance status in rainbow trout. Animal Behaviour 77: 403-410.

Baranzke, H., 2002. Würde der Kreatur? Die Idee der Würde im Horizont der Bioethik. Königshausen and Neumann, Würzburg, Germany.

Barnett, C., Cloke, P., Clarke, N. and Malpass, A., 2011. Globalizing responsibility: the political rationalities of ethical consumption. Wiley-Blackwell, Chichester, UK.

Borkfelt, S., Kondrup, S., Röcklinsberg, R., Bjørkdahl, K. and Gjerris, M., 2015. Closer to nature? A critical discussion of the marketing of 'ethical' animal products. Journal of Agricultural and Environmental Ethics 28: 1053-1073.

Braithwaite, V.A., Huntingford, F. and Van den Bos, R., 2013. Variation in emotion and cognition among fishes. Journal of Agricultural and Environmental Ethics 26: 7-23.

Broom, D.M., 2014. Sentience and animal welfare. Centre for Agriculture and Biosciences International, Oxfordshire, UK. PhD thesis, Gothenburg University/SIK rapport 754, Gothenburg, Sweden.

Diamond, C., 2001. Injustice and animals. In: Elliott, C. (ed.) Slow cures and bad philosophers. Essays on Wittgenstein, medicine and bioethics. Duke university Press, Durham, UK, pp.119-143.

Eisemann, C.H., Jorgensen, W.K., Merritt, D.J., Rice, M.J., Cribb, B.W., Webb, P.D. and Zalucki, M.P., 1984. Do insects feel pain? A biological view. Cellular and Molecular Life Sciences 40: 164-167.

Elwood, R.W., 2011. Pain and suffering in invertebrates? Institute of Research Animals Journal 5(22): 175-184.

Erens, J., Van Es, S., Haverkort, F., Kapsomenou, E. and Luijben, A., 2012. A bug's life: large-scale insect rearing in relation to animal welfare: project 1052. Wageningen University, Wageningen, the Netherlands.

European Commission (EC), 2016. Attitudes of Europeans towards animal welfare. Summary, Special Eurobaromenter 442. EC, Brussels, Belgium. Available at: http://tinyurl.com/y7awv7y7.

European Council (EC), 1976. European Convention for the protection of animals kept for farming purposes (78/923/EEC). EC, Brussels, Belgium.

Food and Agricultural Organisation of the United Nations (FAO), 2013. The contribution of insects to food security, livelihoods and the environment. FAO, Rome, Italy.

Food and Agriculture Organization of the United Nations (FAO), 2012. The State of World Fisheries and Aquaculture (SOFIA) 2012. Fisheries and Aquaculture Department, FAO, Rome, Italy.

Francione, G.L., 2008. Animals as persons: essays on the abolition of animal exploitation. Columbia University Press, New York, NY, USA.

Fraser, D., 1995. Science, values and animal welfare: exploring the inextricable connection. Animal Welfare 4: 103-117.

Fraser, D., 2008. Understanding animal welfare: the science in its cultural context. Willey-Blackwell, Hoboken, NJ, USA.

Gamborg, C., Röcklinsberg, H. and Gjerris, M., in press. Sustainable proteins? Values related to insects in food systems. In: Halloran, A., Roos, N., Flore, R and Vantomme, P. (eds.) Edible insects in sustainable food systems. Springer, New York, NY, USA.

Gjerris, M., Gamborg, C. and Röcklinsberg, H., 2016. Ethical aspects of insect production for food and feed. Journal of Insects as Food and Feed 2: 101-110.

Gjerris, M., Nielsen, M.E.J. and Sandøe, P., 2013. The good, the right and the fair. An introduction to ethics. College Publications, London, UK.

Halloran, A., Roos, N., Eilenberg, J., Cerutti, A. and Bruun, S., 2016. Life cycle analysis of edible insects for food protein: a review. Agronomy for Sustainable Development 36: 1-13.

Hardouin, J., 1995. Minilivestock: from gathering to controlled production. Biodiversity and Conservation 4: 220-232.

Harrison, R., 1964. Animal machines. The new factory farming industry. Vincent Stuart Publishers, London, UK.

Hauskeller, M., 2007. Biotechnology and the integrity of life. Taking public fears seriously. Ashgate Publishing Limited, Surrey, UK.

Huntingford, F., Jobling, M. and Kadri, S., 2012. Aquaculture and behaviour. Wiley-Blackwell, Hoboken, NJ, USA.

Jongema, Y., 2017. List of edible insects of the world. Wageningen UR, Wageningen, the Netherlands. Available at: http://tinyurl.com/mestm6p.

Kellert, S.R., 1993. Values and perceptions of invertebrates. Conservation Biology 7: 845-855.

Kupsala, S., Jokinen, P. and Vinnari, M., 2013. Who cares about farmed fish? Citizen perceptions of the welfare and the mental abilities of fish. Journal of Agricultural and Environmental Ethics 26: 119-135.

Lassen, J and Jamison, A., 2006. Genetic technologies meet the public: the discourses of concern. Science, Technology & Human Values 31: 8-28.

Last, A., 2014. Who's the pest? Imagining human-insect futures beyond antagonism. Science as Culture 23: 98-107.

Lockwood, J.A., 2017. Whatsoever you do to the least of my brethern. Why it is wrong to harm a fly. In: Brotton, M.J. (ed.) Ecotheology and nonhuman ethics in society. A community of compassion. Lexington Books, Lanham. MD, USA, pp. 43-61.

Looy, H., Dunkel, F.V. and Wood, J.R., 2014. How then shall we eat? Insect-eating attitudes and sustainable foodways. Agricultural and Human Values 31: 131-141.

Mather, J.A., 2011. Philosophical background of attitudes toward and treatment of invertebrates. ILAR Journal 52: 205-212.

Pirscher, F., 2016. Consuming for the sake of others: whose interests count on a market for animal-friendly products? Journal of Agricultural and Environment Ethics 29: 67-80.

Pitt, D.B. and Shockley, M., 2014. Don't fear the creeper. Do entomology outreach events influence how the public perceives and values insects and arachnids. American Entomologist 60: 97-100.

Preece, R., 2001. Awe for the tiger, love for the lamb. A chronicle of sensibility to animals. UBC Press, Vancouver, Canada.

Proctor, H.S., Carder, G. and Cornish, A.R., 2013. Searching for animal sentience: a systematic review of the scientific literature. Animals 3: 882-906.

Regan, T., 1983. The case for animal rights. University of California Press, Berkeley, Los Angeles, CA, USA.

Röcklinsberg, H., 2009. Animal ethics in a time of climate changes. In: Millar, K., Hobson, W.P. and Nerlich, B. (eds.) Ethical futures: bioscience and food horizons. Wageningen Academic Publishers, Wageningen, the Netherlands, pp. 92-96.

Röcklinsberg, H., 2015. Fish consumption: choices in the intersection of public concern, fish welfare, food security, human health and climate change. Journal of Agricultural and Environment Ethics 28: 533-551.

Röcklinsberg, H., Gamborg, C. and Gjerris, M., 2014. A case for integrity: gains from including more than animal welfare in animal ethics committee deliberations. Laboratory Animals 48: 61-71.

Röös, E., Patel, M., Spångberg, J., Carlsson, G. and Rydhmer, L., 2015. Limiting livestock production to pasture and by-products in a search for sustainable diets. Food Policy 58: 1-13.

Rose, J.D., 2002. The neurobehavioral nature of fishes and the question of awareness and pain. Reviews in Fisheries Science 10: 1-38.

Rose, J.D., 2007. Anthropomorphism and 'mental welfare' of fishes. Diseases of Aquatic Organisms 75: 139-154.

Rundgren, G., 2016. Food: from commodity to commons. Journal of Agricultural and Environment Ethics 29: 103-121.

Rutgers, B. and Heeger, R., 1999. Inherent worth and respect for animal integrity. In: Dol, M., Van Vlissingen, M.F., Kasanmoentalib, S., Visser, T. and Zwart, H. (eds.) Recognizing the intrinsic value of nature. Van Gorcum, Assen, the Netherlands, pp. 41-53.

Sammet, R., Andres, H. and Dreesmann, D., 2015. Human-insect relationships: an ANTless story? Children's, adolescents', and young adults' ways of characterizing social insects. Anthroz 28: 247-261.

Schetelig, M.F. and Wimmer, E.A., 2011. Insect transgenesis and the sterile insect technique. In: Vilcinskas, A. (ed.) Insect biotechnology. Springer, Dordrecht, the Netherlands, pp. 169-194.

Sherwin, C., 2001. Can invertebrates suffer? Or how robust is argument-by-analogy? Animal Welfare 10, Suppl. 1: 103-118.

Singer, P., 1975. Animal liberation: a new ethics for our treatment of animals. HarperCollins Books, New York, NY, USA.

Smetana, S., Mathys, A., Knoch, A. and Heinz, V., 2015. Meat alternatives: life cycle assessment of most known meat substitutes. International Journal of Life Cycle Assessment 20: 1254-1267.

Smith, J.A., 1991. A question of pain in invertebrates. Institute of Laboratory Animals Research Journal 33(1-2): 25-31.

Smith, R. and Pryor, R., 2013. Work Package 5: pro-insect platform in Europe. Deliverable 5.1 – mapping exercise report with regard to current legislation and regulation: Europe and Africa and China. ProteInsect and Minerva, Brussels, Belgium.

Sneddon, L.U., Elwood, R.W., Adamo, S.A. and Leach, M.C., 2014. Defining and assessing animal pain. Animal Behaviour 97: 201-212.

Taylor, P.W., 1986. Respect for nature: a theory of environmental ethics. Princeton University Press, Princeton, NJ, USA.

Van der Sluijs, J.P. and Vaage, N.S., 2016. Pollinators and global food security: the need for holistic global stewardship. Food Ethics 1: 75-91.

Vilcinskas, A., 2011. Insect biotechnology. Springer, Dordrecht, the Netherlands.

Walter, K.S. and Portmess, L., 1999. Ethical vegetarianism. From Pythagoras to Peter Singer. State University of New York Press, New York, NY, USA.

Webster, J., 2005. Limping towards Eden. Willey-Blackwell, Hoboken, NJ, USA.

Workman, P., Aboagye, E.O., Balkwill, F., Balmain, A., Bruder, G., Chaplin, D.J., Double, J.A., Everitt, J., Farningham, D.A.H., Glennie, M.J., Kelland, L.R., Robinson, V., Stratford, I.J., Tozer, G.M., Watson, S., Wedge, S.R. and Eccles, S.A., 2010. Guidelines for the welfare and use of animals in cancer research. British Journal of Cancer 102: 1555-1577.

Yeates, J., Röcklinsberg, H. and Gjerris, M., 2011. Is welfare all that matters? A discussion of what should be included in policy-making regarding animals. Animal Welfare 20: 423-432.

Ziegler, F., 2006. Environmental life cycle assessment of seafood products from capture fisheries. International Journal of Life Cycle Assessment 12: 61.

20.

Apis mellifera

Creating an enabling environment

P. Vantomme
Via Alessandro Magno 406, 00124 Rome, Italy; paul.vantomme3@telenet.be

Abstract

Although there were important publications on insects as food and feed before 2008, it did not lead towards significant initiatives by government authorities or private enterprises. In the coming years there will be an increasing demand on meat proteins by the ever growing world population. However, the land available for livestock is too constrained to satisfy the increasing demand for meat. In addition, the sustainability of products from conventional production animals is questioned. This prompts the search for alternative protein sources. In 2013, a landmark publication entitled 'Edible insects: future prospects for food and feed security' caught worldwide attention. The reason for such attention was that it was published by the Food and Agricultural Organization of the United Nations and by Wageningen University and Research. The first organization assured global legitimacy and the second academic validity. Enabling conditions at the country level include: (1) a tight collaboration between government, private sector and academic institutions; (2) a strong message that convinces private and public sector of the potential of the endeavour; and (3) an awareness of urgency to find alternative protein sources. Actors at the national and international level can create enabling conditions, also for consumers. Scientific articles, cookbooks, the press, social media, workshops, and conferences all play a role in creating a conducive environment for the sector of insects as food and feed to develop. Such an enabling environment will ensure that the issue of finding novel, sustainable sources of protein will be effectively communicated among all stakeholders.

Keywords: FAO, enabling environment, insect farming, global outreach, program development

20.1 Introduction

Over the last decade (2008-2017) a rapidly growing global interest in using insects for human consumption and for feeding livestock occurred. It is interesting that much earlier similar initiatives had happened such as: the book by Holt (1995) originally published in 1885 'Why not eat insects?', and the books by Bergier (1941) and Bodenheimer (1951). Gene Defoliart was instrumental by publishing 'The food insects newsletter' from 1988 to 1995 (DeFoliart *et al.*, 2009). Ample information can also be found in his on-line bibliography 'The human use of insects as a food resource' (DeFoliart, 2002). An international conference in Paris in 2000 had a number of presentations on edible insects (Motte-Florac and Thomas, 2003), and a book highlighting insects as mini-livestock appeared in 2005 (Paoletti, 2005).

These and other previous initiatives or documentation of existing entomophagy practices did definitely increase our knowledge and scientific literature base to better understand the use of insects for food. However, none of these earlier initiatives, papers, or activities led towards increasing any significant awareness on the contribution of insects to our food security by civil society or by any major public agency dealing with agriculture (research) and/or food security programmes worldwide.

The term 'enabling conditions' is rather general and not to be considered as a unique formula of ingredients, drivers, and stakeholders applicable in a similar way all over the world. They vary also along the many and different levels from subsistence to industrial scale farming. Enabling conditions are not only to be seen as a set of institutional (formal or informal) arrangements among stakeholders within the national context of a given country or region, but can also apply to the (formal and informal) collaboration between and within a group of stakeholders in a given country.

We can raise the question why only in the last decade the interest in, and awareness of, the fact that insects can really be part of our food systems has witnessed such a sudden growth? What were the enabling conditions for a wide range of different stakeholders?

20.2 Changing context and appreciation of food

A dramatical change in our food production systems took place over the last several decades from more traditional family-based farming towards large scale industrial agriculture. For livestock production, the latter is characterized by a dense population of animals raised on limited land and requiring large amounts of food, water and medical inputs. This raises concerns about the impact of such practices on the environment, animal welfare, soil and water quality, and food safety. For that reason, a fast growing number of consumers worldwide are now questioning the quality and sustainability of our food and its production systems.

The Food and Agricultural Organization (FAO) of the United Nations estimated about 68% of the total global land area under farming is designated to livestock (FAOSTAT, http://www.fao.org/faostat). In addition, raising incomes in many regions of the world results in an even higher demand for animal proteins: from 2005/2007 to 2050 this is estimated to be a 76% increase (Alexandratos and Bruinsma, 2012). To sustain our livestock, massive amounts of protein feed are produced and traded globally. Soy cultivation is encroaching on forest lands in South America, raising concerns not only about the environmental impacts, but also for commercial and trade dependency reasons as only three countries make up 90% of the world's total soy protein feed exports (FAOSTAT, http://www.fao.org/faostat). Godfray *et al.* (2010) explored the major drivers affecting the food system between now and 2050, indicated a new way of food production is required as higher yields from the same acreage cannot be achieved without severely impacting the environment. Livestock is also an important contributor to climate change, and to curb greenhouse gas emissions a number of mitigation measures have been proposed (Herrero *et al.*, 2016). Other strategies are increasing yields on underperforming lands (Van Ittersum *et al.*, 2013), reducing waste (Papargyropoulou *et al.*, 2014) and shifting human diets (Hedenus *et al.*, 2014; Tilman and Clark, 2014).

The success 'Insects as food and feed' campaigns received lately has much to do with the perceived potential benefits farming insects for food could provide compared to the 'business as usual' of continuing the industrial scale conventional farming methods. 'Farming insects for food' emerged as a successful message of hope that yes, we can increase the global food production while at the same time reducing the environmental and climate change impact of producing it, creating short and local production cycles and still having safe and healthy food available for nine billion people by 2050. Apparently, these messages launched during the last decade that insects can solve simultaneously several of the actual problems or challenges in our food production has been highlighted in many scientific publications and in the popular press. Society is now much more receptive to accept and support such messages because the perception that something is going wrong with our present food production systems is now shared by more and more people worldwide.

The recent insect messages especially address a dramatically changed context and societal appreciation of meat and the way it is produced. This was not the case at the time of the earlier and above-mentioned initiatives from the 1970s, 1980s, or earlier. At that time, society at large did not perceive a major problem with our food production and was ignorant that agriculture, mainly livestock, was a major contributor to climate change. At that time the key challenge was on increasing production, and knowledge or concerns about associated health risks and environmental impacts was limited. As such reorienting our food production systems or addressing their challenges were not perceived as an issue. Therefore, they were not considered by the agriculture and food sectors and definitely not by the general public. The context for building up enabling conditions to promote insects as food was not ripe yet, particularly not in the western world characterized by a strong aversion of insects in general, and as food in particular.

20.3 Global awareness on insects as food since 2008

It is useful to recall quickly some of the major facts, drivers, and the enabling conditions that emerged during the last 10 years in order to understand why there was such an acceleration in the intensity and impact of insect promotion campaigns and their related activities. In 2008, Wageningen University and Research (WUR) and the FAO entered into a formal collaboration to promote the use of insects for food and feed. This joint collaboration is recognized as the key global driver by most stakeholders in the insect sector, and it has boosted international networking in support of the promotion of using insects for food.

Since 2008, this WUR-FAO initiative quickly became a source of information and inspiration of global relevance. While both WUR and FAO, prior to 2008, had already undertaken, each on its own, activities to promote the use of insects in food systems, combining their resources and outreach capabilities together in a single and well-coordinated campaign proved to have a global catalytical lasting impact. The author has chosen arbitrarily the year 2008, not only because it reflects mathematically the last decade under review (2008-2017), but also because 2008 is the year that the collaboration between WUR and the FAO was signed and took off.

It is interesting to take a closer look at the 'enabling conditions' that existed and/or were created to engage in this joint collaboration in the first place, to start and develop it, and consequently inspiring so many stakeholders in so many countries. It is even more remarkable that it reached such a global success in only six years (if we consider starting from 2008 till the landmark 'First International Conference on Insects to Feed the World', held in the Netherlands in May 2014).

Within the FAO one of the major conditions that enabled the development of the programme on edible insects (started in 2000) was the close collaboration and mutual cross inspiration between staff in headquarters (HQ) and the regional office in Bangkok. The initial work done at FAO HQ to document the role of edible insects in Central Africa (Vantomme *et al.*, 2004), inspired staff at the Regional Asian office to investigate how insects contribute to food security in South East Asia. The FAO report 'Insects bite back' (based on the 2008 regional workshop held in Thailand) was widely noted by the international press (Durst *et al.*, 2010). At the same time FAO funded a two year project in Lao People's Democratic Republic (2009-2011) on the role of insects as food (Hanboonsong and Durst, 2014). Both this FAO report and the project in Laos received a very high visibility by a wide range of national and international stakeholders in the agriculture sector as well as by the (inter-) national media.

In the Netherlands, the first publication about the edible insects in Africa was written in 2003 based on a survey among more than 300 people on the cultural significance of insects in 27 African countries (Van Huis, 2003). In the same year, the first project proposals were written by Wageningen University and Research. Only in 2008 the first project entitled 'The sustainable production of insect proteins' was funded by the Ministry of Economic Affairs of the Netherlands (see Box 20.1). Every two years since 1996, the Laboratory of Entomology of WUR organizes a course entitled 'Insects and society' for students and the general public. The lecture series is announced nationally. The topic 'Edible Insects' always attracts the largest audience, maybe also because insect snacks are made available during these lectures. In 2006, the Laboratory of Entomology transformed Wageningen city during one week into 'The city of insects'. Edible insects featured prominently on the agenda, which was highlighted by the (inter)national public media.

In 2008, a joint collaboration framework between WUR and the FAO was officially signed to cover a wide range of some 50-agricultural research and development topics, including work to promote edible insects. Since 2008, a close collaboration between staff from the

WUR Laboratory of Entomology and the FAO Forestry Department (institutional home of the Edible Insects Programme within the FAO) unfolded and led to the FAO-WUR organized international expert consultation on Insects for Food and Feed, held at FAO HQ in January 2012 (Vantomme *et al.*, 2012).

The expert consultation was attended by 70 global experts and helped to draw up an action plan to advice the FAO-WUR initiative how to move forward. The meeting already inspired participants to elaborate and submit insect-based project proposals to donors or to include insects in their programmes, such as the public support (national projects but international in scope, such as GREEiNSECT from Denmark (2014-2017) and Flying Foods from the Netherlands (2013-2017)). This further inspired and supported the European Commission to issue a call for a research project dealing with insects as a novel source of protein for animal feed (PROteINSECT project 2013-2016), and led the EC asking the European Food Safety Authority (EFSA) for a safety review on insects, published by EFSA (2015). Also during this 2012 expert consultation meeting in Rome, representatives from the private sector started the base work leading to the establishment of an International Platform of Insects for Food and Feed (IPIFF) in 2012 with an office in Brussels. This was followed up in 2016 with the ASEAN Food and Feed Insects Association (AFFIA) and the North American Edible Insects Association (NAEIC). These organizations are now forming a network of stakeholders including all the important insect industry players.

The expert consultation recommended FAO-WUR to: launch a web portal on edible Insects in 2012 (FAO, 2017a), assemble a substantial on-line FAO-WUR report on the present status of knowledge of using insects for food (Van Huis *et al.*, 2013), and prepare for an international conference, which was held in May 2014 (Van Huis and Vantomme, 2014). This sequence of events with global impact and output by WUR-FAO reached a rapidly growing and wide global community as is indicated by the very high number of visits at the FAO edible insects web portal (http://www.fao.org/edible-insects/en), reaching some 100,000 visits each year (by 2015). Even more spectacular was the launching of the FAO-WUR report with some 2,3 million downloads and more than 10 million tweets in the first 24 hours of its launch on 13 May 2013 (Van Huis *et al.*, 2013). Something like this had never happened before at FAO or WUR for any of their previous publications. It clearly indicated how interested and 'ready' the international community was to receive and act upon such a message.

20.4 Enabling conditions for international agencies

What were the key elements for the success of the FAO-WUR collaboration and its achieving worldwide impact? First, the institutional set up of the FAO-WUR collaboration itself as characterized by the following critical 'enabling conditions':

- Formal recognition and documentation of this collaboration officially as part of a larger FAO-WUR partnership programme. In the official document describing the content of this collaboration there was a high level of flexibility and freedom for each partner. In this way staff, both at FAO and at WUR were institutionally backed and officially supported in their activities by senior management of FAO-WUR. In addition, they received high levels of freedom and confidence to conceive, plan, fund and implement the actions they had conceived together. Under this coverage, staff of the WUR supported the FAO HQ staff during several consultancy assignments at the FAO HQ for the elaboration of the FAO action plan on edible insects (FAO, 2017b).

- Full trust and confidence, through a direct and open communication channel between staff from FAO HQ, the FAO regional office for Asia and The Pacific in Bangkok and WUR and with one focal point in FAO HQ as well as in WUR to facilitate and speed up exchange of information and planning, coordination, approval and execution of joint actions. This avoided any 'red tape', lengthy delays in decision making, or heavy bureaucratic procedures so often hampering activities by a United Nations agency.
- Achievement of strong synergy by combining the complementary skills and assets of each: FAO (mostly global communication and development, global networking and outreach) and WUR (knowledge base, political support by the Dutch government, academic review and experiences, strong network, and contacts with the private sector).

However, a good functioning and institutionally sound collaboration programme running fast and efficiently, also needs a strong and well formulated message, scientifically well documented and skilfully presented to targeted audiences. The WUR-FAO collaboration elaborated such a strong message that included a holistic view on what 'insects for food' encompasses. The WUR-FAO was able to elaborate a strong message based on facts of interest to many sectors such as food, feed, environmental protection, wildlife, nutrition, trade, animal health, and policy, which ultimately led to the interest of a very wide range of stakeholders in various sectors, to the media and to the public and people (food consumers) in general.

In addition, the joint WUR-FAO collaboration programme was able to approach, unite, engage with and strengthen an impressive and very active global network of interested governmental and non-governmental agencies, stakeholders, civil society, chefs, restaurants and food catering actors, food distribution, media and individuals from various sectors anywhere in the world. The full and active, voluntarily participation of all these actors was an essential 'enabling condition' in further disseminating and multiplying the WUR-FAO message and insect promotion campaign so widely and fast in all countries of the world. The list of stakeholders at the FAO web platform (http://www.fao.org/edible-insects/en) received most hits on the edible insect's web portal.

The enabling conditions of the WUR-FAO collaboration as described above, resulted two partners (WUR and FAO) were able to obtain output through synergy and multiplication effects to get their 'message' across so quickly, something that neither could have done alone.

Of the other international and regional players that have impacted and inspired national governments, the most active is the Directorate General Health and Consumers Affairs (DG SANCO) of the European Commission. In addition, several civil groups and individuals have also played a significant international role, such as 4Ento, the Insect Centre in the Netherlands, Little Herds in the USA, Essento and Grimian in Switzerland.

20.5 Enabling conditions at country level

The enabling conditions contributing to the success of the WUR-FAO collaboration are unique for large (international) institutions dealing with global issues. However, they may not exactly be a template to follow to facilitate actions at the level of a country. One may even wonder what may have been the enabling conditions that resulted in insects being part of traditional diets for many people worldwide? Partly that is caused by the better availability of insects in temperate regions. However, stakeholders can also play an important role, such as the promotion in Thailand by the government to consume the Bombay locust instead of using pesticides (Hanboonsong, 2010). Many stakeholders can play different roles in maintaining and promoting insects for food or feed. These may be quite different within and across countries and cultures.

Farming insects for food and feed was a new activity emerging, but it is/was not clear in the countries how 'farming insects' should be considered and incorporated into the agricultural and food sector. In particular which regulations would apply, as for most countries no relevant legislation existed on farming insects and using insects in the food chain. As a result, in many countries an enabling environment for insect farming still had to be created first.

Within one country, the enabling conditions evolve over time as soon as the insect sector starts, develops, and matures. For example, the private sector stakeholders and their associations, may take over the lead from academia-led initiatives. As they start to rear insects as an industrial scale, they are most in need for an adjustment of the existing legal framework. Achieving inclusive food and feed regulations is their main concern. This requires policy lobbying and can be achieved easier and more efficiently by them then by governmental or academic agencies or civil society groups.

In many countries, all over the world, the following three components are essential to build strong and lasting enabling conditions at various levels: local, national and regional.

- **Actors**. Having a range of different actors interested and fully participating: this includes private sector, civil society, government, and media. It is best when several actors from different sectors are involved, to provide multiplier effects and synergy. The 'golden triangle', which refers to a collaboration between Government, Private sector, and Academic institutions in The Netherlands, is a nice example of a good coalition for setting up enabling conditions for cross-disciplinary interactions among relevant sectors (see Box 20.1). However, such collaborations will thrive best when they have a jointly agreed common message to focus on.
- **Convincing arguments**. A strong message or a programme of (commercial) interest to many sectors and the public at large, appeals to the audience. For example, the FAO-WUR report (Van Huis *et al.*, 2013) was a source of inspiration to kick off the set-up of working groups in several countries to investigate if and how insect farming could be beneficial for the country.
- **Context**. The context needs to be ripe or ready for the message. The context from one country to another is often quite different in their perception if insects can be included or not in the national food chain. An interesting example here is that in East African countries the fishmeal used for feed is very expensive and of poor quality and therefore

creates an economical profitable set-up for farming insects to produce locally made protein for use as animal feed. The lack of quality protein feed does not exist in most South American countries and as such does not trigger the need to start farming insects to produce animal feed.

Within and among different governmental (and non-governmental) agencies and ministries within a country effective communication is required. Ministries or stakeholders from health, agriculture, food, environment, industry, and trade should talk to each other. The catalytic role of the media and civil society groups can play an important role in a country, such as in the USA, where civil society groups play a major role in promoting the use of insects for food. An example is also the strong organization of the edible insect industry (such as the Dutch Association of Insect Producers, VENIK) with connections to the national government and local public representatives, and with knowledge institutes in the Netherlands (see Box 20.1).

The huge impact of the FAO-WUR report 2013 (Van Huis *et al.*, 2013) is a good example of how in many countries the awareness and subsequent setting up of enabling conditions towards including insects in the local food and feed chains was inspired and driven by events at the international level. In several countries, inspired by this report, actors from government, private sector and others, set in motion activities that would lead to inter-ministerial working groups for the promotion of insects for food. For example, in Belgium, inspired by the FAO report, the Federal food agency approved a list of 10 insects for human consumption in 2014. In addition, an inter-ministerial disciplinary working group was set up for insects for animal feed in 2015 by the line ministry of agriculture. During the 17th FAO/WHO Codex Alimentarius Coordinating Committee for Asia (CCASIA) in November 2010, the Lao People's Democratic Republic (Lao PDR) delegation proposed to create a Codex food standard for crickets for human consumption. The proposal was not accepted at that time but a Codex food standard for human-grade crickets is planned for a future session of the Codex committee.

In larger countries like China, Thailand, India, South Africa, or Brazil, there are often significant differences within the country. In China, the local context and appreciation of using insects for food increases considerably from north to south and from west to east. This is reflected by the different levels of priority given to insect farming by the different governmental and non-governmental, and private sector initiatives among the provinces in China. This does not facilitate the setup of enabling conditions uniformly across all provinces and stakeholders in China.

Cooperation started between producers, academics and the government (green deals): for example, UK (Woven); Switzerland (Grimian, 4Ento and Essento); United States (Bill Gates, FDA); and the Netherlands (see Box 20.1).

Private companies in the food and feed sector are playing a key role in supporting enabling conditions (as part of cross national corporations). The lobbying skills and power of IPIFF towards the European Commission has been crucial in setting a process in motion to review existing food and feed legislations valid in the EU for the inclusion of insects. Legislative hurdles were the main reason that producers created IPIFF. On the other hand, edible-insect

Box 20.1. The edible insect sector in the Netherlands.

In 2008, Wageningen University and Research (WUR) and the Dutch Association of Insect Producers (VENIK) jointly submitted a proposal entitled 'Sustainable production of insect proteins for human consumption' (SUPRO2) to the Ministry of Economic Affairs. This one-million-euro, three-year project was approved in 2010 and three research groups of WUR implemented the project: the Laboratory of Entomology (WUR) working on insect farming, food safety and environmental aspects, the Food Quality and Design (WUR) group and the Biobased Research (WUR) group worked on the isolation and characterization of insect proteins. All research carried out in this project has been published (Oonincx and De Boer, 2012; Oonincx *et al.*, 2010, 2015a,b; Van Broekhoven *et al.*, 2014, 2015, 2016, 2017; Verhoeckx *et al.*, 2014; Yi, 2013, 2015, 2016a,b). Besides technical meetings, the project brought together twice a year all societal stakeholders interested in edible insects among which the private industry, the Ministry and research organizations.

VENIK was founded in 2008 by three insect-rearing companies. A family business specialized in poultry and game birds, marketed the insects in freeze-dried form. Because the branch was so new, they self-imposed protocols of farming to ensure food safety of the marketed products. They also made a green deal with the Ministry of Economic Affairs for a large number of stakeholders: feed and food companies, non-governmental organizations, WUR and other knowledge institutions. The green deal focused on bottlenecks regarding legislation: using waste streams as substrates, insect welfare, storage, killing and processing of insects, feed applications, and the novel food regulation.

In the last five years (2012-2017), several networks have been set up in the Netherlands: (1) The International Insect Centre (2012) organizing workshops; and (2) The Insect Protein Innovation Platform (2015) bringing together government, and education to make the southeast of the Netherlands a hub of insect innovation.

Three large companies on insects as food and feed emerged: (1) Proti-Farm, founded as Kreca in 1978 (originally a company producing insects as pet food), focuses on producing large quantities of insect ingredients for the food industry (mainly lesser mealworm); (2) Protix Biosystems founded in 2009 focuses on recycling organic waste and to produce insect based proteins and lipids (mainly black soldier fly); and (3) Insect Europe in Lelystad, founded in 2013, incorporated the webshop Delibugs (mainly crickets). Protix Biosystems and Proti-Farm are important members of the International Platform of Insects for Food and Feed in Brussels, founded in 2014, that advocates appropriate legislative frameworks in the European Union.

The interest in the Netherlands in edible insects has increased during the last 10 years: the number of newspaper articles per year increased from zero in 2005 to about 50 in 2015. In 2015, several supermarket chains included insect food products on their shelves. The legitimation of the edible insect sector in the Netherlands has been analysed in a thesis by Marberg (2016). She concluded that insects as food and feed are a potential solution to economic, social and environmental problems and have therefore gained traction with national policy makers, business leaders, academic institutions and the public in the Netherlands.

Another boost in the Netherlands was two Dutch research scientists spent a number of months at the Food and Agricultural Organization of the United Nations headquarters with three objectives: (1) to write the book 'Edible insects: future prospects for food and feed security', published in 2013 and up till now downloaded 7 million times and translated into French, Italian and Korean (Van Huis *et al.*, 2013); (2) to organize the conference 'Insects to Feed the World' in May 2014 in the Netherlands and attended by 450 people from 45 countries (Van Huis and Vantomme, 2014); and (3) to start mapping worldwide the stakeholders on edible insects (www.fao.org/edible-insects/en).

producers often do not collaborate technically, as other producers are considered competitors (Dobermann, 2017). Innovations are considered to be in need of protection and company secrecy is imposed.

Consumers also need enabling conditions, in particular because insects in the western world have never been a part of the diet. During the last few years quite a number of scientific publications on consumer attitudes have appeared and a number of enabling conditions can be derived from those publications.

Insects need to be available. Through the internet a variety of edible insect products can be bought. However, the best way would be to make them available in shops, as is the case in the Netherlands and Belgium where insect products can be bought in the supermarket. Preferably those products with insect ingredients should be familiar to the consumer, such as burgers, schnitzels, nuggets, meat balls, flower, pasta, etc. The disguising of insects is proposed as a strategy to convince consumers (Ayieko *et al.*, 2010). Insects as an ingredient are processed and ground into the product. This is the case with crickets, which are processed into a number of food products, such as protein bars, noodles, chips, etc. Consumers need to know how to prepare the insects they purchase. That is why in the cookbook that appeared in the Netherlands (Van Huis *et al.*, 2012), only insects are used which can be bought in shops in the Netherlands (yellow mealworm, lesser mealworm and migratory locust). Similarly, in Switzerland the cookbook: 'Grillen, heuschrecken & Co' (Bärtsch and Kessler, 2016) provides recipes with mealworms, crickets and grasshoppers. Consumers also need to be informed how sustainable the insect products are and whether it is safe to use them (Ruby *et al.*, 2015). For example mealworms, crickets and grasshoppers are available at the Belgian Colruyt group through its premium 'Bioplanet' retail network, known for its high quality and bio products. Role models will legitimize the product even more. Information was seen as trustworthy when provided by scientific researchers, persons familiar with using the product, the government and well known relatives (Lensvelt and Steenbekkers, 2014). For example, In the 'The insect cookbook' (Van Huis *et al.*, 2014), Kofi Anan, former secretary general of the United Nations was interviewed as well as René Redzepi, whose restaurant NOMA was voted four times the best restaurant in the world (San Pellegrino Awards). Last but not least of course the insect product has to be tasty (Deroy *et al.*, 2015). According to 'The Nordic Food Lab', an essential requirement for insects to be commercially successful among consumers is to make tasty and delicious food from them.

20.6 Role of public and private media and of civil society groups, including donor and investment communities

As experienced during the implementation of the WUR FAO partnership activities, an excellent joint collaboration was established with a large number of reporters. Many interviews were made by all staff concerned for television, newspapers, scientific journals, films, website, Facebook, and twitter. This further boosted the dissemination and raised to the public at large.

20.6.1 Academia

Academia plays a special role, particularly in generating facts and knowledge. A new journal was launched in 2015 by Wageningen Academic Publishers: 'Journal of Insects as Food and Feed'. The number of scientific articles on edible insects has grown exponentially. Using 'edible insects' in Web of Science and Google Scholar (consulted 23 May 2017) yielded in the five-year period (2008-2012) before the FAO-WUR book was launched (Van Huis *et al.*, 2013) 36 and 793 hits, respectively, while from 2013 up to June 2017 (less than five years) there were 140 and 1,840 hits, respectively. One book 'Insects as sustainable food ingredients' was published in 2016 (Dossey *et al.*, 2016). A French book appeared in 2016 entitled 'Savoureux insectes: d'aliment traditionnel à l'innovation gastronomique' (Tasty insects: from traditional food to gastronomic innovation) (Le Gall and Motte-Florac, 2016). A number of book chapters on this topic have been written (Finke and Oonincx, 2014; Mishra and Omkar, 2017; Van Huis, 2017; Van Huis and Dunkel, 2016; Yen *et al.*, 2012). In August 2016, the European Federation of Animal Science launched the Insect Study Commission, which addresses all questions concerning insects and insect products in animal feed (http://www.eaap.org/insectsc).

20.6.2 Press

Several cookbooks have appeared: 'Entertaining with insects' (Taylor and Carter, 1996); 'Eat a bug cookbook' by David George Gordon (Gordon, 1998); 'Creepy, crawly cuisine' (Ramos-Elorduy, 1998); 'The insect cookbook' (Van Huis *et al.*, 2014); and 'Eat grub: the ultimate insect cookbook' (Radia *et al.*, 2016). There are also cookbooks in national languages such as in Dutch 'Het insectenkookboek' (Van Huis *et al.*, 2012) and in German 'Das Insekten-Kochbuch' (Dammann and Kuhlenkamp, 2015), 'Das Bienenkochbuch (Ambühl, 2016) and 'Grillen, Heuschrecken & Co' in Switzerland (Bärtsch and Kessler, 2016). Recently, a cookbook was published by the Nordic Food Lab *et al.* (2017) entitled 'On eating insects: essays, stories and recipes'.

A large number of articles appeared in the popular press. To mention just a few: 'For more protein, filet of cricket' in Science (Vogel, 2010), 'Six-legged protein' in the Wall Street Journal (Dicke and Van Huis, 2011); 'Why insects should be in your diet' by Aaron T. Dossey in The Scientist (1 February 2013); 'Why insect eating makes sense' in The Economist (25 September 2014), 'How insects could feed the world' in the Guardian (30 October 2014); 'Centipede vodka and fried crickets: is this the future of food?' by CNN (27 July 2015); 'Chefs make a tasty case for eating bugs' in National Geographic (3 June 2016); and 'Can eating insects save the world' BBC four (24 January, 2017). Daniella Martin wrote a popular book entitled 'Edible: an adventure into the world of eating insects and the last great hope to save the planet' (Martin, 2014). The topic of farming and using insects for food and feed also featured in many TV programmes, shows, news and documentaries all over the world.

20.6.3 Conferences, workshops, symposia and public/private development programmes

A number of conferences were organized by a range of actors (academia, insect producers, civil society groups, chefs, donor agencies, ministries, food or feed companies) after the international conference 'Insects to Feed the World' (Ede, the Netherlands, 2014) which specifically addressed edible insects:

- Two conferences, Insecta2015 and Insecta2016, were held in Magdeburg, Germany and were attended by more than 100 persons.
- The first conference dedicated solely to edible insects in the USA was held in 2016 in Detroit.
- PROteINSECT had their final conference in April 2016 in Brussels, Belgium (PROteINSECT, 2016).
- International Conference on Enhancing Legislation and Policy on Use of Insects as Food And Feed in East Africa held in March 2016 in Kisumo, Kenya; see also Niassy and Ekesi (2016).
- The international conference 'Insectinov' in Paris 2014, and with a second one foreseen for October 2017.
- The Expo 2015 was held in Milan, Italy, from May 1 to October 31, with the theme 'Feeding the Planet, Energy for Life.' Belgium and the Netherlands presented insects in their vision for the future, and Angola as traditional cuisine (Shelomi, 2016).
- 'Taste the change: INSECTS as food' was the theme of a side event at the Paris Climate Conference COP21 in December 2015.
- Several workshops and meetings at the national level have been organized over the last few years in many countries worldwide, such as in Italy, Belgium, Germany, Switzerland, Spain, Sweden, Denmark, China, Thailand and Laos to name a few. Also the Insect Centre organized several seminars/workshops on edible insects in the Netherlands.

The promotion of insects in the food chain is gradually also becoming part of public developmental and aid agencies in donor agencies such as: Denmark's development cooperation (Danida) (Kenya), Finnish International Development Agency (FINNIDA) (Uganda), the Swiss Agency for Development and Cooperation (SDC) (4 countries in West Africa), United States Agency for International Development (USAID) (Nigeria), Food and Agriculture Organization of United Nations (Democratic Republic of the Congo and Lao People's Democratic Republic) and the World Bank (6 million US$ grant to a consortium of East African universities). They mostly have a socially oriented focus to support small scale farming of insects to support livelihoods. Recently, also interest from private investors is coming forward, mainly for funding proposals for setting up large scale insect farms.

20.6.4 Social media and films

Twitter and Facebook played important roles as well as television programmes and films. In 2016, the film 'Bugs' (http://www.bugsfeed.com) featured chefs Ben Reade and Roberto Flore and researcher Josh Evans from the Nordic Food Lab in Copenhagen Denmark, who travelled the world in search of the practices of edible insect consumption around the world. They were accompanied by the filmmaker Andreas Johnsen. On YouTube there are a large number of video's dealing with 'edible insects'.

20.7 Conclusions

'Enabling conditions' are set up or strengthened with the ultimate objective of stimulating consumer acceptance for insects as food or feed. Consumers are driving the demand for insect products. Therefore consumers need to be well informed about the consequences of their food/feed choices and made aware that insect based products are healthier and with a much lower environmental footprint than traditional protein sources such as red meat.

The FAO-WUR publication 'Edible insects: future prospects for food and feed security' has been a landmark in disseminating this message and most scientific articles and the popular press refer to this book as a major event in raising public awareness and interest for insects as food and feed (Van Huis *et al.*, 2013). The combined reputation, force, authority of the WUR (as among the world's leading agriculture universities for innovation), and the FAO as the world summon food and agriculture agency made the report a must read in almost all the relevant agriculture and food related agencies worldwide. The increasing interest for insects as food and feed during the last ten years was probably triggered by the realization that business as usual in food production is not an option, and we need to find alternative protein sources. So, the time was appropriate for attention to alternatives to products from our conventional production animals.

Academia is needed to assess sustainability, food safety, nutritional value, processing techniques, and consumer preferences. Crucial is the collaboration between the government, public enterprise, and the academic world. Government is needed for the legal framework and legitimizing the endeavour. The private enterprise is necessary for bringing the products on the market in such a way that food safety is guaranteed, and the product is tasty. Also in upscaling, private enterprise needs to automate the production process and for processing and marketing the product.

A number of international and national organizations have emerged bringing stakeholders together. As it is a new sector, there are new challenges, such as a legislation, which never considered insects as food and feed, or consumer acceptance for which a number of strategies are proposed. Actors with varying interest in this sector need to work together in order to solve the challenges that emerge.

The media play an extremely important role. The topic seemed to catch their interest probably because of its novelty, and sustainability, in particular as it contributes to a circular economy.

As insects are new food items, a number of hurdles, in particular legislative ones as previously mentioned, became apparent. For the sector to move forward enabling conditions need to be created at the national and international level. This may require a tailor-made approach depending on the country and the insect products concerned.

References

Alexandratos, N. and Bruinsma, J., 2012. World agriculture towards 2030/2050: the 2012 revision. Global Perspective Studies Team. ESA Working Paper No. 12-03. Agricultural Development Economics Division, Food and Agriculture Organization of the United Nations, Rome, Italy.

Ambühl, D., 2016. Beezza! Das Bienenkochbuch. Eine anleitung zur Ernte und Zubereitung von Drohnenwaben der Honingbiene. Skyfood Verlag, Unterterzen, Switzerland.

Ayieko, M.A., Oriamo, V. and Nyambuga, I.A., 2010. Processed products of termites and lake flies: improving entomophagy for food security within the Lake Victoria region. African Journal of Food, Agriculture, Nutrition and Development 10: 2085-2098.

Bärtsch, C. and Kessler, A., 2016. Grillen, Heuschrecken und Co. Kochen mit Insekten. Grundlagen, Rezepte and Hintergrundinformationen. AT Verlag, Aarau, Switzerland.

Bergier, E., 1941. Peuples entomophages et insectes comestibles: étude sur les moeurs de l'homme et de l'insecte. Imprimerie Rullière Frères, Avignon, France.

Bodenheimer, F.S., 1951. Insects as human food; a chapter of the ecology of man. Dr. W. Junk Publishers, The Hague, the Netherlands, 352 pp.

Dammann, F. and Kuhlenkamp, N., 2015. Das Insekten-Kochbuch. Kosmos Verlag, Stuttgart, Germany.

DeFoliart, G., Dunkel, F.V. and Gracer, D., 2009. The Food Insects Newsletter: chronicle of a changing culture. Aardvark Global Publishing, Salt Lake City, UT, USA, 414 pp.

DeFoliart, G.R., 2002. The human use of insects as food resource: a bibliographic account in progress. Department of Entomology, University of Wisconsin-Madison, Madison, WI, USA. Available at: http://tinyurl.com/jna64w5.

Deroy, O., Reade, B. and Spence, C., 2015. The insectivore's dilemma, and how to take the West out of it. Food Quality and Preference 44: 44-55.

Dicke, M. and Van Huis, A., 2011. The six-legged meat of the future. Wall Street Journal, February 19, 2011. Available at: http://tinyurl.com/yby7ctv4.

Dobermann, D., 2017. Insects as food and feed: can research and business work together? Journal of Insects as Food and Feed 3: 155-160.

Dossey, A.T., Morales-Ramos, J. and Rojas, M.G., 2016. Insects as sustainable food ingredients: production, processing and food applications. Elsevier, Amsterdam, the Netherlands.

Durst, P.B., Johnson, D.V., Leslie, R.N. and Shono, K. (eds.), 2010. Forest insects as food: humans bite back. In: Proceedings of a workshop on Asia-Pacific resources and their potential for development, 19-21 February 2008, Chiang Mai, Thailand. Food and Agriculture Organization of the United Nations RAP Publication 2010/02. Available at: http://www.fao.org/docrep/012/i1380e/i1380e00.pdf.

European Food and Safety Authority (EFSA), 2015. Risk profile related to production and consumption of insects as food and feed. EFSA Journal 13: 4257.

Food and Agriculture Organisation (FAO), 2017a. Edible insects stakeholder directory. FAO, Rome, Italy. Available at: http://www.fao.org/edible-insects/stakeholder-directory/en.

Food and Agriculture Organisation (FAO), 2017b. FAO's role in the development of edible insect programmes. FAO, Rome, Italy. Available at: http://www.fao.org/edible-insects/84802/en.

Finke, M.D. and Oonincx, D.G.A.B., 2014. Insects as food for insectivores. In: Juan Morales-Ramos, M.G.R. and Shapiro-Ilan, D. (ed.) Mass production of beneficial organisms. Academic Press, San Diego, CA, USA, pp. 583-616.

Godfray, H.C.J., Crute, I.R., Haddad, L., Lawrence, D., Muir, J.F., Nisbett, N., Pretty, J., Robinson, S., Toulmin, C. and Whiteley, R., 2010. The future of the global food system. Philosophical Transactions of the Royal Society of London B: Biological Sciences 365: 2769-2777.

Gordon, D.G., 1998. The eat-a-bug cookbook: 33 ways to cook grasshoppers, ants, water bugs, spiders, centipedes, and their kin. Ten Speed Press, Berkeley, CA, USA, 101 pp.

Hanboonsong, Y., 2010. Edible insects and associated food habits in Thailand. FAO Regional Office for Asia and the Pacific, Bangkok, Thailand.

Hanboonsong, Y. and Durst, P.B., 2014. Edible insects in Lao PDR: building on tradition to enhance food security. FAO Regional Office for Asia and the Pacific, Bangkok, Thailand.

Hedenus, F., Wirsenius, S. and Johansson, D.A., 2014. The importance of reduced meat and dairy consumption for meeting stringent climate change targets. Climatic Change 124: 79.

Herrero, M., Henderson, B., Havlik, P., Thornton, P.K., Conant, R.T., Smith, P., Wirsenius, S., Hristov, A.N., Gerber, P., Gill, M., Butterbach-Bahl, K., Valin, H., Garnett, T. and Stehfest, E., 2016. Greenhouse gas mitigation potentials in the livestock sector. Nature Climate Change 6: 452-461.

Holt, V.M., 1995. Why not eat insects? Thornton's, Oxford, UK, 67 pp. Text reset from the original 1885 edition by Daniel H. Meeuws, Oxford, July/August 1993.

Le Gall, P. and Motte-Florac, E., 2016. Savoureux insectes: d'aliment traditionel à l'innovation gastronomique. Presses universitaires de Rennes/Institut de Recherche pour le développement, Tours, France, 500 pp.

Lensvelt, E.J.S. and Steenbekkers, L.P.A., 2014. Exploring consumer acceptance of entomophagy: a survey and experiment in Australia and the Netherlands. Ecology of Food and Nutrition 53: 543-561.

Marberg, A., 2016. Tales from the road to organization legitimacy: navigating institutional highways for a licence to operate. PhD thesis, Radboud University, Nijmegen, the Netherlands.

Martin, D., 2014. Edible: an adventure into the world of eating insects and the last great hope to save the planet. Houghton Mifflin Harcourt, Boston, MA, USA.

Mishra, G. and Omkar, 2017. Insects as food. In: Omkar (ed.) Industrial entomology. Springer, Singapore, Singapore, pp. 413-434.

Motte-Florac, É. and Thomas, J.M.C., 2003. Les 'insectes' dans la tradition orale [Insects' in oral literature and traditions]. In: Proceedings of the International Symposium Colloque Internationale Les 'insectes' dans la tradition orale, October, 3-6, 2000, Villejuif, France, 633 pp.

Niassy, S. and Ekesi, S., 2016. Contribution to the knowledge of entomophagy in Africa. Journal of Insects as Food and Feed 2: 137-138.

Nordic Food Lab, Evans, J., Flore, R. and Frøst, M.B., 2017. On eating insects: essays, stories and recipes. Phaidon, London, UK.

Oonincx, D.G.A.B. and De Boer, I.J.M., 2012. Environmental impact of the production of mealworms as a protein source for humans – a life cycle assessment. PLoS ONE 7: e51145.

Oonincx D.G.A.B., Van Itterbeeck, J, Heetkamp, M.J.W., Van den Brand, H., Van Loon, J.J.A. and Van Huis, A., 2010. An exploration on greenhouse gas and ammonia production by insect species suitable for animal or human consumption. PLoS ONE 5: e14445.

Oonincx, D.G.A.B., Van Broekhoven, S., Van Huis, A. and Van Loon, J.J.A., 2015a. Feed conversion, survival and development, and composition of four insect species on diets composed of food by-products. PLoS ONE 10: e0144601.

Oonincx, D.G.A.B., Van Huis, A. and Van Loon, J.J.A., 2015b. Nutrient utilisation by black soldier flies fed with chicken, pig, or cow manure. Journal of Insects as Food and Feed 1: 131-139.

Paoletti, M.G., 2005. Ecological implications of minilivestock: potential of insects, rodents, frogs and snails. Science Publisher Inc., Enfield, NH, USA, 648 pp.

Papargyropoulou, E., Lozano, R., K. Steinberger, J., Wright, N. and Ujang, Z.B., 2014. The food waste hierarchy as a framework for the management of food surplus and food waste. Journal of Cleaner Production 76: 106-115.

PROteINSECT, 2016. Insect protein, feed for the future. In: Proceedings of the final PROteInsect conference, April, 27, 2016. Minerva Communications Ltd, Andover, UK.

Radia, S., Whippey, N. and Holmes, S., 2016. Eat grub: the ultimate insect cookbook. Frances Lincoln Limited, London, UK.

Ramos-Elorduy, J., 1998. Creepy crawly cuisine: the gourmet guide to edible insects. Park Street Press, Rochester, VT, USA, 150 pp.

Ruby, M.B., Rozin, P. and Chan, C., 2015. Determinants of willingness to eat insects in the USA and India. Journal of Insects as Food and Feed 1: 215-225.

Shelomi, M., 2016. The meat of affliction: insects and the future of food as seen in Expo 2015. Trends in Food Science and Technology 56: 175-179.

Taylor, R.L. and Carter, B.J., 1996. Entertaining with insects. Salutek Publishing Company, Yorba Linda, CA, USA.

Tilman, D. and Clark, M., 2014. Global diets link environmental sustainability and human health. Nature 515: 518-522.

Van Broekhoven, S., Bastiaan-Net, S., De Jong, N.W. and Wichers, H.J., 2016. Influence of processing and *in vitro* digestion on the allergic cross-reactivity of three mealworm species. Food Chemistry 196: 1075-1083.

Van Broekhoven, S., Doan, Q.H.T., Van Huis, A. and Van Loon, J.J.A., 2014. Exposure of tenebrionid beetle larvae to mycotoxin-contaminated diets and methods to reduce toxin levels. Proceedings of the Netherlands Entomological Society Meeting 25: 47-58.

Van Broekhoven, S., Gutierrez, J.M., Rijk, T.C.D., Nijs, W.C.M.D. and Van Loon, J.J.A., 2017. Degradation and excretion of the *Fusarium* toxin deoxynivalenol by an edible insect, the yellow mealworm (*Tenebrio molitor* L.). World Mycotoxin Journal 10: 163-169.

Van Broekhoven, S., Oonincx, D.G.A.B., Van Huis, A. and Van Loon, J.J.A., 2015. Growth performance and feed conversion efficiency of three edible mealworm species (Coleoptera: Tenebrionidae) on diets composed of organic by-products. Journal of Insect Physiology 73: 1-10.

Van Huis, A., 2003. Insects as food in sub-Saharan Africa. Insect Science and its Application 23: 163-185.

Van Huis, A., 2017. New sources of animal proteins: edible insects. In: Purslow, P.P. (ed.) New aspects of meat quality: from genes to ethics. Elsevier, Amsterdam, the Netherlands.

Van Huis, A. and Dunkel, F.V., 2016. Edible insects: a neglected and promising food source. In: Nadathur, S.R., Wanasundara, J.P.D. and Scanlin, L. (eds.) Sustainable protein sources. Elsevier, Amsterdam, the Netherlands, pp. 341-355.

Van Huis, A., Van Gurp, H. and Dicke, M., 2012. Het insectenkookboek. Atlas, Amsterdam, the Netherlands, 190 pp.

Van Huis, A., Van Gurp, H. and Dicke, M., 2014. The insect cookbook. Columbia University Press, New York, NY, USA.

Van Huis, A., Van Itterbeeck, J., Klunder, H., Mertens, E., Halloran, A., Muir, G. and Vantomme, P., 2013. Edible insects: future prospects for food and feed security. Food and Agriculture Organization of the United Nations (FAO), Rome, Italy, FAO Forestry Paper no. 171, 187 pp. Available at: http://www.fao.org/docrep/018/i3253e/i3253e.pdf.

Van Huis, A. and Vantomme, P., 2014. Insects to feed the world. Conference report. Food Chain 4: 184-192.

Van Ittersum, M.K., Cassman, K.G., Grassini, P., Wolf, J., Tittonell, P. and Hochman, Z., 2013. Yield gap analysis with local to global relevance – a review. Field Crops Research 143: 4-17.

Vantomme, P., Göhler, D. and N'Deckere-Ziangba, F., 2004. Contribution of forest insects to food security and forest conservation: the example of caterpillars in Central Africa. Odi Wildlife Policy Briefing 3. Odi, London, UK. Available at: http://tinyurl.com/y8e3l5ln.

Vantomme, P., Mertens, E., Van Huis, A. and Klunder, H., 2012 Assessing the potential of insects as food and feed in assuring food security: summary report. Technical Consultation Meeting, January 23-25, 2012, Forestry Department, FAO, Rome, Italy.

Verhoeckx, K.C.M., Van Broekhoven, S., Den Hartog-Jager, C.F., Gaspari, M., De Jong, G.A.H., Wichers, H.J., Van Hoffen, E., Houben, G.F. and Knulst, A.C., 2014. House dust mite (Der p 10) and crustacean allergic patients may react to food containing yellow mealworm proteins. Food and Chemical Toxicology 65: 364-373.

Vogel, G., 2010. For more protein, filet of cricket: could an African caterpillar be the new beefsteak? Science 327: 811.

Yen, A., Hanboonsong, Y. and Van Huis, A., 2012. The role of edible insects in human recreation and tourism. In: Lemelin, R.H. (ed.) The management of insects in recreation and tourism. Cambridge University Press, Cambridge, UK, pp. 169-185.

Yi, L., Van Boekel, M.A.J.S.., Boeren, S. and Lakemond, C.M.M., 2016a. Protein identification and *in vitro* digestion of fractions from *Tenebrio molitor.* European Food Research and Technology 242: 1285-1297.

Yi, L., Van Boekel, M.A.J.S. and Lakemond, C.M.M., 2016b. Extracting *Tenebrio molitor* protein while preventing browning: effect of pH and NaCl on protein yield. Journal of Insects as Food and Feed 3: 21-31.

Yi, L., Lakemond, C.M.M., Sagis, L.M.C., Eisner-Schadler, V., Van Huis, A. and Van Boekel, M.A.J.S., 2013. Extraction and characterisation of protein fractions from five insect species. Food Chemistry 141: 3341-3348.

21.

Freeze-dried insects for human consumption

Product development and promotion

M. Shockley[1]*, R.N. Allen[2] and D. Gracer[3]

[1]Department of Entomology, University of Georgia, 120 Cedar St., Athens, GA 30602, USA; [2]Little Herds, Austin, TX 78757, USA; [3]Community College of Rhode Island, Providence, 400 East Ave, Warwick, RI 02886, USA; entomolo@uga.edu

Abstract

This chapter considers the modern state of edible insects, primarily from the perspective of product development and promotion. Within traditional markets, a certain level of product acceptance is required for the investment in product development to be recouped. However, societal acceptance is only one criterion for progress. Product development also involves agricultural, technological, and economic advances in the field or product line. Insects as food are regulated differently from country to country. This variation in clarity of regulations may delay the acceptance of insects as food and ultimately may stifle innovation in insect based product development. Looking beyond human and animal applications, the individual value streams from insects extends beyond the protein. The fats and the chitin, provide extensive opportunities for scaling up existing and promising research. Upon execution of design, the product must be promoted through various marketing strategies. Perceptions of insects in general, and more particularly as a food source are integral to advancing the edible insect industry. In the field of insects as food the promotion of edible insect products may include company campaigns, K-16 outreach and education programs, non-profit organizations, research organizations, as well as individual proponents. A global network of supporting researchers and other participants contributes to the recent growth of this industry through insect-based food product development and promotion. Trade organizations and research initiatives have been established on several continents (Africa, Asia, Europe and North America) to work closely with food insect-based businesses and regulatory agencies to provide insight and guidance on incorporating edible insects into local, regional, national, and international food systems. Thousands of popular press and peer reviewed articles pertaining to insects as food have been published increasing the overall exposure to the edible insect industry. Product development and promotion of insect based food products will move the edible insect industry forward around the world.

Keywords: product development, promotion, edible insects, insects as food products, entopreneur

21.1 Introduction

In the edible insect industry, varying terms are used including the terms insects as food, edible insects, and insect cuisine. Insects as food refers to specific use for humans rather than livestock, edible insects refer to insects specifically which are edible and insect cuisine refers to the preparation of edible insects. From a marketing perspective moving insect-eating beyond a small group of consumers will require a variety of strategies of which food appropriateness may play larger role than sensory qualities (Gracer, 2010; Tan *et al.*, 2016b). Other strategies may include products having excellent taste, processed formats (versus raw material), attractive packaging, consistent branding and terminology, and consumer education (McFarlane and Pliner, 1997; Ruby *et al.*, 2015). For many countries, edible insects are a normal or traditional food and require little market validation. In countries without a traditional use of insects as food, this critical level of acceptance is increasingly becoming evident. Research addressing consumer acceptance to the idea of insects as food has expanded over the last five years (Balzan *et al.*, 2016; Cicatiello *et al.*, 2016; DeFoliart, 2012; Hartmann *et al.*, 2015; Lensvelt and Steenbekkers, 2014; Megido *et al.*, 2014, 2016; Shelomi, 2015; Sogari *et al.*, 2017; Tan *et al.*, 2015, 2016a). The science of insects as a food-grade ingredient may lead to tasty, healthy, appealing dishes. As we are still in the early stages of exploring the diverse uses of insects and insect derived products, we may not currently be using the most efficient, nutritious or economically beneficial insects now. The field of edible insects has seen recent and encouraging technological and entrepreneurial developments, but other aspects of the field remain unexplored. It is very important that we acknowledge and discuss our own referential frames including our food habits. Edible insects are a global subject, yet most recent consumers are in western societies of North America and Europe.

Product development and the marketing of any product, commodity, or service involves efforts to convince members of the public that a given item or service will solve problems or help the consumer in some form or fashion. We observe this on magazine pages, television channels, billboards, and many other methods of communication. Insects as food may provide solutions to anthropogenic problems such as environmental degradation, nutritional inequities and human health. As with any new or novel food or food ingredient, insects as a solution are challenged by obstacles and barriers. Humankind cannot indefinitely maintain the current status quo regarding the production and consumption of vertebrate-animal-proteins such as beef, pork, and chicken (Gahukar, 2011) and experts from many disciplines and countries have predicted shortfalls in protein production worldwide (Van Huis, 2013).

Numerous themes run throughout the literature on entomophagy. One such theme is that writers acknowledge that their ideas are likely to be met with resistance. In 1885 Vincent Holt published 'Why not eat insects?' which many journalists and researchers have identified as the first book investigating the advantages and disadvantages of entomophagy. We see this in Holt's introduction (Holt, 1885):

> In entering upon this work I am fully conscious of the difficulty of battling against a long-existing and deep-rooted public prejudice. I only ask of my readers a fair hearing, an impartial consideration of my arguments, and an unbiased judgment. If these be granted, I feel sure that many will be persuaded to make practical proof of the expediency of using insects as food.

21.2 Product development

21.2.1 Acceptance

Developing insect based food products presents unique challenges. Edible insects and insect consumption is not readily accepted by that members of many western communities. However, the acceptance of edible insects as food invites us to engage in dialogue with various stakeholders about the nutritional, environmental, and economic potential of this industry (Oonincx *et al.*, 2010). Research and evidence generated over the past decade posit that many societies are acknowledging the role insects may contribute as an alternative protein-source either for animal feed or for direct human consumption (Caparros Megido *et al.*, 2016; Shockley and Dossey, 2014). Environmental impacts of our current agricultural practices (including land use, water, and carbon emissions) may limit the amount of vertebrate protein we can produce (Smil, 2000). It is estimated that the global population will reach 9 billion by 2050 (United Nations, 2015). An overall increase in protein-demand will necessitate a shift in current agricultural production systems as well as the utilization of non-traditional protein-sources, such as insects. Similarly, it is possible that insects may only achieve acceptance when preferred food-sources are no longer available in sufficient supply and price-points.

In areas where insects are a traditional food source whole insects are most often preferred due their availability, abundance, and familiarity. In western societies, however, the idea of eating 'whole foods' has become more esoteric. However, there is simultaneously a resurgence of people who are looking at the whole food while focusing on farm to table concepts and edible insects may be posed for either of these groups. This is part of a general expansion of innovations in many aspects of the human experience, and may manifest in a disconnection between the whole animal and the food item itself: filets of fish versus whole fish, chicken nuggets as pieces of chicken, steaks and hamburgers versus whole cows. People want animal and vertebrate protein but do not necessarily want to know the origins or their protein. In many countries, whole insects are not being marketed or accepted as commonly as powders or flours (Appendix 1). In addition, some research points to disgust and cultural factors as a hurdle to acceptance (Durst, 2010; Looy *et al.*, 2014; Yen, 2009, 2010). This can be observed in the availability of food products where insects are an ingredient not the single food source. The acceptance of insect cuisine and the readiness of consumers to the idea of insects as food is still a challenge, but with research, education and outreach programs the diffusion of entomophagy is possible (Balzan *et al.*, 2016; Cicatiello *et al.*, 2016; DeFoliart, 2012; Hartmann *et al.*, 2015; Lensvelt and Steenbekkers, 2014; Looy *et al.*, 2014; Megido *et al.*, 2014, 2016; Sogari *et al.*, 2017; Tan *et al.*, 2015, 2016a,b).

An entopreneur in an insect based food or feed entrepreneur. Entopreneurs expand the supply chain from farmers, processers, and food development companies and have greatly contributed to the expansion of the edible insect industry. In addition to product acceptance other relevant factors may include efforts to increase the availability of insect foods (Shelomi, 2015). This could be accomplished by edible insects being sold at a wider range of stores (both physical and virtual) and restaurants (Appendix 1). Another factor may be the cultural relevance of insect-based foods (Tan *et al.*, 2016a). Due to the diversity of insect consumers and cultures world wide a product development strategy for one country or region is likely to be very different from another country or region. In addition, another factor may be the degree insect-based foods fit in to and individuals dietary practices (House, 2016). Lastly, the taste of insects or insect-based foods must be palatable and gastronomically pleasing. The importance of cooking with insects and using recipes with insects while combining them with a multitude of ingredients must be an integral part of insect based food in culinary practice, preparation and presentation (Deroy *et al.*, 2015).

21.2.2 Food products

As a new ingredient for individual home use or commercial industrial uses, human ingenuity and entopreneurs have only just begun to explore the myriad of applications of edible insect products. The first step of abstraction of cricket or mealworm powders was refined by entopreneurs and researchers in many countries between 2010 and 2016 (Appendix 1). We have now seen an increasingly competitive push towards value-added ingredients. The first insect powder products to reach the market in western countries were made using fairly low-tech processes; insects were frozen when harvested, then washed and cleaned, dry roasted using convection heat, then ground using grinders similar to food processors found in home kitchens. This provided a powder that was fine enough to prevent visual identification of the insects themselves, and had broad enough applications for baking uses and making consumer products like protein bars, chips, cookies and crackers. The flavours and aromas, often described as nutty, earthy, seedy or umami, lent themselves easily to baking applications.

Over the past few years, however, additional processing methods have been able to refine the powder for more specific uses. One process provided a much finer grain-size and less pronounced flavour and aroma profile by slurring the crickets in water, spray-drying the mix and dehydrating or freeze drying the resulting droplets leaving a fine powder. Another process used fermentation techniques to create versatile ingredients like a cricket miso or cricket tofu. Still other proprietary processes have begun to tweak insect proteins to behave in cooking and product formulation applications like an egg or dairy replacement. Emulsification has also been used by food scientists to create 'puffed' products like various snack foods. As food scientists and recipe creators continue down this path we expect to see an increasingly diverse array of processing methods and desired specifications for insect ingredients, from grain size, to shelf life, to nutritional density and bioavailability.

There has been a great deal of commentary regarding the advantages of processing insects, as opposed to utilizing them intact. This stratagem has been useful in the marketing of standard-vertebrate-proteins worldwide as well. Utilizing food-preparation machinery to transform insect ingredients, and marketing products that incorporate insect powder, may make edible insects more acceptable to the public particularly in western societies. Enhancing culinary techniques may only be a singular method for encouraging the acceptance of new or novel foods such as insects. In addition, industrial-scale insect-production is working towards reaching parity regarding price-points for a given unit of processed final product.

21.2.3 Global regulations

As a newer food industry in the West, insects as food are regulated differently from country to country. This variation in clarity of regulations may delay the acceptance of insects as food and ultimately insect based product development. This immense variation in regulations and understanding of insects as food may create confusion and delays from local, state, and federal/national food agencies, health departments and international/custom trade agents. Questions pertaining to insect farming, selling or marketing insect food products is either being embraced, delayed or halted altogether. Massimo Reverberi, the Italian founder of Bangkok-based Bugsolutely (https://www.bugsolutely.com/yellow-bug-pages), manufactures pasta from cricket flour. In a February 2017 article in Food Navigator Asia, Reverberi detailed the current position of regulatory affiliates concerning insects as food in many Western countries (Reverberi, 2017). Various issues may include farming, production, marketing, and international trade.

Reverberi (2017) described three legal trends concerning regulatory affairs. The first legal trend can be seen within countries where governmental regulations are lenient; insects must be produced for humans and according to all food safety and handling procedures. The next legal trend includes non-English-speaking Western countries and the European Union. In this second trend these groups have opted to establish rules and an approval process before allowing any marketing. The final legal trend can be observed by Non-Western countries where insects are readily consumed at the local level including markets, homes, and restaurants.

21.2.4 Australia and New Zealand

Food Standards Australia New Zealand (FSANZ) is a shared agency between the two nations and is charged with maintaining food safety. Recently FSANZ addressed insect cases including *Zophobas morio* (super mealworm), *Acheta domesticus* (house cricket) and *Tenebrio molitor* (mealworm). It was determined that these insects are a non-traditional food in Australia and New Zealand and not novel food. There were no safety concerns identified, however, labelling of true nature of food required.

21.2.5 North America

Canada

Insects as food are evaluated by the Canadian Bureau of Microbial Hazards in the Food Directorate. Crickets and mealworms are the primary insects currently farmed and utilized as a food ingredient. In the future if different insects are introduced as a food ingredient which lack a history of safe consumption they may fall into the novel food category to then be evaluated by the Bureau of Microbial Hazards in the Food Directorate.

United States

In the United States, there are no specific set of standards for edible insects. The US Food and Drug Administration (FDA) has stated for insects to be allowed for human consumption and to enter the marketplace, the insects must have been reared for human consumption and follow the standards required by the FDA including bacteriological tests and good manufacturing practice certification. In addition, the label on the product must include the common name and the insect's scientific name, as well as noting the potential risks of allergy.

Mexico

Mexico has a long history of insect consumption. Most of the insects consumed are wild harvested and hand collected, farming, raising, and marketing of edible insects is unregulated.

The most common insect consumed in Mexico are chapulines and there are current concerns with overharvesting and resource depletion. However, organic agriculture is regulated by the *Acuerdo por el que se dan a conocer los Lineamientos para la Operación Orgánica de las actividades agropecuarias* (Lähteenmäki-Uutela *et al.*, 2017).

21.2.6 Europe

The European Parliament and the Food Safety Authority (EFSA) considers all insects as novel foods and are subject to an in-depth approval processes. However, Belgium, the Netherlands and Denmark do not accept this interpretation and explicitly permit and (in one case) regulate the marketing and consumption of insects. Other countries, such as France, express a degree of tolerance. Other countries, such as Italy and Germany, have a zero tolerance. Due to the complexity (and cost) of the approvals process, no applications under the novel food regulations have been submitted to date. October 2015, in the European Parliament the novel food law was revised to simplify the steps and reduce the extensive timing for approval which will go into effect January 1, 2018.

Belgium

The Federal Agency for the Safety of the Food Chain (FASFC) has created specific regulations for edible insects. The FASFC approved ten insects to date and have explanative rules for rearing and sales. Currently no insects reared outside of the European Union are accepted. Marketing of edible insects is legal, but the import or export is not (Belgium does not currently accept insects from non-EU countries).

United Kingdom

The Food Standards Agency has a positive position on the sale, utilization and import of edible insects. Edible insects are considered outside the context of the European regulation on novel foods. They require Novel Foods Approval from the EFSA but have given companies leave to continue operating until 2018 by when they should submit the application. However, with Brexit, the future is unclear due to the possibility that the UK will have to comply with EU directions for edible insects.

Denmark

The Danish Veterinary and Food Administration takes the stance that whole insects are not included in the EU novel food legislation. If there is Danish interest, imports from non-EU countries may be possible.

France

As a part of the European Union, France, like other EU countries has not formally recognized the trade.

Germany

Basic regulations of the German food law are laid down in 61 articles of the Lebensmittel- und Bedarfsgegenstaendegesetz (LMBG), last amended September 1997. As an EU member, insects are considered a novel food and cannot be sold in Germany until a procedure for novel foods approval has been established (USDA Foreign Agricultural Service, 2009).

The Netherlands

The Netherlands is a worldwide leader in edible insect research and promotion and is among the most progressive on the European continent. The expanding global edible insect industry can be greatly contributed to the advocacy and research of the academics from the Laboratory of Entomology at Wageningen University and the collaborations with the Food and Agricultural Organization (FAO) on conferences such as the 'Insects to feed the world' conference in 2014, the first international conference on insects for food and feed, bringing together academics, the feed industry, insect farmers and NGOs. Seminal publications, such as the 2013 FAO report published in partnership by Wageningen University scientists and the FAO, 'Future prospects for food and feed security', arguably launched the edible insect industry into the mainstream media (Van Huis *et al.*, 2013). In addition, in 2015 Wageningen Academic Publishers released the first volume of the 'Journal of Insects as Food and Feed' (www.InsectsasFoodandFeed.com).

Norway

The Norwegian Food Safety Authority is Norway regulatory arm, however, Norway, is not an EU member, but belongs to the European Economic Area. The current interpretation of edible insects is limited to whole insects which do not fall under the novel food law.

Switzerland

The Federal Council has worked diligently on the legislation of insects. In December 2016, the council passed a law allowing the sale and consumption of three species. As a part of the requirements, the insects must have be reared for human consumption and must be treated according to the criteria of food security (high temperatures, freezing, etc.).

21.2.7 Asia

China

In many regions of China, insects are a highly marketed and utilized food ingredient. China is the world's largest producer of silk and silkworms are available in very large quantities. They are also exported for food consumption to other Asian countries such as Thailand. However, insects, excluding silkworms which were added in 2014, are not currently listed as food in food laws in China. In China, the feed materials catalogue contains an entry for 'insects and processed insect products'. There is a listing for 'insect meal' composed of dried insects that have been crushed as well as a listing for 'de-fatted insect powder' composed of dried insects that have been crushed and de-oiled. In both cases, only insects that do not affect animal or public health can be used and the specific insect species must be included in the product name.

South Korea

In 2011, South Korea's government sought to legalise some edible insects. On the list were mealworms and crickets (*Gryllus bimaculatus*). In 2016, after the preliminary process, the Korean Food and Drug Administration classified crickets and mealworms as normal foods, without any further restrictions. Other insects are likely to be added to the eligibility list soon.

21.2.8 The Association of Southeast Asian Nations countries

Thailand

There is a tradition of edible insects and insects as food amongst Southeast Asian countries, however, there are no regulations relating to the rearing, sale and export of insects. The world's largest breeder of crickets is Thailand which is working on the creation of a set of rearing and breeding guidelines. The National Bureau of Agricultural Commodity and Food Standards (ACFS) is the Thai government agency for the safety of agricultural products and is expected to release agricultural practices guidelines for the rearing of crickets by the end of 2017. A preliminary set of these guidelines for good agricultural practices (GAP) for cricket farm standards was made public by the University of Khon Kaen. Thailand will be the first country in the world to develop a GAP for cricket farm standards.

Lao People's Democratic Republic

The Lao government has identified the development of the non-wood forest product area as a national priority of which edible insects are important of as nutritious food for villagers and as a source of household cash income. The Ministry of Health, Food and Drug Department is the Lao PDR government agency for food laws and regulations.

India

Insects play an important role in the dietary system of various ethnic groups of India and the number of edible insects eaten in a state varies with the land and tribes (Purkayastha, 2016). Insects provide nutritional security as well as curing various body ailments. The Food Safety and Standards Authority of India is the government agency for food safe and nutritious food.

21.2.9 Africa

Insects are widely consumed in many countries in Africa, however, there is no pan-African legislation that impacts the production of insect protein and its use in human food.

21.2.10 Insect advances beyond food

Advances in many different areas of insect mass-rearing have begun to provide further inroads into possible uses beyond direct food purposes. Looking beyond human and animal applications the individual value streams from an insect beyond the protein, the fats (insect based oils) and the chitin, provide extensive opportunities for scaling up existing and promising research (Sosa and Fogliano, 2017). Chitin is processed for medicinal purposes including bandages and surgical threads. However, the processing of chitin exposes many negative aspects of the mass production in equatorial countries with high amounts of cheap labour and little protection for workers (Feng *et al.*, in press; Gahukar, 2011; Liu, 2011; Van Huis *et al.*, 2013). By extracting chitin from insects, rather than shrimp and seafood shells, we have a 'cleaner' option for scaling up those uses. Some other non-traditional and specific uses of insects include their use in the medical profession as bandages and for cancer treatments (Ratcliffe *et al.*, 2011; Tonk and Vilcinskas, 2017). In addition, insects can be seen in the industrial sector as edible films and as an additive to thicken and stabilize foods. Finally, insects can be observed in the biotechnology purposes (biodegradable plastic) (Cheung *et al.*, 2015; Fernandez and Ingber, 2014; Shahidi *et al.*, 1999). Other uses of insects may include water filtration, seawater and uranium conversions, and their use as polymers (Manjunatha *et al.*, 2016).

21.3 Promotion

21.3.1 Marketing

The term marketing has an inherent commercial aspect (selling a service or product). Strategies will vary and a clear distinction must be made between marketing as business-to-business and 'marketing' as business-to-consumer. In this chapter, we focus primarily on business-to-consumer marketing strategies beginning with the terminology already employed to reference the use of insects as food. While the term entomophagy is accurate for an insect eater this applies to all types of animals not just humans. The term entomophagy may be cumbersome to the layperson. Alternatively, an entopreneur in an insect based entrepreneur. We propose the term insectarian as anyone that eats insects or possibly and more specifically a vegetarian who eats insects like a pescatarian (fish), a pollotarian (chicken), or a flexitarian (occasional meat). GeoEntomotarian is a term that the Goldin brothers of EntomoFarms in Canada have proposed to connote an emphasis on sustainability using insects in the diet. The term insect cuisine was proposed by Kiah Brasch of Rustle Up Some Grub (USA) at the 2016 International Congress of Entomology Edible Insect Symposia as a possible alternate to the term edible insects. To avoid confusion and to place insects as food in a more positive light, it was suggested that symposium attendees modify their vernacular as a marketing strategy for businesses as well as during outreach and educational activities.

Commercial and humanitarian agendas alike depend upon understanding people. Many researchers have noted the intensity with which most people throughout the world prefer their inherited, group-normative food-sources. There is little question that the commercial edible insect industry has undergone a remarkable expansion in the past several years (Appendix 1). Many observers might posit that this development may indicate a societal shift in favour of edible insects. There have been multiple programs designed to encourage the adoption of insect domestication in areas of the world where insects are traditionally harvested from the wild, including but not limited to FAO projects, *icipe* projects, MealFlours in Guatemala, Aspire Food Group in Ghana, MIGHTInsect in Zambia and notably EntoCube in Northern Europe. They design climate controlled shipping containers of various sizes that can be deployed individually or in multiples to provide an 'insect farm-in-a-box,' that can be utilized in multiple different climates to begin farming insects as food quickly and efficiently. When considering the applications for quick deployment in malnourished or resource deficient areas, or areas where the agricultural sector has been compromised by natural disasters, the utility of this drop-and-go strategy and subsequent training in insect rearing bears much potential. Conversely, the Aspire Food Group project or MealFlours project takes a low-tech approach and builds momentum from within the community. Aspire Food Group started in 2014 with ten shareholder farmers refining the domestication of the palm weevil larva in a simple bucket system. The palm weevil larvae were already considered a normal, if not preferred, food source in rural areas of Ghana, but were only wild-crafted seasonally and opportunistically. By growing year-round, in households or small sheds, this system provided the means for women, children, elderly and disabled to access additional food for their families and to sell in the local markets. The reaction from the local communities was overwhelmingly positive and the program has since expanded into four additional villages with over 500 farmers participating. MealFlour began their program in 2016 in Guatemala, teaching women in rural communities to farm, harvest, process and cook mealworms as a food fortification tool. This project stands out as the mealworm is not commonly considered a food insect in Guatemala. Despite this and regulatory challenges, the women in the program have been highly receptive to the idea, and they plan to expand the program to 'train the trainers,' arming women with the tools and knowledge to take the idea back to their own villages to train other women.

Technological and entrepreneurial progress are aspects within the edible insect industry working towards and in conjunction with general acceptance. The broader global industry of insects as food has seen increasing investment in the past few years. Many companies and start-ups have received funding through pitch challengers, accelerator programs and grants. Angel and venture capital investors have also begun to enter the market, as well as serial investors where several insect based product pitch-challenged the TV show SharkTank. This platform not only secured investor resources but also as a large public platform could help with normalizing the idea to a broad swath of the public that might not otherwise know about the potential of insect foods.

Insect foods are diversely represented. Regarding how best to represent and market the products themselves there have been various strategies utilized by entopreneurs. Would showing images of insects help create appeal for insects as food? Or would it be better to hide the animal and just work on concepts? The field of consumer psychology best tackles this issue and provides additional opportunities for research and development. In western societies, for example, it is common to transform animal-proteins from their original appearance (whole chicken into chicken nuggets, and cow into hamburgers, whole fish into breaded filets, etc.). One strategy may be to alter the physical appearance of insects and market that final product without direct images of insects. On the other hand with the recent farm to table sustainability movement there could likely be a market that is fully aware and purchasing insect based food products because of the insect ingredient itself. In this case, sustainably minded consumers may not be deterred from insect related imaging on the packaging or in any related marketing or media promotions.

Abstracting insects in food products may not have been as successful as many had assumed, as the average Western consumer may not associate these types of products with protein. Because the companies that manufacture consumer packaged goods like insect snack foods have, to date, been the primary customer base for insect farmers, may be difficult for new and small insect farmers to grow. Deroy *et al.* (2015), along with numerous culinary experiments (Deroy *et al.*, 2014; Piqueras-Fiszman and Spence, 2014; Spence *et al.*, 2014) suggests Western consumers are amenable to eating insects in whole or lightly coated forms, provided context and positioning, particularly in association with education and outreach events.

These findings corroborate antidotal and informal surveys among edible insect industry stakeholders that a portion of their customers are less interested in insect ingredients in snack foods, and are instead more willing to try insects in their whole form either when cooking at home or when dining at a restaurant. Much like sushi's popularization by *avant garde* chefs, insects are increasingly featured on the menus at restaurants. When consumers eat at a restaurant where they have had a previous positive experience, they trust the chef, and are more likely to try novel foods. Similarly, when consumers cook the food themselves, they are more likely to have a positive experience eating insects. Because a large portion of the early-adopters of edible insects are already more likely to incorporate locality, sustainability, nutrition, and traceability into their purchasing decisions, they may tend to buy and consume whole insects purchased from a local insect farmer at a farmer's market than an abstracted insect protein product from a manufacturer. At this point, very few insect farmers have successfully leveraged farmer's markets to sell directly to consumers in whole form. Consequently, there may be large mutual benefits both to the insect farmer and farmer's markets. Insect farmers my find new retail avenues, and the farmers market that now draws in new customers and chefs who are coming specifically for the insects.

To honour the first edition of World Edible Insect Day (the 23rd of October 2015) Bugburger composed a list of entopreneurs around the world (http://tinyurl.com/zyotzcy). That is companies, organizations and individuals who either are working on an edible-insect product, are planning to deliver one or are just advocating the benefits of eating insects. See Appendix 1 for a modified list of these entopreneurs, restaurants, books and cookbooks highlighting the expansive creativity and ingenuity of insect cuisine.

When used in whole form, insects stand apart, and their perceived benefits are amplified for consumers looking for products that fit their specific dietary preference, rather it be concern for the environment, personal health or animal welfare. Similarly, House (2016), found that abstracted insect products, while successful at garnering first-time purchasers, struggled to accrue the desired amount of repeat purchases. When insects were abstracted, consumers tended to weigh them against similar snack food alternatives and conventional products known to the consumer, using criteria like flavour, familiarity and price. He also concluded that rather than making insect products appealing to wider audiences through abstraction, the industry should focus on the early-adopters who value the inherent 'insect-ness,' of the whole form.

21.3.2 Perceptions

When considering promoting and marketing insect based food products perceptions must be a part of discussion. While true in most aspects of our life, and especially so with insects, we do not always think with our head. In fact, it may be that our head rarely plays a role beyond rationalizing what our 'gut' has already told us. Building on psychologist Daniel Kahneman's notion that we have two distinct parts to our brain, System 1 and System 2, each reacting to external stimuli differently, and consequently how this affects our decision-making process. To illustrate, say we present our average person with a donut. While System 1 is analysing the situation and thoughtfully weighing the outcomes and options, System 2, has already eaten the donut and moved on. They find, through their experiments with behavioural nudges, that people act on their System 1 brain much more often than we would like to admit, sometimes up to 95% of the time. This goes a long way towards understanding why insects can seem to be such a hard sell for anybody who looks at the rational argument and still cannot get past that dreaded 'ick factor' (Sallows, 2015).

Psychologist Paul Rozin *et al.* (2008) finds, through experiments on disgust and revulsion, although not directly insect related, that the foods we eat are often influenced, not only by our learned behaviours from parents and caregivers, but also from an evolutionary instinctive response. On the one hand, we have our biological reaction to potentially harmful foods. When we eat foods that are toxic or rotting there is a putrid taste that biologically tells us it's harmful. This has helped humans learn through history when something is edible or inedible, and is incredibly useful for keeping us from eating foods that could hurt or kill us (Rozin and Fallon, 1987). Simultaneously, we also learn about which foods are 'inedible,' by viewing the responses of those around us as we grow. When a child puts dirt into their mouth, adults have a disgust reaction that can be identified through particular facial expressions found universally around the world (Rozin *et al.*, 1994).

In western cultures the idea of eating insects may be undesirable. The American entomologist Charles T. Brues comments upon a perceived association between two concepts – entomophagy and human development – in the first paragraph of his book 'Insects, food, and ecology' (Brues, 1946):

> Now that we have detailed at such length the great success of the insects in developing food habits that exploit practically everything in their living environment, it may be of interest to view the matter from another angle. Their food relationships with other living organisms are by no means restricted to the categories so far mentioned, for many other animals and even some plants have discovered in the insects themselves a large and nutritious, if not wholly savoury, food supply. We have already devoted one earlier chapter to predatory insects and another to parasitic ones and found these to subsist almost entirely on other insects. Many other groups of carnivorous animals feed, at least to some extent, on insects both as predators and true parasites. Even man himself does not completely exclude insects from his diet although the consumption of insects by members of our western civilizations is at the present time entirely unintentional on their part. Such is not true of other peoples, and if space permits we may return to this as it serves royally to bolster up the feeling of race superiority, Nordic or otherwise.

How entrenched is this culturally acquired aversion to insects? Rozin (2008) conducted an experiment to test this idea; how our gut reaction and learned revulsion exert its power over the rational part of our brain. By acknowledging that our rational brain often takes a backseat to our 'gut,' when it comes to foods with cultural stigmas, the difficulty in marketing insects to western cultures becomes much more understandable (Rozin, 2008).

Sallows (2015) addressed this issue in a collaboration with an edible insect company, Eat Grub. One common advertising and marketing strategy that we have seen work for insects as it does for other products is the 'fear of missing out' (FoMO) concept. Przybylski *et al.* (2013) describe this concept 'as a pervasive apprehension that others might be having rewarding experiences from which one is absent, FoMO is characterized by the desire to stay continually connected with what others are doing'. We can see this exemplified through the OgilvyChange experiment with Eat Grub, in how the average consumer reacts to the idea of insect products, and what marketing messages resonate most effectively with their subconscious. The Alfresco

Lab, a collaboration between marketing agencies OgilvyChange, Kinetic, and Eat Grub, a start-up edible insect company in the UK, set up a stall serving roasted crickets. They designed three different advertisements with images and messaging specifically to draw in pedestrian traffic to sample the insects. The first one provides the generic rational argument of nutritional and environmental benefits; a second focused on normalizing the idea; and a third used the FoMO, a well-known and long used marketing technique saying 'get yours before they're all gone.' Without ads, 34 people were enticed. With the first advertisement, the 'rational argument,' 67 people were attracted. The second ad garnered 126 people while the third ad, using FoMO, reaped an astounding 154 people, double the rational argument and more than four times the number with no ads.

21.3.3 Education and advocacy

Education and outreach events pertaining to insects as food, edible insects, and insect cuisine occurs in diverse settings. The primary function of education, outreach and advocacy are to create new opportunities for learning. Entomology outreach and education programs focused on teaching and engaging the community about insects as food could include insect-based cooking demonstrations or baking competitions. In general, people in educational settings and at public events may be more receptive to new ideas and foods such as edible insects. Edible insect experiences have been available at colleges and universities in multiple disciplines and courses for more than two decades. Various outreach and educational programs worldwide have served insects to the public thousands of times.

Little Herds is an educational non-profit based in Texas, United States that has done hundreds of educational events across the United States since 2013. One important aspect of education and outreach identified by Little Herds was the receptivity of children to the idea of eating insects. By providing the first opportunity to try insect based foods, whether whole insects like roasted mealworms and crickets, or abstracted products like cookies and protein bars made with cricket powders, in a fun and approachable environment, both children and adults were more willing to try insects and leave with a generally positive outlook on the idea. Providing fun, engaging, experiential education around sampling and tasting provides the right environment to dispel misconceptions and encourage kids and adults alike to break down those internal psychological barriers.

21.3.4 Trade organizations and research initiatives

The development of the edible insect industry on a global scale will depend on marketing, advertising and the development of trade organizations to assist companies, researchers, educators, and regulators in defining common language, standard operating procedures and best practices. In North America, including Canada, the United States, and Mexico, the North American Coalition for Insect Agriculture (NACIA), formerly known as the North American Edible Insect Coalition (NAEIC) was created in 2015. The mission of the NACIA 'is to foster collaboration amongst stakeholders and create a consolidated voice to encourage the growth of insects as both feed and food in North America' (http://www.edibleinsectcoalition.org).

Other countries have also created trade organizations to assist with the fast-growing pace of the edible insect industry including the International Platform of Insects for Food and Feed (IPIFF) based in Belgium. IPIFF is a non-profit organisation founded in 2013 which represents the interests of the breeders associated with it and connects the private organizations in the insect industry. Another Belgium based organization Belgian Insect Industry Federation (BIIF) works to unite all active participants in Belgium on the rearing and/or use for commercial purposes of insects and other arthropods. In addition, the BIIF exchanges and spreads knowledge regarding the rearing of insects and different commercial applications of insects and other arthropods to the general public, public authorities, the academic world and to corporate affiliates.

In the United Kingdom, The Woven Network C.I.C is created to directly benefit, through a set of membership services, entrepreneurs and researchers working in the role of insects in the human food chain. Indirectly, through supporting its members, the organization will also benefit the wider public across the Globe that need sustainable, alternative sources of protein.

Association of Southeast Asian Nations (ASEAN) set up Food and Feed Insects Association (AFFIA) for the countries in the ASEAN region. These member states include Brunei, Cambodia, Indonesia, Laos, Malaysia, Myanmar, Philippines, Singapore, Thailand, and Vietnam. With AFFIA, those involved in insect application for feed and food join forces. 'AFFIA aims bringing industry and research stakeholders from the insects' sector in a collaborative movement towards the development of entomoculture, entomophagy and their related activities' (http://affia.org).

Although not a trade organization, The European Parliament adopted a research initiative to address the EU's protein deficit. The PROteINSECT project, completed in 2016, focused on five key areas in order to evaluate insects as a novel source of protein for and to ensure that methodologies are sustainable and economically viable: development and optimization of fly larvae both large and small scale, safety and quality criteria for insect protein products, processing methodologies and insect protein extracts in animal feeding trials, optimal design of insect-based animal feed production systems, and to build a pro-insect platform in Europe including examination of the regulatory framework (http://www.proteinsect.eu).

21.3.5 Press and publications

The popularity of edible insects in the public news media has dramatically increased over the last 10 years. More recently, many media representations of edible insects tend to emphasize their potential and taste. Currently, most public-media coverage includes the standard talking points:

- the presence of edible insects in the diets of many human populations worldwide;
- the nutritional attributes and physiological benefits of insect consumption;
- the scientific findings of the FAO and/or other experts;
- concern regarding the future sustainability of vertebrate-protein-production;
- the general public's negative reactions;
- the presence of insect parts in our standard foods.

Scholars, researchers, and the public may interchangeably use the terms 'edible insects' and 'entomophagy' although they have different definitions. Worldwide the term edible insects is more commonly used in both the popular press media and peer reviewed journals than the term entomophagy. Evans *et al.* (2015) discussed the need for reviewing the evolving terminology associated with the use of the term 'entomophagy' and how more precise and contextual language may ultimately aid in its' acceptance and global understanding. Amongst entomologists and anthropologists the term 'entomophagy' is defined as 'practice of eating insects' which can be observed by primates, mammals, birds, reptiles, amphibians, other animals as well as other insects.

A Google Search with the term 'edible Insects' in the news section shows the increase in the number of popular press articles appearing in newspapers, magazines, blogs, and on television globally. In 2007, 100 articles were captured by this search engine, however in 2016, 6,070 articles were published by the media and popular press outlets (Figure 21.1). In comparison, a Google Search with the term 'entomophagy' in the news section shows an increase from zero articles captured by this search engine and keyword in 2007 to 1,230 popular press articles published in 2016 (Figure 21.2). Much of this observed increase in media exposure can be attributed to the FAO 2013 report (Van Huis *et al.*, 2013) and the subsequent conference held at Wageningen University in the Netherlands in 2014 as well as to other edible insect related conferences, workshops, and symposium wold wide. Another possible contribution, primarily in the USA, is companies that have appeared on various reality television shows to spread their ideas and messages to many viewers simultaneously. This multi-media exposure has been positive throughout the global insects as food industry.

While future shifts in reactions to the term 'entomophagy' remain to be seen, it is apparent that some cultures and countries are having greater success than others in gaining public acceptance for insect-based foods by using the term edible insects when referring to insects as human food. Several countries have begun programs designed to popularize edible insects which could be very beneficial in for the future of the edible insect industry.

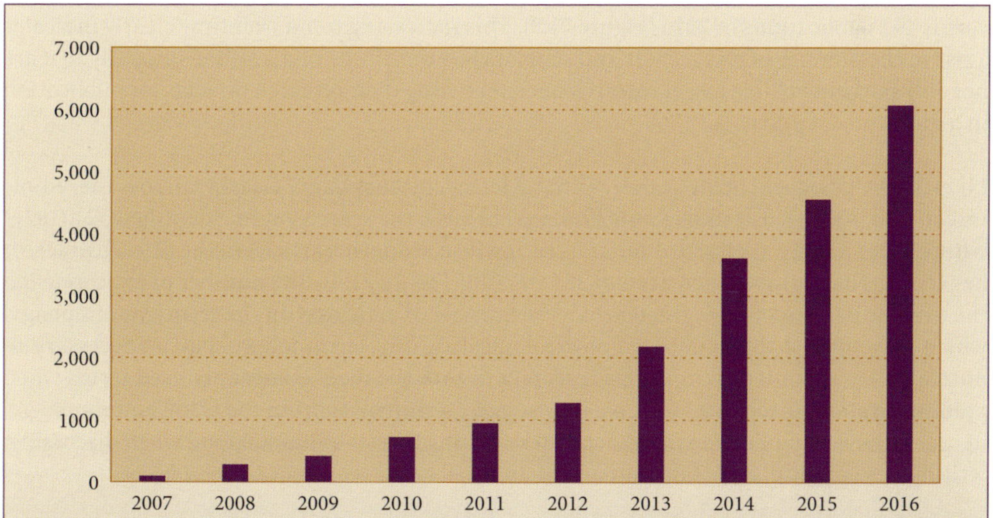

Figure 21.1. Google search 'edible insects' 2007-2016.

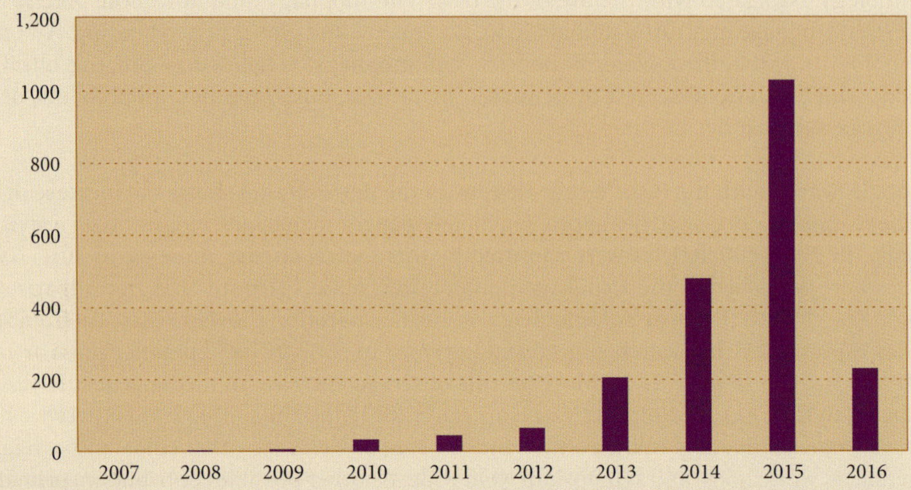

Figure 21.2. Google search 'entomophagy' 2007-2016.

Researchers have also seemingly adopted the idea of edible insects for food which can be observed by the frequency of edible insects and entomophagy in peer-reviewed publications (Figure 21.3 and 21.4). In addition to the thousands of articles published by the popular press and various media outlets there has been a significant increase in peer-reviewed publications around the topic of edible insects and entomophagy. A Google Scholar search with the term 'edible insects' revealed 80 peer-reviewed publications identified by the search engine in 2007 versus 492 publications in 2016 (Figure 21.3). This increasing trend over time clearly indicates a rise in the number of researchers and scholars in a multitude of disciplines globally that are not only pursuing edible insect research but submitting their research projects for publication in a multitude of journals.

Likewise, A Google Scholar search with the term 'entomophagy' clearly mirrors the 'edible insect' peer-reviewed search engine findings. In 2007, the search engine identified 32 articles with 'Entomophagy' (defined as humans eating insects for this article) in the title, abstract, or key terms. That number increased to 285 in 2016 (Figure 21.4). The number of articles using the term 'edible insects' are greater than the number of articles with the term 'entomophagy' which may suggest that a cultural or ideological shift in terminology may be necessary to normalize or attempt to familiarize researchers with the most commonly used terms when publishing about insects as food. We may suggest a shift away from the term 'entomophagy' in peer-reviewed publications specifically researching insects as human food when discussing, writing, researching, or publishing in this area.

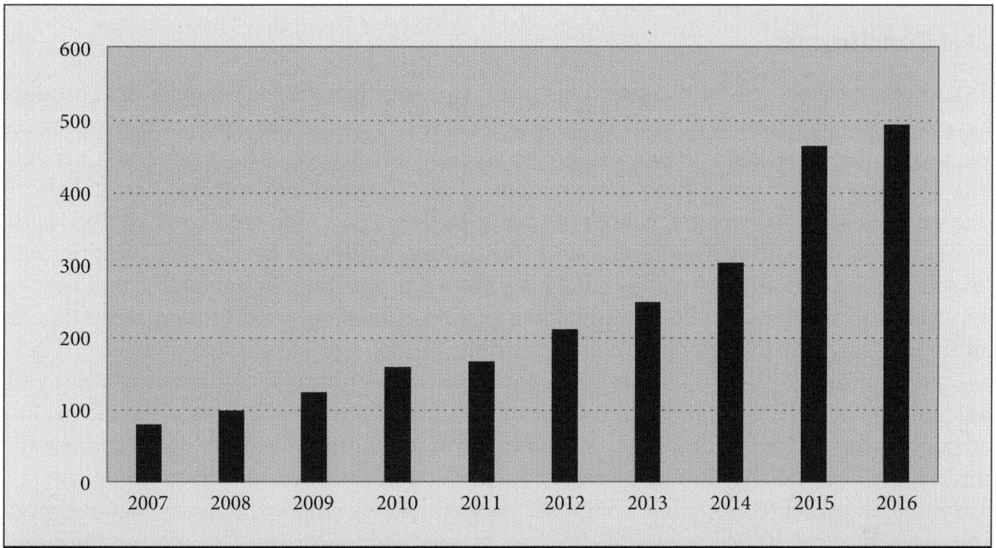

Figure 21.3. Google Scholar search 'edible insects' 2007-2016.

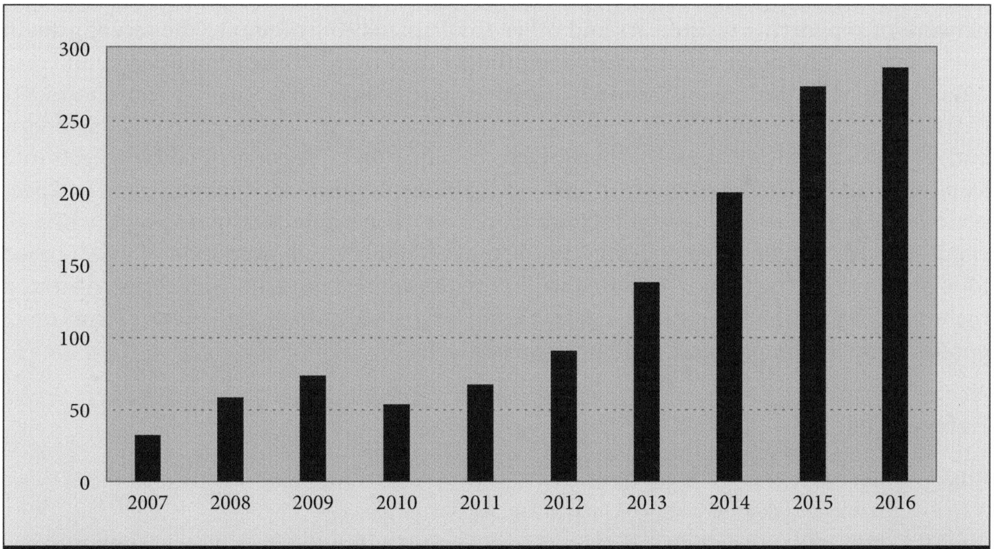

Figure 21.4. Google Scholar search 'entomophagy' 2007-2016.

21.4 Conclusions

This chapter considered the state of edible insects, primarily from the perspective of product development and promotion. Within traditional markets product acceptance is required for the investment in product development to be recouped. However, societal acceptance is only one criterion for progress. Product development also involves agricultural, technological, and economic advances in the field or product line. Insects as food are regulated differently from country to country. This variation in clarity of regulations may delay the acceptance of insects as food and ultimately may stifle insect based product development. Beyond human applications, individual value streams from insects extends beyond protein. The fats and chitin provide extensive opportunities for scaling up existing and promising research. Upon execution of design, the product must be promoted through various marketing strategies. Perceptions of insects in general, and more particularly as a food source are integral to advancing the edible insect industry. We can do much more to recognize and address the underlying reasons that cause many people in Western cultures to reject edible insects. Through education (concerning what insects are; how they enable the movement of energy in ecosystems; the harmlessness of the clear majority of insect species; and humankind's dependence upon the activities of insects for pollination, waste disposal, etc.) we have a chance to positively influence perceptions. In the field of insects as food the promotion of edible insect products may include company campaigns, K-16 outreach and education programs, non-profit organizations, research organizations, as well as individual proponents. A global network of supporting researchers and other participants contributes to the recent growth of this industry through insect based food product development and promotion (Van Huis, 2016). Trade organizations and research initiatives have been established on several continents (Africa, Asia, Europe and North America) to work closely with food insect-based businesses and regulatory agencies to provide insight and guidance on incorporating edible insects into local, regional, national, and international food systems. Thousands of popular press and peer reviewed articles pertaining to insects as food have been published increasing the overall exposure to the edible insect industry. Product development and promotion of insect based food products are the foundations that will move the edible insect industry forward around the world. In addition, with additional small and large scale human grade insect farms more opportunities for entopreneurs will become available.

References

Balzan, S., Fasolato, L., Maniero, S. and Novelli, E., 2016. Edible insects and young adults in a north-east Italian city an exploratory study. British Food Journal 118: 318-326.

Brues, C.T., 1946. Insect dietary. Harvard University Press, Cambridge, MA, USA. Reprinted as: Insects, food, and ecology, 1972. Dover Publications, New York, NY, USA.

Caparros Megido, R., Gierts, C., Blecker, C., Brostaux, Y., Haubruge, É., Alabi, T. and Francis, F., 2016. Consumer acceptance of insect-based alternative meat products in western countries. Food Quality and Preference 52: 237-243.

Cheung, R.C., Ng, T.B., Wong, J.H. and Chan, W.Y., 2015. chitosan: an update on potential biomedical and pharmaceutical applications. Marine Drugs 13: 5156-5186.

Cicatiello, C., Rosa, B., Franco, S. and Lacetera, N., 2016. Consumer approach to insects as food: barriers and potential for consumption in Italy. British Food Journal 118: 2271-2286.

DeFoliart, G., 2012. The human use of insects as a food resource: a bibliographic account in progress. University of Wisconsin-Madison, Madison, WI, USA. Available at: http://tinyurl.com/jna64w5.

Deroy, O., Michel, C., Piqueras-Fiszman, B. and Spence, C., 2014. The plating manifesto (I): from decoration to creation. Flavour 3: 6.

Deroy, O., Reade, B. and Spence, C., 2015. The insectivore's dilemma, and how to take the west out of it. Food Quality and Preference 44: 44-55.

Durst, P.B., Johnson, D.V., Leslie, R.N. and Shono, K., 2010. Forest insects as food: humans bite back. In: Proceedings of a workshop on Asia-Pacific resources and their potential for development, February 19-21, 2008, FAO, Regional Office for Asia and the Pacific, Chiang Mai, Thailand.

Evans, J., Alemu, M.H., Flore, R., Frøst, M.B., Halloran, A., Jensen, A.B. and Eilenberg, J., 2015. Entomophagy: an evolving terminology in need of review. Journal of Insects as Food and Feed 1: 293-305.

Feng, Y., Chen, X., Zhao, M., He, Z., Sun, L., Wang, C. and Ding, W., in press. Edible insects in China: utilization and prospects. Insect Science, DOI: https://doi.org/10.1111/1744-7917.12449.

Fernandez, J.G. and Ingber, D.E., 2014. Manufacturing of large-scale functional objects using biodegradable chitosan bioplastic. Macromolecular Materials and Engineering 299: 932-938.

Gahukar, R.T., 2011. Entomophagy and human food security. International Journal of Tropical Insect Science 31: 129-144.

Gracer, D., 2010. Filling the plates: serving insects to the public in the United States. In: Durst, P.B., Johnson, D.V., Leslie, R.N. and Shono, K. (eds.) Forest insects as food: humans bite back. In: Proceedings of a workshop on Asia-Pacific resources and their potential for development, 19-21 February 2008, Chiang Mai, Thailand. Food and Agriculture Organization of the United Nations RAP Publication 2010/02. Available at: http://www.fao.org/docrep/012/i1380e/i1380e00.pdf.

Hartmann, C., Shi, J., Giusto, A. and Siegrist, M., 2015. The psychology of eating insects: a cross-cultural comparison between Germany and China. Food Quality and Preference 44: 148-156.

Holt, V.M., 1885. Why not eat insects? Leadenhall Press, London, UK. Reprinted as: Why not eat insects?, 1972. Classey, Hampton, Middlesex, UK.

House, J., 2016. Consumer acceptance of insect-based foods in the Netherlands: academic and commercial implications. Appetite 107: 47-58.

Lähteenmäki-Uutela, A., Grmelová, N., Hénault-Ethier, L., Deschamps, M., Vandenberg, G.W., Zhao, A. and Nemane, V., 2017. Insects as food and feed: laws of the European Union, United States, Canada, Mexico, Australia, and China. European Food and Feed Law Review 12: 22-36.

Lensvelt, E.J.S. and Steenbekkers, L.P.A., 2014. Exploring consumer acceptance of entomophagy: a survey and experiment in Australia and the Netherlands. Ecology of Food and Nutrition 53: 543-561.

Liu, Y., 2011. The feasibility study of the new kind crude fat material of biodiesel from insects. 2011 International Conference on Materials for Renewable Energy and Environment 1: 271-279.

Looy, H., Dunkel, F. and Wood, J., 2014. How then shall we eat? Insect-eating attitudes and sustainable foodways. Agriculture and Human Values 31: 131-141.

Manjunatha, S.B., Biradar, D.P. and Aladakatti, Y.R., 2016. Nanotechnology and its applications in agriculture: a review. Karnataka Journal of Agricultural Sciences 29: 1-13.

McFarlane, T. and Pliner, P., 1997. Increasing willingness to taste novel foods: effects of nutrition and taste information. Appetite 28: 227-238.

Megido, R.C., Gierts, C., Blecker, C., Brostaux, Y., Haubruge, E., Alabi, T. and Francis, F., 2016. Consumer acceptance of insect-based alternative meat products in western countries. Food Quality and Preference 52: 237-243.

Megido, R.C., Sablon, L., Geuens, M., Brostaux, Y., Alabi, T., Blecker, C. and Francis, F., 2014. Edible insects acceptance by Belgian consumers: promising attitude for entomophagy development. Journal of Sensory Studies 29: 14-20.

Oonincx, D.G.A.B., Van Itterbeeck, J., Heetkamp, M.J.W., Van den Brand, H., Van Loon, J.J.A. and Van Huis, A., 2010. An exploration on greenhouse gas and ammonia production by insect species suitable for animal or human consumption. PLoS ONE 5: e14445.

Piqueras-Fiszman, B. and Spence, C., 2014. Colour, pleasantness, and consumption behaviour within a meal. Appetite 75: 165-172.

Przybylski, A.K., Murayama, K., DeHaan, C.R. and Gladwell, V., 2013. Motivational, emotional, and behavioral correlates of fear of missing out. Computers in Human Behavior 29: 1841-1848.

Purkayastha, J., 2016. Bioprospecting of indigenous bioresources of North-East India. Springer Singapore, Singapore, 347 pp.

Ratcliffe, N.A., Mello, C.B., Garcia, E.S., Butt, T.M. and Azambuja, P., 2011. Insect natural products and processes: new treatments for human disease. Insect Biochemistry and Molecular Biology 41(10): 747-769.

Reverberi, M., 2017. Exploring the legal status of edible insects around the world. William Reed Business Media Ltd., Crawley, UK. Available at: http://tinyurl.com/ybv7hp22.

Rozin, P. and Fallon, A.E., 1987. A perspective on disgust. Psychological Review 94: 23-41.

Rozin, P., Haidt, J. and McCauley, C.R., 2008. Disgust. In: Lewis, M., Haivland-Jones, J.M. and Barrett, L.F. (eds.). Handbook of emotions (3rd Ed.). Guilford Press, New York, NY, USA, pp. 757-776.

Rozin, P., Lowery, L. and Ebert, R., 1994. Varieties of disgust faces and the structure of disgust. Journal of Personality and Social Psychology 66: 870 881.

Ruby, M.B., Rozin, P. and Chan, C., 2015. Determinants of willingness to eat insects in the USA and India. Journal of Insects as Food and Feed 1: 215-225.

Sallows, J., 2015. Homer vs Spock: using behavioural nudges to persuade people to eat insects. Haymarket Media Group Ltd., Twickenham, UK. Available at: http://tinyurl.com/y7e5a5ly.

Shahidi, F., Arachchi, J.K.V. and Jeon, Y.-J., 1999. Food applications of chitin and chitosans. Trends in Food Science and Technology 10: 37-51.

Shelomi, M., 2015. Why we still don't eat insects: assessing entomophagy promotion through a diffusion of innovations framework. Trends in Food Science and Technology 45: 311-318.

Shockley, M. and Dossey, A.T., 2014. Insects for human consumption. In: Morales-Ramos, J.A., Rojas, M.G. and Shapiro-Ilan, D.I. (eds.) Mass production of beneficial organisms: invertebrates and entomopathogens. Academic Press, Cambridge, MA, USA,764 pp.

Smil, V., 2000. Feeding the world: a challenge for the twenty-first century. MIT Press, Cambridge, MA, USA.

Sogari, G., Menozzi, D. and Mora, C., 2017. Scientific paper: exploring young foodies' knowledge and attitude regarding entomophagy: a qualitative study in Italy. International Journal of Gastronomy and Food Science 7: 16-19.

Sosa, D.A.T. and Fogliano, V., 2017. Potential of insect-derived ingredients for food applications. In: Shields, V.D.C. (ed.) Insect physiology and ecology. InTech, Rijeka, Croatia.

Tan, H.S.G., Berg, E. and Stieger, M., 2016a. The influence of product preparation, familiarity and individual traits on the consumer acceptance of insects as food. Food Quality and Preference 52: 222-231.

Tan, H.S.G., Fischer, A.R.H., Tinchan, P., Stieger, M., Steenbekkers, L.P.A. and Trijp, H.C.M., 2015. Insects as food: exploring cultural exposure and individual experience as determinants of acceptance. Food Quality and Preference 42: 78-89.

Tan, H.S.G., Fischer, A.R.H., Van Trijp, H.C.M. and Stieger, M., 2016b. Tasty but nasty? Exploring the role of sensory-liking and food appropriateness in the willingness to eat unusual novel foods like insects. Food Quality and Preference 48: 293-302.

Tonk, M. and Vilcinskas, A., 2017. The medical potential of antimicrobial peptides from insects. Current Topics in Medicinal Chemistry 17: 554-575.

United Nations, 2015. World population projected to reach 9.7 billion by 2050. United Nations, New York, NY, USA. Available at: http://tinyurl.com/hox2d2y.

USDA Foreign Agricultural Service, 2009. Germany food and agricultural import regulations and standards. FAIRS Country Report. USDA Foreign Agricultural Service, Berlin, Germany. Available at: http://tinyurl.com/y7fp4zqx.

Van Huis, A., 2013. Potential of insects as food and feed in assuring food security. Annual Review of Entomology 58: 563-583.

Van Huis, A., 2016. Edible insects are the future? Proceedings of the Nutrition Society 75: 294-305.

Van Huis, A., Van Itterbeeck, J., Klunder, H., Mertens, E., Halloran, A., Muir, G. and Vantomme, P., 2013. Edible insects: future prospects for food and feed security. Food and Agriculture Organization of the United Nations (FAO), Rome, Italy, FAO Forestry Paper no. 171, 187 pp. Available at: http://www.fao.org/docrep/018/i3253e/i3253e.pdf.

Yen, A.L., 2009. Edible insects: traditional knowledge or western phobia? Entomological Research 39: 289-298.

Yen, A.L., 2010. Edible insects and other invertebrates in Australia: future prospects. In: Durst, P.B., Johnson, D.V., Leslie, R.N. and Shono, K. (eds.) Forest insects as food: humans bite back. Proceedings of a workshop on Asia-Pacific resources and their potential for development, February 19-21, 2008, FAO, Regional Office for Asia and the Pacific, Chiang Mai, Thailand. Available at: http://www.fao.org/docrep/012/i1380e/i1380e00.pdf.

Appendix 1. Entopreneurs.

Protein-bars/insect bars:		
Chapul	http://www.chapul.com	USA
Exo	www.exoprotein.com	USA
Crowbar	www.crowbarprotein.com	Iceland
Gathr foods	gathrfoods.com	UK
BeEnto	https://twitter.com/entofoods	UK
SEQ Foods	www.seq-foods.eu	the Netherlands
Getsharp	www.getsharp.eu	UK
Gryö Bars	www.gryobars.com	France
bodhi	www.bodhiprotein.com	UK
Edible Bug	edible-bug.co	South Korea
Zoic bar	www.zoicbar.com	UK
uKa Protéine	www.ukaproteine.com	Canada
La View Eye	www.lavieweye.com	USA
Crick bar	www.crickbars.com	Sweden
Yumpa	www.yumpabar.co.uk	UK
Sensbar	sensbar.com	
Näak	naakbar.com	Canada
Protein-shakes/nutrition supplements:		
Bugeater Foods	www.bugeaterfoods.com	USA
CrikNutrition	www.criknutrition.com	Canada
Bug Muscle	bugmuscle.com	USA
Grilo	https://www.griloprotein.com.au	Australia
Sustainable gains	www.sustainablegains.net	USA
Premium Cricket Farms		USA
Insect-candy/snacks:		
Hotlix	www.hotlix.com/candy	USA
Crickers	www.crickerscrackers.com	USA
Chirps Chips	https://chirpschips.com	USA
Insectios	www.insectitos.com	USA
Cookies and seasoned dried insects:		
Snack Insects	www.snackinsects.com	Germany
Crispizz	twitter.com/crispizzfood	France
Insecteo	www.insecteo.com	France
Eat:em	www.eatem.se	Sweden
Seek	www.seek-food.com	USA

Insect/cricket flour/powder:		
Bitty Foods	bittyfoods.com	USA
Qvicket	www.qvicket.se	Sweden
Gryllies	gryllies.com	USA
Thinksect	www.thinksect.com	USA
Cricket Flours	www.cricketflours.com	USA
Entolution	twitter.com/entolution	USA
Eco Insect Farming	www.eif-th.com	Thailand
Griopro	cricketpowder.com	USA
Primal Future	www.primalfuture.co.nz	New Zealand
Mophagy	mophagy.com	UK
Acheta	acheta.no	Norway
Insect pasta:		
Nutribug	www.nutribug.com	UK
Going Bugs	twitter.com/GoingBugs	Italy
Goffard Sisters	www.goffardsisters.com	Belgium
Kinjao	www.kinjao.com	France
Bugsolutely	www.bugsolutely.com	Thailand
Insects & vous	on Kickstarter	France
Krick8	www.kric8.co.uk	UK
Special insect products (unique products hard to categorize):		
Nutrient	nutrient.se	Sweden
Critter Bitters	www.critter-bitters.com	USA
Bug Foundation	bugfoundation.com	Germany
C-FU Foods	cfufoods.com	Canada
Gran Mitla	granmitla.us	Mexico
Damhert Nutrition	www.damhert.be/nl/shop/insecta	Belgium
One Hop Kitchen	onehopkitchen.com	Canada
Anti Gin	www.cambridgedistillery.co.uk/antygin	Denmark/UK
Insekt KBH	www.facebook.com/insektkbh	Denmark
Insects as food and farms (generalists):		
Grub	www.eatgrub.co.uk	UK
Bugalicous	www.bugalicious.nl	the Netherlands
ento	www.eat-ento.co.uk	UK
Green Kow	www.greenkow.be	Belgium
EntomoFarms	entomofarms.com	Canada
Next Protein	www.nextprotein.co	France
Dybri – edible insects from Cornwall	dybri.com	UK
Ronzo	www.ronzo.co	Poland

Myldregard	myldregard.no	Norway
Ben's Bugs	www.bensbugs.be	Belgium
Inspiro Foods	www.insprofoods.com	Canada
Bugzz	bugzz.nl	the Netherlands
Entotech	www.entotech.fr	France
Viur	www.fodurskordyr.is	Iceland
Steak TzarTzar	https://hargol.com	Israel
Little food	littlefood.org	Belgium
Spieseplan	www.speiseplan.wien	Austria
Bugs World Solution Food	www.bugsworldsolutionfood.com	Belgium
Entomo Foodafrica	www.facebook.com/entomo.foodafrica	D.R. Congo
Nordic Insect Economy	nie.fi	Finland
Insect cuisine	insectcuisine.jp	Japan
essento	essento.ch	Switzerland
Finsect	www.finsect.fi	Finland
Italibugs	www.italbugs.com	Italy
Imago insect products	www.imago-insects.com	Germany
Detroit Ento		USA
Entosense	www.entosense.com	USA
Entovida!	www.entovida.com	USA
Tiny Foods	www.tinyfoods.nl	the Netherlands
Earth and me	www.earthandme.co.uk	UK
Insekten Essen	insektenessen.at	Austria
Enorm	www.enormfood.com	Denmark

Edible insect books and cookbooks:

Eat a bug cookbook
Creepy crawly cuisine
Edible: an adventure into the world of eating insects and the last great hope to save the planet
Butterflies in my stomach
Entertaining with insects
Insectes à croquer (Insects to munch)
Cuisine des insects: à la decouverte de l'entomophagie
Edible bugs: insects on our plate
Het insectenkookboek/The insects cookbook
Cicada-licious
Man eating bugs: the art and science of eating insects
Why not eat insects
Eat grub
Bug-a-licious
Eating insect: eating insects as food
The Alaskan mosquito cookbook: all time favorite recipes
Let's eat bugs!: a thought – provoking intro to edible insects for adventurous teens and adults
On eating insects: essays, stories and recipes
A survival guide to edible insects

Bugs for lunch
Six-legged livestock: edible insect farming, collection and marketing in Thailand
Top 50 most delicious insect recipes
Bugs appetit
Survival guide: eating insects: how to survive with edible bugs and learning entomophagy
Survival bugs for food

Online stores selling insects for human consumption:

Zocalito Chapulines		USA
EntomoFarms	entomofarms.com	Canada
Thailand Unique	www.thailandunique.com	Thailand
Hakuna mat	www.hakunamat.se	Sweden
Next Food	www.next-food.net	Thailand
BugGrub	www.buggrub.com	UK
Crunchy Critters	www.crunchycritters.com	UK
MiniFood	www.minifood.be	Belgium
Edible Bug Shop	www.ediblebugshop.com.au	Australia
Delibugs	www.delibugs.nl	the Netherlands
Jiminis	www.jiminis.com	France
Insectes comestibles	www.insectescomestibles.fr	France
Haocheng MealWorm Inc	www.hcmealworm.com	China
Cricket Protein Store	cricketproteinstore.com	USA
Entomarket	www.entomarket.com	USA
Incredible Foods		USA

Professional insect farmers:

Enviroflight		USA
Iowa Cricket Farmer		USA
EntomoFarms	entomofarms.com	Canada
Micronutris	www.micronutris.com	France
Proti-Farm	protifarm.com	the Netherlands
Rainbow Mealworms	www.rainbowmealworms.net	USA
Sahakhun Bug Farm Ltd., Part		Thailand
Edible Bugfarm	www.ediblebugfarm.com	UK
Ozark Fiddler Farms	www.ozarkfiddlerfarms.com	USA
Rocky Mountain Micro Ranch	rmmr.co	USA
Zophobas Red	twitter.com/zophobas_red	UK
Coalo Valley Farms	www.coalo.farm	USA
CrickEatz	twitter.com/Crickeatz	UK
Dimini Cricket	diminicricket.com	France
Armstrong Cricket Farm	www.armstrongcrickets.com	USA
Poda foods	podafoods.com	USA
Eco Insect Farming	eif-th.com	Thailand
Chapa farms	www.chapafarms.com	Mexico/Canada

Next Food	www.next-food.net	Thailand
Six legs farms	www.facebook.com/sixlegsfarm	UK
Insagri	www.insagri.com	Spain
Tiny Farms		USA
Home farming insects:		
Open Bug Farm	www.openbugfarm.com	USA
Livin Farms	www.livinfarms.com	Hong Kong/Austria
Insect farming for the poor:		
Aspire FG	www.aspirefg.com	Ghana, Mexico, USA
EntoCube	entocube.com	Finland
Khepri	khepri.eu	France
MealFlour	www.mealflour.org	Guatemala/USA
Education and advocacy:		
4ento	4ento.com	Switzerland
Entomove Project	www.entomoveproject.com	France
Bug Vivant	bugvivant.com	USA
Buggies	buggies.dk	Denmark
Insects for starters	insectsforstarters.com	Sweden
Little Herds	www.littleherds.org	USA
entonote	www.entonote.com	Italy
Girl meets bug	www.girlmeetsbug.com	USA
Bugible	www.bugible.com	USA
Bugs on the menu		Canada
Entoblog	www.entoblog.com	USA
Tastebugs	www.tastebugs.net	Australia
Bugsfeed	www.bugsfeed.com	Denmark
Rocky Mountain Microranch		USA
University of Georgia Bug Dawgs		USA
Audobon Insectarium and Insect Cafe		USA
Food-Insects	http://www.food-insects.com	
Insects are Food	http://insectsarefood.com	
David George Gordon	http://www.davidgeorgegordon.com	
Food Insects Newsletter (2010 through present)	http://www.foodinsectsnewsletter.org	
Ento	http://www.eat-ento.co.uk	
Eating Bugs (from the Manataka American Indian Council)	http://www.manataka.org/page160.html	
Ento Call		

Insects as pet food:		
Hexa Foods	www.hexafoods.com	Canada Produces: Bugbites
EntoBento	entobento.com	USA
Chloes treats	www.chloestreats.com	USA
Conscientious cat	www.conscientious.cat	USA
Cricket Kitchen Pet Treats	www.cricketkitchen.com	USA
Grubbly Farms	grubblyfarms.com	USA

Restaurants:		
Canada	Chirp cuisine	http://chirpcuisine.wixsite.com/chirp
	Jambo Grill	
	Rangoli	
	El Catrin	
	La Mezcaleria	
	Olo Restaurant	
	Tierra Azteca	
	Vij's Restaurant	
Mexico	La Cocinita de San Juan	
	Restaurante Arroyo	
	ElCardenal	http://www.restauranteelcardenal.com
	La Cocina de San Juan	http://www.lacocinadesanjuan.com/home.html
	Corazón de Maguey	http://www.alipus.com/corazondemaguey
	Restaurante Don Chon	
	Limosneros	
	Los Danzantes	
	Azul Condesa	
	Pujol	
United States	The Bug Chef	www.davidgeorgegordon.com
	99 Chicken	
	Bistro LQ, LA	
	Broken Spanish	
	Don Bugito	
	La Cocina Kiosk	
	El Tejate Restaurant	
	El Tule	
	Expression Oaxaquena	
	Guelaguetza	
	Jitlada Restaurant	
	La Sandia, LA	
	La Urbana	
	Mok Maru Jong Sul Jip	
	Monte Alban Restaurant	

	Night Market Song, LA	
	Ortiz Bakery & Restaurant	
	Petty cash Taqueria	
	Real De Oaxaca Restaurant	
	Rivera	
	Sabores Oaxaquenos	
	Typhoon	
	La Sandia Cantina	
	Bug Appetite	
	Johnny Sanchez	
	Tu Y Yo (Somerville)	
	Gringo	
	Toloache	
	The Black Ant	
	Qi Esarn Thai Kitchen	
	El Rey Del Sabor	
	La Slowteria	
	Pocha 32	
	Casa Mezcal	
	Kirakuya	
	Wicked Spoon Yogurt	
	Playground Authentic Thai Cuisine	
	Villa Cemita	
	Sushi Mazi	
	Woods Creek Grill	
	La Condesa	
	Hugo's	
	Xochi	
	Guajillo	
	La Carta de Oaxaca	
	Restaurant El Tio	
	Poquitos	
	Oyamel	
Belgium	Bugs in Mugs	
	Entomochef	www.entomochef.be
Denmark	Noma	
France	Le Festin Nu	
	Livreur de Saveurs	
	Ban Thai	
Germany	Espitas Erfurt	

The Netherlands	Tante Truus	
	Bugzz (Insecuisine)	
	Insect-o-shi	twitter.com/insectoshi
Poland	Co To To Je	
United Kingdom	Archipelago Restaurant	
	Chewed Arms	
	Grub Kitchen	
	Wahaca	
	Rentokil Pestaurant	
	The Bug Banquet	
China	Delicious Inn Restaurant	
Thailand	Seven Spoons	
Cambodia	Romdeng	
	Bugs Cafe	
	Friends	
Australia	El Topo	

Future prospects

22.

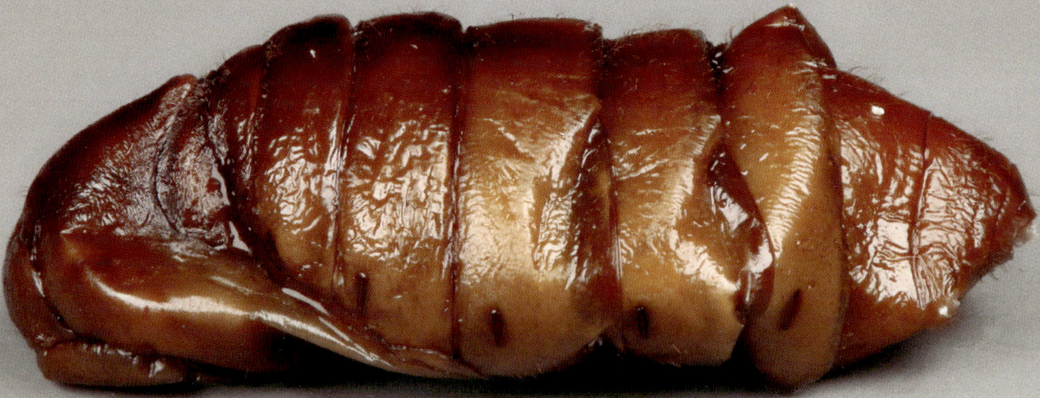

Bombyx mori (canned pupa)

Future prospects of insects as food and feed

A. van Huis[1] and J.K. Tomberlin[2]

[1]Laboratory of Entomology, Wageningen University, P.O. Box 16, 6700 AA Wageningen, the Netherlands; [2]Department of Entomology, Texas A&M University, TAMU 2475, College Station, TX 77843-2475, USA; arnold.vanhuis@wur.nl

Abstract

When considering the rearing of insects for food or feed there are many issues to consider. In this chapter, an overview is given of the insect species used for either animal feed or human food. A number of issues that need to be taken into account when farming insects are discussed. When initiating a colony of a particular insect species for mass-production, the founder population and the strain to be used are critical decisions. Further considerations include the substrate chosen as this material can come at price and impact the nutritional quality of the resulting insects. Rearing conditions and facility design have often been developed by trial and error, but are for that reason company secret in many cases. Insect welfare when rearing edible insects is increasingly becoming an issue, in particular now that emotionlike states of invertebrates seem to be possible. After killing the insects, protein, fat, and chitin need to be recovered while safeguarding food safety and high quality. Also the shelf life of insects or their by-products as food and feed need to be considered. To market edible insect products, a conducive legislative framework is needed and this is a major challenge for the industry. The creation of an enabling environment for the sector is decisive as well. Also strategies need to be developed to make edible insects a mainstream food item. With worldwide need to find alternative sustainable protein sources increasing, the interest for insects to be used either as food and feed is growing. The market prospects are very promising.

Keywords: edible insect species, farming insects, insect industry, edible insect market, consumers

22.1 Introduction

Insects as food and feed is a very new sector in the food and agriculture industry and in full development with many start-up companies. Scientifically, the interest in edible insects continues to increase each year with a growing number of new articles being published dealing with a large variety of topics in disciplines such as biology (e.g. biosystematics, entomology), agrotechnology and food sciences (food quality, human nutrition, food safety), animal science (e.g. breeding and genetics, animal nutrition), environmental sciences (e.g. forest and nature conservation, resource ecology), and social sciences (e.g. marketing and consumer behaviour, law and governance).

The public appreciation for insects in general is low and according to Looy *et al.* (2014) insects are 'deeply embedded in the Western psyche and view of insects as dirty, disgusting, and dangerous.' They are often associated with diseases (insects as vector of human diseases), with filth (cockroaches), and considered a nuisance (house flies and stinging wasps). In Chapter 2 of this book, Dicke (2017) outlines another picture of insects and highlights the important ecosystem services that insects provide. Some insect species are exploited. Examples are the biological control programmes, where insects are used as natural enemies of agricultural pests. Very well known to western society is the use of insect products such as honey and silk. In this book, we focus on using insects both as food for humans and as feed for animals.

With the increasing interest to use insects as food and feed, we need to farm the insects in order to produce the quantity and quality needed to have an impact as a commodity; otherwise, growth of the industry could be stymied. This is an enterprise where many issues need to be considered. We will deal with the different topics in a sequential manner.

22.2 Which insect species

The first question to address is whether the production is either for human consumption or for animal feed. The number of insect species that can be eaten by humans is very large. There are more than 2,000 species to consider (Jongema, 2017). However, most of the listed species are collected from the wild in tropical regions. Some insect species are semi-domesticated such as mentioned in Chapter 1 (Van Huis and Tomberlin, 2017) and Chapter 4 (Van Huis, 2017a).

The different edible insect species reared by companies is often historically based. Companies rear a number of insects species to be used as feed for pet animals such as birds, reptiles, and fish. Customers are pet holders, zoos, and shops that sell fish bait or bird food. When insect rearing companies decided to start producing insects for human consumption, they used the same species as those reared for animal feed. It was often a side-activity, but in order to comply with food laws they had to modify rearing techniques and procedures. This entailed mainly producing under more stringent hygienic conditions and to install a track and tracing system. There are also start-ups that had to begin from scratch. A problem is that the methods to rear insects are often company secret, making it for start-up rearing companies difficult to achieve the same kind of sophistication than the existing companies who became efficient through trial and error.

So, there may be other species than those that are currently used as food and feed which could have potential to be mass produced. This should be further explored.

22.2.1 Insects for human consumption

The most common insect species reared for human consumption are the following.

Beetles

Three mealworm species of the family (Tenebrionidae) are often used: the yellow mealworm *Tenebrio molitor*, the lesser mealworm (*Alphitobius diaperinus*), and the superworm (*Zophobas morio*). In this book, Chapter 14 deals specifically with mealworms (Berezina, 2017). Commercial mealworm producers sometimes include juvenile hormone into the feed to prevent yellow mealworm larvae from moulting into adults, resulting in 'giant' mealworms that can achieve a length of 200 mm or more, and weigh more than 300 mg (Finke, 2002), compared to 20-32 mm and 130-160 mg, respectively, of normal larvae. Mealworms have generally been considered pests, because their larvae feed on stored grains (or damage poultry facilities), hence their name. However, the lesser mealworm, often present in poultry houses, is often considered a pest as a vector of microbial pathogens (Crippen *et al.*, 2009), rather than as a food source with a high nutritional value. Other tenebrionid beetles have also been considered as food such as the confused flour beetle (*Tribolium confusum*), a common pest insect known for attacking and infesting stored flour and grain (Kok *et al.*, 1988). Also the drugstore beetle, also known as bread beetle or biscuit beetle (*Stegobium paniceum*) has been considered, in particular for long term space flights, as the insect is able to convert low quality plant material such as cellulose into high quality proteins biomass (Katayama *et al.*, 2008; Kok, 1983). Palm weevils, curculionid beetles, are popular food throughout the tropics, and in Chapter 6, Muafor *et al.* (2017) give an example of the rearing of the African palm weevil.

Crickets, locust and grasshoppers

The most commonly reared cricket is the house cricket (*Acheta domesticus*). However, other cricket species can also be considered such as the tropical house cricket (*Gryllodes sigillatus*) and the two-spotted cricket (or African or Mediterranean field cricket) (*Gryllus bimaculatus*). We refer to two chapters in this book on crickets, Chapter 5 (Caparros Megido *et al.*, 2017) and Chapter 15 (Mott, 2017). The tropical house cricket is also considered a good alternative for the house cricket which may suffer from a denso virus (Weissman *et al.*, 2012). The migratory locust (*Locusta migratoria*) and the desert locust (*Schistocerca gregaria*) are both reared for human consumption as well as for animal feed. For more information reference is made to Chapter 7 on grasshoppers by Haldar and Malakar (2017).

Caterpillars

The greater wax moth or honeycomb moth (*Galleria mellonella*) is considered by some a delicacy (Martin, 2014; p. 35 and 232). Recently it was discovered that this insect species is able to contribute to bio-degradation of plastics (Bombelli *et al.*, 2017): polyethylene is very resistant to biodegradation and so could be a possible solution for lessening the environmental impact of these materials. Also, silk worms (such as *Bombyx mori*) are produced for silk but can be used as the same time, in particular the pupae, as food.

Bees and wasps

In the mountainous areas of the Central region of Japan, the wasp *Vespula flaviceps* is a seasonal delicacy, which can be captured in the wild but they can also be raised in people's backyard (Nonaka, 2007; Payne and Evans, 2017). Also bee drones have been proposed to be used as food (Ambühl, 2016). Varroa mite infestation is a major problem in beekeeping. In Denmark many beekeepers trap the mites in drone brood frames and then remove those comb frames amounting to many tonnes of drone brood every year (Evans *et al.*, 2016). This can be fed to chickens but another way is to use it for human consumption.

22.2.2 Insects as animal feed

For animal consumption three species are mostly considered: black soldier fly (*Hermetia illucens*), the house fly (*Musca domestica*) and the yellow mealworm. All species do have the capacity of converting low organic side streams into high quality protein products. An overview of which side streams can be used for these insects is provided in Table 1.5, while an overview of which insect species can be used for which production animal, fish or crustacean species see Table 1.6 (Van Huis and Tomberlin, 2017).

The black soldier fly is the major insect species proposed as animal feed, mainly because this insect species is able to convert low quality organic side streams into high quality proteins. Chapter 12 in this book deals specifically with the black soldier fly (Tomberlin and Cammack, 2017).

The house fly has been widely studied as potential feed source but with an emphasis on aquaculture production. Commercially produced larvae and pupae compare well to protein sources such as soya bean meal and fishmeal (Pieterse and Pretorius, 2014). A chapter dealing with the production of house fly is given in Chapter 13 (Tomberlin *et al.*, 2017).

The yellow mealworm can be used as feed for chickens, pigs, fish and shrimps as listed in Table 1.6 (Van Huis and Tomberlin, 2017). Other species such as grasshoppers can also be used, see Chapter 7 (Haldar and Malakar, 2017).

22.3 Genetics

Besides the choice of species, the strain to be used is important. The only literature references referring to strains relate to the yellow mealworm (Urs and Hopkins, 1973a,b) and to the black soldier fly (Zhou *et al.*, 2013). Much more research needs to be carried out to exploit the variation in strains. The other question is how to start a population in order to avoid genetic problems. We would like to refer to Chapter 9 of this book (Jensen *et al.*, 2017).

22.4 Nutrition

One of the major issues is whether insects can be produced with a cost price, which is acceptable and makes the insect product competitive to other food or feed products. Price has to do with the production process and whether economies of scales can be achieved. Certainly when insects are used as feed, huge quantities are required and then saving on labour becomes an important issue. However, another important cost factor is the substrate used and whether low value organic side streams or by-products can be used. It should be noted not all organic by-products are always suitable. This was experienced by Lundy and Parrella (2015) when they fed crickets minimally-processed, municipal-scale food waste or diets composed largely of straw. There was high mortality and the crickets hardly reached a harvestable size. They concluded for the house cricket that 'the potential of crickets to sustainably supplement the global protein supply…will depend on capturing regionally scalable organic side-streams of relatively high-quality that are not currently being used for livestock production.'

The substrate also determines very much the nutritional quality of the insect product. This was among others shown by Van Broekhoven *et al.* (2015) who used for three mealworm diets composed of organic by-products originating from beer brewing, bread/cookie baking, potato processing and bioethanol production. Larval protein content was stable on diets that differed 2-3-fold in protein content, whereas dietary fat did have an effect on larval fat content and fatty acid profile. This was also the case when the black soldier fly larvae was fed only 24 h with fish offal: a higher concentration of omega-3 fatty acids was achieved (St-Hilaire *et al.*, 2007). Finke and Oonincx (2014, 2017) reviewed a large number of variables which can influence the chemical composition of the insects: gender; stage of development; and environmental factors such as temperature, day length, humidity, light intensity and spectral composition. Further information on nutrients and insects is given in Chapter 16 (Finke and Oonincx, 2017).

There is quite some research done on the effect of diet on the chemical composition of insects. However, very little is known about the effect of diet on the sensorial qualities of the insect. Can the insect be made tastier by using certain feed ingredients?

22.5 Rearing conditions and facility design

How to achieve ideal rearing conditions for insects is important in order to get optimal development and a low mortality. This has a lot to do with abiotic conditions such as temperature, humidity and light regime. However, it is also important to get synchronized development. Temperature in a tray may differ as the insects themselves produce metabolic heat. Questions then are how to achieve homogenous temperature throughout the tray, how much space between the trays in order to achieve adequate ventilation? What would be ideal dimensions of the trays and how thick does the layer of substrate need to be? For example, in the case of the black soldier fly, Brits (2017) studied both depth and particle size of the larval feed and how this effects waste reduction, bioconversion, feed depth access, final larval mass, survival, and development. Another issue with the black soldier fly is how much substrate needs to be added during development, in which way and how many times.

All these techniques are often a matter of trial and error and some of the companies have a graveyard of equipment that was tried and afterwards discarded. Therefore, companies are very secretive about their facilities, procedures, and techniques used to rear the insects. Innovative rearing is often based on very simple techniques, which can easily be copied by competitors. Nevertheless companies do file patents to protect their inventions. Secrecy can be a problem when cooperating with universities who have an obligation to publish

(Dobermann, 2017). What is then often negotiated is a time delay between achieving the results and publishing. Another option is not to disclose all the information in a publication. For example, when it concerns the type of feed substrate a company is using, is not to mention the exact composition, but just the ingredients; see the study on the life cycle analysis of mealworms by Oonincx and De Boer (2012).

Chapter 8 deals with 'entotechnology' of edible insects, and describes techniques and methods used and being developed by industrial farming of insects as food and feed (Kok, 2017).

22.6 Diseases and pests

One of the major concerns of the commercial rearing companies is the occurrence of pests and diseases in their rearing. One example is the epizootic *A. domesticus* densovirus (AdDNV) which devastated entire colonies in the Americas and Europe (Weissman *et al.*, 2012). There is quite some knowledge when it concerns how to use other organisms to kill pests. This is called biological control, which is a method of controlling pests such as insects, mites, weeds and plant diseases using other organisms. Those organisms can be predators (insects, mites, and nematodes), parasitoids (most are wasps and flies) and entomopathogens (viruses, bacteria, and fungi). However, when it comes to the insect rearing there is not much knowledge available about the causing pathogenic agents. There are several general publications dealing with this topic such as Shapiro (1984), Inglis and Sikorowski (2009), and Gouli *et al.* (2011), and publications specifically dealing with pests and diseases in insects reared as food and feed, such as Eilenberg *et al.* (2015) and Chapter 10 in this book (Eilenberg *et al.*, 2017). Until know the veterinary sector has not really considered it as their responsibility. However, with the growing interest to rear insects on an industrial scale, insect should be considered as mini-livestock and as such be an expertise belonging to the discipline of veterinary science.

22.7 Insect welfare

Insects are animals, which means the five freedoms for animals, formulated by the Farm Animal Welfare Council (FAWC, 2009) are also valid for insects, viz freedom: (1) from hunger and thirst; (2) from discomfort; (3) from pain, injury and disease; (4) from fear and distress; and (5) to express normal behaviour. Mather (2011) stressed the humanness in caring about animals, considering that 99% of animals are invertebrates.

Can insects experience pain? Pain is distinct from nociception. Pain is an unpleasant sensory and emotional experience associated with actual or potential tissue damage. Nociception has more to do with protective reflexes to mediate noxious stimuli and lacks the negative emotional response or feeling associated with pain. Considering the number of neurons present, the question is whether this would make a cognitive ability of pain experience possible. Van Huis (2017b) assumed that the more complex the brain, the more chance that an animal can experience pain, and the number of neurons present in the brain is for a fruit fly 100,000, a cockroach 1,000,000, a mouse 75,000,000, an octopus 50,000,000, and a human 85,000,000,000. EFSA (2005) considered insects (not social ones) not to be able to experience pain, and concluded that insects can be used for experimental purposes. Because of the uncertainty whether insect can experience pain, several authors recommended to use the precautionary principle (Cooper, 2012; Eisemann *et al.*, 1984). That may be justified as Mendl and Paul (2016) showed that bumblebees show decision-making that reflects emotionlike states, although they were not sure whether this proves that it is accompanied by emotional feelings. Hakman *et al.* (2013) recommended to the Dutch government that killing methods should be used which are quick and effective, such as freezing (insects are cold-blooded), heating (cooking or blanching) and shredding. A review of this topic is given in Chapter 19 by Röcklinsberg *et al.* (2017).

22.8 Processing

After killing there is the whole issue of processing. What to do with the leftover substrate (use as fertilizer), and will the whole animal be used or will it be processed into protein, fat, and chitin? Then also the question of how long the insect product can be conserved and under which kind of conditions? The aim of processing is to obtain raw materials and ingredients from edible insects that can be used for food and feed applications, while safeguarding food safety and ensuring high quality. Chapter 17 provides information on decontamination, drying, and comminution methods and subsequent fractionation methods to recover protein, fat or chitin (Rumpold *et al.*, 2017). Examples for whole insect processing pathways are presented. Also issues like how to prevent enzymatic browning during extractions of mealworms.

22.9 Food safety and legislation

When the word 'animals' is used in legislation, insects are often automatically part of that legislation. However, at the time the legislation was made, nobody considered insects as either food or feed. An example is the EU Regulation 1099/2009 (EC, 2009), which according to Charlton *et al.* (2015) was not designed to apply to insects, i.e. animals in the EU have to be slaughtered in certified slaughterhouses. According to the International Platform of Insects for Food and Feed (IPIFF; http://www.ipiff.org), this is due to 'legal gaps' in the EU legislation. The slaughterhouse issue seems to be solved as the EU commission voted in December 2016 to open the aquaculture feed market for insect-derived protein from July 2017 onwards. IPIFF calls for a revision of EU legislation (i.e. EU TSE & Animal ByProducts legislations) in order to allow insect proteins (insect PAPs) to be used as source of proteins for all farmed animals. The potential market is enormous, 1 billion metric tons a year, worth 460 billion US$, of which 70% goes to poultry (44%) and pigs (26%) (Alltech, 2017).

Food safety issues of edible insects are immediately related to legislative issues (Belluco *et al.*, 2013; Van der Spiegel *et al.*, 2013). There are a number of issues that will have an impact on biological and chemical hazards in insect food and feed: production methods, substrate used, stage of harvest, insect species and developmental stage (EFSA, 2015). Possible hazards are contaminants, like heavy metals, mycotoxins, pesticide residues, and pathogens. According to Eilenberg *et al.* (2015), insect pathogens are specific for invertebrates and do not harm vertebrates. With the use of organic by-products or organic side streams the issue of contaminants and so food safety becomes a real challenge. How do insects deal with contaminants: do they retain, sequester, excrete or detoxify contaminants? Van Broekhoven *et al.* (2017) studied how the yellow mealworm dealt with the mycotoxin deoxynivalenol (DON) and detected no DON or DON-derivatives in the insect after harvest, pointing to degradation by the larvae. However, it has also been shown that bioaccumulation of heavy metals in the black soldier fly and the yellow mealworm is possible (Van der Fels-Klerx *et al.*, 2016).

Very likely insects as food may cause allergies when people are allergic to house dust mites and crustaceans (Van Broekhoven *et al.*, 2016; Verhoeckx *et al.*, 2014) and this means that proper labelling of the insect product is required. The legislative and food safety issues are covered in Chapter 18 (Meijer and Van der Fels-Klerx, 2017).

22.10 How to convince consumers

There have been an impressive number of publications during the last five years dealing with consumer attitudes. It is quite a challenge to introduce a food item in the western world which has never been considered as such before. And this is certainly true for insects often considered to be dirty, disgusting, and dangerous (Looy and Wood, 2006). This notwithstanding the fact insects provide many benefits to human mankind (Dicke, 2017). Also, the eating of insects often is considered as a poor man's diet (DeFoliart, 1999; Van Huis, 2013). According to Yen (2009) globalization may also be responsible as cultural systems based on western values are adopted and insects as food for westerners is not only an alien concept but even taboo. Yet, insects are considered a tasty food in many cultures around the world: Africa (Van Huis, 2003), China (Luo, 1997), India (Chakravorty *et al.*, 2011), and Latin America (Costa-Neto, 2015). However, also in tropical countries the preferences vary from one ethnic group to the other, probably based on taste, abundance, and custom (Ramandey and Mastrigt, 2010). Even in Europe there are differences. Piha *et al.* (in press) determined that in Northern Europe people are more prepared to eat insect food than in Central Europe, and this was explained by a lesser degree of food neophobia in northern Europe and more reliance on subjective and objective knowledge. This was confirmed by a study in Poland, where consumers did not show open-mindedness towards incorporating insect-based food into their diet (Kostecka *et al.*, 2017). In Italy, the majority of the people had a negative opinion about the eating of insects. Issues that were considered important are: availability on the market, product category (whether whole or processed) and the type of communication to the consumers (Sogari, 2015).

Why should we want to convince western consumers to eat insects? The main reason is that it is possible to produce high quality protein from insects with a lower environmental impact than conventional livestock as explained in Chapter 3 (Oonincx, 2017). Consumers in Europe that value sustainability of food may consider insects as an option (Hartmann and Siegrist, 2017). However, than the question is still if this is enough to convince western consumers to incorporate insects into their diets. It certainly helps as providing this information increased the liking of insect burgers (Schouteten *et al.*, 2016). Low environmental impact was also considered important by consumers in the USA and India, but nutrition scored higher (Ruby *et al.*, 2015).

The consumer need not only to be convinced that the product is safe to eat (Ruby *et al.*, 2015), but it needs also to be tasty (Deroy *et al.*, 2015). A negative sensorial experience will not incite to try it once more (Tan *et al.*, 2017). People should be provided with a taste experience (Looy and Wood, 2006), as people who had eaten insects before were more likely to eat insects again than those who had not (Caparros Megido *et al.*, 2016; Lensvelt and Steenbekkers, 2014; Tan *et al.*, 2016). It may also be important to stress to consumers (Caparros Megido *et al.*, 2014) that the familiar crustaceans are taxonomically closely related to insects (Pennisi, 2015).

People are also more inclined to consume food products with a low content of insect flour than those containing whole insects (Ruby *et al.*, 2015). Consumers in Kenya also seem to accept food products easier when processed into familiar products (Ayieko *et al.*, 2010). The acceptability of mealworm products among Dutch consumers was also higher when the mealworms were made invisible (Tan *et al.*, 2016).

Also the promotion by persons that can be considered trustworthy seems to make a difference: e.g. recommendations by scientists, persons familiar with using the product, health officials, and well known relatives famous persons (Lensvelt and Steenbekkers, 2014; Pambo *et al.*, 2016). According to Verbeke (2015) the persons to target in western societies are the younger males with weak attitudes towards meat, open to trying new foods, and who are concerned about the environmental impact of food.

However, giving the assurance that it is safe to eat, stressing the environmental benefits, and making the product delicious, is probably still not enough to convince the consumer. How to achieve consumption on a regular base because 'curious tasting does not imply acceptance' (Tan *et al.*, 2016). Tan *et al.* (2017) recommend to explore the barriers to buying and preparing an insect dish, and those probably relating to availability, pricing and a conducive social environment.

22.11 Future of the edible insect sector

There have been a number of market studies on edible insects, which indicates the potential of growth of the insect sector (Figure 22.1). The latest report (Arcluster, 2017) indicates that the value of the insect feed market will reach US$ 1.07 billion (of which Europe US$ 404 million and North America US$ 250 million) in 2022 with an annual growth of more than 100% from 2017 (Anderson, 2017). The number of start-up companies has now been estimated as more than 200 as of March 2017 (http://tinyurl.com/zyotzcy). Because it is a new field, many of those start-ups work with trial and error, which also means that there are many drop-outs.

Because companies fear competition it is difficult to get an insight in production methods and procedures (Dobermann, 2017). The future market will very much depend also on the enhancing conditions as explained in Chapter 20 (Vantomme, 2017). The development and promotion of insect products is mentioned in Chapter 21 (Shockley *et al.*, 2017).

In order to tackle legislative issues, insect producers have to organize themselves. National stakeholders are increasingly getting organized. More and more conferences and workshops are being organized. Public interest is high considering the many newspaper articles that are being published on this topic. The number of scientific articles, books and book chapters on this topic are increasing almost exponentially. With all this interest it is believed that insects will be increasingly marketed. Producing sustainable food and feed is more and more considered a necessity. So, very likely under these circumstances there will be a bright future for insects as food and feed.

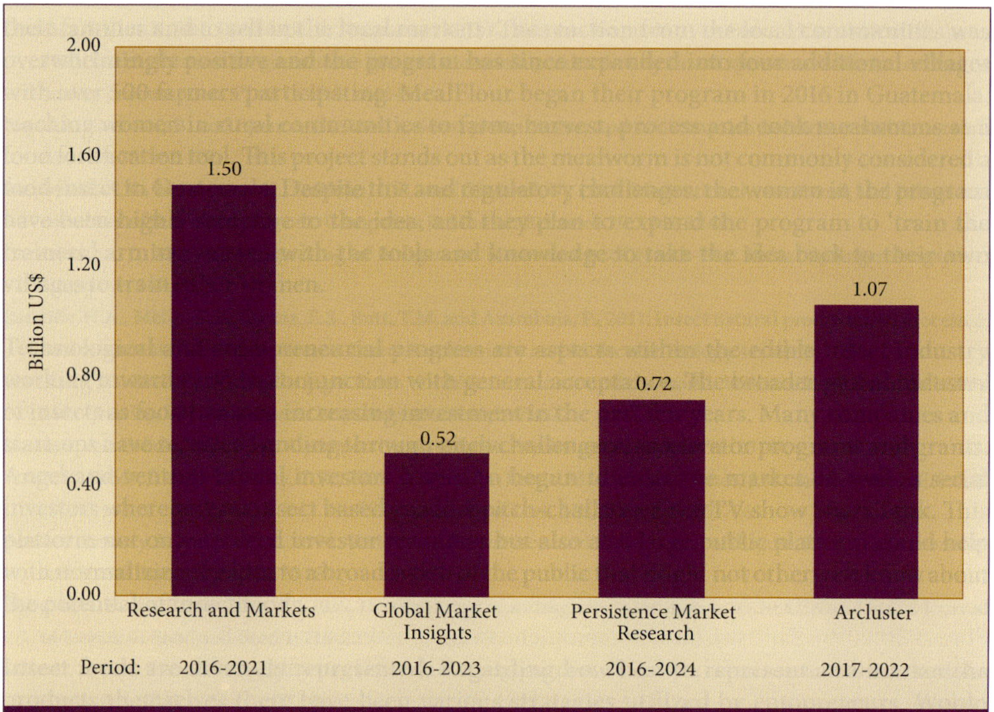

Figure 22.1. Prediction of the value of the market of edible insects (Arcluster, 2017; Global Market Insights, 2016; Persistence Market Research, 2016; Research and Markets, 2017).

References

Alltech, 2017. 2016 global feed survey. Available at: http://go.alltech.com/feedsurveydata.

Ambühl, D., 2016. Beezza! Das Bienenkochbuch. Eine anleitung zur Ernte und Zubereitung von Drohnenwaben der Honingbiene. Skyfood Verlag, Unterterzen, Switzerland.

Anderson, D., 2017. Insect feed market forecast to top $1bn by 2022. Feed Navigator newsletter, 20 May 2017. William Reed Business Media Ltd, Crawley, UK. Available at: http://tinyurl.com/y95l9swn.

Arcluster, 2017. Insect feed market (2017-2022): market analysis and forecasting report. Arcluster, Singapore, Singapore.

Ayieko, M.A., Oriamo, V. and Nyambuga, I.A., 2010. Processed products of termites and lake flies: improving entomophagy for food security within the Lake Victoria region. African Journal of Food, Agriculture, Nutrition and Development 10: 2085-2098.

Belluco, S., Losasso, C., Maggioletti, M., Alonzi, C.C., Paoletti, M.G. and Ricci, A., 2013. Edible insects in a food safety and nutritional perspective: a critical review. Comprehensive Reviews in Food Science and Food Safety 12: 296-313.

Berezina, N., 2017. Mealworms, promising beetles for the insect industry. In: Van Huis, A. and Tomberlin, J.K. (eds.) Insects as food and feed: from production to consumption. Wageningen Academic Publishers, Wageningen, the Netherlands, pp. 258-269.

Bombelli, P., Howe, C.J. and Bertocchini, F., 2017. Polyethylene bio-degradation by caterpillars of the wax moth – *Galleria mellonella*. Current Biology 27: R292-R293.

Brits, D., 2017. Improving feeding efficiencies of black soldier fly larvae, *Hermetia illucens* (L., 1758) (Diptera: Stratiomyidae: Hermetiinae) through manipulation of feeding conditions for industrial mass rearing. MSc thesis, Stellenbosch University, Stellenbosch, South Africa.

Caparros Megido, R., Gierts, C., Blecker, C., Brostaux, Y., Haubruge, É., Alabi, T. and Francis, F., 2016. Consumer acceptance of insect-based alternative meat products in western countries. Food Quality and Preference 52: 237-243.

Caparros Megido, R., Haubruge, E. and Francis, F., 2017. Small-scale production of crickets and impact on rural livelihoods. In: Van Huis, A. and Tomberlin, J.K. (eds.) Insects as food and feed: from production to consumption. Wageningen Academic Publishers, Wageningen, the Netherlands, pp. 100-111.

Caparros Megido, R., Sablon, L., Geuens, M., Brostaux, Y., Alabi, T., Blecker, C., Drugmand, D., Haubruge, É. and Francis, F., 2014. Edible insects acceptance by Belgian consumers: promising attitude for entomophagy development. Journal of Sensory Studies 29: 14-20.

Chakravorty, J., Ghosh, S. and Meyer-Rochow, V., 2011. Practices of entomophagy and entomotherapy by members of the Nyishi and Galo tribes, two ethnic groups of the state of Arunachal Pradesh (North-East India). Journal of Ethnobiology and Ethnomedicine 7: 5.

Charlton, A.J., Dickinson, M., Wakefield, M.E., Fitches, E., Kenis, M., Han, R., Zhu, F., Kone, N., Grant, M., Devic, E., Bruggeman, G., Prior, R. and Smith, R., 2015. Exploring the chemical safety of fly larvae as a source of protein for animal feed. Journal of Insects as Food and Feed 1: 7-16.

Cooper, J.E., 2012. Insects. In: Lewbart, G.A. (ed.) Invertebrate medicine. Wiley-Blackwell, Hoboken, NJ, USA, pp. 267-283.

Costa-Neto, E.M., 2015. Anthropo-entomophagy in Latin America: an overview of the importance of edible insects to local communities. Journal of Insects as Food and Feed 1: 17-23.

Crippen, T.L., Sheffield, C.L., Esquivel, S.V., Droleskey, R.E. and Esquivel, J.F., 2009. The acquisition and internalization of *Salmonella* by the lesser mealworm, *Alphitobius diaperinus* (Coleoptera: Tenebrionidae). Vector-Borne and Zoonotic Diseases 9: 65-71.

DeFoliart, G.R., 1999. Insects as food: why the western attitude is important. Annual Review of Entomology 44: 21-50.

Deroy, O., Reade, B. and Spence, C., 2015. The insectivore's dilemma, and how to take the West out of it. Food Quality and Preference 44: 44-55.

Dicke, M., 2017. Ecosystem services of insects. In: Van Huis, A. and Tomberlin, J.K. (eds.) Insects as food and feed: from production to consumption. Wageningen Academic Publishers, Wageningen, the Netherlands, pp. 60-76.

Dobermann, D., 2017. Insects as food and feed: can research and business work together? Journal of Insects as Food and Feed 3: 155-160.

European Food and Safety Authorisation (EFSA), 2005. Opinion on the 'aspects of the biology and welfare of animals used for experimental and other scientific purposes'. EFSA Journal 292: 1-46.

European Food and Safety Authorisation (EFSA), 2015. Risk profile related to production and consumption of insects as food and feed. EFSA Journal 13: 4257.

Eilenberg, J., Gasque, S.N. and Ros, V.I.D., 2017. Natural enemies in insect production systems. In: Van Huis, A. and Tomberlin, J.K. (eds.) Insects as food and feed: from production to consumption. Wageningen Academic Publishers, Wageningen, the Netherlands, pp. 200-223.

Eilenberg, J., Vlak, J.M., Nielsen-LeRoux, C., Cappellozza, S. and Jensen, A.B., 2015. Diseases in insects produced for food and feed. Journal of Insects as Food and Feed 1: 87-102.

Eisemann, C.H., Jorgensen, W.K., Merritt, D.J., Rice, M.J., Cribb, B.W., Webb, P.D. and Zalucki, M.P., 1984. Do insects feel pain? A biological view. Experientia 40: 164-167.

European Commission (EC), 2009. Council Regulation (EC) No 1099/2009 of 24 September 2009 on the protection of animals at the time of killing. Official Journal of the European Union L303: 1-30.

Evans, J., Müller, A., Jensen, A.B., Dahle, B., Flore, R., Eilenberg, J. and Frøst, M.B., 2016. A descriptive sensory analysis of honeybee drone brood from Denmark and Norway. Journal of Insects as Food and Feed 2: 277-283.

Farm Animal Welfare Council (FAWC), 2009. Farm animal welfare in Great Britain: past, present and future. FAWC, London, UK. Available at: http://tinyurl.com/njpmacf.

Finke, M.D., 2002. Complete nutrient composition of commercially raised invertebrates used as food for insectivores. Zoo Biology 21: 269-285.

Finke, M.D. and Oonincx, D., 2014. Insects as food for insectivores. In: Morales-Ramos, M.G.R.J. and Shapiro-Ilan, D. (ed.) Mass production of beneficial organisms. Academic Press, San Diego, CA, USA, pp. 583-616.

Finke, M.D. and Oonincx, D.G.A.B., 2017. Nutrient content of insects. In: Van Huis, A. and Tomberlin, J.K. (eds.) Insects as food and feed: from production to consumption. Wageningen Academic Publishers, Wageningen, the Netherlands, pp. 290-316.

Global Market Insights, 2016. Global edible insects market size worth $522mn by 2023. Global Market Insights Inc., Ocean View, DE, USA. Available at: https://www.gminsights.com/pressrelease/edible-insects-market.

Gouli, V., Gouli, S. and Marcelino, J., 2011. Common infectious diseases of insects in culture: diagnostic and prophylactic methods. Springer, Dordrecht, the Netherlands.

Hakman, A., Peters, M. and Van Huis, A., 2013. Toelatingsprocedure voor insecten als mini-vee voor het plaatsen van nieuwe insectensoorten op de lijst voor productie te houden dieren. Wageningen University, Wageningen, the Netherlands.

Haldar, P. and Malakar, C., 2017. Farming of grasshoppers for feed and food. In: Van Huis, A. and Tomberlin, J.K. (eds.) Insects as food and feed: from production to consumption. Wageningen Academic Publishers, Wageningen, the Netherlands, pp. 126-139.

Hartmann, C. and Siegrist, M., 2017. Insects as food: perception and acceptance. Findings from current research. Ernahrungs Umschau 64: 44-50.

Inglis, C.D. and Sikorowski, P.P., 2009. Entomopathogens and insect rearing. In: John, C. and Schneider, J.C. (eds.) Principles and procedures for rearing high quality insects. Mississippi State University, Starkville, MS, USA, pp. 224-288.

Jensen, K., Kristensen, T.N., Heckmann, L.H.L. and Sørensen, J.G., 2017. Breeding and maintaining high-quality insects. In: Van Huis, A. and Tomberlin, J.K. (eds.) Insects as food and feed: from production to consumption. Wageningen Academic Publishers, Wageningen, the Netherlands, pp. 174-198.

Katayama, N., Ishikawa, Y., Takaoki, M., Yamashita, M., Nakayama, S., Kiguchi, K., Kok, R., Wada, H. and Mitsuhashi, J., 2008. Entomophagy: a key to space agriculture. Advances in Space Research 41: 701-705.

Kok, R., 1983. The production of insects for human food. Canadian Institute of Food Science and Technology Journal 16: 5-18.

Kok, R., 2017. Insect production and facility design. In: Van Huis, A. and Tomberlin, J.K. (eds.) Insects as food and feed: from production to consumption. Wageningen Academic Publishers, Wageningen, the Netherlands, pp. 142-172.

Kok, R., Lomaliza, K. and Shivhare, U.S., 1988. The design and performance of an insect farm/chemical reactor for human food production. Canadian Agricultural Engineering 30: 307-317.

Kostecka, J., Konieczna, K. and Cunha, L.M., 2017. Evaluation of insect-based food acceptance by representatives of Polish consumers in the context of natural resources processing retardation. Journal of Ecological Engineering 18: 166-174.

Lensvelt, E.J.S. and Steenbekkers, L.P.A., 2014. Exploring consumer acceptance of entomophagy: a survey and experiment in Australia and the Netherlands. Ecology of Food and Nutrition 53: 543-561.

Looy, H., Dunkel, F.V. and Wood, J.R., 2014. How then shall we eat? Insect-eating attitudes and sustainable foodways. Agriculture and Human Values 31: 131-141.

Looy, H. and Wood, J.R., 2006. Attitudes toward invertebrates: are educational 'bug banquets' effective? Journal of Environmental Education 37: 37-48.

Lundy, M.E. and Parrella, M.P., 2015. Crickets are not a free lunch: protein capture from scalable organic side-streams via high-density populations of *Acheta domesticus*. PLoS ONE 10: e0118785.

Luo, Z.-Y., 1997. Insects as food in China. Ecology of Food and Nutrition 36: 201-207.

Martin, D., 2014. Edible: an adventure into the world of eating insects and the last great hope to save the planet. Houghton Mifflin Harcourt, Boston, MA, USA.

Mather, J.A., 2011. Philosophical background of attitudes toward and treatment of invertebrates. Institute for Laboratory Animal Research (ILAR) Journal 52: 205-212.

Meijer, N. and Van der Fels-Klerx, H.J., 2017. Health risks and EU regulatory framework. In: Van Huis, A. and Tomberlin, J.K. (eds.) Insects as food and feed: from production to consumption. Wageningen Academic Publishers, Wageningen, the Netherlands, pp. 346-362.

Mendl, M.T. and Paul, E.S., 2016. Bee happy. Science 353: 1499-1500.

Mott, G., 2017. Cricket rearing. In: Van Huis, A. and Tomberlin, J.K. (eds.) Insects as food and feed: from production to consumption. Wageningen Academic Publishers, Wageningen, the Netherlands, pp. 270-287.

Muafor, F.J., Gnetegha, A.A., Dounias, E., Le Gall, P. and Levang, P., 2017. Palm weevil farming contributing to food security in sub-Saharan Africa. In: Van Huis, A. and Tomberlin, J.K. (eds.) Insects as food and feed: from production to consumption. Wageningen Academic Publishers, Wageningen, the Netherlands, pp. 112-125.

Nonaka, K., 2007. Hebo. Yellow jackets: from the fields to the dinner table – a delightful culinary experience. Tamasaya, Tokyo, Japan.

Oonincx, D. and De Boer, I., 2012. Environmental impact of the production of mealworms as a protein source for humans – a life cycle assessment. PLoS ONE 7: e51145.

Oonincx, D.G.A.B., 2017. Environmental impact of insect production. In: Van Huis, A. and Tomberlin, J.K. (eds.) Insects as food and feed: from production to consumption. Wageningen Academic Publishers, Wageningen, the Netherlands, pp. 78-93.

Pambo, K., Mbeche, R., Okello, J., Kinyuru, J. and Mose, G., 2016. Consumers' salient beliefs regarding foods from edible insects in Kenya: a qualitative study using concepts from the theory of planned behaviour. African Journal of Food, Agriculture, Nutrition and Development 16: 11366-11385.

Payne, C.L.R. and Evans, J.D., 2017. Nested houses: domestication dynamics of human-wasp relations in contemporary rural Japan. Journal of Ethnobiology and Ethnomedicine 13: 13.

Pennisi, E., 2015. All in the (bigger) family revised arthropod tree marries crustacean and insect fields. Science of the Total Environment 347: 220-221.

Persistence Market Research, 2016. Global edible insects market value to increase from US$ 423.8 mn in 2016 to US$ 722.9 mn by 2024, due to low production cost of insect products as compared to beef, chicken, or pork. Persistence Market Research, New York, NY, USA. Available at: http://tinyurl.com/yackzu5w.

Pieterse, E. and Pretorius, Q., 2014. Nutritional evaluation of dried larvae and pupae meal of the housefly (*Musca domestica*) using chemical- and broiler-based biological assays. Animal Production Science 54: 347-355.

Piha, S., Pohjanheimo, T., Lähteenmäki-Uutela, A., Křečková, Z. and Otterbring, T., in press. The effects of consumer knowledge on the willingness to buy insect food: an exploratory cross-regional study in northern and Central Europe. Food Quality and Preference, DOI: https://doi.org/10.1016/j.foodqual.2016.12.006.

Ramandey, E. and Van Mastrigt, H., 2010. Edible insects in Papua, Indonesia: from delicious snack to basic need. FAO, Regional Office for Asia and the Pacific, Bangkok, Thailand.

Research and Markets, 2017. Global edible insects market to reach $1.53 billion in 2021 – analysis by segments, types of insect and regions – Research and Markets. Research and Markets, Dublin, Ireland. Available at: http://tinyurl.com/y8vm8f74.

Röcklinsberg, H., Gamborg, C. and Gjerris, M., 2017. Ethical issues in insect production. In: Van Huis, A. and Tomberlin, J.K. (eds.) Insects as food and feed: from production to consumption. Wageningen Academic Publishers, Wageningen, the Netherlands, pp. 364-379.

Ruby, M.B., Rozin, P. and Chan, C., 2015. Determinants of willingness to eat insects in the USA and India. Journal of Insects as Food and Feed 1: 215-225.

Rumpold, B.A., Bußler, S., Jäger, H. and Schlüter, O., 2017. Insect processing. In: Van Huis, A. and Tomberlin, J.K. (eds.) Insects as food and feed: from production to consumption. Wageningen Academic Publishers, Wageningen, the Netherlands, pp. 318-342.

Schouteten, J.J., De Steur, H., De Pelsmaeker, S., Lagast, S., Juvinal, J.G., De Bourdeaudhuij, I., Verbeke, W. and Gellynck, X., 2016. Emotional and sensory profiling of insect-, plant- and meat-based burgers under blind, expected and informed conditions. Food Quality and Preference 52: 27-31.

Shapiro, M., 1984. Micro-organisms as contaminants and pathogens in insect rearing. In: King, E.G. and Leppla, N.C. (eds.) Advances and challenges in insect rearing. United States Department of Agriculture, Agricultural Research Service, Southern Region, New Orleans, LA, USA, pp. 130-142.

Shockley, M., Allen, R.N. and Gracer, D., 2017. Product development and promotion. In: Van Huis, A. and Tomberlin, J.K. (eds.) Insects as food and feed: from production to consumption. Wageningen Academic Publishers, Wageningen, the Netherlands, pp. 398-427.

Sogari, G., 2015. Entomophagy and Italian consumers: an exploratory analysis. Progress in Nutrition 17: 311-316.

St-Hilaire, S., Cranfill, K., Mcguire, M.A., Mosley, E.E., Tomberlin, J.K., Newton, L., Sealey, W., Sheppard, C. and Irving, S., 2007. Fish offal recycling by the black soldier fly produces a foodstuff high in omega-3 fatty acids. Journal of the World Aquaculture Society 38: 309-313.

Tan, H.S.G., Fischer, A.R.H., Van Trijp, H.C.M. and Stieger, M., 2016. Tasty but nasty? Exploring the role of sensory-liking and food appropriateness in the willingness to eat unusual novel foods like insects. Food Quality and Preference 48: 293-302.

Tan, H.S.G., Verbaan, Y.T. and Stieger, M., 2017. How will better products improve the sensory-liking and willingness to buy insect-based foods? Food Research International 92: 95-105.

Tomberlin, J.K. and Cammack, J.A., 2017. Black soldier fly: biology and mass production. In: Van Huis, A. and Tomberlin, J.K. (eds.) Insects as food and feed: from production to consumption. Wageningen Academic Publishers, Wageningen, the Netherlands, pp. 230-246.

Tomberlin, J.K., Holcomb, C. and Van Leeuwen, S.K., 2017. Industrialization of house fly production for livestock feed. In: Van Huis, A. and Tomberlin, J.K. (eds.) Insects as food and feed: from production to consumption. Wageningen Academic Publishers, Wageningen, the Netherlands, pp. 248-257.

Urs, K.C.D. and Hopkins, T.L., 1973a. Effect of moisture on growth rate and development of two strains of *Tenebrio molitor* L. (Coleoptera, Tenebrionidae). Journal of Stored Products Research 8: 291-297.

Urs, K.C.D. and Hopkins, T.L., 1973b. Effect of moisture on the lipid content and composition of two strains of *Tenebrio molitor* L. (Coleoptera, Tenebrionidae). Journal of Stored Products Research 8: 299-305.

Van Broekhoven, S., Gutierrez, J.M., Rijk, T.C.D., Nijs, W.C.M.D. and Van Loon, J.J.A., 2017. Degradation and excretion of the *Fusarium* toxin deoxynivalenol by an edible insect, the yellow mealworm (*Tenebrio molitor* L.). World Mycotoxin Journal 10: 163-169.

Van Broekhoven, S., Oonincx, D.G.A.B., Van Huis, A. and Van Loon, J.J.A., 2015. Growth performance and feed conversion efficiency of three edible mealworm species (Coleoptera: Tenebrionidae) on diets composed of organic by-products. Journal of Insect Physiology 73: 1-10.

Van der Fels-Klerx, H.J., Camenzuli, L., Van der Lee, M.K. and Oonincx, D.G.A.B., 2016. Uptake of cadmium, lead and arsenic by *Tenebrio molitor* and *Hermetia illucens* from contaminated substrates. PLoS ONE 11: e0166186.

Van der Spiegel, M., Noordam, M.Y. and Van der Fels-Klerx, H.J., 2013. Safety of novel protein sources (insects, microalgae, seaweed, duckweed, and rapeseed) and legislative aspects for their application in food and feed production. Comprehensive Reviews in Food Science and Food Safety 12: 662-678.

Van Huis, A., 2003. Insects as food in sub-Saharan Africa. Insect Science and its Application 23: 163-185.

Van Huis, A., 2013. Potential of insects as food and feed in assuring food security. Annual Review of Entomology 58: 563-583.

Van Huis, A., 2017a. Introducing small production systems for edible insects. In: Van Huis, A. and Tomberlin, J.K. (eds.) Insects as food and feed: from production to consumption. Wageningen Academic Publishers, Wageningen, the Netherlands, pp. 96-99.

Van Huis, A., 2017b. New sources of animal proteins: edible insects. In: Purslow, P.P. (ed.) New aspects of meat quality. Woodhead Publishing, Cambridge, UK, pp. 443-461.

Van Huis, A. and Tomberlin, J.K., 2017. The potential of insects as food and feed. In: Van Huis, A. and Tomberlin, J.K. (eds.) Insects as food and feed: from production to consumption. Wageningen Academic Publishers, Wageningen, the Netherlands, pp. 24-58.

Vantomme, P., 2017. Creating an enabling environment. In: Van Huis, A. and Tomberlin, J.K. (eds.) Insects as food and feed: from production to consumption. Wageningen Academic Publishers, Wageningen, the Netherlands, pp. 380-397.

Verbeke, W., 2015. Profiling consumers who are ready to adopt insects as a meat substitute in a western society. Food Quality and Preference 39: 147-155.

Weissman, D.B., Gray, D.A., Pham, H.T. and Tijssen, P., 2012. Billions and billions sold: pet-feeder crickets (Orthoptera: Gryllidae), commercial cricket farms, an epizootic densovirus, and government regulations make for a potential disaster. Zootaxa 3504: 67-88.

Yen, A.L., 2009. Edible insects: traditional knowledge or western phobia? Entomological Research 39: 289-298.

Zhou, F., Tomberlin, J.K., Zheng, L., Yu, Z. and Zhang, J., 2013. Developmental and waste reduction plasticity of three black soldier fly strains (Diptera: Stratiomyidae) raised on different livestock manures. Journal of Medical Entomology 50: 1224-1230.

About the

Dr Arnold van Huis is an emeritus professor of the Laboratory of Entomology of Wageningen University in the Netherlands. He completed his undergraduate studies at the State Horticultural College in Utrecht in the Netherlands and his graduate studies at Wageningen University. He worked from 1974 to 1979 in Nicaragua for the Food and Agriculture Organization (FAO) of the United Nations in a research project on integrated pest management (IPM) in food grains. Based on these studies he got his PhD from Wageningen University. From 1982 to 1985 he coordinated a regional crop protection training project for eight Sahelian countries. From 1985 to 2015 he worked as tropical entomologist at Wageningen University, the Netherlands, and has been responsible for a number of IPM and biological control projects in the tropics.

Arnold has a research background in IPM, biological control, tropical pest management, insects as food and feed, interactive research, and interdisciplinary agricultural research in Africa. On these different topics he supervised 24 PhD theses. In 2013 he published together with FAO the book 'Edible insects: future prospects for food and feed security', which has been downloaded seven million times. He is also the first author of the 'The insect cookbook' published in 2014 by Columbia University Press. In 2014 he organised with FAO the conference 'Insects to feed the world', attended by 450 participants from 45 countries. From 2010 to 2014 he was responsible for a Netherlands' funded project entitled 'Sustainable production of insect protein'.

Arnold has published about 250 papers of which more than half refereed, co-authored three books and edited more than 10 conference proceedings. He is chief editor of the Journal of Insects as Food and Feed. On edible insects, he (co)authored ten book chapters, and 20 refereed and 10 non-refereed publications. Each year he gives numerous presentations in and outside the Netherlands on insects as food and feed.

editors

Dr Jeffery K. Tomberlin is an associate professor and AgriLife research fellow in the Department of Entomology at Texas A&M University (College Station, TX, USA). He is also the director of the Forensic & Investigative Sciences Program at Texas A&M University, and the principle investigator of the Forensic Laboratory for Investigative Entomological Sciences (F.L.I.E.S.) Facility (forensicentomology.tamu.edu) at Texas A&M University.

Research in the F.L.I.E.S. Facility examines species interactions on ephemeral resources such as vertebrate carrion, decomposing plant material and animal wastes in order to better understand the mechanisms regulating arthropod behaviour as related to arrival, colonization and succession patterns. The goals of his program are numerous; however, a major focus of his research is on waste management in confined animal facilities as well as concerns with food waste being placed in landfills. His research efforts for the past 19 years have been developing methods for the production of alternate protein sources for use as livestock, poultry and aquaculture feed from these resources. Predominately, these efforts have been accomplished through his research with the black soldier fly. Since arriving on campus at Texas A&M University in 2007, nine PhD and seventeen MS students have completed their degrees under his supervision. Dr Tomberlin welcomes those that are interested in collaborating or gaining experience in forensic entomology or other areas of his research to visit the F.L.I.E.S. Facility.